AUSTRALIAN REGIONAL FOOD GUIDE

SALLY & GORDON HAMMOND

First published in Australia in 2005 by
New Holland Publishers (Australia) Pty Ltd
Sydney • Auckland • London • Cape Town

14 Aquatic Drive Frenchs Forest NSW 2086 Australia
218 Lake Road Northcote Auckland New Zealand
86 Edgware Road London W2 2EA United Kingdom
80 McKenzie Street Cape Town 8001 South Africa

Copyright © 2005 text: Sally Hammond and Gordon Hammond
Copyright © 2005 New Holland Publishers (Australia) Pty Ltd

All rights reserved. No part of this publication may be reproduced, stored in a retrieval system or transmitted, in any form or by any means, electronic, mechanical, photocopying, recording or otherwise, without the prior written permission of the publishers and copyright holders.

National Library of Australia Cataloguing-in-Publication Data:–
Hammond, Sally,
Australian regional food guide.

ISBN 1 74110 147 6.

1. Food industry and trade - Australia - Directories.
I. Hammond, Gordon. II. Title.

664.002594

Publisher: Fiona Schultz
Project Editor: Liz Hardy
Designer: Joanne Buckley
Production Manager: Linda Bottari
Cartographer: Ian Faulkner
Printed in Australia by Griffin Press, Adelaide

2 4 6 8 10 9 7 5 3 1

The authors and publisher have made every effort to ensure that the information contained in this book was correct at the time of going to press and accept no responsibility for any loss, injury or inconvenience sustained by any person using this book.

Cover image (top) courtesy of Tourism Victoria; cover image (bottom) courtesy of Tourism Queensland

Contents

INTRODUCTION

**NEW SOUTH WALES AND THE
AUSTRALIAN CAPITAL TERRITORY 9**
    Sydney and the Southern Highlands 12
    Blue Mountains and the Hawkesbury 20
    Canberra, Capital Country and the Snowy Mountains 30
    Central Coast 44
    Explorer Country and Riverina 48
    The Hunter 66
    New England and North-West 79
    North Coast NSW 85
    Northern Rivers 99
    Outback and the Murray 110
    South Coast and Illawarra 113

NORTHERN TERRITORY 125
    Central Australia 126
    Top End 131

QUEENSLAND 135
    Brisbane and Surrounds 136
    Bundaberg, Fraser Coast and South Burnett 139
    Capricorn, Gladstone and the Outback 147
    Far North 151
    Gold Coast 167
    Sunshine Coast 171
    Toowoomba and the Southern Downs 181
    Townsville, Mackay and the Whitsundays 188

SOUTH AUSTRALIA 193
 Adelaide and Surrounds 195
 Barossa 217
 Clare Valley and Yorke Peninsula 225
 Eyre Peninsula, Flinders Ranges and the Outback 231
 Kangaroo Island 236
 Limestone Coast 241
 Murray and Riverland 249

TASMANIA 253
 Hobart and Surrounds 254
 Bass Strait Islands 260
 The North 263
 The South 275
 The North-West 288

VICTORIA 295
 Melbourne 298
 Bays and Peninsulas 301
 Goldfields 310
 Goulburn Murray Waters 316
 Lakes and Wilderness 321
 Legends, Wine and High Country 325
 Macedon Ranges and Spa Country 343
 Murray and the Outback 350
 Phillip Island and Gippsland 357
 The Grampians 365
 The Great Ocean Road 369
 Yarra Valley and the Dandenong Ranges 375

WESTERN AUSTRALIA 387
 Perth and Surrounds 388
 The South-West 400
 Australia's North-West 418
 The Outback 424
 Australia's Coral Coast 426

INDEX 428

Introduction

Six years ago when the first edition of the *Australian Regional Food Guide* was released, its scope amazed many. This was a big book, representing the entire country. What was even more fascinating, though, was the people who made it such an interesting book. Like a mosaic, each one was individual, valuable and vitally important to the whole picture.

The first edition came about because there was a need, and the time was right, to bring together all the states and territories and give the reader a book they could slip into the glove box of their car, or into a suitcase, as they travelled around the country. It was also important that there should be no link with advertising of any sort.

The main intention was to awaken people to the wealth of produce and initiatives throughout this great country, so that they would choose to eat food that was grown and prepared locally rather than settling for a mass-produced item they could eat anywhere in Australia, or even worldwide, for that matter.

Obviously the spin-offs from this regional emphasis are many. People come to a region for a variety of reasons, but everyone needs to eat. So why not support the local producers, visit the restaurants and wineries, and maybe extend the stay in order to see what else the area has to offer?

As you can see, this book is not just about restaurants and cafes, although almost half the book is devoted to them. There are also fruit wineries, the occasional food museum, tours, markets and events that feature food, as well as orchards, bakeries, butchers and much more.

This has been a massive task, one that has kept us up late, out often, and away from home a lot. It has tied us each to our computers for longer than we would ever want, yet such is our passion for this concept, we have kept with it. So now it is here for you to use and enjoy.

In this book, space limitations have meant that sometimes difficult decisions have had to be made about which places could be included. And others are missing because, for whatever reason, they did not respond to our invitation to submit information.

Then there are the ones that slipped past because we were not told about them. And this is where you all come in. A book such as this should be kept up to date and the intention is to do this—with a little help from our friends.

If you know someone who should be listed in the next edition, please contact the authors care of shammond@iprimus.com.au. A file is already filling with names and addresses. Please help us add to it.

French gastronome, Brillat-Savarin, once said, 'What a country eats, defines its personality.' The aim of this book is to raise the awareness and appreciation of Australian food. If we can each be inspired and in turn inspire others to value our local, regional and seasonal produce more, and to respect and value the people who grow and produce it, then it will have succeeded.

And if we eat and savour the fresh, healthy, lush, bounteous food this country is so skilled at producing, then—if Mr Brillat-Savarin's premise is correct—just think what Australia's personality could be!

Bon appetit and bon voyage.

Sally & Gordon Hammond

Acknowledgments

Of course this book would not exist if it was not for the hundreds of busy people who took the time to fill in questionnaires and reply by post or email, or who answered the phone and patiently helped us with the research for this book. Thank you for taking the time.

Then there were the local and state tourism officers and those from the Department of Primary Industries (or its equivalent) in each state and region who offered valuable insights, circulated emails requesting information, and answered numerous questions. Thank you for your patience.

Two others deserve thanks. Robynne Millward who commissioned this book, and Angela Handley who saw it to completion a year later.

And finally, thanks are due to Liz Hardy, the editor of this massive work, whose diligence in checking and querying and tidying up the manuscript took enormous dedication and care. Thank you for your tenacity and sharp eye.

What Happened to My City?

Space rules! This is a regional guide, and many city restaurants already feature regional food on their menus, so this time we made the decision to concentrate primarily on places outside the capital cities.

DINING PRICE GUIDE
Main courses
$	Under $15
$$	$15–$25
$$$	$26–$35
$$$$	Over $35

NEW SOUTH WALES

AND THE AUSTRALIAN CAPITAL TERRITORY

Bordering three other states and encircling the Australian Capital Territory, New South Wales was Australia's first state and birthplace of agriculture in this country. It was here that the infant wheat and wool industries began in harsh and primitive conditions, in what is now suburbia. New South Wales is Australia's fourth largest state, yet it is home to one-third of the total population, and accounts for a great proportion of the country's primary produce. The diversity of these industries is enormous.

In the lush, semi-tropical rainforest regions in the north, bananas and pineapples, macadamias and coffee flourish, alongside some of the country's richest beef grazing lands. To the south there is dairying and lamb, and wheat in the west, as well as pecans and olives, and a coastline bristling with fish and seafood.

The wine industry in New South Wales is phenomenal too. The Riverina alone produces seventy per cent of the state's wine grapes, yet the Hunter, Mudgee, Cowra and Orange regions spring more quickly to many people's minds. With wine comes gastronomy, and these areas have fostered fine dining and accommodation, gourmet goodies and many crops.

Mid-state, the Young area is the cherry and stone fruit capital, lovely in spring, even better in summer when the fruit ripens.

Further south, the Riverina spreads fertile and self-sufficient, a tribute to water and hard work.

South coastal New South Wales provides all that is rich and delicious—oysters, cheeses, berries, fruit wines and venison. The theme continues over the mountains onto the southern highlands where cool-climate foods abound.

Big-city Sydney ties it all together with markets and dining options to suit every palate and purse. Cosmopolitan and fast-paced, lacking, some say, the elegance of Melbourne, this city more than makes up for any perceived shortfall with an extravagant coastline of beaches and bays and, in the west, the equally stunning Blue Mountains.

Canberra is the federal capital, created in 1909, and features circular town-planning in many areas.

Canberra's cosmopolitan flavour spills over into its restaurants, shops, and inevitably, foodstuffs. The city is notable for its many markets, and these are a vibrant place to shop at weekends, particularly when they glow with produce—much of it local or from just across the border in New South Wales.

NSW WEBSITES
www.hunterharvest.com
www.tourism.nsw.gov.au
www.winecountry.com.au
www.visitnsw.com.au

QUEENSLAND

Regions:
1 Sydney & the Southern Highlands
2 Blue Mountains & the Hawkesbury
3 Canberra, Capital Country & the Snowy Mountains
4 Central Coast
5 Explorer Country & Riverina
6 The Hunter
7 New England & North-West
8 North Coast NSW
9 Northern Rivers
10 Outback & the Murray
11 South Coast & Illawarra

Bourke

NEW

SOUTH

WALES

SOUTH AUSTRALIA

10

5

Griffith

Young

Wagga Wagga

Albury

Jindab

VICTORIA

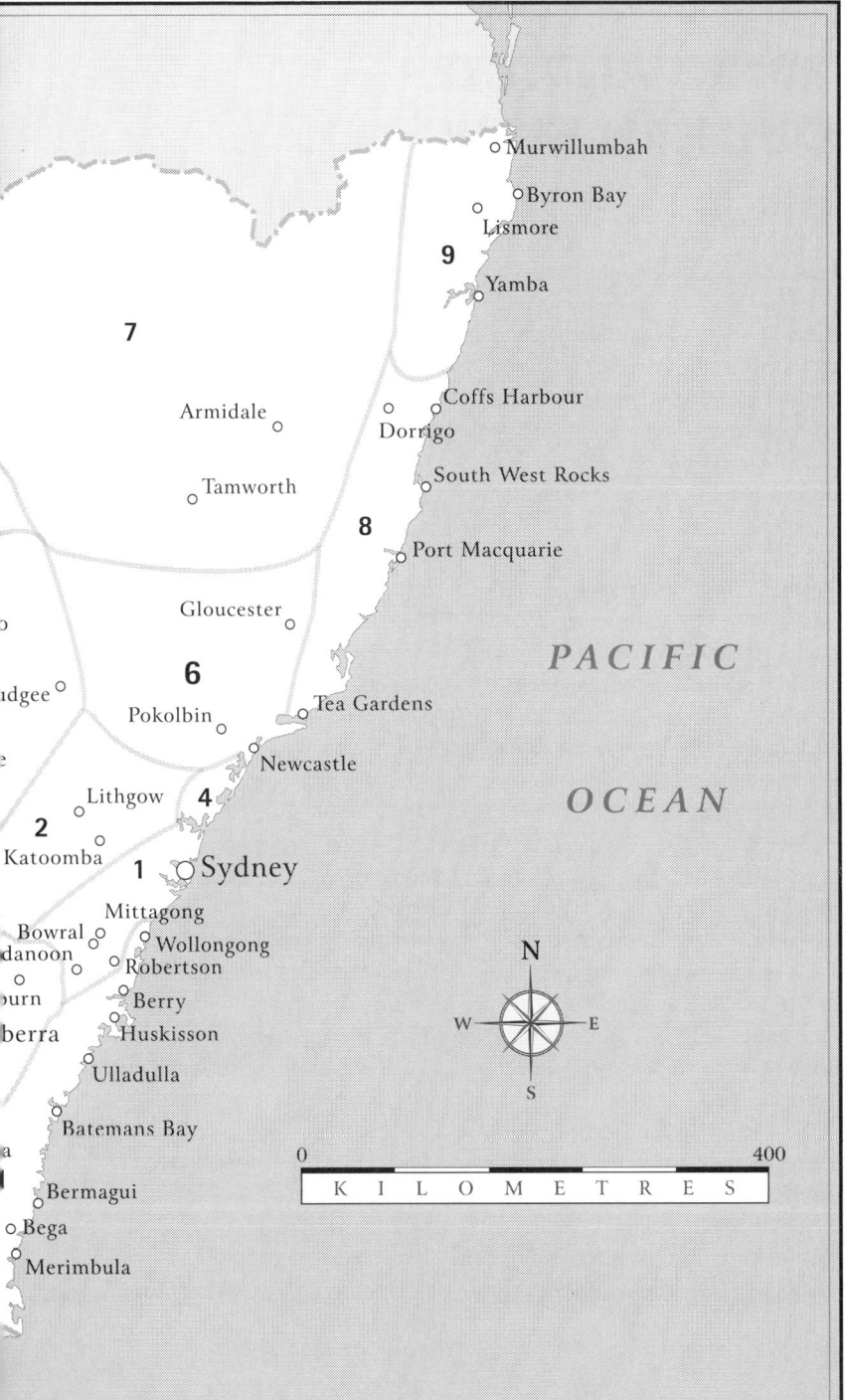

SYDNEY AND THE SOUTHERN HIGHLANDS

Sydney believes it was born lucky. And certainly the tourists who come each year to marvel at the harbour, clean air, and day after day of sunshine, seem to think that could be true. Add fabulous water and bushland views, to food and wine equal to any in the world, and perhaps you can start to see why.

This is a complex city, a cosmopolitan city,, and one where food can vary from ambrosial to abysmal. It is a city, after all. Sydney is best seen from the point of view of a huge showcase, where you can easily locate foods from all over the state — and indeed the country — and also taste them prepared by some of the best chefs in the world. You may not be dining on them at their source, but at least you won't have missed them entirely.

Yet Sydney does have a regional food of its own. Sydney Rock oysters are farmed in many parts of the east coast, and in fact originated here. Sydney's many seafood restaurants are ideal spots to taste the local fish and shellfish, often within metres of the water. Even it that was the only reason to come, it would be enough.

The Southern Highlands is a favourite weekend retreat from Sydney. It contains a wealth of orchards, jam-makers nurseries and cottage industries. The very English climate suits the practise of these rural arts.

SYDNEY

PRODUCERS

BLUE MOUNTAINS HONEY COMPANY
Genevieve and Steve Craig
Unit 6/14–16 Penrith St, Penrith 2750
Phone: 02 4721 2840
Email: bmhoney@acay.com.au

The Craigs are as busy as their little buzzing insect co-labourers. Not content with just managing hives they opened a factory store in Penrith selling everything from giftwrapped beeswax candles to comb honey and, of course, bottled honey. They quickly expanded their range of produce to include mustards, chocolate, nuts in honey, conserves and liqueur fruits. Now they have opened a store in Leura, Mountain Fine Fare (Shop 4, 166–168 The Mall) where they not only sell their honey products, but showcase a wide range of regionally produced foods from all over Australia—organic licorice, herbs, spices, oils, gift hampers and picnic baskets. The Penrith outlet caters for groups and bus tours. Seeing is bee-lieving.

■ **Open:** 9am–4pm weekdays, weekends by appointment

STORES

AC BUTCHERY
Carlo Colaiacomo
174 Marion St (and Norton Street Plaza), Leichhardt 2040
Phone: 02 9569 8687
www.acbutchery.com.au

SYDNEY AND THE SOUTHERN HIGHLANDS—NSW

'Excellence' and 'food' are two words that marry well. When Carlo Colaiacomo was awarded a Jaguar Award for Excellence in 2002, the industry acknowledged what his clients had known for a long time—nobody does it better than this boutique butchery. The son of an Italian butcher, trained in Rome, the tradition lives on in Australia with wonderful gourmet sausages and a range of products sourced from regional NSW. A wide range of regional meats end up here—Wagyu beef from North Queensland, pork from Temora, chemical-free lamb from Young. For meat lovers in Sydney's Eastern Suburbs, he has recently open a branch in 22 Plumer Rd, Rose Bay.
- **Open:** Marion St—6am-6pm weekdays, 6am-3pm Sat, Norton St—6am-6pm weekdays, 7am-5pm Sat, 10am-4pm Sun

BOULEVARD MARKET
Olympic Boulevard,
Sydney Olympic Park, Homebush 2127
Phone: 02 9714 7300
Email: info@sopa.nsw.gov.au
www.sydneyolympicpark.nsw.gov.au
An eclectic mix of fresh food, organic produce, gourmet produce and a few crafts, in the ideal setting. Parking and train and bus connections.
- **Open:** 9am-2pm fourth Sun of the month

CLEAVER'S—THE ORGANIC MEAT COMPANY
Ken Taylor
Shop 6, Grove Arcade,
174-176 Military Rd, Neutral Bay 2089
Phone: 02 8969 6982
Email: nswsales@organicmeat.com.au
www.organicmeats.com.au
This butcher began small, located just in one shop while sourcing organic meats from all over New South Wales. Now the brand is found in major supermarkets and it includes organic chicken, pork, beef, lamb, veal, bacon and ham, as well as sausages. Wholesale from: Unit 13b/12 Jusfrute Dr, West Gosford 2250, phone: 02 4322 4528
- **Open:** 7am-7pm weekdays, 7am-4.30pm Sat

RELISH TASMANIA
Molly Alexander
60 Carrington St, Sydney 2000
Phone: 02 9202 2009,
Freecall 1300 722 555
Email: info@relishtasmania.com.au
www.relishtasmania.com.au
All sorts of wonderful Tasmanian produce—wines, olives and olive oil, jams, beauty products, arts and crafts, gift hampers and baskets—shipped anywhere in the world, including Australia of course. This is the first and only store of its kind, located on the Tourism Tasmania site.
- **Open:** 9am-5pm weekdays, 9am-1pm Sat

SYDNEY FISH MARKET
Bank St, Pyrmont 2009
Phone: 02 9004 1122
Email: fishline@sydneyfishmarket.com.au
www.sydneyfishmarket.com.au
This major fish and seafood market is worth a leisurely visit. There are a number of seafood outlets here where you can stock up on seafood to cook at home, as well as cafes and waterside places to enjoy a meal of fresh seafood or simply fish and chips. From 6am on the first and third Thursday of each month you may join a tour to attend the Dutch auction by which retailers purchase seafood, or see Sydney's top chefs selecting their choices. Group tours are

NSW – SYDNEY AND THE SOUTHERN HIGHLANDS

available on request. The public can view the auction on weekday mornings from the public viewing platform, through reception on Level 1 of the main building, Sydney Fish Market.
- **Open:** 7am–4pm daily

SYDNEY MARKETS
Parramatta Rd, Flemington 2129
Phone: 02 9325 6200
Tour info: 02 9325 6295
Email: info@sydneymarkets.com.au
www.sydneymarkets.com.au,
www.paddys.com.au
This massive fruit and vegetable wholesale market is the only wholesale market of its kind open to the public in Australia. Tours are available for groups or individuals, and there is also a growers' market and flower market on the site. Before 9.30am, weekdays, members of the public must pay an entry fee. Tours on the first Friday of the month, and other times.
- **Open:** From 5am Mon, from 6am other days

TOURS

BRADLEYS HEAD BUSHFOOD TOURS
National Parks and Wildlife Service
Cadman's Cottage,
110 George St, The Rocks 2000
Phone: 02 9247 5033
Led by a National Parks Guide, the tours last 1.5 hours and also includes bush medicine and tasting of bush foods. Bookings necessary. Departs opposite the rear entrance of Taronga Park Zoo in the Ashton Park car park.
- **Open:** 1.30pm–3.30pm first Sun of the month

CADI JAM ORA: FIRST ENCOUNTERS
Donna Osland
Royal Botanic Gardens,
Mrs Macquaries Rd, Sydney 2000
Phone: 02 9231 8134
Community.Education@rbgsyd.nsw.gov.au
www.rbgsyd.nsw.gov.au
Take a journey of discovery with an Aboriginal guide and learn about the Royal Botanic Gardens' rich Aboriginal heritage and bush tucker. Find out what this site meant to the Cadigal, the original inhabitants of the Sydney city area, and what it means to Aboriginal people today. One hour self-guided tours return to the Palm Grove Centre. Simply follow the numbers on the map.
- **Open:** By appointment

GOURMET SAFARIS
Maeve O'Meara
Phone: 02 9960 5675, weekdays
Email: info@gourmetsafaris.com.au
www.gourmetsafaris.com.au
Sydney has such a diverse restaurant and dining mix that you need an expert to take you by the hand and explain it to you. Maeve O'Meara, co-author of the successful *SBS Eating Guide to Sydney*, is just the person to do this. The range of Sydney tours is extensive, and includes tastings, information and sometimes a meal. But there are rural options too: blackberry and mushroom picking in Oberon, three-day weekends in Mudgee to coincide with End of Harvest Festival, including a harvest dinner, or a three-day weekend in Orange.
Tours: Approximately twice a month

SYDNEY AND THE SOUTHERN HIGHLANDS—NSW

BOWRAL
STORES

THE CHEESE STORE AT BOWRAL
Liz Telford and Will Laver
Shop 6B/Corbett Plaza,
14 Wingecarribee St, Bowral 2576
Phone: 02 4862 3749
Email: wlaver@austarnet.com.au
This store, opened in 2003 and described as a 'semi-sea change' by the owners, is now the place to come for local cheeses, olives, oil, and all types of local produce, including Doodles Creek and Poacher's Pantry.
■ Open: 9am–5pm weekdays,
8.30am–4pm Sat, 10am–1pm Sun

 DINING

BISTRO MONT
Mark Chance (chef) and Ranee Monaghan
250 Bong Bong St, Bowral 2576
Phone: 02 4862 2677
Email: ranee@hinet.net.au
Licensed/BYO
Cuisine: Modern Australian
Mains average: $$
Specialty: Salt and pepper squid, ice-creams made on site, soufflés
Local flavours: Herbs, lettuce, rhubarb, berries, zucchini flowers, wines
Extras: Large, noisy, popular, with a good wine list. Wood-fired oven for pizzas and takeaway deep-fried and fish and chips. Homemade ice-cream also for sale.
■ Open: Lunch Tues–Fri and Sun, dinner daily

BRIARS COUNTRY LODGE AND INN
Bob and Sarah Adams
Moss Vale Rd, Bowral 2576
Phone: 02 4868 3566

Email: briars@briars.com.au
www.briars.com.au
Licensed
Cuisine: Bistro
Mains average: $$
Specialty: Briars steak to cook yourself
Local flavours: Prime Angus beef and vegetables
Extras: Overlooking the southern ranges, hearty country food, outdoor dining and children's play area.
■ Open: Breakfast, lunch and dinner daily

CAPEL'S BOWRAL
Alexandra Capel and Simon Leese
238 Bong Bong St, Bowral 2576
Phone: 02 4861 5554
Licensed/BYO wine
Cuisine: Modern Australian
Mains average: $$
Specialty: Confit of Maryland duck, seafood platters
Local flavours: Nearly everything, including local wines
Extras: Winner Best Regional Restaurant, Restaurant and Caterers Association Awards, 2003, parking at rear, disabled access, outdoor heated courtyard.
■ Open: Lunch Mon and Wed–Fri, dinner Mon and Wed–Sat

ESCHALOT
Tony Capps and Richard Kemp
17 Links Rd, Bowral 2576
Phone: 02 4861 6177
Licensed
Cuisine: Modern international
Mains average: $$$
Specialty: Duck leg confit with kipfler potatoes and maple roasted eschalots
Local flavours: Meredith ducks from Thirlmere, mushrooms from Mittagong, Robertson potatoes

15

NSW – SYDNEY AND THE SOUTHERN HIGHLANDS

Extras: Cocktail lounge, extensive wine list, opposite golf links and manicured gardens. On the same grounds as Links House, a seventeen-room boutique hotel, also open to the public.
- Open: Brunch, lunch Sun, dinner Tues–Sun

HORDERN'S RESTAURANT
Wayne Stanley
Milton Park Country House Hotel,
Horderns Rd, Bowral 2576
Phone: 02 4861 1522
Email: dine@milton-park.com.au
www.milton-park.com.au
Licensed
Cuisine: Modern Australian
Mains average: $$$
Specialty: Venison or kangaroo game pies with chef's pastry
Local flavours: Berries, mushrooms, fruit and vegetables, seafood from the coast, farmed barramundi and trout.
Extras: Originally the country mansion of the Hordern family early last century.
- Open: Breakfast, lunch and dinner daily

BUNDANOON

PRODUCERS

BUNDANOON VILLAGE NURSERY
Howard Nicholson and Trisha Arbib
71 Penrose Rd, Bundanoon 2578
Phone: 02 4883 6303
Email: nicholbib@ozemail.com.au
This green-fingered couple grow a large and interesting range of rare and unusual food plants, including some bush tucker plants for cold climates. Their twenty-two year passion for growing things, and for food, has led them to become author-

ities on this branch of gardening. Fruits include medlars, damsons, sloes, raisin tree, Chinese quince, and white alpine strawberries. Vegetables include New Zealand yams (oca), Chinese artichokes, yacon, rakkyo, water chestnuts and samphire, as well as culinary herbs. They also offer a mail-order service for out-of-print cooking and gardening books.
- Open: 10am–5pm daily (occasionally closed Tues or Wed)

DINING

CAFE DE RAILLEUR AND YE OLDE BICYCLE SHOPPE
Christine Cole
9 Church St, Bundanoon 2578
Phone: 02 4883 6043
BYO
Cuisine: Healthy international
Mains average: $
Specialty: Caters for people with coeliac disease, diabetes, and most allergies
Local flavours: Free-range eggs, home-grown fruit and vegetables
Extras: Bikes and cafes hardly seem to go together—until you think about hiring a bike and cycling off to the perfect picnic spot. All food, including pastries, made on the premises. Outdoor area and historic mural.
- Open: 10am–4.30pm weekdays, 9am–5pm weekends

BURRAWANG

STORES

MAUGER'S MEAT
John and Vicki Mauger
Hoddle St, Burrawang 2577
Phone: 02 4886 4327

SYDNEY AND THE SOUTHERN HIGHLANDS—NSW

The Mauger family grow their own beef and lamb for sale in this third-generation country butcher's shop. In addition, they make their own sausages and pies, have a catering business, and smoke hams and bacon using pork from Burrengong.
■ **Open:** 8am–5pm weekdays, 8am–12pm Sat

SCARLETT'S EXOTIC FRUIT AND VEG
David Scarlett
96 Hoddle St, Burrawang 2577
Phone: 02 4886 4240
Email: scarlett@myaccess.com.au
The Scarlett's trademark is the very fresh, organic vegetables they sell, made possible because everything is local, and because suppliers grow specially for them. For example, potato growers keep coming up with different strains and these are brought into this grocery shop with its deli section full of good cheeses, and nice touches like the strings of chillies, garlic and onions hanging from the ceiling. There are also Scarlett's Jams, a brainwave from David's mother.
■ **Open:** 8am–5pm weekdays, 8am–1pm weekends

MITTAGONG
DINING

FITZROY INN
Maria and Cosmo Aloi (chef),
Gai (chef) and Paul Lovell
26 Ferguson Cres, Mittagong 2575
Phone: 02 4872 3457
Email: bookings@fitzroyinn.com.au
www.fitzroyinn.com.au
Licensed/BYO
Cuisine: Italian-influenced modern Australian

Mains average: $$$
Specialty: Homemade pasta and gnocchi, pastries
Local flavours: Berries, ducks and quails
Extras: Alfresco dining, and restored historic cellars dating back to 1835. B & B guests have breakfast provided, and may have dinner any night of the week. Small groups often feast in the kitchen.
■ **Open:** Lunch Thurs–Sun, dinner Wed–Sat

ROBERTSON

ROBERTSON POTATOES
R J Mackey
Hoddle St (Illawarra Hwy),
Robertson, 2577
Phone: 02 4885 1329
Those famous Robertson potatoes are available here, even though the signs says that this is a real estate office. In season there are plenty of potatoes sold by the kilo or the bag, and they are worth stopping for.
■ **Open:** 9am–5pm daily

THE ROBERTSON PIE SHOP
Will and Jenny Bleeker
Illawarra Hwy (cnr Jamberoo Rd),
Robertson 2577
Phone: 02 4885 1330
There are dozens of sorts of pies here at this very popular 'you can't miss it' pie shop. Local beef is used in the meat pies.
■ **Open:** 7am–7pm daily

17

SUTTON FOREST
🍎 PRODUCERS

MONTROSE BERRY FARM
Margaret and Robert Hutchison
Ormond St, Sutton Forest 2577
Phone: 02 4868 1544
Montrose Farm is a working berry farm, surrounding a restored 1861 homestead. The farm shop has ready-packed frozen berries, or you can pick your own. The giftwrapped 'Berries in a Basket' filled with jams, jellies and vinegars make lovely presents. The farm grows gooseberries, blueberries, red raspberries, loganberries, blackberries, strawberries, autumn raspberries and red and blackcurrants, which ripen between mid November and the end of April, and asparagus between October and December. Bed and breakfast is available too (think of it, berries for breakfast!) in a self-contained guest cottage.
- **Open:** 10am–4pm Tues–Sat, 11am–4pm Sun
- **Location:** 1.5km along Exeter Rd, towards Bundanoon

WILDES MEADOW
🏠 STORES

FARMGATE AT STATFORD PARK
Barrie Loiterton
Pearsons Lane, Wildes Meadow 2577
Phone: 02 4885 1101
Email: info@statford.com.au
www.statford.com.au
This recently established place has more than twenty thousand lavender plants, as well as olives and vines. The farmgate shop features lavender products, of course—everything from pet shampoos to perfume, aprons to aromatherapy—but there are also honeys, mustards, olives and soaps. Just a walk away, through the gardens, the cellar shop has daily tastings of Statford Park wines. Lavender blooms all year, but from December through summer this lavender garden is awash with brilliant colour.
- **Open:** 10.30am–4.30pm daily

SYDNEY AND THE SOUTHERN HIGHLANDS—NSW

MARKETS AND EVENTS

✦ **BONDI MALL ORGANIC MARKETS,**
BONDI JUNCTION
02 9999 2226
9am–3pm Fri and Sat

✦ **CHATSWOOD ORGANIC MARKET,**
CHATSWOOD
02 9999 2226
7.30am–1pm Sat

✦ **FOX STUDIOS FARMERS' MARKET,**
MOORE PARK
02 9383 4333
10am–4pm Wed,
10am–5pm Sat

✦ **GOOD LIVING GROWERS' MARKETS,** PYRMONT
02 9282 3606
7am–11am first Sat (except Jan)

✦ **HAWKESBURY HARVEST FARMERS' AND GOURMET FOOD MARKET,**
CASTLE HILL
02 4572 6260
8am–1pm second Sat

✦ **LEICHHARDT FARMERS' MARKET,**
LEICHHARDT
02 9999 2227
7.30am–1pm Sat

✦ **MUDGEE WINE AND FOOD FAIR,**
BALMORAL BEACH
02 6372 7409
Mid Aug

✦ **NORTHSIDE PRODUCE MARKET,**
NORTH SYDNEY
02 9922 2299
8am–12pm third Sat

✦ **NSW FARMERS' ASSOCIATION FARMERS' MARKET,** LIVERPOOL
02 8251 1857
8am–12pm Sat

✦ **ORGANIC FOOD AND FARMERS' MARKET,** HORNSBY
02 9999 2226
9am–5pm Thurs

✦ **ST IVES ORGANIC MARKET,** ST IVES
02 9999 2226
8.30am–2pm Wed

✦ **THE ORGANIC MARKET,**
FRENCHS FOREST
02 9999 2227
9am–1pm Sun

✦ **BOWRAL FARMERS' MARKET,**
BOWRAL
02 4861 6498
8am–1pm second Sat

✦ **MITTAGONG FRUIT MARKET,**
MITTAGONG
02 4871 1370
8am–6pm Mon–Fri,
8am–1pm Sat

✦ **GREAT WINE TASTING AND APPRECIATION EXPO,**
BUNDANOON
02 4883 6005
First Sat in Feb

✦ **SOUTHERN HIGHLANDS OLIVE FESTIVAL,** MITTAGONG
02 9427 9797
www.australianolives.com.au
Mar

BLUE MOUNTAINS AND THE HAWKESBURY

So close to Sydney, yet with a life all of its own, the Blue Mountains region forms the essential lungs for the city.

As a food-producing region, the Blue Mountains has enough altitude for apples, pears and stone fruit to thrive.

Best of all, though, the Blue Mountains are just an hour or so from Sydney, so it's an ideal day trip—but make sure you're armed with a picnic hamper or shopping basket.

BILPIN

 **PRODUCERS**

BILPIN SPRINGS ORCHARD
Cedric and Jenny Leathbridge
2550 Bells Line of Rd, Bilpin 2758
Phone: 02 4567 1294
www.bilpinspringsorchard.com.au
Cross your fingers this place is still open, as, at time of going to press, drought had almost crippled it. A great place to come to pick your own fruit (although PYO is not available for the cherries and peaches).

There are cherries, berries and citrus in spring, a stone fruit smorgasbord in summer, and apples and nashi in autumn. Check what's picking at the time by visiting the website or phoning for a recorded message.
■ **Open:** Afternoons on weekends (in season—from Australia Day to Queen's Birthday weekend)

KARAMBI ORCHARD
Neville and Shelley Julian
2814 Bells Line of Rd, Bilpin 2758
Phone: 02 4567 1200
Unlike some other roadside stalls in the area, everything the Julians sell is grown on their own property. They have ten varieties of cherries, seventeen varieties of peaches and nectarines and seven types of plum, as well as apples. Only a few of their apples make it off to market, the rest are snapped up by hungry passers-by.
■ **Open:** 9am–6pm Fri–Mon
● **Location:** 2km west of Bilpin, the first fruit stall past the BP service station

SHIELDS ORCHARD
Bill and Julie Shields
2270 Bells Line of Rd, Bilpin 2758
Phone: 02 4567 1206
Email: bilpinbfb@hawknet.com.au
Now seeing the grandchildren of some of their original customers, the Shields have been selling fruit here since 1955. There are nine peach and at least ten apple varieties, and you may pick your own tree-ripened apples in certain varieties.

THE HAWKESBURY HARVEST FARMGATE TRAIL
Kate Faithorn
PO Box 447, Windsor 2756
Phone: 02 4572 6260
Fax: 02 4560 4400
Email: info@hawkesburyharvest.com.au
www.hawkesburyharvest.com.au

BLUE MOUNTAINS AND THE HAWKESBURY—NSW

Some of the older varieties are maintained because of their eating qualities. Peaches ripen January to February then apples come in between February and May, so there is always a fresh variety being picked.
■ **Open:** 9am–5pm daily (early Jan to late May)
● **Location:** On the left, 1.5km east of Bilpin Village

THE LOCAL HARVEST
Louise and Harry Saarinen
Cnr Johnsons Rd and Bells Line of Rd, Bilpin 2758
Mobile: 0414 671 154
Email: saarinen4@bigpond.com
Apple pies, baked fresh daily, are just some of the temptations at this place. The fruit 'year' begins in early December with peaches and plums, which last through until mid February. Apples continue until the end of June, and there are also local berries in season, vegetables, persimmons, walnuts, chestnuts and eggs. The produce is sold from a roadside shed, and the advice is to 'call first to check out what's in harvest' before coming. Some of the produce is certified organically grown, others are grown in farming systems using integrated pest management, which means minimum insecticides (or none at all) are used to grow the fruit.
■ **Open:** 10am–6pm Fri–Sun (Dec–July)

WIRRANINNA RIDGE
APPLE CIDER VINEGAR
Claudia Kindler
134 Kurts Rd, Bilpin 2758
Phone: 02 4567 1240
Email: claudiakindler@hotmail.com
www.wirraninnaridge.com.au

Claudia Kindler made a mistake once when making natural apple cider—and it worked to everyone's advantage. Those ninety litres were so good that she switched to specialising in pure, mountain apple cider vinegar. Visitors can see the process first-hand if they wish. It involves an initial fermentation of whole, ripe Bilpin apples, to turn the juice into cider. The cider is then cultured, and this changes the alcohol to acid. The entire process takes at least two years and the vinegar is available to purchase here or at local outlets. Claudia also has some particularly fine six-year-old vinegar available.
■ **Open:** By appointment
● **Location:** 3km from the Bilpin Fruit Bowl, turn left off the Bells Line of Road into Kurts Road and continue to the end

 **STORES**

BILPIN FRUIT BOWL
Simon and Joe Tadrosse
2093 Bells Line of Rd, Bilpin 2758
Phone: 02 4567 1152
This popular stop in the apple-growing area of the mountains has fresh seasonal fruit, including local apples and stone fruit, local fruit juices, and the small cafe offers light snacks and coffee.
■ **Open:** 8am–6pm daily

BLACKHEATH
STORES

BLACKHEATH CONTINENTAL DELI
Joe and Chris Frazer
32 Govetts Leap Rd, Blackheath 2785
Phone: 02 4787 8984
Email: jc.frazer@pnc.com

21

NSW—BLUE MOUNTAINS AND THE HAWKESBURY

Local produce, including some from Collits' Inn over the mountains, as well as a range of coeliac products, gourmet sandwiches to eat in or takeaway, smallgoods and cheeses.
- **Open:** 7am–6.15pm weekdays, 7.30am–4.15pm Sat, 8am to 4.15pm Sun

🍴 DINING

ASHCROFTS RESTAURANT
Mary-Jane Craig
18 Govetts Leap Rd, Blackheath 2785
Phone: 02 4787 8297
www.ashcrofts.com
BYO
Cuisine: Modern Australian
Mains average: $$$
Specialty: Slow-roasted crispy-skin duck in Persian aromatics and citrus flavours, or baked vanilla bean custard with spun toffee.
Local flavours: All meats, poultry and game from Wallerawang Butchery—forty minutes' drive west of Blackheath. Takeaway available.
Extras: Photography exhibitions
- **Open:** Dinner Wed–Sun

BUSH ROCK CAFE
Mark (chef) and Annette Kern
198 Evans Lookout Rd, Blackheath 2785
Phone: 02 4787 7111
Email: bushrockcafe@bigpond.com
BYO
Cuisine: Eclectic traditional Australian
Mains average: $
Specialty: Eggs Jackie, sticky black rice, Devonshire teas
Local flavours: Organic wherever available
Makes for sale: Jams, pickles
Extras: Grows its own herbs for herbal teas, log fires, sunny deck, native birds and three acres of bush setting close to National Park. Some take away food.
- **Open:** 9am–5pm weekends and public holidays

BOWEN MOUNTAIN

🍎 PRODUCERS

HONEYSUCKLE COTTAGE
Dr Judyth and Keith McLeod
Lot 35, Bowen Mountain Rd, Bowen Mountain 2753
Phone: 02 4572 1345
Email: kamcleod@zeta.org.au
Honeysuckle Cottage claims to have the largest and rarest herb collection in Australia, and when you consider the 850 herb varieties, you would have to agree. Add to these a hundred or more heirloom tomato varieties, 180 chilli pepper varieties, and hundreds of heirloom vegetables and fruits. In the 'Still Room' there are ornamentals, and a range of seasonally produced herb products and preserves, made to old-fashioned or traditional standards. Honeysuckle Cottage began in 1977, and many plant varieties were first released or have been developed here and are still only available on site, or by mail order. Mouthwatering products include sweet and spicy chilli jelly, Tahitian lime jelly, and brandied peach sauce, but there are many more available as well as dried or potted herbs, potted tomatoes and chillies. It's apparent that the McLeods have made propagation of these rare and heirloom plants an artform and a life-work, and although plants are for sale, their expertise is freely shared with visitors, and their enthusiasm for the task is infectious.

- **Open:** 10am–4.30pm Fri and Mon, 9am–5pm Sat-Sun, other times by appointment

CASTLEREAGH
🍎 PRODUCERS

PENRITH VALLEY ORANGES
Ivan and Lyn Glover
800 Castlereagh Rd, Castlereagh 2749
Phone: 02 4776 2332
Email: penvalorange@bigpond.com.au
www.penrithvalleyoranges.com.au
How good is this? You can plan a picnic at this orange orchard and even pick your own fruit—navel oranges in spring, then Valencias until Christmas. 'We always receive the comment "These are so sweet and juicy!"' say the Glovers. Tours, including school tours, are free, but you will need a minimum of ten people. These will demonstrate how the Glovers manage the orchard, picking, grading and packing the fruit for sale. Presently there are 5000 trees with new plantings all the time and no insecticides are used. The farm also sells local vegetables in season, local pure honey and free-range eggs.
- **Open:** 9am–5pm daily (12 June–24 Dec)
- **Location:** Midway between Penrith and Richmond

EBENEZER
🍴 DINING

TIZZANA WINERY
Peter and Carolyn Auld
518 Tizzana Rd, Ebenezer 2756
Phone: 02 4579 1150
Email: enquiries@tizzana.com.au
www.tizzana.com.au

Licensed
Cuisine: Country style
Platters: $$
Local flavours: Uses mainly Hawkesbury Harvest produce on cheese and antipasto platter available at the cellar door—Willowbrae chevre cheese, Windsor Smokehouse salmon, Kurrajong Kitchens lavosh biscuits, Tizzana olives and wine.
Extras: A restored 1887 National Trust classified historic winery, with views overlooking vineyards, and a lake with waterlilies. Catering also available for functions.
- **Open:** 12pm–6pm weekends, other days by arrangement

GROSE VALE
🍎 🍴 PRODUCERS/DINING

ENNISKILLEN ORCHARD
John and Patricia Maguire
753 Grose Vale Rd, Grose Vale 2753
Phone: 02 4572 1124
Mobile: 0412 649 651
Email: maguires@hawknet.com.au
John Maguire is a man who cannot be stopped. With part of his packing shed a market for fresh local fruit and vegetables, and the other half of the shed a small cafe, he has also planted a herb garden behind it, aided by herb guru, Ian 'Herbie' Hemphill. The veranda overlooks the orchard with the magnificent Grose Valley beyond, making this a great coffee stop for the traveller—or the herb lover.
- **Open:** 10am–6pm weekdays, 9.30am–6pm weekends

HARTLEY VALE

🍴 DINING

COLLITS' INN
Cyrillia Van Der Merwe (manager), Laurent Deslandes (chef), Christine and Russell Stewart (owners)
Hartley Vale Rd, Hartley Vale 2790
Phone: 02 6355 2072
Email: info@collitsinn.com.au
www.collitsinn.com.au
Licensed
Cuisine: French country
Two- or three-course menus: $$$
Specialty: Goats cheese galette, degustation menu that changes monthly
Local flavours: As much as possible
Extras: Since opening in early 2002 Collits' Inn has received many awards. One of Australia's earliest inns, built in 1823, this is the Blue Mountains' oldest building. There is a history room, early cemetery, shop and accommodation for up to sixteen people. A range of homemade jams and chutneys are available.
■ **Open:** Breakfast daily (guests), lunch Fri–Sun, dinner Thurs–Sun

KATOOMBA

🏠 STORES

HOMINY BAKERY
Jenny Ingall and Brent Hersee
185 Katoomba St, Katoomba 2780
Phone: 02 4782 9816
Sourdough traps the wild yeasts from the air, so you could hardly find a more 'regional' food. Hominy's starter dough is sixteen years old and has complex flavours. The baker trained with Poilane in Paris, and the range of authentic, handmade sourdough breads is extensive and well liked by locals and visitors. Look for plain or wholemeal sourdough every day, and pumpkin or potato sourdough several days a week. Then there are daily specials such as walnut or fruit loaves.
■ **Open:** 6am–5.30pm daily

🍴 DINING

DARLEY'S
Hugh Whitehouse (chef)
Lilianfels Blue Mountains Resort and Spa
Lilianfels Ave, Katoomba 2780
Phone: 02 4780 1200
Email: reservations@lilianfels.com.au
www.orient-express.com
Licensed
Cuisine: Australian
Mains average: $$$
Local flavours: Meat, fowl, salad greens, and herbs
Extras: Darley's is located in a heritage-listed house on Lilianfels' grounds, with over one hundred years of history behind it, and stunning views of the mountains from the veranda. There is also a Mediterranean cuisine restaurant, Tre Sorelle, and the Lobby Bar at this highly acclaimed boutique accommodation.
■ **Open:** Dinner daily

ECHOES BOUTIQUE HOTEL AND RESTAURANT
Colin and Gai O'Brien (owners), Peter van Kruyssen and Mark McCluskey (chefs)
3 Lilianfels St, Katoomba 2780
Phone: 02 4782 1966
Email: enquiries@echoeshotel.com.au
www.echoeshotel.com.au
Licensed
Cuisine: Country French, northern Italian
Two courses: $$$

BLUE MOUNTAINS AND THE HAWKESBURY—NSW

Local flavours: As much as possible, as well as veal from Cowra
Extras: Located near Echo Point on the escarpment with a terrace and conservatory overlooking the Jamison Valley. Luxury boutique accommodation.
- **Open:** Breakfast daily, lunch and dinner Thurs–Tues

KURRAJONG HILLS
DINING

PATRICE'S TABLE AT LOXLEY ON BELLBIRD HILL
Paul Maher (owner), Patrice Falantin (chef)
993 Bells Line of Rd, Kurrajong Hills 2758
Phone: 02 4567 7711
Email: patricefalantin@froggy.com.au;
loxley@iprimus.com.au
www.loxleyonbellbirdhill.com.au
Licensed
Cuisine: French
Mains average: $$$
Specialty: Fresh herbs, and produce from the Farm Gate Trail producers
Extras: An award-winning restaurant and accommodation, with sweeping views towards Sydney. Rainforest, outdoor reception area, spring-water pool, two hundred-year-old kauri tree, hundred-year-old Loxley Cottage.
- **Open:** Breakfast, lunch and dinner daily

LEURA
STORES

LEURA BUTCHERY
Kane and Janine Hughan,
176 The Mall, Leura 2780
Phone: 02 4782 4411

This butcher specialises in two things: highest quality country-killed meats, mainly from Cowra, and a large range of gourmet sausages produced on the premises, as well as free-range poultry.
- **Open:** 7.30am–5.30pm weekdays, 7.30am–12pm Sat

DINING

CAFE BON TON
Ralph Potter
192 The Mall, Leura 2780
Phone: 02 4782 4377
Email: bonton@ozemail.com.au
Licensed/BYO
Cuisine: Mod Oz bistro
Mains average: $$$
Specialty: Braised pork cheeks (dinner only)
Local flavours: Goats cheese, local wines
Extras: Large outdoor courtyard under shady trees.
- **Open:** Breakfast and lunch daily, dinner Fri–Sun

SILKS BRASSERIE
Stewart Robinson (owner),
David Waddington (chef)
128 The Mall, Leura 2780
Phone: 02 4784 2534
Email: enquiries@silksleura.com
www.silksleura.com
Licensed
Cuisine: Modern Australian, with a strong French influence
Mains average: $$$
Local flavours: Varies seasonally, including local wines, hazelnuts from Oberon, Jannei goat curd
Extras: An elegant, high-ceilinged dining room with a Parisian-style bar. Fireplace for winter and fully airconditioned in summer.
- **Open:** Lunch and dinner daily

LEURA FALLS
🍴 DINING

SOLITARY
Georgia Shepherd and John Cross
90 Cliff Dr, Leura Falls 2780
Phone: 02 4782 1164
Email: info@solitary.com.au
www.solitary.com.au
Licensed
Cuisine: Modern Australian
Mains average: $$$
Specialty: Dessert and entree soufflés
Local flavours: Produce and wines of the Central West, including Genoa figs from Orange, Mudgee venison, and in autumn forest mushrooms collected in and around the mountains
Extras: The highly acclaimed Solitary was established in 2000 and is one of only two remaining original Blue Mountains kiosks. The heritage-listed building is set in a landscaped garden on its own promontory with magnificent views over the Jamison Valley.
■ **Open:** Lunch weekends, dinner Wed–Sun, kiosk open 10am–4pm daily

LITHGOW
🍎 PRODUCERS

ARCHVALE TROUT FARM
Jill Cutcliffe
Hughes Lane, Marrangaroo, Lithgow 2790
Phone: 02 6352 1341
A great day out for the family—or maybe just a mad keen fisher-person or two. If you can't fish, there are fresh trout, smoked trout, and barbecue facilities, as well as a kiosk. If you do catch your own fish, you may cook it on the barbecue, and there are fingerlings available too.
■ **Open:** Wed–Mon 9am–5pm, (9am–5pm daily during school holidays)

LITTLE HARTLEY
🍴 DINING

HARTLEY VALLEY TEAHOUSE
Duncan and Barbara Wass
5 Baaners Lane, Little Hartley 2790
Phone: 02 6355 2043
Email: dunbar@pnc.com.au
www.hartleyvalley.com.au
BYO
Cuisine: Teas and light lunches
Mains average: $
Specialty: Warm chicken salad, waffles with own brand of liqueur sauces
Local flavours: Pies, free-range eggs, breads
Extras: Eccentric teapot collection, veranda dining, gardens with views of the escarpment, and boutique nursery. A range of homemade gourmet products, sauces and dressings is for sale.
■ **Open:** 9am–5pm Thurs–Mon (breakfast and lunch)
● **Location:** Cnr of Great Western Hwy and Baaners Lane, 6km after Mt Victoria. Look for a 1934 Bedford truck parked near the dam.

MEGALONG VALLEY
🍴 DINING

MEGALONG TEAROOMS AND KIOSK
Christine Bundy
Megalong Rd, Megalong Valley 2785
Phone: 02 4787 9181

Cuisine: Home-cooked light meals
Mains average: $
Specialty: Freshly baked scones and blackberry jam, apple pie (made to grandma's recipes)
Local flavours: Blackberries, apples, apple juice
Extras: Cosy log fires, veranda dining with stunning views towards the Hydro Majestic and the Blue Mountains cliffs.
■ Open: Breakfast, lunch, morning and afternoon teas daily
● Location: 10km from Blackheath, through rainforest

MEGALONG VALLEY HERITAGE FARM
Gary and Glenda Lane
Megalong Rd, Megalong Valley 2785
Phone: 02 4787 8688
Email: megalong@megalong.cc
www.megalong.cc
Licensed/BYO
Cuisine: Traditional Australian
Local flavours: Steaks, local farmed fish, and gourmet pies
Extras: A unique way to dine with horse and carriage meals and great views over the valley. These picnics with a difference include a carriage and staff to attend you in spectacular surroundings.
■ Open: Lunch and dinner daily. Bookings essential for dinner.

MOUNT IRVINE
 PRODUCERS

KOOKOOTONGA
Bill and Ruth Scrivener
247 Mount Irvine Rd, Mount Irvine 2786
Phone: 02 4756 2136
You can pick your own chestnuts and walnuts in season at Kookootonga, a property which is over a hundred years old. There are three varieties of chestnuts here and the walnuts are Irvine walnuts. Home-grown pickled walnuts are also available.
■ Open: 8am–4pm daily (20 Mar–25 Apr only)
● Location: 20km from Richmond on Hwy 40

MOUNT TOMAH
🍴 DINING

MT TOMAH RESTAURANT
Jason Saville
Mount Tomah Botanic Gardens,
Bells Line of Rd, Mount Tomah 2758
Phone: 02 4567 2060
Email: mttomah@dnd.com.au
www.rbgsyd.nsw.gov.au
Licensed/BYO
Cuisine: Modern Australian

MARKETS AND EVENTS

✦ **CAMDEN PRODUCE MARKET,** CAMDEN
7am–12pm, second and fourth Sat

✦ **THE BILPIN BITE,** BILPIN
02 4567 1287
10am–4pm first Sun after Easter

✦ **FESTIVAL OF HERBS,**
WENTWORTH FALLS
02 4751 5163
March

✦ **YULEFEST,** KATOOMBA
02 6355 6200
www.wonderland.bluemts.com.au/yulefest
July–August various venues

Mains average: $$$
Specialty: Rabbit and wild mushroom pies
Local flavours: Olive oil, Mudgee Blue Wren verjuice, apple juice
Extras: Outdoor deck dining with fantastic views across the botanic gardens and beyond to the Brokenback Ranges. Regular exhibitions in the gardens.
■ Open: 9.45am–4.45pm daily (during daylight saving time), 9.45am–3.45pm daily (Mar–Sept)

MOUNT VICTORIA
DINING

MAGNOLIA RESTAURANT
Jason Saville (owner),
Robbie Stapleton (head chef)
Park West Boutique Hotel,
84 Great Western Hwy,
Mount Victoria 2786
Phone: 02 4787 1112
Email: parkwest@dnd.com.au
www.parkwest.com.au
Licensed
Cuisine: Modern Australian
Mains average: $$$
Specialty: Double-roasted duck with honey and coriander, desserts
Local flavours: As many as possible
Extras: Private garden available for wedding ceremonies. Outdoor dining with mountain views. Cooking classes also available. Nine guestrooms.
■ Open: Breakfast daily (in-house guests), lunch weekends (bookings only), dinner Fri–Sun (bookings only), dinner also Mon–Thurs for in-house guests

SOUTH BOWENFELS
DINING

BLUE MOUNTAINS PARADISE
Victoria Choo
Lot 3, Sir Thomas Mitchell Dr,
South Bowenfels 2790
Phone: 02 6352 3122
Email: theteashack@ozemail.com.au
www.bluemountainsparadise.com
Cuisine: Japanese teahouse
Mains average: $
Specialty: Thirty-two types of organic herbal teas, produced on site
Extras: The tea room is located in the bath house, facing the lake and mountains. Indoor and outdoor mineral hot spring pool, indoor cold spring pool, herbal steam bath, reflexology in the Zen garden (as a foot massage), Japanese deep tissue massage. Teas are also sold at city produce markets to naturopaths, herbalists and five-star restaurants.
■ Open: Lunch weekends, dinner Fri–Sun (2pm–9pm)
● Location: Take Magpie Hollow Rd, then turn right at the yellow teapot sign into Sir Thomas Mitchell Dr.

THE OAKS
DINING

POSSUM FRUITS GOURMET FOODS
David Constantine
Shop 2/83 John St, The Oaks 2570
Phone: 02 4657 1455
Email: possumfruits@bigpond.com
www.possumfruits.com.au

BLUE MOUNTAINS AND THE HAWKESBURY—NSW

> **GARDEN SPECIALS**
> The Garden Shops at Mount Annan, Mount Tomah and the Royal Botanic Gardens, Sydney all stock bush foods as part of their normal retail range. Items range from lilly pilly jam to native hibiscus flowers in syrup. There are guided walks through both Mount Annan and the Royal Botanic Gardens in Sydney that specifically relate to indigenous foods.

Licensed/BYO
Cuisine: Modern Australian
Mains average: $$
Specialty: Bruschetta, lamb shanks, pancakes
Local flavours: Fresh herbs, free-range eggs, mushrooms, local bread
Extras: Intimate dining in a heritage building with a small but private outdoor garden area. Chef's cooking classes on Wednesday and Thursday nights. Homemade olive tapenade and infused oils available for sale. Takeaway available.
■ **Open:** Breakfast and lunch Thurs–Sun, dinner Fri and Sat
● **Location:** Turn off toward Camden along Camden Bypass, after the bridge, turn right to The Oaks

WENTWORTH FALLS
DINING

CONSERVATION HUT CAFE
Ralph Potter
End of Fletcher St, Wentworth Falls 2782
Phone: 02 4757 3827

BYO
Mains average: $
Specialty: Great sandwiches, hearty affordable lunches and afternoon teas
Local flavours: All produce and bread from local suppliers
Extras: Information on Blue Mountains bushwalks. Stunning views of the Jamison Valley. Enjoy great coffee in front of an open fire on a crisp winter's day. Takeaway available.
■ **Open:** 9am–5pm, daily

WILBERFORCE
 PRODUCERS

WILLOWBRAE CHEVRE CHEESE
Karen Borg
143 Singleton Rd, Wilberforce 2756
Phone: 02 4575 1077
Email: willowbrae@bigpond.com
It all began with a love of goat's cheese, two lactating nannies and a book on how to make goat's cheese. Between them, Karen and the goats must have done something right because there is now a herd of one hundred goats on a six-acre block on the outskirts of Sydney's metropolitan area. This boutique cheese producer now makes ten varieties of cheeses from super-fresh cheese aged for only eight hours, to nine-month-old mature blues and whites. At present, Borg is experimenting with other mature cheeses and washed rinds. She milks twice a day, and in the warmer months makes cheese every day. Available from the farm and found at a number of growers' markets around Sydney.
■ **Open:** By appointment

29

CANBERRA, CAPITAL COUNTRY AND THE SNOWY MOUNTAINS

The Australian Capital Territory covers just 2400 square kilometres, much of which is mountainous and sparsely populated, yet somehow the territory manages to produce a wide range of fruit and other foods and is nurturing a growing wine industry.

Canberra, the Federal Capital, and newish by Australian standards, was begun in 1909, and features circular town-planning in many areas.

Base for many overseas embassies and consulates, Canberra's cosmopolitan flavour spills over into its restaurants, shops and, inevitably, foodstuffs. The city is notable for its many markets, and is a vibrant place to shop at weekends, particularly when, even on the bleakest Canberra day, they glow with produce—much of it local.

Nestled within New South Wales, it would be impossible to enforce too strictly the territory-state borders. It is inevitable that much of Canberra's 'local' produce stems from nearby mountain and highland regions in New South Wales.

CANBERRA
🍎 PRODUCERS

GOURMET TASTE BUD
Marion Lea
33 Bettington Circuit, Charnwood 2615
Phone: 02 6258 6300
Email: gtb@austarmetro.com.au

Here's a saucy solution. Marion's husband makes the boxes that protect and display the sauces, chutneys, jams and marmalades that Marion lovingly creates from local produce. 'I use as much local fruit and vegetables as I can,' she says. She makes twenty varieties, including a chilli plum sauce, mellow mango chutney, apricot chilli jam, spicy lime pickle, cumquat ginger marmalade or the ever-popular mint sauce. Sold in assorted gift boxes, singly, or in hampers, they make great gifts, and are available at the weekly Old Bus Depot markets in Kingston. Handmade Christmas puddings and a special range of chocolate orange, chocolate rum and raisin and chocolate macadamia Christmas puddings are available late each year.
■ **Open:** By appointment

PIALLIGO APPLES
Robyn and Jonathan Banks
10 Beltana Rd, Pialligo 2609
Phone: 02 6248 9228
Email: apples3@bigpond.com
As owner of an orchard bursting with over sixty different varieties of organic apples, as well as quinces and summer vegetables, Dr Jonathan Banks admits to finding some of the older apple varieties hard to resist—hence, his growing collection. There is an apple stall on the roadside, open in season, and here you will also find homemade cider vinegar, fresh apple juice, jams and preserves (classic preserves, Jonathan calls them)

as this organic BFA-certified orchard also grows stone fruits and some blackberries.
- **Open:** 9am–5pm weekends in apple season (Feb–Apr)
- **Location:** Turn at the last roundabout before Canberra Airport

STORES/CLASSES

AS NATURE INTENDED
Michael Sirr
Shop 40A, Belconnen Fresh Food Markets, Lathlain St, Belconnen 2617
Phone: 02 6253 0444
www.asnatureintended.com.au

As you'd expect from the name, this store features a great selection of top-quality, fresh, 100 per cent certified-organic and biodynamic dairy goods, bulk natural foods, sugar-free natural snacks, gluten-free specialty products, organic breakfast cereals, soymilks and grain milks. There are daily deliveries of quality breads, including organic sourdoughs from a number of small, specialty, organic bakers. Only fresh fruit, vegetables and herbs that have been certified organic or biodynamic are allowed in the door, and there's a continual supply of fresh produce throughout the week and a close relationship with the growers. There is also an organic and vegetarian cafe on site.
- **Open:** 8.30am–6pm Tues–Sun

CANBERRA FRESH FRUIT MARKET
Frank and Joe Spagnolo
Shop 17, City Markets, Bunda St, Canberra City 2601
Phone: 02 6248 8269
Email: canberrafresh@bigpond.com

If Joe Spagnolo finds time to sleep, you can bet he dreams about how to make Canberra Fresh even better. This squeaky-clean place could be a model for any aspiring fruit-and-vegetable vendor. Apart from an almost entire range of everything edible from the plant kingdom, there are neat additions such as nutritional information on the name tags over each glowing heap of produce. You'll find perhaps the largest range of greens and mushrooms you've ever seen. Joe buys local blackberries and raspberries, accounting for entire crops in some cases, and freezes them. Recently the gourmet lines have expanded, adding long-life juices, more mushrooms and sun-dried tomatoes, and an even wider selection of herbs.
- **Open:** 8am–8pm Mon–Thurs, 8am–9.30pm Fri, 8am–7pm weekends

COOKING COORDINATES
Jan Martin-Brown and Liz Posmyk
Shops 39/40, Belconnen Fresh Food Markets, Lathlain St, Belconnen 2617
Phone: 02 6253 5133 (shop), 02 6253 5132 (school)
Email: markets@dynamite.com.au
www.belconnenmarkets.com.au

Cooking Coordinates is a veritable treasure house of kitchen gadgets and cook's equipment, together with a range of gourmet ingredients from around the world including the best of award-winning Australian olive oils, Herbie's spices, Maggie Beer products and Charmaine Solomon's pastes and marinades. Next door is a state-of-the-art cooking school with an auditorium seating up to ninety guests, offering a diverse program of cookery classes and events. Well-known chefs such as Rick Stein, Margaret Fulton, Belinda Jeffery, Gabriel Gaté, Ian Parmenter, Geoff Jansz, Kurma

NSW—CANBERRA, CAPITAL COUNTRY AND THE SNOWY MOUNTAINS

Dasa, Charmaine Solomon, Maggie Beer, Stephanie Alexander, Ian Hemphill and Elizabeth Chong are in the high profile line-up of those who have presented here, plus there are school holiday cooking classes for children aged between five and fifteen years.
- **Open:** 9am–5.30pm Wed–Sun, classes as advertised

ECO MEATS
Gino D'Ambrosio
Shop 41, Belconnen Fresh Food Markets,
Lathlain St, Belconnen 2617
Phone: 02 6251 9018
Email: ecomeats@bigpond.com
'The preservative-free, gluten-free organic sausages are a favourite with our customers,' says Gino D'Ambrosio, owner of Eco Meats. 'We carry an excellent range of organic beef and lamb cuts including porterhouse, rump and T-bone steaks, fat-reduced mince, fillet and kebabs, fat trimmed cutlets and chops, as well as organic pork and organic free-range chickens.' For those whose culinary skills venture into the wild side of things, Gino supplies a fabulous range of game meats. There's wild rabbit, kangaroo, crocodile, emu, ostrich, venison and even Tasmanian possum and wallaby! Drawing on his Italian background, Gino now offers some of the best peperoni, cabanossi and kransky, made on site. There's also trout, salmon, ham, bacon, chicken and turkey fresh from Gino's smokehouse, as well as a range of fine cheeses, homemade pasta, organic foods, Italian and Turkish bread, bagels, dips, pestos and pizzas.
- **Open:** 8am–6pm Wed–Sun

FOOD LOVERS
Peter Turner, Michele Menchin and Brian Lawler
Shop 36, Belconnen Fresh Food Markets,
Lathlain St, Belconnen 2617
Phone: 02 6253 5079
Email: foodlovers@austarnet.com.au
www.foodlovers.com.au
The Food Lovers story is that of three friends with a passion for food and an enthusiasm for spending many hours in the kitchen. Together they create over 500 handmade products in small batches, including sauces, mustards, dressings, jams, marmalades, jellies, chutneys, relishes, biscuits, cakes, biscotti, oils, chilli, vinegars, preserved fruits and vegetables...the list goes on! Fresh tapenades and spreads, pestos, and pasta sauces, gourmet bush food products, hampers—it's all here.
- **Open:** 8.30am–6pm Wed–Sun

GRIFFITH BUTCHERY AND BAKERY
Richard Odell
10 Barker St, Griffith 2603
Phone: 02 6295 9781
Email: info@griffithbutchery.com.au
www.griffithbutchery.com.au
If you have limited time in Canberra, make sure you squeeze in a visit to this place. Richard Odell is determined that his shop only sell the best of all things. And that includes a wide range of bio-dynamic beef, which he hangs to achieve optimum tenderness and flavour, as well as lamb, pork and veal, organic chicken and other organic meats. The shop also sells free-range eggs from hens fed on organic grains. The bakery makes breads as well as pasties and pies using Demeter biodynamic flour.
- **Open:** 7.30am–6.30pm weekdays, 7.30am–1.30pm Sat

CANBERRA, CAPITAL COUNTRY AND THE SNOWY MOUNTAINS—NSW

ORGANIC ENERGY
Karen Medbury
Shop 8A, Barker St, Griffith 2603
Phone: 02 6295 6700
You can stock up on all your organic fruit and vegetables here, in a very foody street. A whimsical touch is the gardening glove in the crate of potatoes, a Christmas present from a customer to help others keep clean.
■ **Open:** 9am–7pm weekdays, 9am–5pm Sat

 MARKETS

OLD BUS DEPOT MARKETS
Morna Whiting
27 Wentworth Ave, Kingston 2604
Phone: 02 6239 5306 or 02 6292 8391
www.obdm.com.au
Free admission, ample parking—and those inducements are just for starters, before you even get in the door. Once there, you will be bowled over by the wealth and variety of the stands in this old industrial building, which has been splendidly revived for these weekly markets. Come here for everything from honey and jam to bread and coffee, fruit and vegetables, cheese, smoked trout, gifts and arts and crafts. It's the ideal place to have a quick taste around the region, as producers compete for a spot to display their goods.
■ **Open:** 10am–4pm Sun

ARALUEN
 PRODUCERS

HARRISON AND SONS ORCHARD
Keith and Mary Harrison, Ken and Tracey Harrison
'Araglen', Araluen 2622
Phone: 02 4846 4017

The legendary Araluen peaches and nectarines have been grown here for forty-five years or more. Ken Harrison's grandfather started the business, which now sells to both the domestic and overseas markets. The family is more than happy for customers to drop in, even though the orchard is a little off the beaten track. Follow the directions, and you can see what Limousin cattle look like, at the same time.
■ **Open:** 8am–5pm daily (mid Nov–mid Jan) or by appointment at other times.
● **Location:** From Moruya, follow signs to Araluen for 60km, unsealed road follows Deua National Park and River. From Braidwood, follow signs to Araluen for 25km; watch for 'peach' sign.

BATLOW
 PRODUCERS

SPRINGFIELD ORCHARD
Peter and Enid Wilkinson
Batlow Rd, Batlow 2730
Phone: 02 6949 1021
Email: sfield@dragnet.com.au
Here you'll find nineteen varieties of apples (including Fuji and Pink Lady) available year-round at the fruit stall. There is a full range of tree-ripened stone fruits from mid December to Easter, cherries (mid-to-late December), and chestnuts in season, and there is an orchard shop and picnic area in which to enjoy them.
■ **Open:** 8am–5pm daily
● **Location:** 6km north of Batlow

WILGRO ORCHARD ROADSIDE STALL
Ralph Wilson
4066 Batlow Rd, Batlow 2730
Phone: 02 6949 1224
Email: ralphjw@bigpond.com

The Wilsons sell a wide range of both new and old varieties that have been tree-ripened to taste the way an apple should taste. At the roadside stall there are high quality apples (March to May), pears, peaches, nectarines, cherries (mid December to mid January), plums, apricots and quinces as well as berry fruits like blackberries, raspberries, strawberries and currants in season.

- **Open:** 8.30am–5pm daily (Dec–Jan, Mar–May)
- **Location:** 4km north of Batlow

STORES

BATLOW FRUIT COOPERATIVE LTD
Lachlan Moore
Forest Rd, Batlow 2730
Phone: 02 6949 1408
lachlan.moore@batlowapples.com.au
www.batlowapples.com.au
Primarily concerned with the storage, packing and distribution of fresh fruit such as peaches, nectarines and the well-respected Batlow apples, this co-op commenced in 1922 as a small facility in response to the rapid growth of the local fruit industry. Today it has a turnover of more than 800 000 cartons, which is around eleven per cent of Australia's annual apple and stone fruit production. Batlow apples include Red Delicious, Granny Smith, Bonza, Red Fuji, Pink Lady, Braeburn and Royal Gala varieties and are sold nationwide. Apples are sorted using a sophisticated computer video scanner that views the apples from many angles before sorting them. Some apples are stored in controlled atmosphere (CA) storage that does not affect their constitution.

- **Open:** 8am–5pm, daily

BERRIDALE

PRODUCERS

DELEGATE RIVER DEER FARM
Walter Widmer
'Ruetihof', Hilltop Rd, Berridale 2628
Phone: 02 6456 7152
Email: widmer@acr.net.au
The Widmers are happy to sell frozen Denver legs and their excellent venison steaks, if ordered in advance. They run 300 red deer. When the farm began nearly twenty years ago, it was the first in the area, but today their venison is used in top restaurants. Eddie Hope, at the nearby Nimmitabel Butchery in Cooma, makes sausages with their venison.

- **Open:** By appointment

BINALONG

DINING

THE BLACK SWAN
Kevin and Marita Gallimore
Stephens St, Binalong 2584
Phone: 02 6227 4236
Email: blackswanrest@aol.com
Licensed
Cuisine: Modern Australian
Mains average: $$
Specialty: Garlic prawns, duck, kangaroo
Local flavours: Wattleseed bread
Extras: The warmth and hospitality that is special to the country helped along by open fires in the colder months. This charming old-world building is a fully restored Cobb and Co Inn built in 1847. The town's name is pronounced 'bine-along' and has a wonderful rustic feel.

- **Open:** Lunch Sun, dinner Fri–Sun (bookings for groups)

BOOROWA

🍎 PRODUCERS

WALSH'S COUNTRY KITCHEN
Roland and Cathy Walsh
33 Campbell St, Boorowa 2586
Phone: 02 6385 3251
Email: walshshomemade@bigpond.com
Take home the flavours of the region in a jar or a packet. Here there are homemade jams, preserves, rich fruitcakes, and Christmas puddings, made with fruit from Young. A butcher by trade, Roland Walsh always wanted to live in the country, and ten years ago he saw this business as the ideal way to combine his dream with a career in the food industry. The fruit comes straight from the growers and the Walshs have a range of jams such as strawberry, raspberry, blackberry, fig, quince jelly, and apricot, as well as marmalades, chutneys, relishes and sauces. Available in the local region from shops in Boorowa, Young, Yass and Canberra.
- **Open:** By appointment

BRAIDWOOD

🍴 DINING

THE DONCASTER SMALL LUXURY HOTEL
Jon Davies and Chris Seaman
1 Wallace St, Braidwood 2622
Phone: 02 4842 2356
Email: thedoncaster@bigpond.com
www.doncasterinn.com.au
Licensed
Cuisine: Australian
Mains average: $$
Local flavours: Trout, organic chicken, seafood, cheeses, wine

Extras: The restaurant is located in The Doncaster's chapel, originally used by the Order of Good Samaritan Nuns from the early 1900s. Now a comfortable B & B with lovely gardens.
- **Open:** Breakfast daily (for guests), lunch weekends, dinner Fri-Mon

COLLECTOR

🍴 DINING

BUSHRANGER HOTEL
24 Church St, Collector 2581
Phone: 02 4848 0071
www.gdaypubs.com.au/collector
Licensed
Cuisine: Pub meals during the week and à la carte on Saturday nights
Mains average: $$
Specialty: Pizzas, pub grub
Local flavours: All meats are local
Extras: A very popular and welcoming spot with good quality food. Open fireplace and accommodation. A memorial to fallen policeman, Samuel Nelson, stands next to the pub. Nelson was killed by Johnny Dunn, one of bushranger Ben Hall's cobbers. A good example of what a NSW country pub should be like.
- **Open:** Lunch and dinner Wed-Sun

LYNWOOD CAFE
Robin and Alan Howard
1 Murray St, Collector 2581
Phone: 02 4848 0081
Email: lynwood@goulburn.net.au
www.lynwoodcafe.com.au
Licensed/BYO
Cuisine: Modern regional Australian
Mains: $$
Specialty: Jams, preserves and chutneys
Local flavours: Fruits and vegetables, meat

Extras: The jam room, plus locally produced cakes, sweets, ceramics and gifts, outdoor dining, open fireplaces. Robbie's jams, preserves and chutneys are made in individual copper-pot batches. The produce is sourced from local orchards and properties and often features unusual and old varieties.

- **Open:** 10am–5pm Wed–Sun, dinner Fri–Sat

COOMA

🍎 PRODUCERS

HERB OF GRACE ORGANIC HERBS
Margaret Reynolds
'Langrove', Pine Valley, Cooma 2630
Phone: 02 6452 4511
Here you'll find dried culinary herbs and teas, some of them medicinal, as well as the packaging plant where you can see how the herbs are processed and packed. There is a large garden of herbs, with twenty to thirty varieties. The herbs are also sold at stalls and functions locally and through naturopaths and other outlets.

- **Open:** 10am–4pm Thurs–Fri (sometimes Mon, Sat or Sun)
- **Location:** 6km west of Cooma on Snowy Mountains Hwy, towards snowfields

🏠 STORES

NIMMITABEL BUTCHERY
Eddie Hope
225 Sharp St, Cooma 2630
Phone: 02 6452 7800
Email: nimbu@acr.net.au
Everyone in the region tells you about Eddie Hope. No wonder, as his tasty gourmet sausages have won prizes for many years. He makes twenty-six types of gourmet sausages, including tasty beef—thick and thin—chicken, pork and lamb, herb and garlic, and Mexican beef curry, and gets his meat from Cowra. Hope also makes kangaroo sausages and venison sausages for Delegate Deer Farm. His smallgoods have won many awards at the Sydney Royal Easter Show, including the 2002 first prize for most innovative sausage—a skinless lamb roulette wrapped in bacon. In 2003 he won third place in New South Wales with a red meat innovation product, 'beef rouladen'—schnitzel spread with French mustard, filled with spinach and cheese, then tied. His latest invention is a sweet beef sausage using honey.

- **Open:** 9am–6pm weekdays (Wed to 5pm, Fri to 8pm), 9am–1pm Sat, 10am–5pm Sun

🍴 DINING

ELEVATION RESTAURANT
David McDougall and Marianne Webb (owners), Ellen Webb (chef)
Alpine Hotel, 170 Sharp St, Cooma 2630
Phone: 02 6452 5151 or 02 6452 1466
Email: alpinehotel@bigpond.com.au
Licensed
Cuisine: Modern Australian
Mains average: $$
Specialty: Snowy Mountains trout, regional beef
Local flavours: Trout, beef, berries, cheese, wines
Extras: Opened in 2003, the restaurant, decorated in art deco style, was that year's winner of Best Restaurant in a Hotel/Motel/Guest House Category (NSW Restaurant and Catering Association).

- **Open:** Lunch and dinner daily (counter meals), Restaurant—lunch daily, dinner Tues–Sat

CROOKWELL

🍎 PRODUCERS

DOONKUNA ORCHARD
Gary and Sue Armstrong
Binda Rd, Crookwell 2583
Phone: 02 4832 1127
Email: gs.armstrong@bigpond.com
You could picnic on the good food from this orchard—apples and pears (several varieties of each), plus cherries and potatoes. The produce is sold from the shed on their property, at supermarkets and fruit shops in the town, and markets in Canberra and Goulburn.
- **Open:** 9am–5.30pm daily (Mar–Sept)
- **Location:** 1km west of Crookwell towards Bathurst

THE POACHER'S TRAIL
Yass Tourist Information Centre, Yass
Phone: 02 6226 2557
Email: tourism@yass.nsw.gov.au
www.poacherstrail.com.au
The Poacher's Trail is a wonderful food and wine trail and introduces you to winemakers, galleries and Poacher's Pantry—the superb smokehouse situated on a farm with a cafe and cellar door.

GOULBURN

🏠 STORES

BRYANTS HERITAGE BAKERY CAFE
Louise and Shane Hobbs
170 Auburn St, Goulburn 2580
Phone: 02 4821 2561
Email: shobbs@tpg.com.au
In 1928, Syd Bryant started making pies in the Orange and Nyngan districts. Later his son Keith opened the Goulburn shop on the site of the first wool sale held in the area. The business was taken over in May 2000 by Shane and Louise Hobbs, maintaining the staff and the tradition. Today their twenty varieties of prize-winning meat and fruit pies made from local meats and western wheat are still hugely popular. Louise's father Bob Thompson runs the wholesale side of the business. He is a former Australian champion baker, who has passed on his expertise to the Bryants' retail store.
- **Open:** 6am–5pm weekdays, 6am–4pm weekends

🍴 DINING

FIRESIDE INN RESTAURANT
Julian and Sue de Cseuz
23 Market St, Goulburn 2580
Phone: 02 4821 2727
www.firesideinn.com.au
Licensed/BYO wine
Cuisine: Modern Australian country
Mains average: $$
Specialty: Duckling with honey and lime
Local flavours: Lamb shanks, berries, vegetables
Extras: Alfresco dining, across the road from the park.
- **Open:** Lunch Tues–Sun, dinner Tues–Sat

GUNDAROO

🍴 DINING

THE CORK STREET CAFE
Gina Collins
Cork St, Gundaroo 2620
Phone: 02 6236 8217
Email: ginacollins@ozemail.com.au
www.corkstreetcafe.com

BYO
Cuisine: Pizza
Mains average: $$
Specialty: Pizza, smoked lamb salad and a range of breads made on the premises (including pesto and goats cheese)
Local flavours: Poachers Pantry products and Allsun Farms organic produce used extensively
Extras: The restaurant is in the former horse stables of the old police station, which was built in 1850. The original jail sits empty, waiting for the 'customer who is not right!' It hasn't been used for years. Winner of Best Pizza in Regional NSW, Restaurant and Catering Association Awards, 2003, and nominated again in 2004.
■ Open: Brunch Sun, lunch weekends, dinner Fri–Sun

HALL
PRODUCERS/DINING

POACHER'S PANTRY, WILY TROUT VINEYARD, SMOKEHOUSE CAFE
Susan and Robert Bruce
Marakei, Nanima Rd, Hall 2618
Phone: 02 6230 2487
Email: unwind@poachers.com.au
www.poacherspantry.com.au
Licensed
Cuisine: Mod Oz smokehouse-style
Mains average: $$
Specialty: Smoked meat, poultry and vegetables
Local flavours: Poachers Pantry
Extras: Poachers Pantry grew out of a farming venture run for more than thirty years in the area by the Bruce family. By 1991, diversification seemed the key to economic survival, so Poachers Pantry was formed, and it is still situated on the property. Sensing a need in the market for hot- and cold-smoked meats, poultry and game they set about to meet that need with industry awards following, and a place in most quality delis and gourmet shops throughout Australia. Their range includes pork, kangaroo and emu prosciutto, cold-smoked herb-cured peppered sirloin, honey-cured double-smoked ham, smoked tomatoes, hot-smoked rack of lamb, kangaroo and native pepper chipolatas and many other mouth-watering products. The Smokehouse Cafe showcases Poachers Pantry's smoked goods and the Bruces' own Wily Trout Wines, grown on the hillsides around the cafe. A meal in the cafe is an ideal way to enjoy the product in a relaxed environment. The cafe is located in a restored farm cottage with a terrace overlooking an 1870s slab woolshed and rolling hills. No takeaway, but picnic hampers are available.
■ Open: Breakfast and lunch Fri–Sun and public holidays, tastings available 10am–5pm daily
● Location: 5km north of the ACT/NSW border, turn right into Nanima Rd and continue for 5km

JINDABYNE
PRODUCERS

HOBBITT FARM GOAT CHEESE
Mike Corbett
Barry Way, Jindabyne 2627
Phone: 02 6457 8171
Email: hobbittfarm@hotmail.com
In 1987 Mike Corbett went cycling in France, working on farms, and became hooked on goats cheese. Back home he set

up his own dairy herd of Toggenburg and Saanen goats, which now numbers around sixty, and began making fine cheeses such as the fresh curd quark, waxed and fresh chevre, or ashed chevre. He coated some cheeses with cracked or alpine pepper, mixed pepper into the cheeses, or even put the cheese into a marinade of olive oil and garlic *à la Provence*. Mike also makes a garlic and dill herbed chevre which he describes as 'somewhere between spiced quark and Boursin', and another matured one with a white, brie-like mould crust. These lovely artisan cheeses are available in the local area, or from The Essential Ingredient in Sydney. Corbett also runs ski tours in the area, and there is B & B accommodation available on the property. Of course, guests are spoiled with dishes such as goats cheese soup or quiche.

- **Open:** 10am–5.30pm daily, by appointment
- **Location:** 11km south of Jindabyne

KINGSVALE
PRODUCERS

JUST PRUNES
JC Granger and Sons
Ventnor Rd, Kings Vale 2587
Phone: 02 6384 4240
Email: jeffgranger@ozemail.com.au
Third-generation orchardists on this land, the Grangers are justly proud that they grow and dry all the fruit they pack. It's a big business now, much of it wholesale, but depending on the season, you will find prunes, peaches, nectarines, apples, pears and apricots, some of them sulphur-free which are sold in bulk to health food shops. But the real arm of the business, as the name implies, is prune processing—dried, as spreads, pulp and also juice.

- **Open:** 8am–5pm weekdays, holidays and weekends by appointment
- **Location:** Approximately 20km from Young. Take Woodlands Rd and follow the sealed road for 2.5km, turn right into Ventnor Rd

LAGGAN
DINING

WILLOWVALE MILL
Graham and Takako Liney
Mill Rd, Laggan, via Crookwell 2583
Phone: 02 4837 3319
Licensed/BYO
Cuisine: Eclectic Japanese–Australian blend
Three-course meal: $$
Specialty: Home-grown potatoes
Local flavours: As much as possible, local meats, fruit and vegetables, complemented by local wines
Extras: Accommodation in on-site cottages adjacent to a charming, old stone flour mill—Tuscany transplanted! Chef/owner Graham Liney has turned raising heirloom potatoes into an art-form and they are always on the menu.

- **Open:** Lunch Sun, dinner Fri–Sun

LAKE GEORGE
DINING

GRAPEFOODWINE
Darren Perryman and Meaghan Pidd
Westering, The Vineyards, Federal Hwy, Lake George 2581
Phone: 02 4848 0026
Email: gfw@madewwines.com.au,

thefirstfloor@netspeed.com.au
www.madewwines.com.au
Licensed
Cuisine: Modern Australian
Mains average: $$
Extras: A balcony at the front of the restaurant offers fantastic views of the vineyards and Lake George.
Open: Lunch (brunch) Thurs–Sun, dinner Fri–Sat
Location: Look for Madew Wines signs at Gurney VC Rest Area

QUEANBEYAN
DINING

BYRNE'S MILL RESTAURANT
Basil Smith
55 Collett St, Queanbeyan 2620
Phone: 02 6297 8283
Licensed/BYO wine
Cuisine: Classic international
Mains average: $$
Specialty: Braised duck
Local flavours: All meats
Extras: A century-old flour mill with elegance and character and comfortable, upholstered captain's chairs. Often chosen for small weddings and functions. Christmas in July is very popular. Basil Smith, the chef, has been here for two years and must have made an impression as the telephone directory has mistakenly listed the restaurant 'Basil's at Byrne's Mill'.
- Open: Lunch Tues–Fri, dinner Tues–Sat

SNOWY MOUNTAINS
This is high country. The seasons are sharp, the weather sometimes brutal, and the visitors come seasonally, so the food outlets and producers somehow have to fit around these conditions. They manage it, and do so brilliantly. Make sure you go—winter or summer. The weather may get cold, but the welcome is always warm.

THREDBO VALLEY
DINING

CRACKENBACK COTTAGE AND FARM
Sarah Young
Alpine Way, Thredbo Valley 2627
Phone: 02 6456 2198
Email: info@crackenback.com
www.crackenback.com
Licensed
Cuisine: Australian with traditional influence
Mains average: $$$
Specialty: Lamb and rosemary pie, 'high country pie'
Local flavours: Trout, goats cheese
Extras: The gourmet shop in the loft above the restaurant sells homemade preserves, jams, relishes, mustards and sauces. The vineyard on the property is soon to produce its own vintage. Guesthouse on site (seven rooms) with indoor swimming pool, day spa, massage and body treatments. A wedding chapel for weddings, functions and product launches. A giant wooden maze for those who want to get lost. A produce market is planned, and there is a crafts shop, to round out the mix of attractions.
- Open: Lunch and dinner daily (July–Sept and school holidays), Thurs–Sun (other months)

CANBERRA, CAPITAL COUNTRY AND THE SNOWY MOUNTAINS—NSW

TUMBARUMBA

🍴 DINING

LAUREL HILL BERRY FARM
Cor Smit
2150 Batlow Rd, Laurel Hill,
Tumbarumba 2649
Phone: 02 6949 1717
Email: cormary@bigpond.com
The farm produces a range of berries including blueberries, raspberries, boysenberries, blackberries and gooseberries and makes blueberry pancakes, jams and sauces. These are served in the Pancake Pavilion in a tranquil garden setting. Bed and breakfast accommodation available.
■ **Open:** 10am–6pm Thurs–Sun (Nov–Apr), 10am–6pm every day during school holidays, other days by arrangement
● **Location:** 20km from Batlow, halfway between Batlow and Tumbarumba on Tourist Drive 6

YOUNG

🍎 PRODUCERS

ANE'S CHERRYGROVE
Barry and Ane Apps
Olympic Way, Young 2594
Phone: 02 6384 3333
There are pick-your-own cherries, peaches and strawberries here as well as plums, nectarines, and berries. A range of homemade products such as cherry and strawberry pies, pickles, jams, sauces, chutneys, oils and vinaigrettes is also available. There are also tastings of fruit wines and local cheeses, or you may simply relax with a Devonshire tea.
■ **Open:** 8am–6pm daily (Oct–May)
● **Location:** 6km from Wombat

BIT O' HEAVEN ORCHARD
Charles and Yolanda Mullany,
Michael and Bernadette Mullany,
Vince and Michelle Fernon
240 Fontenoy Rd, Young 2594
Phone: 02 6384 3353
Mobile: 0418 659 819
Email: bitoheaven@hotmail.com
Just come and pick as much as you like (eat as much as you like too, is the offer) but you do have to pay for what you pick to take home. It's a family-run business and PYO is seen here as a fun family activity. There are eight varieties of cherries, and no entry fee to this seventy-acre orchard. Or just buy the cherries already picked from the orchard stall. Plums are available too, from early February onwards.
■ **Open:** Daily from mid Nov to Christmas (cherry season)

CHERRYHAVEN ORCHARDS
Ian and Arna Hay
Cowra Rd, Olympic Way, Young 2594
Phone: 02 6382 4023
Email: cherry@dragnet.com.au
The Hay family have been here since 1862, and it was the Hays's great-grandfather who planted this, the first commercial cherry orchard in Australia. Today's generation sells twenty varieties of fresh cherries (November to January) and you may watch fresh cherries and plums being packed for the domestic and export market. There are also plums (December to March), peaches, nectarines, jam, chutneys and sauces in the coffee shop, and out of season you can always buy cherries which have been canned in the local cannery.
■ **Open:** 9am–5pm daily
● **Location:** 4km from Young—look for the big red cherry

41

EQUITAS ORCHARDS
Scott Cormack
20 Stanley Park Rd, Young 2594
Phone: 02 6382 1479
Email: office@equitas.com.au

In season this place is really busy, as you would imagine, but the owners still find time to open the cherry shop packed with fresh fruit (cherries, plums, stone fruit such as nectarines and peaches) plus macadamias, jams, sauces (under the Hilltops Fruit Co label) and souvenirs. No pick your own, but there are two or three tours of the packing facilities from a viewing platform each day in season.
- **Open:** 9am–4pm daily (Nov–Jan)

JD'S JAM FACTORY
Jan and Lester Donges
Lot 1, Grenfell Rd, Young 2594
Phone: 02 6382 4060
Email: jdsjam@yol.net.au
www.jdsjamfactory.com.au

Walk into another world—a world of cherries, cherries, and more cherries. Which is appropriate, because Young is the country's cherry capital. Look closer and you will see that other stone fruits also get a good airing here, as this jam factory produces 118 varieties of Young Maid products, canned fruit, jams, pickles, chutneys, butters, jellies, and sauces, which are sold in supermarkets throughout the country. This sweet 'take' on recycling was developed twenty years ago to make use of waste fruit—good fruit, but too ripe to survive transport to the city or interstate markets. Lester Donges's Aunty Coral, using her mother's old-style recipes, began cooking cherry jam in her own kitchen. Soon the kitchen migrated to adjoin the packing shed and now takes care of around 100 tonnes of the area's fruit seconds. You may tour the streamlined jam factory, but most people find the Devonshire teas and cherry pie irresistible, and few can resist tasting the cherry port or wine.
- **Open:** 8am–6pm daily
- **Location:** 2km from Young

VERITY PRUNES
Kelvin Cronk
Young District Producers Co-op,
43 Nasmyth St, Young 2594
Phone: 02 6382 2656
Email: verity@dragnet.com.au

Ever wondered how a prune shrivels? Or why they taste so different to plums? The staff at Verity will answer these and many other questions for you. Commercial prune growing commenced in the Young area after World War I, with the establishment of soldier settlement orchards. Dehydration plants followed, and an industry was born. The shop sells choc-coated prunes, prunes in port, liqueur fruits, figs, brandy prunes—and bagged prunes, of course, to snack on as you go.
- **Open:** 9am–5pm weekdays

DINING

ZOUCH CAFE-RESTAURANT
Susie (chef) and Andy Forrest
26 Zouch St, Young 2594
Phone: 02 6382 2775
Email: zouch@bigpond.com.au
Licensed/BYO
Cuisine: Modern Australian
Mains average: $$
Specialty: Roast of Dutton Park duck with local poached apricots
Local flavours: As much as possible, including meat, poultry, vegetables, stone fruit, olives, oils and bread.

Extras: In the former Masonic Hall with courtyard area and fountain. Winner of the Restaurant and Catering Award for Excellence, Modern Australian, 2003. Fresh and frozen meals, homemade pesto, caramel sauce and biscuits available.
- **Open:** Lunch Wed–Mon, dinner Thurs–Sat

MARKETS AND EVENTS

✦ **CANBERRA GROWERS' & PRODUCE MARKET,** SYMONSTON
02 6232 7968
8am–2pm first and third Sun

✦ **CANBERRA REGION FARMERS' MARKET,** EPIC (Exhibition Park in Canberra)
0400 852 227
8am–11am Sat

✦ **DAYS OF WINE AND ROSES FESTIVAL,** CANBERRA REGION
02 6205 0044
Nov, participating wineries

✦ **FESTIVAL OF THE FALLING LEAF,** TUMUT
02 6947 7025
Last weekend in Apr

✦ **COOLAC FESTIVAL OF FUN SHOWCASE,** DINNER, COOLAC
02 6954 3240
First Sat in Mar

✦ **SNOWY MOUNTAINS REGIONAL FOOD FAIR,** DALGETY
www.dalgetychamber.org.au
Oct

✦ **NATIONAL CHERRY FESTIVAL,** YOUNG
02 6382 3394
End Nov–early Dec

CENTRAL COAST

Like the Blue Mountains, because it's so close to Sydney, the Central Coast region is the city's playground.

When it comes to food, the Central Coast has plenty of fish and seafood to go round, including some of the best oysters you'll find.

At just an hour or so from Sydney, it makes an ideal day trip. So what are you waiting for? Grab your picnic hamper and start exploring.

AVOCA BEACH
DINING
FEAST RESTAURANT
André Chouvin
Shop 3, 85 Avoca Dr, Avoca Beach, 2251
Phone: 02 4381 0707
Email: tchouvin@aol.com
www.feast.com.au
Licensed
Cuisine: Modern French
Mains average: $$$
Local flavours: Lobster, snapper
Extras: This fine-dining waterfront restaurant has outdoor seating with amazing views.
■ **Open:** Breakfast Sun, lunch and dinner daily

BROOKLYN
STORES
JJ'S AT THE MOUTH OF THE HAWKESBURY RIVER
Joshua Smith
8 Dangar Rd, Brooklyn 2253
Phone: 02 9985 7106
Mobile: 0413 050 861
This is a tiny dot of a place with a serving window that opens onto the footpath, but it has some of the best fresh seafood and oysters around. There are also burgers and takeaways that you can eat at the river's edge. The fish comes via local fishermen, the Newcastle Fish Co-op and Sydney Fish Market, and the oysters from Bayou Bill. The oysters come in jars, and you can also get fresh seafood and live, local mud crabs to cook at home.
■ **Open:** Daylight hours Fri–Sun (summer), 10am–7pm Mon–Thurs, 10am–8pm Fri–Sun (from Easter)

DINING
LIFEBOAT SEAFOODS
Tom and Sheree Cosgrove
1 Dangar Rd, Brooklyn 2253
Phone: 02 9985 7510
Mobile: 0410 554 777
BYO
Cuisine: Seafood
Mains average: $$
Specialty: Seafood platter, green Thai curry
For sale: Oysters, crabs, prawns, lobster
Extras: Garden setting, the charm of Brooklyn itself, so near yet so far from

CENTRAL COAST—NSW

Sydney. This place has been serving seafood to appreciative locals and visitors for around fifteen years. The seafood couldn't be fresher—it comes from the Hawkesbury Fish Co-op next door. Takeaway available.
- **Open:** 10am–5pm Mon–Thurs, 9am–6pm Fri–Sun

🚌 TOURS

CRAB'N'OYSTER CRUISE
Catherine Cigneguy
Hawkesbury Princess Cruises,
5 Bridge St, Kangaroo Point, Brooklyn 2253
Phone: 02 9985 8237
Mobile: 0415 184 191
Email: hawkesburyriver@ozemail.com.au
www.hawkesburyprincess.com.au
The Crab'n'Oyster cruise is one of the most popular cruises offered by these operators. An oyster farmer opens oysters and talks about how they are raised, and passengers can pull up traps containing live mud crabs. The one-and-a-half to two-and-a-half hour cruises include morning or afternoon tea or lunch, which of course features some of the mud crab harvest, oysters, prawns, and salads. The licensed vessels can take one hundred passengers, making them popular for charters, weddings, and barbecue cruises as well.
- **Open:** Morning tea cruises Tues–Thurs and Sun, lunch cruise Sun

ERINA HEIGHTS

🍴 DINING

FLAIR RESTAURANT
Lawry Gordon
1/488 The Entrance Rd, Erina Heights 2260
Phone: 02 4365 2777
Email: tmic@tac.com.au

BYO
Cuisine: Smart modern
Mains average: $$$
Specialty: Duck confit
Local flavours: Redgate Farm duck, Sydney rock oysters from Mooney Mooney, wild rocket
Extras: Alfresco dining, winner of Best New Restaurant Central Coast, 2003, and Best New Restaurant Regional NSW, 2003 (Restaurant and Caterers Association). Takeaway available.
- **Open:** Lunch Wed–Fri, dinner Tues–Sat

GOROKAN

🏠 STORES

COMMERCIAL FISHERMEN'S COOPERATIVE LTD
Wallarah Rd, Gorokan 2263
Phone: 02 4392 1603
This cooperative began in 1945 when a group of fishermen joined together to sell their daily catches. Each day there are many fish species available as well as lobsters and crabs, school and king prawns—basically whatever is caught that day. Perfectly situated on the lake, this branch sells mostly lake fish (whiting, flathead and mullet) with regular deliveries of ocean fish.
- **Open:** 9am–7pm daily
- **Location:** Beside Toukley Bridge

SOMERSBY

🍎 PRODUCERS

LA TARTINE
Nick and Laurence Anthony
Unit 2/111 Wisemans Ferry Rd,
Somersby 2250

Phone: 02 4340 0299
Email: latartine@ozemail.com.au
Look for this wonderful bread at various outlets on the Central Coast and at Sydney farmers' markets and delis. When La Tartine is baking, the scents of wood-fired sourdough waft from the huge traditional oven that Nick Anthony built himself. This bread is organic through and through and if your taste buds can't tell, the NASAA certification proves it. The Anthonys use organic, Junee wheat flour from Green Grove Flour Mills. Nick Anthony learned his sourdough skills in a bakery in France. Sourdough is a less precise method of baking, and can depend on the weather, so the locals call him to see when the bread will be done, then pop down for their wholemeal, white, sesame, raisin or black rye loaf. Lucky locals.
- **Open:** Phone for details of opening hours

TACOMA
STORES
COMMERCIAL FISHERMEN'S COOPERATIVE LTD
77 Wolseley Ave, Tacoma 2259
Phone: 02 4353 2344
This cooperative began in 1945 when a group of fishermen joined together to sell their daily catches. Each day there are many lake and ocean fish species available as well as lobsters and crabs, school and king prawns—basically whatever is caught that day.
- **Open:** 8.30am–4.30pm weekdays, 8.30am–1pm Sat

TERRIGAL
DINING
LETTERBOX
Karl and Nicole Kard
Old Terrigal Post Office,
4 Ash St, Terrigal 2260
Phone: 02 4385 4222
Email: kard@integritynet.com.au
www.letterboxrest.com.au
Licensed/BYO
Cuisine: Modern Australian
Mains average: $$$
Local flavours: Organic produce where possible
Extras: Private garden room, open kitchen, airconditioned in summer, log fire in winter.
- **Open:** Lunch and dinner Mon–Sat

SEASALT RESTAURANT
Julie Donohoe (chef)
Level 1, Crowne Plaza Terrigal,
Pinetree Lane, Terrigal 2260
Phone: 02 4384 9133
seasalt@crowneplazaterrigal.com.au
www.terrigal.crowneplaza.com.au
Licensed
Cuisine: Modern Australian
Mains average: $$$
Local flavours: Seasonal, regional produce
Extras: Weekend seafood buffet, waterfront views, terrace dining.
- **Open:** Breakfast, lunch and dinner daily

UMINA BEACH
🍴 DINING

FISH HEADS CAFE AND FINE FOODS
Lisa and Richard Zammit (owners),
Chrissy Juno (chef)
471 Ocean Beach Rd, Umina Beach 2257
Phone: 02 4342 2879
BYO
Cuisine: Eclectic
Mains average: $$
Specialty: Asian-style duck
Local flavours: Free-range eggs, seafood
Extras: Private courtyard, art gallery, restaurant has a 'fishy' theme.
- **Open:** Breakfast and lunch daily, dinner Thurs–Sat

WOY WOY
🍴 DINING

FISHERMEN'S WHARF
Merv Clayton
The Boulevarde, Woy Woy 2256
Phone: 02 4341 1171
Email: wwfw@hotkey.net.au
BYO
Cuisine: Seafood
Mains average: $$
Local flavours: Oysters, prawns, squid, snapper, whiting, flathead, kingfish, mullet and bream
Extras: Restaurant over the water on a pier located on the Woy Woy Channel. Most tables have a window seat. Most of the seafood is supplied by local fishermen, though some may come from the Sydney Fish Market when supply exceeds demand. Buy whole fish or they can either fillet or steak it for you. Takeaway available.
- **Open:** 9am–5pm Mon–Sat, 9am–4pm Sun

YARRAMALONG
🍎 PRODUCERS

YARRAMALONG MACADAMIA NUT FARM
Philip Davis
RMB 1253, Yarramalong Rd,
Yarramalong 2259
Phone: 02 4356 1170
Email: yarramac@bigpond.com
Philip Davis took over this macadamia plantation a few years ago and it is one of the more southerly ones in the country. At the cafe you will find superb chocolate-coated nuts as well as the plain, cracked ones (and check out the macadamia-cracker—one of the most simple and effective you'll find) as well as sandwiches made with homemade 'Mac' bread and salads and pies crunchy with nuts. Macadamia trees have rich dark foliage and there are several hundred trees here in this already delightful valley.
- **Open:** 9am–4.30pm Tues–Sun
- **Location:** 12km from Wyong

EXPLORER COUNTRY AND RIVERINA

This region is a massive arc of prime growing land, one of the richest food and wine belts in the state. You be the explorer of this region—one which is possibly the most satisfying for the independent food- and wine-lover. This is cellar-door country with dozens of wineries, many of them still in their infancy, but showing signs of a maturity beyond their years. So important are the ancillary food industries—orchards, olives, aquaculture, honey and many more—that growers and others have branded the region to market it more effectively.

Further south, the Riverina is one of those places that you feel could stand independently even if the rest of the country somehow dropped away. Like an island, it has almost enough resources, agriculturally at least, to be entirely self-sufficient. There are grain crops, citrus, many types of fruits, chicken and eggs, cheeses, meat and rice. And of course wine, for this region produces around seventy per cent of all wine grapes in New South Wales. The percentages go on—forty-two per cent of Australia's rice, thirty per cent of our citrus.

Two things have influenced this region immensely. Irrigation and immigration. One brought the means to grow the food, the other the industry and expertise. The Italian influence is strong in the Riverina, and has affected everything from the restaurant scene to the type of goods produced here. Antipasti reigns, okay!

BATHURST

DINING

RESTAURANT LEGALL
Angele, Philippe and Gwenael Legall
56 Keppel St, Bathurst 2795
Phone: 02 6331 5800
Email: legall@bigpond.net.au
Licensed/BYO
Cuisine: French
Mains average: $$$
Specialty: Crème brûlée
Local flavours: Chicken, lamb, goats cheese, all fruits and vegetables in season, wine list from Central West winemakers
Extras: Several dining rooms in a delightful terrace house.
- **Open:** Lunch Thurs-Fri, dinner Tues-Sat

BORENORE

PRODUCERS

HILLFARM SPRING WATER
Lyster Ormsby and Susan Maitland
Hillfarm, Borenore 2800
Phone: 02 6365 2378
Email: maitlandsm@optusnet.com.au
Crystal-clear spring water bottled at source on this historic property in distinctive bottles that are glass-etched to order. Vintage bottling machinery is still in use, and is a feature of the packing-shed plant.
- **Open:** By appointment

NORLAND (FIG) ORCHARD
Warren Bradley
'Norland', Borenore 2800
Phone: 02 6365 2225
Figs have been grown on this property for close to a hundred years. Bradley is the fourth generation of his family to tend the six acres of black Genoa, white Adriatic, and brown turkey fig trees, which create the same amount of work, Bradley says, as 30 000 apple trees. Bradley has added his own touch to the 2.5-hectare orchard, thinning the plantings, adding some new and old varieties to create the largest and, many say the best, fig orchard in Australia. He has also trialled a very successful process of drying the fruit, and his fresh and dried figs, preserves and jams are sold from the orchard, while the fresh figs are used in local wineries and restaurants as well as Sydney restaurants.
■ **Open:** 7am–7pm daily (Jan–Apr)

COWRA
DINING

NEILA
Jerry Mouzakis and Anna Wong
5 Kendal St, Cowra 2794
Phone: 02 6341 2188
Email: eat@alldaydining.com
www.neila.com.au
BYO
Cuisine: Modern regional
Mains average: $$$
Local flavours: Home-grown Chinese red dates, duck, pork, asparagus, venison, white peaches, white nectarines, strawberries, olives
Extras: Award-winning restaurant, that also sells olive oil and dukkah packs.

■ **Open:** Dinner Thurs–Sat, other times by arrangement for groups

RED CARP
Karen Harding and Jane Reeves
Binni Creek Rd, Cowra 2794
Phone: 02 6342 5222
Licensed
Cuisine: Australian and Japanese
Mains average: $$
Local flavours: Cheeses, pâté, olives
Extras: Children's menu. Takeaway available. Views of the Japanese garden.

■ **Open:** Breakfast and lunch (all day) and dinner daily

THE BREAKOUT BRASSERIE
Ruth Moore
5/37 Macquarie St, Cowra 2794
Phone: 02 6342 4555
BYO
Cuisine: Gourmet lunches
Mains average: $
Specialty: Crunchy chicken salads, prawn and avocado salad
Local flavours: Fruit and vegetables, Parkes chicken
Extras: Courtyard and veranda dining. Homemade cakes and slices for sale.

■ **Open:** Lunch daily, functions by bookings only

THE MILL CELLAR DOOR AND FUNCTION CENTRE
David O'Dea
6 Vaux St, Cowra 2794
Phone: 02 6341 4141
Email: themill@windowrie.com.au
www.windowrie.com.au
Licensed
Local flavours: Sheeps cheese, smoked trout, sun-dried tomatoes, fresh produce in season

Extras: Located in the oldest building in Cowra, restaurant upstairs has outside dining. Caters for booked functions.
- **Open:** 10am–6pm daily (restaurant by appointment only)

THE QUARRY RESTAURANT AND CELLARS
Anne and Paul Loveridge
The Cowra Estate, 7191 Boorowa Rd, Cowra 2794
Phone: 02 6342 3650
Email: quarry@bigpond.net.au
www.cowraregionwines.com
Licensed
Cuisine: Modern Australian
Mains average: $$
Specialty: Use of regional produce
Local flavours: Lamb, beef, pork, chicken, asparagus, olives, olive oil, sheeps cheese, trout, vegetables, fruit
Extras: Tuscan-style shaded courtyard and gardens.
- **Open:** Lunch Wed–Sun, dinner Fri–Sat
- **Location:** 4km from Cowra

DUBBO

PRODUCERS

BEN FURNEY FLOUR MILLS
John Furney
Sarah Jane Fine Food,
101–105 Brisbane St, Dubbo 2830
Phone: 02 6884 4388
Email: bffm@benfurney.com
An independent family-owned business since early last century, Ben Furney Flour Mills has kept abreast of technology. Located in the heart of premium wheat-growing country in the Central West, it is on hand for the pick of the crop. However, the mill in turn supports local growers, and now produces bread pre-mixes; whole, rolled and kibbled grains and cereals; flours and meals. The factory outlet sells a range of home baking supplies—flours, grains, bread pre-mix, and cake mix, as well as snack food.
- **Open:** 9.30am–5pm weekdays

STORES

NEWTOWN PROVIDORES
Peter A Davis
62 Wingewarra St, Dubbo 2830
Phone: 02 6882 0055
Email: service@newtownprovidores.com.au
www.newtownprovidores.com.au
In the heart of the Macquarie Valley food bowl, this gourmet deli has no trouble stocking regional produce—the area is full of great produce and people with innovative ideas on what to do with it. Look for gourmet platters, cheeses, smallgoods, dips, antipasto, and breads.
- **Open:** 10am–6pm weekdays, 10am–2pm weekends

DINING

THREE SNAILS RESTAURANT
Stephen Neale
7/36 Darling St, Dubbo 2830
Phone: 02 6884 9994
Email: snails123@bigpond.com
www.threesnails.com.au
Licensed
Cuisine: Eclectic modern Australian
Mains average: $$$
Specialty: Local produce
Extras: Outdoor area, newly established restaurant owned by the former chef-owner of Echidna, a popular restaurant in Dubbo some years ago.

■ Open: Lunch Wed-Sun, dinner Tues-Sat

FORBES
🍴 DINING

EDA-BULL
Andrew Bruem
137 Rankin St, Forbes 2871
Phone: 02 6852 1000
Email: ed@edabull.com.au
www.edabull.com.au
Licensed
Cuisine: Modern Australian
Mains average: $$
Specialty: American-style pork ribs, Asian beef, salt and pepper prawns
Local flavours: Regional wines, seasonally available vegetables, regional meats
Extras: Alfresco dining, at this multiple award-winning restaurant.
■ Open: Brunch all day Sun, lunch Wed-Fri, dinner Wed-Sat

GRIFFITH
🍎 PRODUCERS

RIVERINA GROVE PTY LTD
Louis Marangon
4 Whybrow St, Griffith 2680
Phone: 02 6962 7988
Email: info@riverinagrove.com.au
www.riverina.grove.com.au
'Turning local produce into speciality foods' is the aim of this producer, which began as a backyard olive preserving business in 1964. Today there are over forty products (all preservative-free) under five labels mostly made from Riverina produce. The Gourmet Grocer section of the factory sells marinated eggplant, sun-dried tomatoes, olives, pasta sauces, and the bestseller—the amazingly named Bum Hummers (pickled onions). There's a host of other antipasto-style products, as well as local brands. You don't even have to hurry away as now there is Riverina Grove House Blend coffee on hand, which you can enjoy as you taste products and decide what to buy.
■ **Open:** 9am-5pm weekdays, 9am-12pm Sat
● **Location:** Mooreville Industrial Estate, east of Griffith

🏠 STORES

RIVERINA GOURMET GIFTS
Christine Ison
432 Banna Avenue, Griffith 2680
Phone: 02 6962 9199
Email: riverh@bigpond.net.au
www.riverh.com.au
Although she runs a busy coffee shop and makes some great cakes to go with the coffee, Christine still finds time to put together hampers based around local produce. These gourmet gift baskets are especially popular at Christmas, but are available all year, and the products—all local brands, representing some of the best local produce—are available for sale separately.
■ **Open:** 8.30am-5.30pm Mon-Sat

🍴 DINING

L'OASIS RESTAURANT
Martin and Suzy O'Donnell
150 Yambil St, Griffith 2680
Phone: 02 6964 5588
Email: martinod@iinet.net.au
Licensed
Cuisine: Modern Australian

NSW—EXPLORER COUNTRY AND RIVERINA

Mains average: $$
Specialty: Slow-cooked duck, tart of potato and lemon thyme, almond brittle ice-cream and meringue
Extras: Monthly regional dinners utilising local products. Winner of Restaurant and Catering awards last few years.
- **Open:** Lunch and dinner Tues–Sat

MICHELIN
Julian Raccanello
72 Banna Ave, Griffith 2680
Phone: 02 6964 9006
Email: jraccanello@nugan.com.au
Licensed/BYO
Cuisine: Modern French
Mains average: $$
Specialty: Char-grilled quail and seared scallops
Local flavours: Lamb, poultry, quail, citrus, vegetables, sweetcorn
Extras: An elegant restaurant with a special 'Riverina' menu.
- **Open:** Brunch Sun, lunch daily, dinner Mon–Sat

GRONG GRONG
🍎 PRODUCERS

UARAH FISHERIES
Bruce Malcolm
Old Wagga Rd, Grong Grong 2652
Phone: 02 6956 2147
Mobile: 0428 696 927
Email: uarahfisheries@austarnet.com.au
Murray cod is sometimes called the Kobe Beef of fish, and Uarah Fisheries grows them out to around 1.5kg to 3kg. If you think that's large, in the wild the biggest one ever caught reeled in at a line-breaking 130kg. This place also raises the other great Australian fishes—silver perch and golden perch, or yellow belly. These are all pond grown and bred. Bruce Malcolm has been in this industry for twenty years, one of the first growers in Australia, and developed the necessary technology for the table-production of Murray cod. He also helped re-establish rivers, returning them to a healthy state, with Queensland DPI (Department of Primary Industries), in the 1980s, and now exports his fish, as well as supplying some of Sydney's top restaurants.
- **Open:** By appointment

HANWOOD
🚌 TOURS

CATANIA FRUIT SALAD FARM TOURS
Sharon and Joe Maugeri
Farm 43, Cox Rd, Hanwood,
via Griffith 2680
Phone: 02 6963 0219
Mobile: 0427 630 219
Email: cataniafruitsaladfarm@bigpond.com.au
One of the oldest farms in the area, a genuine 'fruit salad' farm growing many varieties of fruit, and with an original mud-brick house. See plums drying, taste fresh fruit and purchase homemade, preservative-free wine, olives, jams, prunes in port, and mustard.
- **Open:** Tours at 1.30pm daily

HAY
🍴 DINING

JOLLY JUMBUCK
Brian Gibbs
Riverina Hotel, 148 Lachlan St, Hay 2711
Phone: 02 6993 1137

EXPLORER COUNTRY AND RIVERINA—NSW

Licensed
Cuisine: Country bistro
Mains average: $$
Local flavours: Local steaks, lamb, chicken from Griffith
Extras: A typical country pub, with lots of character and lots of interesting characters.
- **Open:** Lunch and dinner daily

HILLSTON
PRODUCERS

BIJI BUSH QUBES
Tony Upton
106-108 Cowper St, Hillston 2675
Phone: 02 6967 2417
Biji Bush Qubes are a unique Australian concept, a natural product designed to bring the aromas and flavours of a typical Aussie campfire to your own patio or backyard barbecue. Tony Upton has used some lateral thinking and processes the leaves of certain Australian trees such as red or blue gums, native pine, lemon box, black box and ironbark, from natural stands in his local area. He compresses them into cubes so that they can be used with either gas or kettle-type barbecues. A further spin-off is his Biji-Barbie, based on the old plough-disc, traditionally used for bush barbecues. It is made in Hillston and the stand folds flat for easy storage or transport. Available by mail order.
- **Open:** 7:30am-5:30pm daily

HUNTLEY
PRODUCERS

HUNTLEY BERRY FARM
Brian McCarthy
Huntley Rd, Huntley, via Orange 2800
Phone: 02 6365 5282
The list of berries is as long as your arm—strawberry, raspberry, blackberry, boysenberry, blueberry, jostaberry and youngberry—at least a dozen seasonal varieties. You can pick your own or buy direct, along with locally made jams and sauces. Keep your eyes open for fresh vegetables also grown on the farm, and for something completely different, ask about their feijoas—related to kiwifruit. The farm is also a supported workplace for people with disabilities.
- **Open:** 8.30am-3.30pm weekdays, possibly weekends during summer holidays (phone for details)

JUNEE
PRODUCERS

GREEN GROVE ORGANICS—JUNEE LICORICE AND CHOCOLATE FACTORY
Neil Druce, Max Reid, Carol Druce
8-18 Lord St, Junee 2663
Phone: 02 6924 3574
Email: greengroveorganics@bigpond.com
www.greengroveorganics.com
If you thought Green Grove was all about organic flour, you haven't been keeping up. In 2001, organic licorice—a medicinal herb that dates back over 4000 years—was introduced in its confectionery form, and in 2003 chocolates made a very natural addition to the line-up available here. The wheat has been stone ground

in a 1920s flour mill that closed in the 1970s but was rebuilt recently. The flour is supplied to major organic bakers throughout the state. It also grinds wheat, rye, barley, oats, maize—anything organic, in fact—to order, and makes bread mixes. Now the product of the sweet arm of the business is sold in all states and overseas. The Cafe De Mill has snacks, and there are tours available.
- Open: 10am–4pm daily

LEETON
PRODUCERS
FRUITSHACK
Michael and Debbie Ierano
Farm 312, Henry Lawson Dr, Leeton 2705
Phone: 02 6953 2451
Mobile: 0429 866 965
Email: fruitshack@hotmail.com
One hundred per cent yummy orange juice, nothing added, is the only product made here. Rising citrus farm costs and lower profits 'encouraged' the Ieranos to diversify, beginning Fruitshack in 1994. The Valencia orange juice is natural, and has a short shelf life because it is unpasteurised. But it's so good, most people can't keep it long enough.
- Open: 9am–5pm weekdays

STORES
MICK'S BAKEHOUSE
Michael Di Salvatore
56 Pine Ave, Leeton 2705
Phone: 02 6953 2212
Email: micksbh@dragnet.com.au
www.micksbakehouse.com.au
When Mick's Bakehouse won nineteen gold and one silver out of a possible twenty at the Great Aussie Meat Pie Competition, Mick thought it was time to retire and let someone else have the glory. So he judged the event the next year. That was between running the Leeton shop and opening a new one at Wagga (phone 02 6925 9599). Both places are bakery cafes where you get the pies, of course, but also coffee, hot food and light meals.
- Open: 6.30am–5pm weekdays, 7am–5pm weekends

TOURS
SUNRICE COUNTRY VISITORS CENTRE
Lyn Wood (Information Manager)
Calrose St, Leeton 2705
Phone: 02 6953 0596
Email: lwood@sunrice.com.au
www.sunrice.com.au
There is a video, plus photographic and product displays, rice milling demonstrations, and tastings at this key centre in the Riverina. Daily presentations (one hour, telling the story of rice and rice tasting) at 9.30am and 2.45pm. All rice products are for sale.
- Open: 9am–5pm weekdays (bookings essential for coaches and groups)

LIDSDALE
PRODUCERS
JANNEI GOAT DAIRY
Neil and Janette Watson
8 View St, Lidsdale 2790
Phone: 02 6355 1107
Email: jannei@lisp.com.au
www.lisp.com.au/~jannei
This busy dairy, one of the first to begin goats cheese making almost ten years

EXPLORER COUNTRY AND RIVERINA—NSW

ago, has fresh waxed and white mould cheeses, goats curd, yoghurt and goats milk. These multi award-winning chevre varieties are sold in cheese shops and specialty delis in Sydney, and elsewhere. Jannei won golds at the Sydney Royal Easter Show, 2004. It's a busy working farm so sound the horn if no-one is in sight. They could be milking or working with the goats.
- **Open:** 9am-4pm Mon-Wed, 10am-5pm Fri-Sat
- **Location:** 15 minutes from Lithgow on the Mudgee side of Lidsdale.

MILLTHORPE
DINING

THE OLD MILL CAFE AND RESTAURANT
Janis Pritchard (owner),
Lisa O'Leary (chef)
12 Pym St, Millthorpe 2798
Phone: 02 6366 3188
Licensed/BYO wine
Cuisine: Modern Australian
Mains average: $$
Specialty: Fresh local produce, homemade desserts, gluten-free food
Local flavours: Lamb, pork, chicken, veal, beef, venison
Extras: Outdoor courtyard dining at this former bakery in a small heritage village. Takeaway available.
- **Open:** Lunch daily, dinner Wed-Sat

TONIC
Cnr Pym and Victoria Streets,
Millthorpe 2798
Tony and Nicole Worland
Phone: 02 6366 3811
Licensed/BYO

Cuisine: Australian
Mains average: $$$
Local flavours: As much as possible and local producers bring in rabbit, venison, lamb, duck, chestnuts, hazelnuts, walnuts
Extras: A rustic restored corner shop.
- **Open:** Brunch and lunch weekends, dinner Wed-Sat

MUDGEE
PRODUCERS

BRISTOWE FARM HAZELNUTS
Vanessa and Clem Cox
8 Court St, Mudgee 2850
Phone: 02 6372 3224
Email: ausnut@hazelnuts.net.au
Bristowe Farm Hazelnuts is a leader in the development of a hazelnut industry in south eastern Australia. Established in 1987 in hills near Mudgee, the Coxes propagate young trees of a unique hazelnut variety that was developed in Australia. These trees and their pollinators are sold to commercial growers, with a buy-back agreement for the nuts produced. The Coxes crack and process the nuts at Mudgee and sell the kernels as raw, dry roast, honey spiced, and roast crumb. They also coat premium, dry-roast nuts with vanilla and butterscotch flavoured white chocolate (Duos), which are sensational with coffee. With the industry still in its infancy, volume produced is small but growing. Bristowe Farm Hazelnuts are available at Mudgee Gourmet, by mail order and at selected local retail outlets, regional and city farmers' markets.
- **Open:** 9am-4pm Thurs-Tues

DI LUSSO ESTATE
Eurunderee Lane, Mudgee 2850
Phone: 02 6373 3125
Email: sales@dilusso.com.au
www.dilusso.com.au
Wine is made and bottled on the premises, and extra virgin olive oil is made from Italian olive varieties, plus there are limited amounts of table olives available. Fresh figs are also sold from the cellar in season, five different varieties, and these are in high demand from top Italian restaurants in Sydney. Tapenades and dried figs and other fig products are available all year round.
- **Open:** 10am–5pm Fri–Sun

FIGTREE RETREAT OLIVES AND FARMSTAY
Richard Lawson
241 Old Grattai Rd, Mudgee 2850
Phone: 02 6372 7237
Mobile: 0400 072 728
Email: figtree-retreat@bigpond.com.au
www.figtree.austasia.net
Figtree Retreat are growers and producers of premium organic table olives, tapenades and organic preserves and pickles, which are sold at the farm gate or at growers' markets in the area. The farm produce is served to B&B guests who stay in two fully self-contained cabins (sleeping six, and four people) with breakfast and linen provided.
- **Open:** 10am–4pm Thurs–Sun

MUDGEE GOURMET @ HEART OF MUDGEE
Vanessa and Clem Cox
8 Court St, Mudgee 2850
Phone: 02 6372 3224
Email: enquiries@mudgeehampers.com.au
enquiries@mudgeegourmet.com.au
www.mudgeehampers.com.au,
www.mudgeegourmet.com.au
Production of gourmet food products in the Mudgee region has come a long way since Clem and Vanessa Cox began Mudgee Hampers nearly a decade ago. Their first hampers featured products from only eight local food suppliers, but the range has grown steadily. Today the shelves of their produce store, set in the town's oldest house, are stocked with goods from more than thirty local suppliers—including wine jelly, tapenade, mustard, jam and sauce, extra virgin olive oil, verjuice and paste—and the tasting table features more than one hundred samples for visitors to savour. The Coxes recently renamed their shop Mudgee Gourmet, reflecting the increased range of locally produced foods supported by the same high standard of service and product knowledge.
- **Open:** 9am–4pm Thurs–Tues

MUDGEE HONEY COMPANY PTY LTD
Gavin Adams
28 Robertson St, Mudgee 2850
Phone: 02 6372 2359
Mudgee Honey is almost a byword with honey lovers. Gavin Adams admits to being 'well into his sixties, close to retiring age', yet he shows no sign of doing so. He has spent his entire working life with bees, in the seventy-eight year old family business. The premises are anything but imposing. A tin shed, a hive to show how it all happens and rows of golden jars—around twenty varieties at any one time—labelled to show their origin (stringybark, canola, blue gum, wildflower, sally gum, or giant goo bush). The honey, as well as royal jelly capsules and beauty cream, is often sold under the Goldvita label.
- **Open:** 9am–4pm daily

THE OLIVE NEST
Sue Clubb, Pam Colqhuoun, Sue Robertson
Ridgeback Park Olives, Pipeclay Lane,
Mudgee 2850
Phone: 02 6373 3119
Email: info@ridgeback.cc
You can tour the olive grove of 4000 olive trees (the first trees were planted in 1996) if you wish, or cut to the chase and visit the 'press door outlet'—that's a cellar door for olive oil. The grove began producing award-winning extra virgin olive oil in 2001 when a state of the art Italian oil press was installed. There are tastings of at least two varietal oils— always extra virgin and cold pressed, as well as Ridgeback Park verjuice, pickled olives, tapenades and soap. Also sold at selected Sydney outlets.
- **Open:** 10am–4pm Fri–Sun
- **Location:** 8km from Mudgee on Cassilis Rd, opposite Steins Wines

DINING

BLUE WREN RESTAURANT
Blue Wren Wines, Cassilis Rd, Mudgee 2850
Phone: 02 6372 6205
Email: james@bluewrenwines.com.au
www.bluewrenwines.com.au
Licensed
Cuisine: Contemporary Australian
Mains average: $$
Specialty: Lamb shanks braised in Blue Wren chardonnay port
Local flavours: Beef, venison, rabbit, olives, pistachios—whatever Mudgee offers
Extras: Alfresco dining with views across vineyards and rolling countryside.
- **Open:** Lunch daily, dinner Wed–Sat
- **Location:** 3km from town near the airport

EXPLORER COUNTRY AND RIVERINA—NSW

CRAIGMOOR RESTAURANT
Peter Marshall
Poet's Corner Winery, Craigmoor Rd,
Mudgee 2850
Phone: 02 6372 2208
Email: peter.marshall@orlando-wyndham.com
www.poetscornerwines.com
Licensed
Cuisine: Regional
Mains average: $$
Specialty: A menu complemented by the wines
Local flavours: Beef, lamb, venison, plus whatever else is grown in this food bowl
Extras: Alfresco dining in the vineyards. Poet's Corner pays respect to local boy turned poet, Henry Lawson.
- **Open:** Lunch daily, dinner Fri–Sat

DEEB'S KITCHEN AT THE SCHOOLMASTER'S HOUSE
Bechora and Sybil Deeb
Cnr Cassilis Rd and Buckaroo Lane,
Mudgee 2850
Phone: 02 6373 3133
www.homestead.com/deebskitchen.index.html
BYO
Cuisine: Homestyle
Mains average: $$
Specialty: Mezze plate, seasonal omelettes
Local flavours: Meat, vegetables, fruits, cheeses
Extras: Small weddings a speciality. A true country welcome and simple fresh food in an alfresco setting close to major vineyards.
- **Open:** Lunch weekends, dinner by arrangement (for eight or more)
- **Location:** Opposite Huntingdon Estate Winery

ELTON'S BRASSERIE
Alan and David Cox
81 Market St, Mudgee 2850
Phone: 02 6372 0772
Email: eltons@hwy.com.au
Licensed
Cuisine: Modern Australian
Mains average: $$
Specialty: Wood-fired pizzas
Local flavours: Farm-fresh eggs, olives and olive oils, local wines
Makes for sale: Dressings, relishes, jams
Extras: Located right in the shopping area. Takeaway available.
■ **Open:** Breakfast and lunch daily, dinner Tues–Sat

LAURALLA GUESTHOUSE AND GRAPEVINE RESTAURANT
Vinh Van Lam and Stuart Horrex
Lauralla Guest House,
Cnr Lewis and Mortimer Streets, Mudgee 2850
Phone: 02 6372 4480
Email: GreatTastes@lauralla.com.au
www.lauralla.com.au
BYO
Cuisine: Modern Australian
Seven-stage degustation: $$$
Specialty: Native foods and regional produce
Extras: Atmospheric alfresco dining under an eighty-year-old vine. Fully enclosed with open fires in winter and open to balmy nights in summer.
■ **Open:** Dinner Fri–Sat (daily for house guests), bookings essential

WINEGLASS BAR AND GRILL
Scott Tracey
Cnr Market and Perry Streets, Mudgee 2850
Phone: 02 6372 3417
Email: info@cobbandcocourt.com.au
www.cobbandcocourt.com.au
Licensed
Cuisine: Regional
Mains average: $$
Specialty: Wineglass antipasto plate, grain-fed char-grilled steaks
Extras: Cobb and Co Court Boutique Hotel offers accommodation. Le Verre Patisserie has takeaway, snacks and street-front dining. Courtyard and loft dining at The Wineglass Bar and Grill.
■ **Open:** Breakfast, lunch and dinner daily

MULLION CREEK
● PRODUCERS

CABONNE COUNTRY HONEY
George and Pearl Butcher, Bradean Mullion Rd, Mullion Creek 2800
Phone: 02 6365 8475
Email: cabonnehoney@bigpond.com
'It may be possible to demonstrate honey extracting,' say the Butchers, 'but you have to arrange that with the bees!!' George Butcher collects honey from hives located within about 100km of his property. Most popular are Paterson's curse, stringybark, yellow box and ironbark. He currently has around seventy hives but is building up, after first being 'stung' by beekeeping enthusiasm at a Canberra Festival bee display years ago.
■ **Open:** By appointment
● **Location:** 20km from Orange off Burrendong Way (3km dirt road to the gate)

NARRANDERA
🍴 DINING

AT THE STAR RESTAURANT
Margaret (chef) and John Britton
Historic Star Lodge, 64 Whitton St,
Narrandera 2700
Phone: 02 6959 1768
enquiries@historicstarlodge.com.au
www.historicstarlodge.com.au
Licensed
Cuisine: Contemporary Australian
Mains average: $$
Local flavours: Murray cod, silver perch, Riverina lamb and wines
Extras: Formerly the Star Hotel, the Historic Star Lodge is now classified by the NSW National Trust and operating as a B&B with ten traditional rooms. The main dining room was formerly the public bar and features original tile work, stained glass windows, arches and pressed metal ceilings. Homemade ice-creams made to order. Takeaway available.
■ **Open:** Dinner Tues–Sat, other meals by arrangement

NARROMINE
🍎 PRODUCERS

LIME GROVE
Peter and Susie Collett
4606 Mitchell Hwy, Narromine 2821
Phone: 02 6889 1962
Email: nlt@tpg.com.au
www.limegrove.com.au
Australia's largest lime orchard has luscious juicy Tahiti limes to buy fresh on the farm from March through June, or purchase Lime Light—Heaven in a Bottle, they call it. These lime products are available at the farm gift shop, as well as pure lime juice cordial, lime and chilli seed-mustard, lime juice dressing, Asian lime marinade, lime infused oil and other products. No chemicals are sprayed at Lime Grove, the Colletts telling customers that 'Limes are a magic green health tablet—full of vitamin C and anti-inflammatory flavonoids'. The products are available in many places locally and in some capital city outlets.
■ **Open:** 9am–5pm weekdays, by appointment on weekends, coach groups by arrangement
● **Location:** 5km west of Narromine

ORANGE
🍎 PRODUCERS

APPLEDALE PROCESSORS COOPERATIVE
David Gartrell
5 Stephen Place, Orange 2800
Phone: 02 6361 4422
Email: appledale@cww.octec.org.au
Appledale apple juice may be clear or cloudy, but it is all natural juice produced from fruit grown in the Orange district. There is also apple and boysenberry juice.
■ **Open:** 9am–5pm weekdays

BORRODELL ON THE MOUNT
Borry Gartrell and Gaye Stuart-Nairne
Lake Canobolas Rd, Orange 2800
Phone: 02 6365 3425
Mobile: 0418 865 217
Email: accommodation@borrodell.com.au
www.borrodell.com.au
Borry Gartrell is not just an orchardist. He is an artist who plants fruit trees so that the blossoms complement each

other. In addition to a wide range of fruits, with a specialty in old apple varieties, there are cellar-door sales of his wine and cider, including nectarine, plum and cherry. Borry is currently planting acorns from two beautiful trees in Orange and hopes that in time they will host truffles. You'll meet him at the Orange farmers' market, or you can stay in one of the two three-bedroom guest cottages or the sleep-six converted railway carriages on the property. What appears at first glance to be a house is actually a cellar door and shop where you can buy and taste Borry's fruit wines and a range of local Orange wines.
- **Open:** Cellar door 11am–6pm daily, fruit sales by arrangement in season, farm tours by arrangement
- **Location:** 12km south-west of Orange

BRITTLE JACKS
David Ogilvy
Lookout Rd, Mullion Creek, Orange 2800
Phone: 02 6365 8353
Email: ogers007@netwit.net.au
David Ogilvy has been growing chestnuts and plums here for twenty-seven or so years simply because he felt there was a need for chestnuts in the marketplace. Sometimes he has chestnut honey available too, and will meet with coach and special interest groups by appointment. After the main harvest, he will allow people to pick their own chestnuts, but this is usually only for a few weeks a year from mid March to the end of May.
- **Open:** By appointment
- **Location:** Take Burrendong Way from Orange to Mullion Creek

🏠 STORES

PROVEN—ARTISAN BREADS AND PASTRIES
Lisa and Paul Wilderbeek
26B Sale St, Orange 2800
Phone: 02 6360 0722
Email: proven@hn.ozemail.com.au
Handmade organic breads, sourdoughs, fine tarts and pastries—can't you almost smell them? Wilderbeek, formerly chef at Highland Heritage Estate in Orange, makes all this and more from scratch with no pre-mixes, additives or preservatives, and using no machines. There are twenty-eight varieties of bread, even spelt, and gluten-free bread, made with ingredients from Gunnedah, Parkes, Manildra and Orange. The products are also available at local farmers' markets.
- **Open:** 8am–4pm Tues–Thurs, 8am–5.30pm Fri, 8am–2pm Sat

🍴 DINING

HIGHLAND HERITAGE ESTATE RESTAURANT
Rex and Jacky D'Aquino (owners),
Keith Wilson (chef)
Mitchell Hwy, Orange 2800
Phone: 02 6361 3054
Email: heritage@netwit.net.au
www.highlandheritageestate.com
Licensed/BYO wine
Cuisine: Modern Australian
Mains average: $$$
Local flavours: Mandagery Creek venison, rabbit, organic herbs, Mardies Olives (Cowra), Dutton Park duck
Extras: Situated in the heart of a vineyard, this restaurant has 360-degree views of the surrounding countryside, and features recycled timber from an old bridge. The

restaurant has won numerous awards. Takeaway is available. They also grow and sell their own raspberries and blackberries (fresh or frozen) as well as jams and Mountain Flame raspberry liqueur.
- **Open:** Lunch Thurs–Sun, dinner Thurs–Sat

LOLLI REDINI
Simonn Hawke (chef)
48 Sale St, Orange 2800
Phone: 02 6361 7748
Email: simhawke@hotmail.com
Licensed
Cuisine: Modern regional with an Italian influence
Mains average: $$$
Specialty: Menu changed daily
Local flavours: Uses as much food and wine from the Orange district as possible.
Extras: Features fine art work on the walls with the occasional exhibition, and relaxed ambience.
- **Open:** Lunch Fri–Sat, dinner Tues–Sat

SELKIRKS RESTAURANT
Josephine Jagger and Michael Manners (chef)
179 Anson St, Orange 2800
Phone: 02 6361 1179
Licensed
Cuisine: Fine dining à la carte
Mains average: $$$
Specialty: Regional produce
Local flavours: Venison, pigeon, rabbit
Extras: A multi award-winning restaurant, in an elegant, older-style building. Outdoor area for drinks only.
- **Open:** Dinner Tues–Sat

PARKES
🍎 PRODUCERS
ALPACA COUNTRY SHOP AND CAFE
Maree Hornery
Valley Crest, Newell Hwy,
Tichborne via Parkes 2870
Phone: 02 6863 1133
Email: hornery@westserv.net.au
www.macusanialpacas.com
This alpaca stud has diversified by opening a farm shop, which includes alpaca products, as well as serving morning and afternoon teas and light lunches in its country cafe. Look for local produce too, such as olive oil, lavender and jams.
- **Open:** 10am–4pm Thurs–Sun, public holidays and school holidays (other times by appointment)
- **Location:** 10km south of Parkes

RYLSTONE
🍴 DINING
BRIDGE VIEW INN BAKERY CAFE
Kim Currie
28–30 Louee St, Rylstone 2849
Phone: 02 6379 0996
Email: kim.currie@bigpond.com.au
Licensed
Mains average: $
Specialty: Danishes, pies, crusty country breads
Local flavours: Berries, free-range chickens, own breads and salads
Extras: Located in Rylstone's most historic landmark building, a two-storey, 1830 sandstone inn with ten open fireplaces. Dining also available in back courtyard or on front veranda, overlooking Rylstone's main street.

■ **Open:** 8am–4pm daily, 6–9pm Sun–Thurs

RYLSTONE FOOD STORE
Kim Currie
47 Louee St, Rylstone 2849
Phone: 02 6379 0947
Email: eat@rylstonefoodstore.com.au
www.rylstonefoodstore.com.au
BYO
Cuisine: Regional, seasonal comfort food
Four-course set menu: $$
Extras: This historic town is close to Wollemi National Park, Mudgee and Rylstone wine and farm tours. Located in the original village general store with minimal contemporary makeover. Shares space with number 47, a contemporary gallery space, and also sells handmade gifts and breads. Saturday dinners are feast like, and Friday dinners feature lighter, simpler food.
■ **Open:** Lunch Sun, dinner Fri–Sat, foodstore 10am–4pm Fri–Sun

SOFALA
DINING

CAFE SOFALA
Ralf-Josef Mueller
42 Denison St, Sofala 2795
Phone: 02 6337 7053
Email: cafesofala@dodo.com.au
Licensed
Cuisine: International
Mains average: $$
Specialty: Great steak, local trout, curries
Local flavours: Wine, trout, honey, jams
Extras: Takeaway available. Vines and gardens surround the cafe.
■ **Open:** Lunch Fri–Mon, dinner Fri–Sun

WAGGA WAGGA
PRODUCERS

RIVERINA OLIVE GROVE
Vici Murdoch and Gerard Gaskin
'Federation', RMB 274,
Wagga Wagga 2650
Phone: 02 6922 9221
Mobile: 0408 694 194
Email: gerardgaskin@bigpond.com
www.riverinaolivegrove.com
The owners of this award-winning olive grove are happy to show people around during the processing season from mid April to late July. Recently they have begun manufacturing a range of all-natural body products made from olives, which can be ordered via the website, and are also found in an increasing number of pharmacies.
■ **Open:** By appointment

STORES

KNIGHT'S MEATS WHOLESALE
Michael and Anne Knight
187 Fitzmaurice St, Wagga Wagga 2650
Phone: 02 6921 3725
Email: knight@riverinatelco.com.au
www.knightsmeats.com.au
The meat products at this shop are described by fourth-generation butcher, Michael Knight, as 'diverse but still evolving after thirty years'. This massive deli has become an outlet for local produce such as Hampden Bridge hams (the sawdust for smoking comes from Adjungbilly in the mountains), Riverina Fresh Products, olive oil, lentils, honey, mustard-seed oil, free-range eggs, antipasto, fresh cheese, marinades, blueberry topping, trout, jams and chutneys and fruit juices.

■ **Open:** 8am-6pm Mon-Sat (Thurs until 8pm, Sat until 4pm)

🍴 DINING

MAGPIE'S NEST RESTAURANT
Christopher and Wendy Whyte
20 Pine Gully Rd, Wagga Wagga 2650
Phone: 02 6933 1523
Email: magpiesnest@bigpond.com
Licensed/BYO
Cuisine: Modern Australian
Mains average: $$
Specialty: Murray cod braised with baby fennel and ginger
Local flavours: Olive oil, regional wines, lamb, beef, cod and silver perch, fruit and vegetables, pistachios
Extras: Oils made from the fruit of wild olive trees throughout the state, and Magpie's Nest's own olive grove, pressed using a traditional press. Separate vegetarian menu. Takeaway available. A wonderful view over the city of Wagga, and a Mediterranean ambience. Large cottage gardens, vegetable and herb gardens, olive grove and vineyard, plus ducks, geese, hens and magpies.
■ **Open:** Lunch Wed-Sun, dinner Wed-Sat

WAGGA WAGGA WINERY
Peter and Kerry Fitzpatrick
'Hillsley', RMB 427, Oura Rd,
Wagga Wagga 2650
Phone: 02 6922 1221
Licensed
Cuisine: Aussie barbecue, cook your own
Mains average: $$
Specialty: Rump steak, marinated chicken, rainbow trout, butterfly lamb steak
Local flavours: Meats, chicken, trout, cherries, apples

Extras: There are two large indoor bistros or enclosed, sheltered gardens.
■ **Open:** Lunch and dinner daily
● **Location:** 13km from Wagga Wagga on Oura-Wantabagery Rd

WALLENDBEEN
🍎 PRODUCERS

YANDILLA MUSTARD SEED OIL
Kaye Weatherall
'Yandilla', Olympic Hwy, Wallendbeen 2588
Phone: 02 6943 2516
Email: kaye@yandilla.com
www.yandilla.com
A trip to India gave Kaye Weatherall and her husband a taste for mustard-seed oil. Returning to Australia, and unable to find it, they discovered that the CSIRO deemed the oil to be too high in erucic acid. When a lower-acid plant was developed, farmers were encouraged to plant this hardy crop, and so, fifteen years ago, the Weatheralls set about producing their own-brand oil. Local farmers grow the mustard seed under contract and Yandilla processes it. Although related to mustard powder, the oil itself is not hot, and may be used like any good cooking or salad oil. Health benefits include thirty per cent Omega-6 and fifteen per cent Omega-3 essential fatty acids, no cholesterol and extremely low saturated fat. Add to this the nutty flavour, and Yandilla feels it has a winning combination. The oil is available through David Jones Food Halls, delicatessens, health food and gourmet shops. A delightful tearoom serves light lunches and afternoon teas during the farm opening hours.
■ **Open:** 10am-5pm daily

- **Location:** Halfway between Young and Cootamundra

WELLINGTON

🛒 STORES

GILLIN'S BUTCHERY
Greg Gillin
6 Nanama Cres, Wellington 2820
Phone: 02 6845 2062
The word is getting around about the great range of gourmet sausages here—including a knockout spicy peach and pork sausage with eighteen ingredients. All the meat used here is local, as Gillin deals direct with local small producers, some of whom attend local growers' markets. Gillin admits he is self taught in the arts of smoking and making small-goods, but his traditionally smoked range of mulga-smoked ham, beef, and chicken, as well as pastrami and bacon (all made on the premises) show he has learned well.
- **Open:** 6am–6pm weekdays, 7am–1pm weekends

🍴 DINING

CACTUS CAFE AND GALLERY
Marilyn Keirle
33–35 Warne St, Wellington 2820
Phone: 02 6845 4647
Licensed/BYO
Cuisine: Simple, fresh, tasty
Mains average: $
Local flavours: Vegetables, herbs, lamb, beef, fruit, olives
Extras: This award-winning cafe-gallery is on the site of the old Catholic infants school where many locals started school, but now it is filled with craft and home wares. Veranda seating, and pergola with a water feature in the garden.
- **Open:** Lunch Wed–Sun, all day snacks

YEOVAL

🍴 DINING

THE FAT CAT CAFE
Gordon Klaare
Newsagency and Information Centre,
27 Forbes St, Yeoval 2868
Phone: 02 6846 4102
BYO
Cuisine: Australian country cafe
Mains average: $
Specialty: Burgers, especially the pork Fat Cat burger
Extras: The Fat Cat Cafe in Yeoval and its plump 'kitten' in Cumnock (37 Obley St, Cumnock, 20km along the road towards Orange) are examples of pit-stop cafes which have got the right mix of atmosphere and food. Each Fat Cat is located at a newsagency and tourist information centre.
- **Open:** 9am–6.30pm daily

MARKETS AND EVENTS

✦ **BATHURST FARMERS' MARKETS**
02 6332 4447
8am–12.30pm fourth Sat

✦ **BATHURST ORGANIC MARKETS**
02 6334 2020
9am–5.30pm Fri, 9am–1pm Sat

✦ **COWRA REGION FARMERS' MARKET**
02 6342 9225
8.30am–12pm third Sat

✦ **DUBBO FARMERS' MARKET**
02 6884 0161
8am–12pm fourth Sat

✦ **MUDGEE REGIONAL FARMERS' MARKETS**
02 6379 0728
9am–1pm first Sat

✦ **ORANGE FARMERS' MARKET**
02 6362 0276
8.30am–12pm second Sat

✦ **PCYC MARKETS,** Griffith
02 6964 2004
7am–12pm Sun

✦ **WAGGA WAGGA FARMERS' MARKETS**
02 6922 9221
8am–1pm second Sat

✦ **TRUNDLE BUSH TUCKER DAY,** TRUNDLE
02 6892 1430
First Sat in Sept

✦ **SUNRICE FESTIVAL,** LEETON
02 6953 6481
Mar–Apr biennial festival (even-numbered years)

✦ **MUDGEE WINE CELEBRATION**
1800 816 304
www.mudgeewine.com.au
Sept

✦ **THE TASTE OF COWRA**
02 6341 4141
Mid May

✦ **COWRA WINE SHOW**
02 6342 4333
Aug

✦ **COWRA FOOD AND WINE WEEKEND**
02 6342 4333
Early Nov

✦ **TUMBAFEST,** TUMBARUMBA
02 6948 2462
Last weekend Feb

✦ **FOOD WEEK,** ORANGE
02 6362 3822
www.orangefoodweek.com.au
Apr

✦ **UNWINED IN THE RIVERINA,** GRIFFITH
1800 681 141
www.unwined-riverina.com
June

✦ **MACQUARIE VALLEY WINE AND FOOD FESTIVAL,** DUBBO
0417 891 962
www.dubbotourism.com.au
Nov, various locations

THE HUNTER

This magnificent region is known for its fine wines and the beauty of the countryside. Every season has a different attraction and a wide range of eateries are ready to cater to your every need.

ANTIENE

🍎 PRODUCERS

HUNTER GROVE, HUNTER OLIVE COOPERATIVE LTD
168 Hebden Rd, Antiene 2333
Phone: 02 6541 3522
www.huntergrove.com.au

This cooperative was formed three years ago to meet the processing and marketing needs of over eighty growers with, between them, 130 000 olive trees in the Hunter Valley. Oils, olives and olive products from various cooperative members throughout the Hunter region are now processed here. The products are available at Hunter Valley Cheese Company, Pokolbin, so look for Hunter Grove extra virgin olive oils there, and at most good delis in Sydney and the Hunter Valley.
■ **Open:** Tours by arrangement

ASH ISLAND

🍎 PRODUCERS

KOORAGANG CITY FARM
Rob Henderson (manager)
Milhams Rd, Ash Island 2322
Phone: 02 4964 9308
Email: rob@kooragang.hcmt.org.au
www.hcmt.org.au/kooragang

Integral to the surrounding wetlands, the Kooragang City Farm, begun eight years ago, is promoting and demonstrating 'farming in harmony with wetlands'. The farm encourages volunteers, casual visitors and large groups (by appointment) to see holistically managed beef cattle, organic fruit and vegetables, permaculture, and bush foods. There are bottled preserves, farm tours, fishing and even frog spotting. Farm products are sold on site and at local farmers' markets.
■ **Open:** 10am–3pm, Mon–Fri, other times by appointment
● **Location:** Via Ash Island Bridge, Pacific Hwy, Hexham

BELFORD

🍷 BREWERY/WINERY

BRAMBLEWOOD FRUIT WINES
Graham and Robyn Renfrew
80 Lindsay St, Belford 2335
Phone: 02 6574 7172
Mobile: 0438 747 172

Graham Renfrew is passionate about this hobby-turned-career, and is delighted with how it is going. After long hours of experimentation, this self-taught fruit winemaker now has four wines—orange, lemon, strawberry, boysenberry, and lemon and passionfruit. Mango is fermenting as we speak, he tells us, and fruit blends are coming soon. You get the feeling it's hard work, but he's having a lot of fun too. Tastings are available. Fruit wines don't cellar for long and are best drunk within twelve months.

THE HUNTER—NSW

■ **Open:** 9am–5pm daily (buses and tour groups by appointment)

BOBS FARM
🍎 PRODUCERS

THE AVOCADO FARM STALL
Henry and Robin Willner
37A Nelson Bay Rd, Bobs Farm 2316
Phone: 02 4982 6037
Basically this is a fruit shop located on an avocado farm which started small around twelve years ago. Of course the avocados are good, and real bargains too. Ask them to show you the difference between the varieties: Fuerte, Haas, Sharwil and Reed. Also here are Anna Bay tomatoes, jams, pickles, chutneys, sauces, marinades, avocado oil and crafts.
■ **Open:** 10am–6pm Wed–Mon (June–Jan)
● **Location:** Fifteen minutes on the Newcastle side of Nelson Bay, between Salt Ash and Anna Bay

BROKE
🍎 PRODUCERS

FORDWICH GROVE
Kjeld Jakobsen
203 Fordwich Rd, Broke 2330
Phone: 02 6579 1179
Email: mail@fordwichgrove.com.au
www.fordwichgrove.com
This is an olive grove that evolved from a berry farm. The previous owners pulled out the berries, and now some of the olive trees are ten years old. Tastings of oil (including infused oil) and sales are available at weekends. In addition there are olives (marinated and plain) tape-nades, wine vinegars and caramelised balsamic vinegar, chutneys, jams, dukkah, and some ceramics. Accommodation is also available in two, two-bedroom, self-catering cottages.
■ **Open:** 9am–5pm weekends, or by appointment

BULGA
🍎 PRODUCERS

HILLSDALE ORANGE ORCHARD
Betty and Harold Harris
Hillsdale Orange Orchard,
Bulga, via Singleton 2330
Phone: 02 6574 5173
Pick your own sweet juicy oranges, then set off on a mapped and marked ten-kilometre walking trail through the rainforest. Betty Harris has lived on this property all her life and she and her husband have recently added a farmstay in a three-bedroom former farmhouse. The Harrises grow navel and Valencia oranges, mandarins, lemons and grapefruit 'and they are the best', they say, adding that buyers at the farmers' market just love their Bulga navels. Many are sold to markets and for juice, but there are plenty for passing visitors, and there are morning teas for coaches by appointment.
■ **Open:** Tours are run all day, every day—phone to arrange a time

CONGEWAI
🍎 PRODUCERS

SNAILS BON APPETITE
Helen Dyball
245 Congewai Rd, Congewai 2325
Phone: 02 4998 0030

Email: snails@hunterlink.net.au
www.snailsbonappetite.com.au
It's a long way from a wild snail to that tasty morsel of escargot that delights the palates of gourmets. Helen Dyball needs to keep 50 000 breeding snails to generate the 30 000 snails needed to supply Hunter and Sydney markets. *Helix Asperasa* is an edible Australian snail which is not only highly nutritious, but low in fat and cholesterol. The young are fed a diet high in calcium, combining powdered milk, dried feed and green herbs. The succulent escargot is ready for the table when it reaches a length of 3.5–4cm. They are raised in pens housed in large igloos. Helen is always looking for new snails for her breeding stocks, so has created a popular fundraising project for local schools—snail hunting. Snails come to her from everywhere, some even through the post (snail mail?). Check her website to learn more about this unusual industry. You will find them on the menus of Roberts at Peppertree and the Quay Grand in Sydney. In all likelihood, any restaurant in this region serving fresh escargot is probably using these little beauties.

■ **Open:** 10am–3pm, by appointment

DUNGOG

DINING

CRAZY CHAIRS RESTAURANT AND GALLERY
Kate Hare
205 Dowling St, Dungog 2420
Phone: 02 4992 3272
Cuisine: Country Australian
Mains average: $
Specialty: 'Crazy' crammed full sandwiches
Local flavours: Smoked Barrington perch

Extras: Eclectic furnishings, and gallery of local artists' works.
■ **Open:** Lunch daily, dinner Fri

EAST MAITLAND

DINING

THE OLD GEORGE AND DRAGON RESTAURANT
Ian (chef) and Jenny Morphy
48 Melbourne St, East Maitland 2323
Phone: 02 4933 7272
Email: oldgeorge@netcentral.com.au
www.oldgeorgeanddragon.com.au
Licensed
Cuisine: Produce-driven Anglo–French
Mains average: $$$
Specialty: Duck
Local flavours: Duck, squab, seafood, seasonal vegetables
Extras: Multi award-winning restaurant in a unique building that dates from 1837, and has been refurbished with quality antiques and 18th-century paintings. Accommodation in delightfully decorated rooms.
■ **Open:** Lunch by arrangement, dinner Wed–Sat

GLOUCESTER

PRODUCERS

ALPACA VIANDE
Phillip and Roma Lahey
2549 Waukivory Rd, Gloucester 2422
Phone: 02 6558 0971
Email: alpacaviande@bigpond.com
Viande is the French word for meat, the Spanish for delicacy and the Latin term for 'grazed on pasture', which should give a hint as to what the Laheys are market-

ing here. The twenty-five acre farm runs fifty or so alpacas, and the aim is to market the meat as smallgoods and also show visitors how the farm operates. Visitors can try alpaca sausages, frankfurters and terrine, watch a shearing video and handle alpaca fleeces.
■ **Open:** By appointment

CAPPARIS
Aled Hoggett and Tania Parkinson, Coral Hoggett and Jim Hoggett
Berllanber, 1333 Bowman River Rd, Gloucester 2422
Phone: 02 6558 5557
Email: capparis@tpg.com.au
www.capparis.com.au
The Capparis goat herd is mostly pure Anglo-Nubian stock as the milk from Anglo-Nubians is noted for its cheese making qualities. Cheese maker Aled Hoggett, is currently making three styles of goats milk cheese—a fresh traditional goats cheese, a brined fetta style and a matured white mould cheese in the foothills of the World Heritage Barrington Tops area, near Gloucester. Capparis goat cheese farm is open for farm visits, and the products may be bought in Sydney at several farmers' markets as well as at Rozelle Fine Food Store and jones the grocer.
■ **Open:** 10am-4pm daily, or by appointment
● **Location:** 26km west of Gloucester, left on Walcha Rd

HILLVIEW HERB FARM
Karen O'Brien
5 Fairbairns Rd, Gloucester 2422
Phone: 02 6558 2369
In this herb garden with magnificent views, visitors can take a tour as well as purchase herbal products. There are herbal scones, honey, cakes, bread, butters, quiches, salads and also massage oils. The O'Briens grow a wide variety of herbs and sell the excess to restaurants. There are a couple of acres under herbs that Karen describes as being 'like a big back yard, which is getting bigger all the time.'
■ **Open:** Morning and afternoon teas and light lunches by appointment
● **Location:** Cnr Bucketts Way and Fairbairns Rd, five minutes south of Gloucester

🏠 STORES

DARREL'S GOURMET BUTCHERY
Darrel Wisemantel
39 Church St, Gloucester 2422
Phone: 02 6558 1009
One of the first NSW shops to receive the Q Award—shop with assurance—Darrel's Butchery prides itself on the fine, fresh local meat it sells, also using local herbs in his products. There is rabbit, goat, quality local beef, pork, ham and bacon cured on the premises, silver perch, lamb and chicken, and you can even hire a spit roast to cook it yourself. Named the NSW North Coast Sausage Kings in 2002, the butchers here say their breakfast sausage is the king of the range. The butchery now sells MSA beef (the brand means 'guaranteed tender'), and also smoke their own hams, and cater for functions.
■ **Open:** 6am-5pm weekdays, 6am-12pm Sat

🍴 DINING

PERENTI
Jane Westcott
69 Church St, Gloucester 2422
Phone: 02 6558 9219

Email: perenti@tpg.com.au
www.perenti.com.au
BYO
Cuisine: Regional
Mains average: $
Specialty: Use of Essentially Barrington products
Local flavours: Gloucester Gold smoked chicken, Barrington Beef steak, Hillview Herb Farm salad mix, Capparis goats cheese
Extras: Founding member of a food alliance called Essentially Barrington. Chic, stylish cafe, Toby's Estate coffee, fresh juices and frappes, frozen take-home meals, vegetarian dishes and own range of chutneys, jams and salad dressings. Caters for special dietary requirements.
■ **Open:** Breakfast and lunch daily

GRESFORD
🍎 PRODUCERS
THE WHOLEFOOD GARDEN
Marc and Caroline Intervera
46 Pound Crossing Rd, Gresford 2311
Phone: 02 4938 9159
Email: thewholefoodgarden@hotmail.com
This wholefood garden evolved as a lifestyle choice for this family. It's all organic and the salad greens and vegetables 'taste the way they should' say the Interveras. They must do, as this seasonal produce is being sought out by top restaurants in the Hunter Valley such as Wyndham Estate and Hungerford Hill.
■ **Open:** Fri–Sun, phone for availability
● **Location:** Pound Crossing Rd is on the left 2km from Gresford on the Singleton–Glendonbrook Rd

GRETA
🍴 DINING
THE TABLE GUESTHOUSE
Malcolm Martin
3 Water St, Greta 2334
Phone: 02 4938 7799
Email: thetable@kooee.com.au
www.thetable.com.au
BYO
Three-course set menu: $$$
Specialty: Duck confit
Local flavours: Beef, veal, vegetables
Extras: A dedicated 'slow-foodist' chef–owner who is committed to serving regional, seasonal foods. Tuscan farmhouse-style guesthouse with leafy terrace for outdoor dining.
■ **Open:** Breakfast and dinner daily (for guests, dinner for non-guests by arrangement)

LAMBS VALLEY
🍎 PRODUCERS
MOUNT HUDSON STRAWBERRIES
David and Janice Campbell
1593 Maitland Vale Rd, Lambs Valley 2335
Phone: 02 4930 6224
Mobile: 0412 608 564
Email: campbell69@bigpond.com
The Campbells began this interesting venture in 1997. The strawberries are grown using hydroponic, NFT (Nutrient Flow Technology), which is a complicated way of saying these are bright red juicy strawberries, sun ripened in the open air for maximum colour and flavour. You can purchase them on the farm or from local outlets (pick your own not available), and

Janice also makes jams and toppings—mixing strawberries with pineapple, apples, passionfruit, or Cointreau. Both David and Janice once had other jobs. In fact the strawberries began as a hobby, and although they live on thirty acres, only half an acre is planted with berries, all at chest height, though, which saves their backs.
- **Open:** 9am–6pm daily (Apr–May and mid Sept to early Jan), phone first

LUSKINTYRE
🍎 PRODUCERS

POKOLBIN CREEK OLIVES
Margaret Yeatman
215 Luskintyre Rd, Luskintyre 2321
Phone: 02 4930 6132
Email: theolivefarm@yahoo.com.au
The Yeatmans grow and press their own olives, and have five varieties of freshly pressed olive oil available, as well as a range of bread dippers, marinated olives, tapenades, and olives to home-cure in season. Their brands are Pokolbin Creek Olives, Get Stuffed Olives, and Popeyes Olive Farm and you may see them in some Hunter wineries, as well as markets and shows, or you can order them by mail order. Everything is made using Hunter Valley oils and olives and they have tastings of all their products. They are also planning cookery classes, demonstrating the use of olive oil and olive products.
- **Open:** By appointment
- **Location:** Second gate on the left over the old Luskintyre bridge

MAITLAND
🏠 STORES

ORGANIC FEAST
Brent Fairns
Shop 3, Maitland Plaza, Bulwer St,
Maitland 2320
Phone: 02 4934 7351
Email: info@organicfeast.com.au
www.organicfeast.com.au
Nearly thirty certified local growers supply organic fruit and vegetables from all over the Hunter region to this shop. There are organic grains and flours, local biodynamic chickens and beef, as well as organic honey. Now you can order online for deliveries (including delivery to the Central Coast—metro areas only). Available Sundays at Honeysuckle Markets, Newcastle, the first Sunday of the month at Gosford Fine Food Market, and the third Sunday of the month at Mount Penang.
- **Open:** 9am–5.30pm weekdays, 9am–1pm Sat

MT VINCENT
🚌 TOURS

PICK-OF-THE-CROP WINE EXPEDITIONS
Heather Jefferys
19 Rodney Rd, Mt Vincent 2323
Phone: 02 4990 1567
Mobile: 0416 187 295
Email: info@wineexpeditions.com.au
www.wineexpeditions.com.au
What a fun way to experience the Hunter! Wine and food tours for small groups in a chauffeur-driven, classic Pontiac sedan. Awarded Best Significant Tour Operator for the Hunter Region in

2003 as well as being a finalist in the 2003 NSW Tourism Awards.
- **Open:** Daily by appointment

MUSWELLBROOK
🍎 🍴 PRODUCERS/DINING

HUNTER BELLE CHEESE
Kate Woodward
Verona Winery,
75 Aberdeen St, Muswellbrook 2333
Phone: 02 6541 5066
Mobile: 0402 838 462
Email: info@hunterbellecheese.com.au
www.hunterbellecheese.com.au
'How Now Swiss Brown Cow?' Very well, says Kate Woodward whose new business Hunter Belle Cheese, producing Swiss-style cheeses, is taking off nicely. Located within Verona Winery (02 6541 4777), on the same premises as the Verona Cafe, you can kill three birds with one stone—watch the cheese making through windows into the dairy, enjoy lunch and buy some wine for later. And you can meet a couple of the cows, residents in a paddock nearby, which provide the milk. Verona also has a gourmet food area that retails many other local food products. In 2002 Kate was awarded the Australian Grand Dairy Awards Young Cheese-maker Scholarship which allowed her to train in several major cheese-making companies. Currently the range of cheese includes fetta, camembert, washed rind, and blue styles.
- **Open:** 9am–5pm daily

NELSON BAY
🍴 DINING

RED BELLIES AT THE PUB
Bruce Sanders (owner),
Matthew Hallcroft (chef)
Seabreeze Hotel, cnr Stockton and Laman Streets, Nelson Bay 2315
Phone: 02 4981 1511
Email: seabreeze@nelsonbay.com
www.seabreezehotel.com.au
Licensed
Cuisine: Pub style
Mains average: $$
Local flavours: Seafood
- **Open:** Breakfast, lunch and dinner daily

ZEST RESTAURANT
Glenn Thompson
16 Stockton St, Nelson Bay 2315
Phone: 02 4984 2211
Email: zestrestaurant@optusnet.com.au
Licensed
Cuisine: Modern European
Mains average: $$$
Local flavours: Seafood, duck, quail, rabbit
Extras: Winner Best New Restaurant and European Restaurant in Hunter (and Regional NSW) Restaurant and Catering Awards.
- **Open:** Dinner daily

NORTH ROTHBURY
🍴 DINING

SHAKEY TABLES
Paula Rengger
The Hunter Country Lodge, Branxton Rd, North Rothbury 2335
Phone: 02 4938 1744

THE HUNTER—NSW

Email: eat@shakeytables.com.au
www.shakeytables.com.au
Licensed
Cuisine: Modern Australian
Mains average: $$$
Specialty: Roast pigeon stuffed with black pudding and prunes on cabbage parcels
Local flavours: Cheeses, ducks, venison
Extras: Grows much of its own produce including wine, verjuice, pigeons, vegetables, duck, chicken and quail eggs, herbs and preserves.
- **Open:** Lunch Sun, dinner daily
- **Location:** 17km from Cessnock, 5km from Branxton

PATERSON

DINING

YABBIES RESTAURANT
Steve Kelly (owner), Daniel Wilks (chef)
Paterson Tavern,
25 Prince St, Paterson 2421
Phone: 02 4938 5835
Licensed
Cuisine: Quality pub food
Mains average: $$
Specialty: Mushroom sauce, casseroles and soups
Local flavours: Hunter Valley steaks, yabbies, breads, chicken, fruit and vegetables
Extras: Country atmosphere with log fires, ideal for the climate in the foothills of the Barringtons. Takeaway available.
- **Open:** Lunch and dinner daily

TOURS

HUNTER VALLEY DAY TOURS
Rob McLaughlin
Paterson 2421
Phone: 02 4938 5031
Email: daytours@hunterlink.net.au
www.huntertourism.com/daytours
Fully escorted wine and cheese tasting tours of Lower Hunter wineries and selected food outlets, or may be tailored to suit your own interests. There is the option of hotel pick-up, and a restaurant lunch is included. Cheese, fudge and chocolate tasting is also available, and there is tour commentary and wine education. Small- or large-group tours and charters available on request.
- **Open:** 8am–9pm daily

THE CB ALEXANDER AGRICULTURAL COLLEGE AT TOCAL
Sandra Earle
Tocal Rd, Paterson 2421
Phone: 02 4939 8888
Freecall: 1800 025 520
Email: homestead@tocal.com
www.tocal.com
Interactive farm tours! Sounds trendy but here is a great opportunity to get away from Sydney for the weekend and see a living farm. The Hunter Harvest Network and Gourmet Trails was the inspiration behind this concept. This is an ideal place to start your tour of the Hunter food and farm trail. Not only can you visit the farm and historic homestead on weekends and public holidays but you can find out everything you want to know about the region with a selection of brochures and even maps from the visitor centre. The Tocal Field Day has become an annual event offering a chance to get up close

73

and personal with beef cattle, sheep and dairy cows on Friday to Saturday coinciding with the first Sunday of May (the weekend before Mother's Day). Self-guided audio tours of the heritage site (not the farm) are available, and there are plans for an increasing range of local produce to be sold from the visitors centre in the future.

- **Open:** 10am–4pm, weekends and public holidays (Mar–Nov), group bookings any time

POKOLBIN

🍎 PRODUCERS

HUNTER VALLEY CHEESE COMPANY
Peter Curtis and Rosalia Lambert
McGuigans Centre, McDonalds Rd,
Pokolbin 2320
Phone: 02 4998 7744
Email:
contact@huntervalleycheese.com.au
www.huntervalleycheese.com.au

Wine and cheese are perfect partners—so what better place for a specialty handmade gourmet cheese company than slap-bang in the middle of one of New South Wales' richest wine-making regions? Hunter Valley Cheeses include grapevine-ashed brie, Hunter Valley Gold washed-rind, and Pokolbin smear-ripened cheeses as well as fresh cheeses. You can buy lunch platters at the cafe to enjoy with local wines outside on the lawns, or for a picnic. A viewing window allows you to watch the cheese-making process. Cheese-making tutorials are conducted in the factory under the tutelage of highly experienced cheese-makers Peter Curtis or Rosalia Lambert.

- **Open:** 9am–5.30pm daily

🍴 DINING

AMANDA'S ON THE EDGE
Amanda North (owner), Karl Avis (chef)
Windsor's Edge Vineyard,
McDonalds Rd, Pokolbin 2320
Phone: 02 4998 7900
Email:
amandasontheedge@hunterlink.net.au
www.amandas.com.au
Licensed/BYO wine
Cuisine: Modern Australian and international
Mains average: $$$
Specialty: Duck, desserts
Local flavours: Red wine vinegar, extra virgin olive oil, fresh baked bread
Extras: Spectacular views from every table overlooking the vineyard and surrounding rolling hills and country. Fires in winter, alfresco in warmer months. Menu changes seasonally.

- **Open:** Lunch Fri–Mon, dinner daily

AUSTRALIAN REGIONAL FOOD STORE AND CAFE
Rosalia Lambert and Peter Curtis
Small Winemakers Centre,
Shop 2, Lot 59 McDonalds Rd,
Pokolbin 2320
Phone: 02 4998 6800
Email:
australianregionalfood@telstra.com.au
Licensed
Cuisine: Modern Australian
Mains average: $
Specialty: Regional products from all states
Local flavours: Bread, wines, cheeses, meat, fruit and vegetables.
Extras: Outdoor seating, views of the dam, wildlife, great for picnics and kid-friendly. Takeaway available. Begun at the end of 2003 to showcase the best of

THE HUNTER—NSW

the best Australian produce to visitors in the Hunter Valley. Products available for tasting include preserves, cordials, flour, pasta, breads, nuts, dried fruits, coffee, tea, cookies, macadamia and oils. These are also used in tasting menus in the cafe. The Lamberts call the stock for sale, 'The best of clean green Australian Produce'.

■ **Open:** 10am–5pm daily (lunch all day)

THE CELLAR RESTAURANT

Mark Hosie and Andy Wright (chefs)
Hunter Valley Gardens Village,
Broke Rd, Pokolbin 2320
Phone: 02 4998 7584
Email: thecellar@idl.com.au
Licensed
Cuisine: Modern Australian
Mains average: $$$
Specialty: Kangaroo, duck, Asian influences
Local flavours: Mostly local, including meats and dairy
Extras: Adjacent to the Hunter Valley Gardens Resort, and gardens.
■ **Open:** Lunch daily, dinner Mon–Sat

CHEZ POK

Jamie Hartcher (chef)
Peppers Guest House Hunter Valley,
Ekerts Rd, Pokolbin 2320
Phone: 02 4998 7596
Email: pghres@peppers.com.au
www.peppers.com.au
Licensed
Cuisine: Country-style with French, Asian, and Italian influences
Mains average: $$$
Specialty: Curried lamb and apple salad with 'Binnorie' mascarpone and curry dressing

Local flavours: Grapes, berries, nuts, root vegetables, salad greens
Extras: Surrounded by beautiful country gardens, the terrace veranda overlooks vineyards and bushland. Grows as much produce as possible in its own organic garden.
■ **Open:** Breakfast, lunch and dinner daily

ESCA BIMBADGEN

Bradley Teale (chef)
790 McDonalds Rd, Pokolbin 2321
Phone: 02 4998 7585
Email: office@bimbadgen.com.au
www.bimbadgen.com.au
Licensed
Cuisine: Modern Australian
Mains average: $$$
Specialty: Wood spit-roasted duck, wine and food matching tasting plates
Local flavours: Cheese, ducks, as much fresh produce as possible
Extras: 'Best restaurant views in the area', they say, plus an outdoor dining balcony overlooking vineyards and the Barrington Ranges. Takeaway available.
■ **Open:** Lunch daily, dinner functions and weddings any day by arrangement

ROBERTS AT PEPPERTREE

Robert Molines
Halls Rd, Pokolbin 2320
Phone: 02 4998 7330
Email: tani@towerestate.com
www.robertsatpeppertree.com.au
Licensed
Cuisine: Modern Australian with French influences
Mains average: $$$
Specialty: Duckling
Local flavours: Uses as much fresh local produce as possible.

Extras: Dining room with large glass windows overlooking the vineyard and sunny terrace dining right beside the vines. Delightful lawns and gardens.

- **Open:** Lunch and dinner daily

ROTHBURY
🍴 DINING

MOJO'S ON WILDERNESS
Adam and Ros Baldwin
84 Wilderness Rd, Rothbury 2320
Phone: 02 4930 7244
www.mojos.com.au
Licensed
Cuisine: Modern Australian–British
Two or three-course menu: $$$
Specialty: Desserts, beef
Local flavours: All local produce used
Extras: A bright cottage with a sunny outdoor courtyard. Vegetables and herbs are grown in an organic garden. Home-grown roasted olives, tapenades and caramelised balsamic vinegar for sale.

- **Open:** Brunch Sun, lunch and dinner Thurs–Mon
- **Location:** Take the Lovedale Rd exit off Main Rd for 8km, then the first left for 500m

SINGLETON
🏠 STORES

TRUNK'S GOURMET MEATS
John and Jenny Trunk
29 John St, Singleton 2330
Phone: 02 6572 2749
Known for its salami and beef jerky, this place has an old-fashioned feel to it, but that's okay. It means they make good local smallgoods in the traditional Polish way (learned from the original salami-maker, Alphonse Kmack) plus gourmet sausages, and a 'best in Australia' beef jerky. The beef is organic from the Upper Hunter and some other locally grown meats are also used. These products are sold here and at the Hunter Valley Cheese Company at Pokolbin.

- **Open:** 7am–5pm weekdays, 7am–12pm Sat

SWAN BAY
🍴 DINING

MOFFAT'S OYSTER BARN RESTAURANT
Lloyd and Michelle Moffat
Moffat Rd, Swan Bay 2324
Phone: 02 4997 5605
Email:
fishermansvillage@bigpond.com.au
Licensed/BYO
Cuisine: Seafood
Mains average: $$
Specialty: Local oysters every which way, Moffat's famous Seafood Platter for two
Extras: Family-owned and operated with self-contained cabin accommodation. The Moffat family have been oyster farmers for four generations and have lived and worked in the area since 1893 (yes, more than one hundred years). The restaurant is located on the edge of Port Stephens, with fantastic views, and takeaway is available.

- **Open:** Breakfast, lunch and dinner weekends, dinner Fri (Mon–Thurs by appointment)
- **Location:** 5km south of Karuah, turn off to Swan Bay from the Pacific Hwy, 12km to Moffat's

VACY
🍎 PRODUCERS

NAREEDA VALLEY OLIVES
Kelly and Mauro Melai
509 Fishers Hill Rd, Vacy 2421
Phone: 02 4938 8258
Email: nareedavalley@bigpond.com
If you want a gourmet treat using the best of two major local industries, try these olives marinated in Hunter Valley wines. There is olive oil too, pressed from an Italian cultivar grown on the farm. The olives are hand-picked early, cold pressed and the oil is unfiltered giving it a distinctive nutty flavour. There is kalamata tapenade (Italian Vegemite, they call it) and pesto from organic basil. Sold at Nelson Bay and Tomaree markets on alternate Sundays, and from selected local outlets.
- **Open:** By appointment
- **Location:** Off Summerhill Rd

WALLAROBBA
🍎 PRODUCERS

CAMELOT LAVENDER FARM
Denise Gale
1312 Dungog Rd,
Wallarobba, via Dungog 2420
Phone: 02 4995 6166
Email: info@camelotlavender.com.au;
denise@camelotlavender.com.au
www.camelotlavender.com.au
If you have a sudden fancy for lavender ice-cream, you'd better hope you are within reach of Camelot Lavender Farm. Here there are lavender chocolates as well as lavender scones and biscuits, jams, vinegars and sauces, all available from the tearooms at this delightful lavender farm. They also have a range of gourmet foods, which are available from the farm and selected retail outlets in the Newcastle, Central Coast and Sydney areas. Great news! Camelot's raspberry vinegar won a gold, and the apricot and almond jam snared a bronze at the Sydney Royal Fine Food Show, 2004.
- **Open:** 9.30am–5.30pm Tues–Sun and public holidays, Mon by appointment
- **Location:** Midway between Paterson and Dungog, 35km north of Maitland

WOLLOMBI
🍴 DINING

GRAY'S INN
Patricia Keane
Lot 2 Maitland Rd, Wollombi 2325
Phone: 02 4998 3475
Licensed/BYO
Cuisine: Country fare
Mains average: $$
Local flavours: Wines, fruit and vegetables
Extras: Delightful rural and mountain views from outdoor dining area on the back patio. This old, country-style guesthouse has two rooms with ensuite, and one double room. Cellar door tastings and sales. Menu varies from week to week but the chef always has a surprise up his sleeve.
- **Open:** Cafe—breakfast and lunch daily, restaurant—dinner Fri–Sat

GUESTHOUSE MULLA VILLA
Francisca and Caroline Maul
Old North Rd, Wollombi 2325
Phone: 02 4998 3254
Email: mullavilla@idl.net.au
www.mullavilla.com.au

NSW — THE HUNTER

BYO
Cuisine: Classic Australian
Mains average: $$
Specialty: Home-grown organic slow-cooked beef and poached china pears
Local flavours: Home-grown beef and vegetables
Extras: A working cattle farm, set in one hundred acres of unspoilt rainforest and also growing olives and grapes, next to the Wattagan State Forest and Yengo National Park. An 1840 convict-built house, originally designed for the district's first police magistrate, Mulla Villa still has three complete convict cells beneath the main living area.
■ **Open:** Lunch and dinner daily (bookings only)

MARKETS AND EVENTS

✦ **BRANXTON-GRETA FARMERS MARKET,** BRANXTON
02 4938 3300
Fourth Sun

✦ **HONEYSUCKLE MARKETS,** NEWCASTLE
02 4927 5366
9am–3pm Sun

✦ **HUNTER HARVEST FARMERS' MARKET,** MAITLAND
02 4939 8864
8.30am–1pm fourth Sat

✦ **POKOLBIN GROWERS' MARKET,** POKOLBIN
02 4998 7550
8am–11am fourth Sat

✦ **WOLLOMBI MARKETS,** WOLLOMBI
02 4900 4477
From 8am Mon on long weekends

✦ **JAZZ IN THE VINES,** POKOLBIN
02 4933 2439
Last Sat in Oct

✦ **BIMBADGEN BLUES,** POKOLBIN
02 4933 2439
Easter Sunday

✦ **MORPETH HONEY FESTIVAL,** MORPETH
02 4933 1407
Mid April

✦ **HUNTER VALLEY HARVEST FESTIVAL,** POKOLBIN
02 4990 4477
www.hunterharvestfestival.com.au
Mar–Apr

✦ **LOVEDALE LONG LUNCH,** POKOLBIN
02 4930 7317
www.lovedalelonglunch.com.au
Mid May

✦ **FEAST OF THE OLIVE,** LOWER HUNTER (various venues)
02 6579 1000
www.hunterolives.asn.au
Last weekend in Sept

NEW ENGLAND AND NORTH-WEST

Big sky, open spaces, wide horizons and huge acreages at the extremities. This is the area where the land has tested people to the limits, where many have had to leave sheep and wheat farming and diversify into trout and olives, emus and ostriches.

In the northern tablelands, the older, colder pursuits continue with apple and stone fruit orchards and berries thriving in the chill mountain air. Also look for fine meat, some of it organic, and biodynamic wheat, for these are healthy and health-conscious places too.

Perhaps the mountain air does something to you. Or maybe it's just having so much sky to enjoy.

ARMIDALE
DINING

RESTAURANT Q
Robert and Kate Finn
Girraween Shopping Centre, Armidale 2350
Phone: 02 6771 1038
Licensed
Cuisine: Modern Australian
Mains average: $$$
Specialty: Crown roast of hare
Local flavours: Wherever possible
Extras: Winner Best Modern Australian Restaurant for Regional NSW, 2003 (Restaurant and Caterers Association). Alfresco dining.
■ **Open:** Brunch Fri–Sat, lunch and dinner Tues–Sat

EBOR
TOURS

LP DUTTON TROUT HATCHERY
Peter Selby
Point Lookout Rd, Ebor 2453
Phone: 02 6775 9139
This place is called 'birthplace of the rainbow trout' and although you can't buy the fish themselves here, you are able to take a self-guided tour. The facility, run by NSW Fisheries, gives a good insight into the history of the fish and how they are raised.
■ **Open:** 9am–4pm daily
● **Location:** 70km east of Armidale towards Coffs Harbour on the Waterfall Way

GLEN INNES
PRODUCERS

THE SUPER STRAWBERRY
David and Cecily Tarrant
9922 New England Hwy, Glen Innes 2370
Phone: 02 6732 1210
Email: supastrawb@hotmail.com
You can see the strawberry farm through the back window of this shop, so it's no wonder your strawberry thick shake is so full of flavour. The Tarrants have been doing this for thirty years now, growing the berries, then making strawberry jam and ice-cream, and serving up strawberries and cream with their irresistible sponges and pavlovas. They also sell local honey.
■ **Open:** 9am–5.30pm daily

🚌 TOURS

NGOORABUL ABORIGINAL TOURISM
Glen Innes Visitors Centre,
The Willows, Glen Innes 2370
Phone: 02 6732 5960
A project of the Cooramah Community Organisation featuring Koori cuisine and bush tucker tours. Bookings essential.

GUYRA

🍎 PRODUCERS

MIDLANDS AQUACULTURE
Tom Sole
Midlands, Guyra 2365
Phone: 02 6779 4206
Email: midlandsfish@bigpond.com
A silver medal in the fine foods section at the Sydney Royal Show for 2003 for rainbow trout smoked fillets is one of the most recent achievements for Midlands. After diversifying from wool and beef to raise trout fifteen years or so ago, the Soles now have twenty-two trout ponds, although the trout are mainly raised in sheds using a recirculation system and filtered water. The Soles taught themselves by trial and error to smoke the fish over hardwood. There was no-one to copy as they pioneered this new industry in their area. Midlands Aquaculture smoked and fresh trout sells to restaurants, seafood centres, and fishermen's co-ops. The fish is smoked and seasoned with either sumac, a garlic mixture, or an Egyptian mix with mint. The Soles travel to southern Queensland farmers' markets and their fish is used in major hotels in Brisbane and the Gold Coast.
■ **Open:** 7am–7pm daily, fishing by arrangement
● **Location:** 20km from Guyra on Inverell Rd

HANGING ROCK

🍎 PRODUCERS

ARC-EN-CIEL RAINBOW TROUT
Ron and Ivy Bishop
'Malonga', Hanging Rock 2340
Phone: 02 6769 3665
Mobile: 0402 277 114
Email: arctrout@ceinternet.com.au
www.nundle.info/arcenciel
Rainbow trout at its best, many say. You will find fresh and smoked trout available here, grown out in fresh spring water, just a picturesque twenty minutes' drive from the old gold-mining town of Nundle. After twenty years, the Bishops' trout is sold to restaurants between Sydney and Brisbane and as far west as Dubbo. As well as making pâté, they also smoke their own trout, and fresh trout fillets, and whole fish are available. Also sold at the local markets.
■ **Open:** 9am–5pm daily by appointment
● **Location:** 20km from Nundle, at the top of the Great Dividing Range

INVERELL

🍎 PRODUCERS

GWYDIR OLIVES
35 Brissett St, Inverell 2360
Phone: 02 6721 2727
www.gwydirolives.com.au,
www.originolives.com.au
Olives have been grown commercially since 1994 in Moree and members of the growers' company have travelled and studied extensively to learn the latest olive oil extraction methods. An underground cellar was constructed to maintain oil at a constant temperature. Now Gwydir Olives are processors of olives for

NEW ENGLAND AND NORTH-WEST—NSW

many growers in the state, and the oil is distributed under the Gwydir Grove label, offering a range of Australian extra virgin olive oils, oils infused with Australian native herbs and spices (such as lemon myrtle), table olives and an olive pâté. A tourist facility allows visitors to see olives being processed using an environmentally friendly olive crusher. Visitors can taste and buy both the olives and oil, which are also available at many farmers' markets in Sydney and Queensland.
- **Open:** 8.30am–5pm weekdays (tours at 11am)

MANILLA
🍎 BREWERY/WINERY

DUTTON'S MEADERY
Ian and Wilga Dutton
59 Barraba St, Manilla 2346
Phone: 02 6785 1148
Ian Dutton was Australia's first mead maker and has been making it now for just on fifty years. He procured the original recipes from monasteries in Europe and taught himself, persevering for more than eighteen months, until he got it right. The Duttons now make a range of meads—dry, medium, and sweet, as well as melamol (a blend of grapes and honey), and one blended with mulberries, plus a spiced mead. They keep their own bees and collect the honey, but mead is the only product for which they offer tastings. They use predominantly white-box honey, but say that any sort of honey works well for mead, which must mature for a minimum of two years.
- **Open:** 8.30am–5pm daily
- **Location:** 2km north of Manilla

🍴 DINING

TOOT'S CAFE
Raelene Turner
The Big Fish Centre,
79 Arthur St, Manilla 2346
Phone: 02 6785 1113
Email: tootscafe@telstra.com
BYO
Cuisine: Homestyle country
Mains average: $
Specialty: Toot's jacket potatoes, pizza with homemade sauce
Local flavours: Eggs, Dutton's honey from Dutton's Meadery, bread, meat
Extras: Doubles as a tourist information and gift shop.
- **Open:** Breakfast, lunch and dinner daily
- **Location:** Just look for the Big Fish (blue, with a top hat)

NIANGALA
🍎 PRODUCERS

KOOLKUNA BERRIES
Lothar and Barbara Kalz
Koolkuna, Nowendoc Rd, Niangala 2354
Phone: 02 6769 2221
Buy or pick-your-own berries, or simply settle in to taste the traditionally fermented wines at this tablelands property. There are jams and vinegars, all homemade, and the Berry Shop on the property sells these and serves Devonshire teas, light lunches, berry ice-cream, smoothies, cakes and muffins. Berries are available in season and include strawberries, raspberries, blueberries, blackberries, and bramble-berries—a term which encompasses a bunch of other berries: boysenberries,

loganberries, tayberries, youngberries, red and black currants, English gooseberries, yosterberries (a cross between blackcurrant and English gooseberry) and silvanberries. All these are superb wineberries, say the Kalzes, who become almost lyrical when describing their tayberry liqueur, which they are able to make when there is enough fruit. Brandy wine is also popular. If you can't get to Niangala, the wines are sold at Liquor Stax in Tamworth.

- Open: 10am–4pm Thurs–Mon, closed August and major public holidays
- Location: 75km from Tamworth

NUNDLE

🍎 PRODUCERS

NUNDLE YABBY FARM
Joy Burton
Happy Valley Rd, Nundle 2340
Phone: 02 6769 3363

If you have a yearning for yabbies, you can come here and catch them in season, or else buy them most of the year. It's best to order two or three days ahead, if possible, so they can be purged. There are also tours of the shed and you can see yabbies on display. The farm has an eating area, built much like an old woolshed, and large groups can arrange a country lunch complete with good old-fashioned stews and damper, and a pig on a spit.

- Open: 10am–4pm daily, longer hours in summer, groups by appointment
- Location: About a kilometre from Nundle, on the road to Hanging Rock

🍴 DINING

CHA CHA CHA
Judy and Peter Howarth (owners)
Jenkins Street Guesthouse,
85 Jenkins St, Nundle 2340
Phone: 02 6769 3239
Email: ghnundle@northnet.com.au
www.nundle.info

Licensed
Cuisine: Local Nundle produce
Mains average: $$
Specialty: Hanging Rock trout, smoked or fresh
Local flavours: Local, Nundle 'Blue Stripe Beef', trout
Extras: Located within the elegant Jenkins Street Guesthouse in a remote country village. Knitting lessons available in conjunction with the local Nundle Woollen Mills. Grows own vegetables and fresh eggs are collected every morning.

- Open: Breakfast daily (in-house guests), lunch weekends, dinner Thurs–Sat (daily for in-house guests and private functions), picnics anytime

PALLAMALLAWA

🍎 PRODUCERS

STAHMANN FARMS INC
Jon Craven
Trawalla MSF 2058, Pallamallawa 2399
Phone: 02 6754 9259
Email: marketing@stahmann.com.au
www.stahmannfarms.com.au

An old German proverb quoted by Stahmann Farms, goes: 'God gives the nuts, but He does not crack them'. You could ask, why would he when Stahmann Farms does it so well? Thirty years of

NEW ENGLAND AND NORTH-WEST—NSW

experience growing pecans has resulted in product lines that include fruit cakes, puddings, pecan oil, kitchen-ready pieces and roasted flavoured nuts, as well as whole nuts, halves and even sixty per cent mono-saturated pecan oil (pecans have a low glycemic index of ten). Currently these farms supply over ninety-five per cent of the Australian pecan crop from around 75 000 trees. To complement the pecan pie recipes, Stahmann Farms have devised the obvious accessory—a baking and topping syrup made from Australian wheat. The pecans are widely available in Australia, and exported to the United States to supply the off-season market.

Tour bookings: Moree Visitors Centre (02 6757 3350) 9am–5pm weekdays, 9am–1pm weekends and public holidays

TAMWORTH

PRODUCERS

MANDALONG GRAIN-FED LAMB
Mark Taylor
14 Bass St, Tamworth 2340
Phone: 02 6765 3653
Mobile: 0418 610 324
Email: mandalong@bigpond.com
www.mandalonglamb.com.au

Grain-fed lamb is the signature of this tablelands farm at Walcha, and it is available from the company's butcher's shop at Tamworth, which also sells other cuts of grain-fed beef and a recently added range of gourmet sausages. If you can't get to Tamworth, the meat is regularly available at Sydney and Brisbane farmers' markets.

■ **Open:** 8.30am–5.30pm weekdays

STORES

THE ESSENTIAL INGREDIENT
Anna Madgwick
15B White St, Tamworth 2340
Phone: 02 6766 5611
Email: essential@mcsonline.com.au

One of a chain of franchised stores now in most capital cities, this one, opened late 2003, is presently the only one in a country town. Local lines include olive oils, olives, local barramundi (cryovaced or smoked) as well as condiments, chutneys, local meat and trout.

■ **Open:** 9am–6pm weekdays, 9am–2pm Sat

DINING

SSS BBQ BARNS (STETSON'S STEAKHOUSE AND SALOON)
Graham and Marlene Manvell
Cnr Craigend Ln and Sydney Hwy, Tamworth 2340
Phone: 02 6762 2238
Email: info@sssbbq.com.au
www.sssbbq.com.au
Licensed
Mains average: $$

Local flavours: Beef, buffalo, wine, beer, apples, berries, asparagus, olives, herbs and pumpkins, all grown on SSS Ranches
Extras: A microbrewery, Ironbark Brewery, makes ginger tawny ale. SSS BBQ Buffalo and Beef ranches supply pasture and grain-fed prime buffalo and Dexter beef. SSS BBQ Barns has outlets in Tamworth, Coffs Harbour, Cessnock (Hunter Valley), Lake Munmorah (Central Coast) and Brisbane.

■ **Open:** Dinner daily

MARKETS AND EVENTS

✦ **GUYRA LAMB AND POTATO FESTIVAL,** GUYRA
02 6779 1577
Jan

✦ **NATIONAL TOMATO CONTEST,** GUNNEDAH
02 6742 0400
Week before Jan long weekend

✦ **ORANGE PICKING DAY,** BINGARA
02 6724 0006
From end of June on—whenever the oranges are ripe!

✦ **ORANGE FESTIVAL,** BINGARA
02 6724 0006
Late July–early Aug

✦ **SUMMERLAND OLIVE FESTIVAL,** CASINO
02 6662 3390
Second Sun in Feb

NORTH COAST NSW

Once called The Holiday Coast, the name doesn't quite say it all. The coastal strip is certainly a great place for everyone from the completely hooked fisherman to a young family wanting a place in the sun for their vacation, but there is even more—much more—on offer here.

Inland, the mountains rise up, hiding tiny communities, timber-getting towns, and some of the state's best dairy lands. Here the rivers are wide and lazy, the soil in their valleys so good it's almost edible. Here you will get natural milk and cheese, bananas, avocados and meat, for this is an organic-grower's paradise.

Beachside, the eating just gets better with seafood so fresh it almost winks at you, straight from the fishing boats at towns along the coast. Holiday coast, or gourmet escape. You name it.

BELLINGEN
DINING

LODGE 241 GALLERY CAFE
Laurence Crooks
117–121 Hyde St, Bellingen, 2454
Phone: 02 6655 2470
Email: kalaurie@ozemail.com
www.bellingen.com/thelodge
BYO
Cuisine: Modern Australian
Mains average: $
Specialty: Ricotta and basil gnocchi with fresh tomato and pesto sauce
Local flavours: Dorrigo-pepper smoked tuna

Extras: Award-winning cafe housed in the old Masonic Lodge (a heritage landmark building) in Bellingen, the gallery sells local paintings and has great views of the Bellingen Valley from every window, the terrace and the veranda. Award-winning coffee too.
■ **Open:** 8am–5pm daily

NO 2 OAK ST
Toni and Ray Urquhart
2 Oak St, Bellingen, 2454
Phone: 02 6655 9000
Email: urquhart@ozconnect.net
Licensed/BYO wine
Cuisine: Modern Australian
Mains average: $$$
Specialty: Local meats
Extras: Veranda dining at this multi award-winning restaurant in a delightfully restored house.
■ **Open:** Dinner Tues–Sat

BOBIN
PRODUCERS

DINGO CREEK RAINFOREST NURSERY
Kim and Peter Gollan
82 Schneiders Rd, Bobin 2429
Phone: 02 6550 5167
Email: dingock@midcoast.com.au
A biodynamic rainforest nursery is unusual enough, but when it has 150 different species, and specialises in raising indigenous bush-tucker plants, then it turns into a one-of-a-kind marvel. Located on the mid north coast, some of the more common plants grown here

include pepper bush, black apple, native tamarind, wombat berry, lemon aspen and plum pine.
- **Open:** By appointment
- **Location:** On the Tourist Drive to Ellenborough Falls, 27km from Wingham

BONVILLE
PRODUCERS/DINING

KIWI DOWN UNDER ORGANIC FARM AND TEAHOUSE
Tom and Marguerite Hackett
430 Gleniffer Rd, Bonville 2441
Phone: 02 6653 4449
Email: tom@kiwidownunder.com
www.kiwidownunder.com

The philosophy behind Kiwi Down Under Farm is that it is not just a farm or a teahouse but a lifestyle, where a message about the environment is delivered by example. So says Tom Hackett, one of the owners of this certified-organic/biodynamic fruit and nut farm. The family is zealous in educating customers about the non-use of chemicals, care of the environment and good food. In 1982, the Hacketts began growing safe, healthy foods for themselves, and establishing a kiwifruit farm producing the Delicious Dexter variety. Before they knew it people started to flow in, now at the rate of around 1000 a month, to see the property. You'll understand why when you visit their Top Shed Teahouse and sample the freshly made meals, and see the jams, dried fruits and herbal body products. On Thursdays the Top Shed organic market sells fresh fruit, meats, dairy products, vegetables and sauces, as well as a host of other organic products.

- **Open:** 9am–4pm Thurs (organic vegetable market), group bookings of twenty or more people at other times
- **Location:** 15km south of Coffs Harbour. Turn off south of Bonville Post Office and follow signs along Gleniffer Rd for 4km

DINING

FLOODED GUMS RESTAURANT
Stephen Seckold (chef)
Bonville International Golf Resort,
North Bonville Rd, Bonville 2441
Phone: 02 6653 4002
Email: info@bonvillegolf.com.au
www.bonvillegolf.com.au
Licensed
Cuisine: Modern Australian
Mains average: $$$
Local flavours: A wide range of local fresh produce
Extras: Set in a Federation-style clubhouse overlooking the golf course this was awarded Best Restaurant in a Resort 2003, by the Restaurant and Catering Association for the Northern Division. There is accommodation, and the restaurant grows much of its own produce in the orchard and extensive herb and vegetable garden.
- **Open:** Breakfast, lunch and dinner daily

BOWRAVILLE
DINING

BOWRA HOTEL BISTRO
Stuart (chef) and Kathy Oliver
33 High St, Bowraville 2449
Phone: 02 6564 7739
Licensed
Cuisine: Modern Australian bistro

Mains average: $$
Specialty: Slow-cooked lamb shanks, red roasted Thai chicken with lychees
Local flavours: Seafood, beef, local fruit
Extras: Located in an old pub with an open fireplace in the restored dining room and with mountain and valley views. Takeaway available.
- **Open:** Dinner Thurs–Sun

BYABARRA
DINING

BLUE POLES CAFE AND GALLERY
Deborah Collins and Miranda Mills
1086 Comboyne Rd, Byabarra 2446
Phone: 02 6587 1167
Email: info@bluepoles.com.au
www.bluepoles.com.au
Licensed/BYO
Cuisine: Modern cafe
Mains average: $
Local flavours: Hastings milk and cheese products, avocados, salad vegetables, pecans, macadamias, local jams
Extras: Modern art exhibitions, as well as crafts from mainly Hastings-based artists, change monthly. Large deck for dining overlooking a lovely valley. Cabins will be available soon.
- **Open:** Lunch Thurs–Sun, dinner functions by appointment
- **Location:** 20km west of Wauchope

COFFS HARBOUR
PRODUCERS/DINING

THE BIG BANANA
Kevin and Marie Rubie
Pacific Hwy, Coffs Harbour 2450
Phone: 02 6652 4355
Email: info@bigbanana.com
www.bigbanana.com
Bananas are so important in this area that you should understand how much effort goes into raising them. This place has it all. There are plantation tours by train, and the grand tour takes an hour as you travel amongst the bananas to the summit lookout that offers a stunning all-round view and there are toboggan rides down the mountain. There are three types of large bananas, pawpaws, exotic fruits (which are not harvested and are just for show), macadamias and avocados, and a display to explain them. The cafe serves banana smoothies, banana muffins, cakes, even a banana burger, ice-creams, and the must-have choc-coated bananas.
- **Open:** 8.30am–4.30pm daily, 8.30am–5pm during school holidays

STORES

COFFS HARBOUR FISHERMEN'S CO-OP
Phillip Neuss
69 Marina Dr, Coffs Harbour 2450
Phone: 02 6652 2811
Email: phil@coffsfishcoop.com.au
Fresh and hot seafood made with the day's catch from the fishing fleet, are available here as well as prawns, lobster and fish. The co-op has won three Sydney Fish Market awards.
- **Open:** 9am–6pm daily, hot takeaways available later in summer

ESSENTIAL INGREDIENTS ON THE MALL
Liz and Tony Locandro
Shop 1, Boulevard Arcade,
15–21 Harbour Dr, Coffs Harbour 2450
Phone: 02 6651 1770
Email: pmbooks@ozemail.com.au
Sourcing from approximately eighty individual suppliers, this shop has most of the best food from around the country as well as local products including de Paoli coffee, grown on a former banana plantation in Coffs Harbour, and gluten-free products from another local supplier.
- **Open:** 8.30am–5.30pm weekdays (Thurs until 6pm), 8.30am–2.30pm Sat, 9.30am–1.30pm Sun

DINING

THE OCEAN FRONT BRASSERIE
Matthew Donovan and Arnold Underwood
Coffs Harbour Deep Sea Fishing Club,
Jordan Esplanade, Coffs Harbour 2450
Phone: 02 6651 2819
Email: nic@ozconnect.net
Licensed
Cuisine: Seafood
Mains average: $
Specialty: Ocean Front seafood platter for two or four
Extras: Balconies on three sides of the Club allow spectacular views of the ocean, harbour, beach and mountains.
- **Open:** Lunch and dinner daily

THE TIDE AND PILOT LOWER DECK AND UPPER DECK
Cindy McCarthy
International Marina, Marina Dr,
Coffs Harbour 2450
Phone: 02 6651 6888
Email: tideandpilot@aol.com

Licensed
Cuisine: Modern Australian seafood
Mains average: $$
Specialty: local seafood and seafood platters
Local flavours: Seafood, fruits, vegetables, game, beef
Extras: Oyster and crustacean bar, harbour views, outdoor dining on the deck, and seafood, straight from the trawler to the plate. The Lower Deck is a smaller daytime cafe (not a takeaway—why would you want to take your food away from such a superb location anyway?) while the Upper Deck is for serous dining.
- **Open:** Breakfast, lunch and dinner daily

COMBOYNE

DINING

THE UDDER COW CAFE
Maureen Sulis
1 Main St, Comboyne 2429
Phone: 02 6550 4188
Mobile: 0428 665 200
Cuisine: Wholesome country food
Mains average: $
Local flavours: Wherever possible, including also rhubarb, avocado and citrus
Extras: Located in the original bakery built in the 1920s, with lovely rural views.
- **Open:** 9.30am–5.30pm daily, dinner Sat (bookings essential)

DORRIGO

PRODUCERS

DORRIGO WOODFIRED BAKERY
Kim and Rick Frewen
39 Hickory St, Dorrigo 2453
Phone: 02 6657 2159

The Frewens purchased a run-down bakery more than a decade ago and have turned it around, using traditional recipes for pies, pastries, bread and rolls and baking them in the wood-fired ovens. The Dorrigo-pepper steak pies, which use native pepper grown by local farmer Leo Pollard, are very popular, and of course only local beef is used in the pies.
- Open: 4.30am–5.15pm weekdays, 4.30am–12.30pm Sat

BREWERY/WINERY

THE LONELY PALATE WINERY
Susie Snodgrass and David Scott
Hickory St, Dorrigo 2453
Phone: 02 6655 1714
Email: goodfood@optusnet.com.au
www.bellingen.com/lonelypalatewinery
How do you describe a business that has outgrown itself and has to uproot and relocate? The Lonely Palate wines are too good to omit, so even though at the time of going to press they are in transit, we will give you an email address and tell you that they are moving to Dorrigo. These unique fruit wines which are made from the best quality regionally grown fruits—lime, guava, boysenberry, strawberry, mango, raspberry—will continue production and no doubt expand once Susie and David have settled into their new premises and established the new cellar door. This has been the only winery in the Bellingen Shire and the wines are made using traditional wine-making techniques, the same as those employed in the grape wine industry.
- Open: 10am–5pm weekdays, 10am–12pm Sat, other times by appointment

STORES

LICK THE SPOON
Jo Sweeney
53 Hickory St, Dorrigo 2453
Phone: 02 6655 1714
Brand new in 2004, this shop is packed with regional produce and unique, locally produced gourmet treats, the aim being, as Jo puts it, 'To bring a piece of "foody heaven" to a small country town and make gourmet products accessible to country people'. So look for cakes, spices, take-home meals, preserves, olive oils, fine chocolates, coffees, mayonnaise, antipasto, homemade ice-cream, organic meats including game, cheeses, condiments and sauces. And labels such as Bespiced, Lick the Spoon, Whisk and Pin, Doodles Creek, Limelight, Toby's Estate coffee, Simon Johnson, Cippango, Bellatta, Aunty Joans toffee, and The Cake Lady panforte.
- Open: 10am–5pm weekdays, 10am–12pm Sat

DINING

MISTY'S RESTAURANT AND ACCOMMODATION
33 Hickory St, Dorrigo 2453
Phone: 02 6657 2855
Email: mistys@dorrigo.com
www.dorrigo.com/mistys
BYO
Cuisine: Contemporary regional
Mains average: $$
Specialty: Dorrigo-peppered duck breast
Local flavours: Dorrigo beef, farmed eastern cod, organic produce
Extras: Located in a cottage with leadlight windows. The Booma Hyland eastern cod is a freshwater fish now being farmed locally.

Open: Lunch Sun, dinner Wed–Sat (with extended holiday trading)

FREDERICKTON
🏠 STORES

FREDO FAMOUS PIES AND ICE-CREAMS
Nola Turnbull
75 Macleay St, Frederickton 2440
Phone: 02 6566 8226
Email: nolar@midcoast.com.au
www.fredopies.com.au

You could be forgiven for thinking that Frederickton's most colourful building is a zoo. But the murals equate with the fillings of Fredo's 148 types of pies—crocodile, buffalo, rabbit and more. 'Only fifty-six sorts are on display at any one time', says Nola apologetically, although no-one minds, as even that number makes it very hard to choose. The bakery recently won two gold, two silver and a bronze medal in the 2003 Great Aussie Pie Competition. Turnbull gets the beef from a local butcher who also raises the meat, and rabbits from a local supplier. Game meats come from Brisbane, and local vegetables are used in the pies. And the good news for Sydney rail commuters is that a new shop has just opened in the concourse of Wynyard Railway Station.

Open: 7am–7pm daily

HARRINGTON
🍴 DINING

SEASHELLS CAFE
Catherine McDonald and Wolfgang Zichy
19 Beach St, Harrington 2427
Phone: 02 6556 0220

Email: elegant_epicure@bigpond.com.au
BYO
Cuisine: Modern Australian with Asian influence
Mains average: $
Specialty: Breakfasts, especially 'Eggs with Attitude'
Local flavours: Whenever possible, ingredients are sourced from local organic farms
Extras: The owners formerly ran The Bank Guesthouse and Tellers Restaurant at Wingham. Stunning views of the river, beaches and the ocean. Takeaway available.

Open: Breakfast and lunch daily
● **Location:** Ten minutes off the Pacific Hwy, between Port Macquarie and Taree

KENDALL
🍎 PRODUCERS

LORNE VALLEY MACADAMIA FARM
Ray and Joanne Scott
1181 Lorne Rd, Lorne, Kendall 2439
Phone: 02 6556 9653

City escapees, the Scotts were in search of tranquillity, and a chance to work together. Having achieved that, they now offer to share their country hospitality and beautiful views as well as home-baked food and fresh farm produce with all their visitors. Macadamias are flavoured and roasted on the premises in small batches to ensure sweetness and there are now seven flavours including honey-roasted and sweet chilli. There are homemade jams and a variety of other macadamia products, such as a super-yummy honey and macadamia butter and macadamia oil. These products are also sold at local

weekend markets. The farm cafe offers coffee and light meals.
- **Open:** 10am–4pm Wed–Sun, closed for the two weeks before Christmas and two weeks in mid Feb, by appointment during summer
- **Location:** 15km west of Kew. Take the Laurieton–Comboyne Rd exit off the Pacific Highway, halfway between Port Macquarie and Taree

NORFOLK PUNCH (AUSTRALIA)
Blair and Rhonda Montague-Drake
Batar Creek Rd, Kendall 2439
Phone: 02 6559 4464
Email: montague-drake@bigpond.com
www.earthimages.com.au
Even the scenic drive to this place is worth the effort, but the destination is even better. Norfolk Punch, the drink that is made here, is a unique combination of over thirty herbs and spices, made to a recently-discovered 700-year-old recipe. The shop is crammed with all things herbal. There's the punch of course, and liqueur-style Celtica, honeys, Kendall Ginger Bliss, herbal waters, dried herbs, insect repellents and herb pillows. Outside, the gardens are amazing, with herbs set out in various categories, groves of gums and the Bower Bird Cafe, serving Norfolk Punch's own organic coffee along with scones and cakes flavoured by native herbs and spices. There is also a window to view the bottling process and a kitchenware museum.
- **Open:** 9am–4pm weekdays and the first and third weekend of each month, daily during school holidays

KORORA
🍎 PRODUCERS

CAROBANA CONFECTIONERY
Ian and Jean Hamey
125 James Small Dr, Korora 2450
Phone: 02 6653 6051
Email: info@carobana.com.au
www.carobana.com.au
'Fruit bats', replies Ian Hamey, when asked how his business began. His family has farmed this land, growing bananas, for sixty years and eventually started drying the excess bananas to prevent the fruit bats getting into them. From there it was a longish step to dipping the fruit in carob, and finally moving into carob confectionery in a big way. Today they produce a large range of unsweetened carob treats with forty fillings. Visitors may watch a video and learn more about this unusual chocolate alternative. Coach groups also may tour the factory.
- **Open:** 10am–4pm Mon–Sat
- **Location:** 5km north of Coffs Harbour

LORNE
🍎 PRODUCERS

BARBUSHCO PTY LTD
Bruce and Barbara Barlin
50 Gills Rd, Lorne 2439
Phone: 02 6556 9656
Email: barlinb@bigpond.com
www.barbushco.com.au
All sorts of bush foods are produced here— herbs and spices, sauces, chutneys and syrups made from the leaves and fruit of this property's 26 000 trees. Davidson plums, lemon myrtle, Dorrigo pepper, and lilly pillies are used to produce spices, teas,

jams, fettuccine, nut pastes and satays. Sold in local farmers' markets, and special food fairs, as well as delis and health food stores. All Barbushco produce is chemical-free. The Barlins use organic practices, and are in the process of organic certification.
- **Open:** By appointment

MACKSVILLE
STORES
DANGEROUS DAN'S BUTCHERY
David and Donna Hoffmann
13 Princess St, Macksville 2447
Phone: 02 6568 1036
www.here.com.au/dans
The beef sold in this butchery is all from the Hoffmanns' own property, where they raise beef. They also grow Australian native herbs for the Bushman range of gourmet sausages made in the butchery—flavours such as native pepper, lemon myrtle, macadamia nuts, and lilly pilly. A range of sauces from nearby Valley of the Mists is the ideal complement. If you can't get to Macksville, look for these products at North Sydney farmers' market on the third Saturday of each month.
- **Open:** 7am–5pm weekdays, 7am–1pm Sat

MANNING POINT
TOURS
MANNING VALLEY RIVER CRUISES
Darren and Marie Ryan
Main Rd, Manning Point 2430
Phone: 02 6553 2683
Mobile: 0428 532 683
Email: darrenryan@bigpond.com.au
www.manningrivercruises.com.au

Four-hour tours of the local oyster-growing areas, as well as a crab-pot cruise and seafood cruises, full roast dinners and seafood baskets. Functions also catered for. Bookings essential.

MEDOWIE
PRODUCERS
MEDOWIE MACADAMIA GROWERS
Scott and Stacy Leech
32 Medowie Rd, Medowie 2318
Phone: 02 4982 8888
What set out as a simple change of lifestyle—a tree-change, if you like—has turned into a macadamia farm that sells macadamia nuts and products including oil, spread, soaps and honey, and a range of flavoured macadamias. The products are available from the shop on the farm, as well as at local markets, restaurants and wineries.
- **Open:** 9am–5pm daily
- **Location:** Between Newcastle and Karuah on the coast side of the highway

MOORAL CREEK
PRODUCERS
MOORAL CREEK FARMS
Jennifer Johnstone (owner), Standish Kemmis and Lia Szokalski (operators)
1597 Mooral Creek Rd,
Mooral Creek 2429
Phone: 02 6550 5847
Email: sandlando@iprimus.com.au.
This farm began four years ago because the owners had a belief in organic agriculture as 'a method of growing food that sustains people and is not detrimental to the environment'. The business and

the vegetables have grown and now about thirty varieties are raised here and mixed boxes are delivered weekly direct to individuals and families in Sydney as well as local farmers' markets.
■ **Open:** By appointment

NABIAC
🍴 DINING

AMISH COUNTRY STORE
Ferdinand Weerheim
Hancock Bldgs, Nabiac 2312
Phone: 02 6554 1113
Mobile: 0438 150 955
Email: amish@ceinternet.com.au
www.amishcountrystore.com.au
Cuisine: Wholesome country-style
Mains average: $
Local flavours: All local produce, own-baked breads
Extras: A unique and relaxed place selling a range of Amish memorabilia—books, tea, coffee, jam and pickles—plus outdoor veranda dining
■ **Open:** Lunch Mon–Sat

KIT AND KABOODLE
Jan and Colin Whittaker
64 Clarkson St, Nabiac 2312
Phone: 02 6554 1160
BYO
Cuisine: Eclectic contemporary Australian
Mains average: $$
Specialty: Dishes using duck, mangoes, olives and cherries
Local flavours: Lamb, barramundi, macadamias, avocado, fruits and vegetables
Extras: Quaint old building full of character with beautiful leadlight windows

and heritage photographs in the hallway. Finalists for the past six years in the Restaurant and Catering Association Awards for the Northern Region.
■ **Open:** Lunch Fri–Sun, dinner Thurs–Sun

NORTH ARM COVE
🍎 PRODUCERS

CRAYHAVEN AQUACULTURE
Robert McCormack
6408 Pacific Hwy, North Arm Cove 2324
Phone: 02 4997 3002
Email: yabby@nobbys.net.au
www.crayhaven.com.au
Most of the live blue-claw yabbies from this oldest and largest privately owned yabby farm in New South Wales go straight to restaurants, but there are gate sales for bait, yabby stock and, yes, even yabbies as pets. And now there are live fish and eels as well. Robert admits he bought the property as a fishing spot for retirement. 'We never intended to get as big as we are today,' he says, 'it has just grown over the last twenty years and we have never been able to produce enough to meet demand—and we are still growing.'
■ **Open:** 9am–4pm weekdays, 11am–4pm Sat
● **Location:** 11km north of Karuah River Bridge, 2km south of Ayers Rock Roadhouse

---

CHOCOLATE-COATED BANANAS
Travelling north of Port Macquarie be on the alert for signs advertising these addictive delights made with local bananas—a regional specialty.

PORT MACQUARIE

🍎 PRODUCERS

HASTINGS RIVER OYSTER SUPPLIES
Mark Bulley
Shed 1, Sandfly Alley,
Port Macquarie 2444
Phone: 02 6583 2444
Email: marbul@tsn.cc
'Just oysters', says Mark Bulley, when asked what he grows. But these aren't just any oysters. Sydney rocks, he'll tell you, don't grow best north of Moreton Bay in Queensland, or south of the Victorian border, but they do very well indeed in this area, thank you. Mark grows them in the river and has done so for more than twenty years. A butcher by trade, he came from a rural background, and realised that the oyster industry had great potential. Now the industry can't keep up with demand, he reckons. The oysters are all sold closed-shell. Recently he has begun selling seed stock to other farmers. Because of the volume of oysters harvested, it is usually impractical to supply small orders, so ring him to negotiate your order.
■ **Open:** 9am–3pm weekdays during harvest (Oct/Nov to Feb/Mar)

🏠 STORES

HASTINGS RIVER FISHERMENS' COOPERATIVE
Clarence St, Port Macquarie 2444
Phone: 02 6584 7399
Email: hrfc@optusnet.com.au
Straight from the co-op's own fishing trawlers, you'll find snapper, flathead, whiting, mullet, blackfish, bream, prawns, crabs, lobsters, bugs and oysters. If there are shortages, they will also buy from other local co-ops, but one thing is sure, you'll never find fresher.
■ **Open:** 9am–5pm daily

🍴 DINING

PORTABELLO'S CAFE
Craig and Janine Smith
Shop 6, 124 Horton St,
Port Macquarie 2444
Phone: 02 6584 1171
Email: janinesmith1@bigpond.com
BYO
Cuisine: Modern Australian
Mains average: $$
Specialty: Duck dishes, seafood chowder, home-cooked bread, pasta and desserts
Extras: Outdoor area, wheelchair access and children's menu. Homemade pestos and dips for sale.
■ **Open:** Breakfast and lunch Tues–Sat, dinner Thurs–Sat

SAWTELL

🍴 DINING

THE BLUE FIG
Donna (chef) and Mark O'Sullivan
23 First Ave, Sawtell 2452
Phone: 02 6658 4334
Email: madb@tpg.com.au
BYO
Cuisine: French-influenced modern Australian
Mains average: $$$
Specialty: Sauces, freshly baked bread, relishes and chutney.
Local flavours: Fruit and vegetables, meat, Coramba venison, Stuarts Point rabbit, quail, and spatchcock
■ **Open:** Dinner Tues–Sat

NORTH COAST NSW—NSW

SHOAL BAY

🍴 DINING

BIG FISH RESTAURANT
Warren Thompson
Shoal Bay Resort and Spa, Beachfront,
Shoal Bay 2315
Phone: 02 4984 8130
Email:
warren.thompson@shoalbayresort.com
www.shoalbayresort.com
Licensed/BYO
Cuisine: Contemporary Australian
Mains average: $$
Specialty: MSA Hereford beef and Port Stephens Sydney rock oysters
Local flavours: Oysters, blue swimmer crabs, whole snapper and flathead fillets
Extras: Views of magnificent Shoal Bay and Mount Tomaree. The adjacent Cafe Sandyfoot is open all day for lighter meals and lunches on weekends.
■ Open: Breakfast and dinner daily, lunch weekends

SOUTH WEST ROCKS

🍎 PRODUCERS

BARNETT'S RAINBOW REACH OYSTERS
John Barnett
551 Rainbow Reach Rd,
South West Rocks 2440
Phone: 02 6565 0050
Freshly opened Sydney rock oysters, grown on the river, are sold to Sydney markets, but you will find them too at Trial Bay Kiosk, Crescent Head Tavern, and many other places in the local area. You can buy them direct from the barn itself, freshly opened on the half-shell, unopened or bottled.
■ Open: 8am–4pm weekdays, 8am–12pm Sat
● Location: Off Pacific Hwy, 2km south of BP Clybucca, left into Suez Rd, then 5.3km to blue oyster barn

🍴 DINING

THE ROCKS RESTAURANT
Wayne Gornall and Sherie Brown
Rockpool Motor Inn,
45 McIntyre St, South West Rocks 2431
Phone: 02 6566 7755
Email:
bookings@rockpoolmotorinn.com.au
www.rockpoolmotorinn.com.au
Licensed
Cuisine: Modern Australian
Mains average: $$
Specialty: Wok-seared green prawns and baby squid
Local flavours: Oysters, seafood, beef, duck, goats cheese
Extras: Makes marmalade, sweet tomato jam, lilly pilly and cinnamon jelly for sale.
■ Open: Dinner Mon–Sat

TAYLORS ARM

🍴 DINING

BILLABONG RESTAURANT
Sam Aboud
Bakers Creek Station,
Greenhills Rd, Taylors Arm 2447
Phone: 02 6564 2165
Email: info@bcstation.com.au
www.bcstation.com.au
Licensed
Cuisine: Modern Australian
Mains average: $$

95

Specialty: Rack of lamb with Davidson plum sauce and grilled pork neck
Local flavours: Aabenraa prime beef, Hoffman's sausages and sauces, Valley of the Mist bush food products and preserves
Extras: The restaurant has a westerly outlook and is built out over an eighteen-acre lake. Blinds on three sides can be opened to create a virtually open-air restaurant. Small country resort in a valley surrounded by rainforested mountains.
■ **Open:** Breakfast, lunch and dinner daily (for in-house guests), dinner Fri–Sun (general public)
● **Location:** 8km from Taylors Arm

TEA GARDENS

🍎 PRODUCERS

KORE FARM PRODUCE
Helmut and Vivien Panhuber
Tea Gardens 2324
Phone: 02 4997 0488
Email: helmut@myallcoast.net.au
Almost twenty years ago, Helmut, a CSIRO scientist, and Vivien, a forensic biologist, were looking for a holiday home. They bought a small property on Kore Creek in the Port Stephens area, and over the years planted many exotic fruits there on their regular visits. Several years ago, when they moved to Tea Gardens full-time, the place turned into a working farm, growing all sorts (around 130 varieties) of raspberries, tamarillos, custard apples, citrus, berries, and rare and unusual fruits. In this microclimate they can grow many fruits out of season, so you could find Tahitian limes, carambola, guavas, strawberry guavas and mangoes too. Of course the next step was to use the excess fruit to make things like jams and cordials, butters, fruit pastes, even chocolate truffles with fruit centres. This wide range of products is sold at wholesale markets and retail outlets in Sydney.
■ **Open:** By appointment (best road directions will be provided at the time)

🍴 DINING

WATERFRONT RESTAURANT
Lee and Ros Anderson (owners), John Berriball (head chef)
Tea Gardens Hotel and Motel, cnr Marine Dr and Maxwell St, Tea Gardens 2324
Phone: 02 4997 0203
Email: tghotel@myallcoast.com.au
www.tghotel.myallcoast.net.au
Licensed
Cuisine: Pub food and seafood
Mains average: $$
Specialty: Fish cones (local fish and chips in paper cones), Guinness pies, local flathead with lime mayonnaise
Local flavours: Fish, oysters, prawns
Extras: Beer garden, deck overlooking the Myall River and tropical garden. Takeaway available.
■ **Open:** Bistro—lunch and dinner daily, restaurant—dinner Fri–Sat

URUNGA

🏠 STORES

THE HONEY PLACE
Jeff and Julie Daley
Pacific Hwy, Urunga 2455
Phone: 02 6655 6160
Email: honey@midcoast.com.au
www.honeyplace.com.au
This place is a honey-lover's paradise. From the moment you enter, as if into a

huge yellow beehive, you are confronted with masses of honey products—jams, and chutneys, pickles, honeyed macadamia nuts, and sparkling honey nectar. A cafe serves Devonshire teas with scones made on the premises, and of course, honey and local cream and jams. Unusual items include a ginger flavoured honey, and powdered honey from Superbee in Queensland.
- **Open:** 8.30am–5.30pm daily
- **Location:** 30km south of Coffs Harbour

WAUCHOPE
PRODUCERS

HASTINGS VALLEY OLIVES
Graeme and Kathy Booker
780 Forbes River Rd, Birdwood, Wauchope 2446
Phone: 02 6587 7225
Email: info@hastingsvalleyolives.com.au
www.yarrasgrove.com.au
This is the largest non-irrigated olive grove in Australia, established in 1995. The extra virgin olive oil, olives, cosmetics using olive oil are sold through retail outlets in many states. The oils and olives include infusion products with a taste of Australia, such as smoky fried garlic, wild Thai twist, wild mountain pepper, lemon myrtle, garlic and gumleaf.
- **Open:** 8am–5pm daily
- **Location:** About forty minutes west of Wauchope off the Oxley Hwy

WICKHAM
STORES

COMMERCIAL FISHERMEN'S COOPERATIVE LTD
97 Hannell St, Wickham 2293
Phone: 02 4965 4221
This cooperative began in 1945 when a group of fishermen joined together to sell their daily catches. Each day there are many fish species available as well as lobsters and crabs, school and king prawns—basically whatever is caught that day, so it's a bit of a lucky dip. Just like fishing itself. Fish preparation classes also available, and a restaurant is planned.
- **Open:** 7am–5pm weekdays, 8am–3pm Sat, 9am–2.30pm Sun

WINGHAM
DINING

THE BANK GUEST HOUSE AND TELLERS RESTAURANT
Anne Relph
48 Bent St, Wingham 2429
Phone: 02 6553 0006
Email: relliesretreat@bigpond.com
www.bankguesthouse.com.au
Licensed
Cuisine: Country style
Mains average: $$
Specialty: Seafood
Local flavours: Seafood, fruit and vegetables, beef, cheeses
Extras: Catering, courtyard dining, outside barbecue area. A former 1920s bank, with comfortable accommodation.
- **Open:** Breakfast weekends, lunch and dinner Wed–Sat

WOOLGOOLGA
🍴 DINING

POSSUM'S CAFE
Helen Milne
Shop 4/53, Beach St, Woolgoolga 2456
Phone: 02 6654 2807
BYO
Mains average: $
Specialty: Homemade cakes and muffins, Aussie burgers
Local flavours: Chutneys, Stotts macadamia dressings
Extras: Relaxed atmosphere, outdoor seating, takeaway available.
■ **Open:** Breakfast and lunch Mon–Sat, daily in holiday periods

MARKETS AND EVENTS

✦ **BELLINGEN ORGANIC MARKETS,**
BELLINGEN
02 6655 8720
8am–1pm second and fourth Sat

✦ **COFFS COAST GROWERS' MARKET,**
COFFS HARBOUR
02 6648 4084
Fortnightly, 4pm–7pm Thurs

✦ **GREAT LAKES GREAT PRODUCE MARKET,** FORSTER
02 6554 4184
8am–12pm third Sat

✦ **MARKETS AT THE PALMS,**
FORSTER–TUNCURRY
02 6554 0348
9am–1pm last Sun

✦ **HASTINGS FARMERS' MARKETS,**
WAUCHOPE
02 6581 8633
8am–12pm fourth Sat

✦ **THE ADVOCATE COFFS COAST FOOD AND WINE FESTIVAL,**
COFFS HARBOUR
02 6648 4240
Last weekend in Oct

✦ **HASTINGS HARVEST PICNIC AND CULTURAL FESTIVAL,**
PORT MACQUARIE
02 6581 8000
Early Nov www.hastings.nsw.gov.au

✦ **BOUNTY OF THE SEA FESTIVAL,**
FORSTER–TUNCURRY
02 6555 5527
Feb

NORTHERN RIVERS

The beauty of the northern part of New South Wales will take your breath away. Lush and volcanically fertile, it seems almost anything will grow here, especially on the alluvial flood plains. Everything, from tropical fruits and coffee to dairy cattle, macadamias, beef, sugar cane, blueberries and chillies, thrives here and seems to be brighter and bigger than elsewhere.

From the waters off the dramatic and scenic coast, some of the country's finest seafood is harvested or caught. This is picnic land, either overlooking stunning ocean scenery or tucked away in a green and grassy valley. With such a wealth of produce, you don't need to bring your own food—simply pick it up as you travel through.

ALSTONVILLE

PRODUCERS

FATHER MAC'S HEAVENLY PUDDINGS
Our Lady of the Rosary Catholic Church
9 Perry St, Alstonville 2477
Phone: 02 6628 5474
Email: fathermc@nor.com.au
www.fathermac.org.au
You could call this story: Local Boy Makes Pud; or even Father McCarthy's Heavenly Brainwave. This speciality business came about because the local Catholic school needed new classrooms. It all began in 1981 with various fundraising activities when puddings were sold from a cake stall outside the church. Serious pudding production, staffed mainly by volunteers, commenced in 1985–87. Father Darcy McCarthy was behind it all—hence the name. They became so popular that they are still made, and with the same quality ingredients (no salt, artificial flavours or preservatives). They are sold in selected IGA, Franklins and Action stores in New South Wales, Queensland, and Victoria.
■ **Open:** By appointment

BANGALOW

DINING

COUNTRY FRESH RESTAURANT
Bangalow Hotel, 1 Byron St,
Bangalow 2479
Phone: 02 6687 1711
Licensed
Cuisine: Contemporary Australian
Mains average: $$
Local flavours: Seafood
Extras: Dine on the veranda amongst the Bangalow palms. Takeaway available.
■ **Open:** Lunch and dinner daily

BENTLEY

PRODUCERS

FAIRBRAE MILK CO
Phil and Jan Denniston
2025 Kyogle Rd, Bentley 2480
Phone: 02 6663 5356
Email: phild@nor.com.au
www.fairbraemilk.com
Jersey A2 milk from the Dennistons' herd of around 200 cows is processed at the

farm. Visitors may buy homogenised and unhomogenised full-cream milk, skim milk and cream from the factory or from outlets in the town. The Dennistons pasteurise the milk and visitors may also watch milking if they are there at the right time (from 5am–7am, or 4pm–6pm), or watch calves being fed at other times. The milk is also available in Sydney.

■ **Open:** 6am–7pm daily, by appointment

BYRON BAY

🍎 PRODUCERS

RAINFOREST FOODS
Greg Trevena
Unit 2, 12 Bayshore Dr, Byron Bay 2481
Phone: 02 6685 8097
Email: rainforest@rainforestfoods.com.au
www.rainforestfoods.com.au

Greg Trevena grows a number of northern NSW rainforest trees in a plantation but, more importantly, he produces a range of rainforest jams, macadamia butter, a rainforest sauce, a chutney and a marmalade, as well as a unique lemon myrtle honey. The Davidson plum is very local and very rare. In fact it only grows naturally in two valleys in the world, both of which are behind Byron Bay. Trevena also makes lemon myrtle cosmetics—soap, body lotion and essential oil which works like tea-tree oil—under the Rainforest Delights label, and grows riberry, Davidson plum, small leafed tamarind, lemon myrtle, and finger limes. Try all these in the new showrooms.

■ **Open:** 11am–4pm weekdays, by appointment

WILDBITE
Roe Ritchie and Trace Gordon
5 Ti-Tree Place, Byron Bay 2481
Phone: 02 6680 9622
Email: wildbite@nrg.com.au
www.wildbite.com.au

Roe Ritchie and Trace Gordon, formerly organic greengrocers in Melbourne, call their product 'biscuits that bite with wild food flavour'. Indeed, the Lismore-grown lemon myrtle and macadamia biscotti, Buderim ginger and pistachio biscotti, and double chocolate and peanut cookies are great examples. Committed to using organic and free-range ingredients where possible, sourced from around Byron Bay, Wildbite has bitten off a good hunk of the local market and the biscuits are widely available in Lismore, Ballina, Byron Bay and Bangalow at cafes, delis and gift shops. They also sell a range of packaged and bulk biscuits, gift bags and boxed biscotti, regional foods, hampers, cakes and slices from the factory, where you can see them baking.

■ **Open:** 8am–4pm Mon–Thurs

🏠 STORES

BYRON BAY WHOLLY SMOKED GOURMET FOODS
John Garret
Shop 7, 130 Jonson St, Byron Bay 2481
Phone: 02 6685 6261

Wholly Smoked now has a wholly new shop—bigger, better, brighter—and with more organic products. Here the specialty is in combining fresh organic food and traditional methods. So while the ingredients are totally certified organic, fresh beef jerky is dry-aged, and local meats such as beef, lamb, poultry and tuna are treated traditionally, and macadamia-smoked.

■ **Open:** 7am–5pm weekdays, 7am–1pm Sat

🍴 DINING

DISH RESTAURANT RAW BAR
Ben and Belinda Kirkwood (owners), Luke Southwood (chef)
Shop 4, cnr Marvel and Jonson streets, Byron Bay 2481
Phone: 02 6685 7320
Email: dish@nor.com.au
Licensed
Mains average: $$$
Specialty: Oysters
Local flavours: Seafood
Extras: This multi award-winning restaurant is one of the most attractive restaurant spaces around, featured in many design, food and travel magazines.
■ **Open:** Dinner daily

FINS SEAFOOD RESTAURANT
Steven Snow
Beach Hotel, Bay St, Byron Bay 2481
Phone: 02 6685 5029
Email: finssnow@norex.com.au
www.fins.com.au
Licensed
Cuisine: Seafood
Mains average: $$$
Specialty: Cataplana—local seafood and potatoes poached in a saffron and star anise flavoured broth
Local flavours: Fish, prawns and herbs from the chef's garden
Extras: An outside Balinese style area, called 'the jungle room'. Winner of Best Seafood Restaurant in Australia, Restaurant and Catering Association Awards 2003. Steven Snow sometimes teaches cooking classes at the Sydney Seafood School.
■ **Open:** Dinner daily

FISHHEADS@BYRON RESTAURANT
Mark Matheson
Byron Bay Swimming Pool,
Jonson St, Byron Bay 2481
Phone: 02 6680 7632
BYO
Cuisine: Modern Australian seafood
Mains average: $$
Specialty: Hot and cold tapas platter
Local flavours: Local seafood
Extras: Seaside location with beachy views. Takeaway available.
■ **Open:** 8am–late daily

OLIVO
James Lancaster (owner), Anthony Telford (chef)
34 Jonson St, Byron Bay 2481
Phone: 02 6685 7950
Email: olivo_byron@hotmail.com
Licensed/BYO wine
Cuisine: Modern Australian
Mains average: $$
Local flavours: Seafood, fruit and vegetables
Extras: Located in the middle of Byron—great for people-watching!
■ **Open:** Dinner daily

RAE'S ON WATEGOS
Vincent Rae and Sean Jarrett
8 Marine Pde, Wategos Beach, Byron Bay 2481
Phone: 02 6685 8246
Email: raes@wategos.com.au
www.raes.com.au
Licensed
Cuisine: Mediterranean influence (lunch), Royal Thai (dinner)
Mains average: $$$$
Specialty: Local seafood—salt and pepper squid, large mud crabs, champagne lobster
Extras: On the waterfront overlooking

Wategos Bay, complex includes a seven-room boutique hotel.
- **Open:** Lunch and dinner daily

THE BOWER ROOM
Max and Rowen King (owners),
Gavin Hughes (chef)
Upstairs, cnr Jonson and Lawson Streets,
Byron Bay 2481
Phone: 02 6685 7771
Email: info@thebowerroom.com.au
www.thebowerroom.com.au
Licensed
Cuisine: Modern Italian
Mains average: $$$
Specialty: Pan roasted duck confit with sweet potato puree and fresh mustard fruits
Local flavours: Alstonville free-range chickens, Bangalow pork, herbs, fruit and vegetables when available
Extras: The restaurant and cocktail lounge is set on extensive Queenslander-style verandas overlooking the busy streets of Byron Bay, and faces west, with sunsets over the Byron hills. There is a log fire for cosy winter dining.
- **Open:** Dinner daily

DURANBAH
PRODUCERS

TROPICAL FRUIT WORLD
Bob Brinsmead
Duranbah Rd, Duranbah 2487
Phone: 02 6677 7222
Email: info@tropicalfruitworld.com.au
www.tropicalfruitworld.com.au
In 1983, horticulturalist Bob Brinsmead started an idyllic farm on which to raise his family. It grew into a massive tourist venture, growing the world's largest collection of tropical fruits, and far outpacing even the energetic Brinsmead's dreams. This sixty-five hectare farm grows over 500 varieties of tropical and rare fruits. Guests can safari through it on Adventure Buggies and see fruits they never knew existed. They can ride a miniature railway, wonder at the thirteen 'gardens of the world', or relax over a fruit platter or cool juice in the cafe. Ideal for families, the Adventure Park will keep the youngsters amused for hours. Fruit World sells its fruit to local markets, at its own market on-site, and of course uses the produce in the on-site licensed Rainforest Cafe, as well as selling a wide range of jams, dressings and pickles in the gift shop.
- **Open:** 10am–5pm daily
- **Location:** Ten minutes south of Gold Coast airport

FEDERAL
DINING

POGELS WOOD CAFE AND RESTAURANT
Fay Niederhauser (owner),
Rupert Stevens (chef)
Lot 1, Federal Dr, Federal 2480
Phone: 02 6688 4121
Email: sales@pogelswood.com
www.pogelswood.com
Licensed/BYO
Cuisine: Informal modern Australian
Mains average: $$
Specialty: Salt and pepper calamari, zarzuela (Spanish seafood dish)
Local flavours: Macadamias, rainforest fruits, beef
Extras: Finalist three years running in Restaurant and Catering Association

NORTHERN RIVERS—NSW

awards. Located in a century-old building, with covered courtyard and tropical gardens. Takeaway available.
- Open: Breakfast and lunch Wed–Sun, dinner Thurs–Sat

GRAFTON

DINING

GEORGIES AT THE GALLERY
Mark (chef) and Judy Hackett
Grafton Regional Gallery,
158 Fitzroy St, Grafton 2460
Phone: 02 6642 6996
Licensed/BYO wine
Cuisine: Modern Australian
Mains average: $$
Specialty: Yamba prawns
Local flavours: Fish sourced from Iluka and Coffs Harbour, Chiquita raspberries and blueberries
Extras: Courtyard surrounded by the heritage gardens of the Grafton Regional Gallery. The restaurant has access to one of the Gallery Rooms for indoor dining surrounded by artwork. Takeaway available. Grafton Show champion pickles and homemade ice-cream are also made for sale.
- Open: Lunch Tues–Sun, dinner Tues–Sat

KNOCKROW

PRODUCERS

KNOCKROW RIDGE COFFEE
Barry Heffernan
Lot 3, Pacific Hwy, Knockrow 2479
Phone: 02 6687 8080
Barry Heffernan had a farming background, but a few years ago he was looking for a new challenge. He got it when he moved into producing Australian-grown and processed coffee, free of harmful sprays. Knockrow Ridge grows the coffee, then produces medium roast, drip/plunger, medium dark, and espresso blends which, because they are organically based, contain fifteen per cent less caffeine. Available in northern New South Wales and the Gold Coast, by mail order, and in selected local restaurants and supermarkets.
- Open: 9am–5pm daily

STORES

THE MACADAMIA CASTLE
Jerome Hensen
Pacific Hwy, Knockrow 2479
Phone: 02 6687 8432
Email: castle@nrg.com.au
www.macadamiacastle.com.au
This is more like a showcase for the local macadamia industry. The store features a large range of roasted and seasoned macadamia nuts—gently heated for maximum taste sensation. There is also a unique range of fine-food products featuring this magnificent nut as well as locally grown coffees, which are naturally low in caffeine thanks to the volcanic soil of the area, as well as fresh coffee and home-baked macadamia specialty cakes.
- Open: 8.30am–5pm daily

LENNOX HEAD

DINING

SEVEN MILE CAFE RESTAURANT
Jason and Julie Gilmor, Tracey Craft and Marcel Verhage
41 Pacific Pde, Lennox Head 2478

103

NSW—NORTHERN RIVERS

Phone: 02 6687 6210
Email: sevenmilecafe@bigpond.com
www.lennoxhead.au.nu/sevenmile.htm
Licensed
Cuisine: Modern Australian
Mains average: $$
Specialty: Slow-roasted duck, lemon and passionfruit cheesecake
Local flavours: Finger limes, Bangalow pork, fruits, oysters and seafood
Extras: Undercover front deck overlooking the Pacific Ocean. Located in a bright red beach house. Wheelchair access and catering for children. The second Sunday of the month is a Jazz Sunday, featuring a seven-piece Dixie band. Outstanding wine list; winners of the Tucker Seabrook Australia-wide award for Best Regional Wine List on two occasions in recent years.
- Open: Brunch Sun, lunch Fri–Sun, dinner Wed–Sat

LIMPINWOOD
DINING

LIMPINWOOD TEAHOUSE
Peter Merel (owner), Gary Jackson and Tom Twyford (chefs)
538 Zara Rd, Limpinwood 2484
Phone: 02 6679 3246
Email: teahouse@limpinwood.com
www.limpinwood.com
Licensed/BYO
Cuisine: Fusion world cuisine and bush tucker
Mains average: $
Specialty: Tea (Twyford Teas grown and milled in Limpinwood Valley, Madura teas, Clothier's Creek, and locally milled herbal extracts) with scones, fresh clotted cream, local rosella and blueberry jams, rindless marmalade. Also signature ethnic dishes with bush tucker added to them.
Local flavours: Milk, cream and butter from Holstein's ranch in Limpinwood valley, meats all locally caught/grown, most of the bush tucker from native gardens at Banana Cabana, Chillingham.
Extras: Limpinwood honey jars for sale year-round, as well as various jams, chutneys, and pickles and raw honey made by own bees from exotic flowers in the Teahouse garden. Takeaway and B & B accommodation available. Natural amphitheatre used for performances. The teahouse itself is a work of art, with stained glass, unique paintings and sculptures, and imported slate floors full of fossils.
- Open: Breakfast and lunch Fri–Sun, dinner Fri (with show)
- Location: Take Tourist Route 34 from Murwillumbah to Chillingham. Continue three blocks to Zara Road (Tourist Route 36)

LISMORE
PRODUCERS

MEDICINE GARDEN AUSTRALIA
Jodie and Stephen Barnier
13 Arthur St, North Lismore 2480
Phone: 02 6622 2322
Email: admin@medicine-garden.com.au
www.medicine-garden.com.au
This gardener's delight, which has taken twelve years to build up, has a public display of traditional waterlilies, botanical carnivores and medicinal plants. For non-gardeners, the Barniers make herb teas, and it's so simple—just choose the herb you want from the garden, and they will brew up your tea. You then sit on the veranda, or in the garden under shadecloth, and enjoy this rare touch of pampering.

■ **Open:** 9am–5pm daily, tours by appointment

NORCO COOPERATIVE
Union St, South Lismore 2480
Phone: 02 6627 8000
Email: markw@norco.com.au
www.norco.com.au

All Norco cheeses are made without animal rennet and all use the rich creamy milk from this area the Nor(th) Co(ast), hence the name. The co-op store stocks all products including Norco Nimbin low-fat and low-salt cheeses, cheddars, local milk, Cape Byron Supreme ice-cream (including the Australiana range) and the sinfully rich Prestige ice-cream, although it is potluck as the range is usually what remains after the sales runs.

■ **Open:** 8.30am–5pm weekdays

🍴 DINING

PAUPIETTES
David Forster
56 Ballina St, Lismore 2480
Phone: 02 6621 6135
Licensed
Cuisine: Modern Australian
Mains average: $$
Specialty: Braised duck, roast pork belly
Local flavours: Raspberries, limes, produce sourced directly from local markets
■ **Open:** Dinner Tues–Sat, other times by appointment

MOOBALL
🍎 PRODUCERS

RAINFOREST SECRETS
Marshall Family (Shane Marshall)
Pottsville Rd, Mooball 2483

Phone: 02 6677 1215
Email: info@rainforestsecrets.com.au
www.rainforestsecrets.com.au

This property has been in the family for four generations, but now is a tourist attraction with admission charged. There's a family restaurant specialising in local products such as Madura tea and Fernleigh coffee, and the farm offers tours and a wildlife walkabout area, as well as a waterbird lake and rainforest walks. There is also a regeneration trail, a bush tucker trail, and bush medicine trail, as well as farm and native animals.

■ **Open:** 9.30am–4pm daily during DST
● **Location:** On Tourist Drive 38, 4km north of Mooball

THE SCOTTISH FOOD OF MACLEAN

The town of Maclean prides itself on its Scottish connections. Look for shortbread and sweeties at The Chocolate Box, Palace Arcade; Haggis sausage at Towells Tender Meats in River St; sliced sausage at Walters Butchers, River St; Irn-Bru, Tizer and Haggis-burgers at The Tartan Lounge, Palace Arcade; and rumbletythump at Ferry Park Restaurant, Pacific Highway.

MURWILLUMBAH
🍎 PRODUCERS

MADURA TEA ESTATES
Ray Fien
753 Clothiers Creek Rd,
Murwillumbah 2484
Phone: 02 6677 7215
Email: info@maduratea.com.au

Madura Tea has been producing premium

quality tea for over twenty years. Naturally lower in tannin and caffeine, it is an ideal healthy alternative to mainstream teas, and is sold in many health food shops. The tea is processed on the plantation.
- **Open:** 9am–5pm weekdays
- **Location:** 2km from the off ramp from the new freeway

THE RIGHT FOOD GROUP
Anni Brownjohn
Factory 5/2 Lundberg Dr,
Murwillumbah 2484
Phone: 02 6672 5882, 02 6672 8504 (sales)
Email: anni@ozganics.com.au
www.ozganics.com.au
This company manufactures bottled organic pasta sauce, jam, cholesterol-free dressings, Australian olives and tapenade. Tastings are available.
- **Open:** 7am–4pm weekdays

NASHUA
PRODUCERS

SPRINGVALE HILLS
Janet and Ian Fraser
Lot 4, Taylors Rd, Nashua 2479
Phone: 02 6629 1263
Email: janetfraser@ozemail.com.au
The Frasers grow several varieties of bush food on a small farm between Bangalow and Lismore. They grow lemon myrtle, cinnamon myrtle, Davidson plums, finger limes, riberries, lemon and silver aspens, cut leaf native mint as well as cabinet timbers. Although the plantation is not generally open to the public the Frasers say they could show people the farm by special arrangement, but only a few at a time, no bus loads.
- **Open:** By appointment

- **Location:** 8km from Bangalow, towards Lismore

NEWRYBAR
PRODUCERS

PLANTATION LORNA MACADAMIAS
Geoff Dorey
144 Old Byron Bay Rd, Newrybar 2479
Phone: 02 6687 2122
Email: plantlorna@ozemail.com.au
www.plantationlorna.com
Plantation Lorna's (named for grandma Lorna) macadamias have been in the works for a while. You have to start early with this nut tree, as macadamias take ten or so years to fruit. The plantation has been producing nuts for the past couple of years and there are plans for a purpose-built tour facility soon.
- **Open:** 9am–5pm weekdays

NIMBIN
STORES

NIMBIN ORGANICS
Gerhardt Weihermann
50A Cullen St, Nimbin 2480
Phone: 02 6689 1445
Nimbin may not be everybody's cup of (herbal) tea, but if you stumble upon this village tucked away in the fertile valleys near Lismore, spare a glance for more than just the psychedelic graffiti on the buildings. There is organic produce—subtropical fruits, bananas, pineapples, pawpaw, avocados, mangoes, stone fruit, juices and breads—all locally produced. Nimbin Organics make dried organic bananas on the premises, and sell local coffee. These sales support local growers. Delicious and

seasonal, all this has sprung naturally from the fertile volcanic soils of the region.
- **Open:** 9am–5pm daily

TWEED HEADS
🚌 TOURS
CATCH A CRAB TOUR
Lee Eyre
Drydock Jetty, Drydock Rd,
Tweed Heads 2385
Phone: 07 5599 9972
Email: info@catchacrab.com.au
www.catchacrab.com.au
Join a group and catch mud crabs on the Terranora Inlet of the Tweed River. You can feed pelicans too, go yabby pumping or just fish, and then top it off with a great crab meal. Or join an eco-tour to see how Terranora rock oysters are farmed.
- **Open:** 9am daily, bookings essential

WEST BALLINA
🍴 DINING
GARDEN RESTAURANT
Shirley Carruthers (owner),
Peter Ilich (chef)
Best Western Ballina Island Motor Inn,
Pacific Hwy, West Ballina 2478
Phone: 02 6686 8866
Email: ballinaisland@bigpond.com
www.bestwestern.com
Licensed
Cuisine: Modern Australian
Mains average: $$
Specialty: Macadamia crusted whiting with lemon myrtle aioli
Local flavours: Seafood, beef, pork, macadamias, dairy, tropical fruits, native foods, coffee

Extras: Multi award-winning chef, outdoor dining available.
- **Open:** Breakfast and dinner daily

WOLLONGBAR
🍴 DINING
BELLOWING BULL RESTAURANT
Coral Martin
Bruxner Hwy, Wollongbar 2477
Phone: 02 6628 0715
Email: bellowingbull@bigpond.com
www.bellowingbull.com.au
Licensed/BYO
Cuisine: Modern Australian steakhouse
Mains average: $$
Specialty: Marinated veal shanks and steaks
Local flavours: Prime Northern Rivers beef
Extras: Housed in a hundred-year-old restored Federation-style building, winner of the Restaurant and Catering Award for Best Steakhouse in Regional NSW.
- **Open:** Dinner daily

WOOLI
🍎 PRODUCERS
WOOLI OYSTER SUPPLY
Ron and Kim Guinea
Riverside Dr, Wooli 2462
Phone: 02 6649 7537
Halfway between Grafton and Coffs Harbour, head east until you hit the coast and there you will find Wooli. Ron Guinea specialises in oysters but sells anything he can get his hands on, providing it has come straight out of the water. It is all local and all fresh—snapper, pearl perch, parrot fish, flathead, crabs, prawns,

NSW—NORTHERN RIVERS

scallops, shark, even mackerel in season. Mention this book and you will get a 10 per cent discount, says Ron.
- **Open:** 8.30am–4pm weekdays, 11am–4pm weekends

WOOMBAH

🍎 PRODUCERS

WOMBAH COFFEE PLANTATION
Joan Dibden and Joy Phelps
51 Middle St, Woombah 2469
Phone: 02 6646 4380
Email: wombah@hotkey.net.au
www.wombahcoffee.com.au
Here's another retirement enterprise that became a runaway success. Wombah coffee is mild, low in caffeine, not blended and is 100 per cent Australian. There is a cafe in a cottage on the property, and conducted tours of the processing factory. All the coffee is grown on the property or in the Richmond and Clarence rivers area, and is processed on site, where there is a small gift shop for handcrafts and of course it also sells the coffee. Coffee flowers November to December and the berries ripen around May or June.
- **Open:** 10am–5pm weekends, school and public holidays
- **Location:** 4km along the Iluka Rd from the Pacific Hwy

YAMBA

🍴 DINING

GORMAN'S RESTAURANT
Shirley and Gaye Gorman
Yamba Bay, Yamba 2464
Phone: 02 6646 2025
Licensed

Cuisine: Seafood
Mains average: $$
Specialty: Seafood platter for two—bugs, crays, prawns, oysters, calamari, fish and chips, prawn cutlets
Extras: Wholesale shop downstairs for fresh seafood, and Yamba Bay oysters. Restaurant overlooks Yamba Bay and the mouth of the Clarence River. Takeaway available.
- **Open:** Lunch and dinner daily

RESTAURANT CASTALIA
Margaret Matthews and Gregory Gray-Matthews (owners), Andrew Causley (chef)
Shop 1, 15 Clarence St, Yamba 2464
Phone: 02 6646 1155
Email: restaurantcastalia@yambansw.com.au
www.yambansw.com.au/content/pages/businesses/castalia/castalia.htm
Licensed/BYO wine
Cuisine: Modern Australian
Mains average: $$
Specialty: Crispy skin duck with sesame Asian greens with orange and chilli glaze
Local flavours: Seafood, chemical and hormone-free meats, herbs, vegetables, fruit and bush foods whenever possible.
Extras: This award-winning restaurant is perched on the top of the hill in the coastal village of Yamba with sweeping views of the Clarence River mouth in the east to the hinterland in the west. Four separate dining areas, both indoor and outdoor.
- **Open:** Breakfast Fri–Sun, lunch Wed–Sun, dinner Wed–Sat

MARKETS AND EVENTS

✦ **BANORA POINT FARMERS' MARKET,** BANORA
07 5590 4862
7am–12pm first, third and fifth Sun

✦ **BYRON FARMERS' MARKET,**
BYRON BAY
02 6629 1059
8am–11am Thurs

✦ **LISMORE FARMERS' MARKET,**
LISMORE
02 6621 5961
8am–12pm Sat

✦ **LISMORE RAINBOW REGION ORGANIC MARKET,** LISMORE
02 6628 1084
8am–11am Tues

✦ **MACLEAN FARMERS' MARKET,**
MACLEAN
02 6645 1980
8am–12pm first Sat

✦ **GRAFTON FARMERS' AND GROWERS' MARKET,** GRAFTON
02 6643 1967
7am–1pm second Thurs

✦ **AVOCADO FESTIVAL,**
DURANBAH
02 6677 7222
June

✦ **BEEF ON BARKER,** CASINO
02 6662 3566
Last Sun in May

✦ **NORTHERN RIVERS HERB FESTIVAL,**
LISMORE
02 6622 8147
Third week in Aug

✦ **A TASTE OF BYRON,** BYRON BAY
02 6687 8692
Oct

✦ **TWEED VALLEY BANANA FESTIVAL AND HARVEST WEEK,**
MURWILLUMBAH
02 6672 6186
Last weekend in Aug

✦ **PRIMEX,** CASINO
07 5531 4600
www.exhibitionmarketing.com.au
June

OUTBACK AND THE MURRAY

The mighty Murray symbolises the size of this country perhaps better than any other natural feature. The waters from this river system have helped the desert blossom.

Today orchards and vineyards thrive because of irrigation, yet mere kilometres out of range, the land reverts to wide, flat pastureland, dependent for its success on the fickle climate.

Australia's greatest river it certainly is, but when seen in the light of its contribution to our food-lands, you could also call it 'Australia's greatest giver'.

ALBURY

🍎 PRODUCERS

BUTT'S GOURMET SMOKEHOUSE
Graham Hulm
417 Tribune St, Albury 2640
Phone: 02 6021 3987
If you love the subtle taste of smoke in your food, make straight for this place. Local smoked trout, triple smoked hams, smoked beef, oxtongue, smoked loin or rack of lamb, smoked poultry, kangaroo, rabbit when available, turkey and quail all happen here. The smokehouse also carries Mrs Whitehead's vacuum-packed olives, hand-picked from hundred-year-old trees in Urana. Products are available from the shop or by mail order, and overnight delivery is available outside the Riverina and north-east Victoria.
■ Open: 8.30am–5.30pm weekdays, 8.30am–12pm Sat

🏠 STORES

PETERS AND SONS SELECTED MEATS AND DELICATESSEN
Lutz and Joan Peters
317A Urana Rd, Lavington,
North Albury, 2641
Phone: 02 6025 1796
Email: ljpeters@albury.net.au
The beef here is locally sourced, and Lutz Peters is sure of that, as his brother and other regional producers grow it. Other meats, such as pork, lamb, veal and buffalo are also local. The shop, established fifty years ago by Lutz's father, Paul, a German master butcher, makes its own gourmet sausages and a wide range of smallgoods—all gluten free—and hams are made the traditional way, too, using nothing artificial. The deli also stocks some local produce, as well as free-range poultry from Milawa.
■ Open: 7.30am–5.30pm weekdays, 7am–1.30pm Sat

🍴 DINING

THE GLOBE HOTEL
Brian Grenfell
586 Dean St, Albury 2640
Phone: 02 6021 2622
Email:
briangrenfell@globehotelalbury.com
www.globehotelalbury.com
Licensed
Cuisine: Modern Australian pub food
Mains average: $
Specialty: Risotto, seafood lasagne
Local flavours: Smallgoods, chicken, meat, seasonal vegetables and fruits, olives, olive oil, breads

Extras: Alfresco dining area, barbecue breakfast cooked outside on weekends.
- **Open:** Breakfast weekends, lunch and dinner Tues–Sat

> **MURRAY REGIONAL FOOD AND WINE DIARY**
> Noelle Quinn,
> PO Box 515, Albury 2640
> **Phone:** 03 6058 2996
> **Email:** noelle.quinn@tafensw.edu.au
> This is the way to keep in touch with this region and its produce all year.

BOURKE

DINING

MORRALL'S BAKERY AND CAFE
Mark and Alan Morrall
37 Mitchell St, Bourke 2840
Phone: 02 6872 2086
Email: morralls@auzzie.net
Mains average: $
Specialty: Meat pies
Extras: This family business, established in 1905, won a bronze medal in the Great Aussie Meat Pie Competition for their Back'o'Bourke lamb pie, and also makes breads, sandwiches, cakes and slices.
- **Open:** Breakfast, lunch, morning and afternoon tea, Mon–Sat

TOURS

BACK OF BOURKE FRUITS
Bourke Visitor Information Centre
Old Railway Building,
Anson St, Bourke 2840
Phone: 02 6872 2280
Email: tourinfo@ozemail.com.au
www.backobourke.com.au

The local tourism office conducts around nine, guided, four-hour field and shed tours of citrus and grape farms and other places of food interest each week in this rich area. Tours may also include a visit to a jojoba plantation, the airport (with its memorabilia of Nancy Bird Walton), and a cotton farm. In season, tour members may sample and take fruit home.
Tours: Mon–Sat, morning or afternoon

BURONGA

PRODUCERS

ORANGEWORLD
Maria and Mario Mammone
Silver City Hwy, Buronga 2739
Phone: 03 5023 5197
www.orangeworldmildura.com.au
They call this 'the land of the living orange', and certainly it's quite an experience to see thirty different types of citrus and avocados, learn how to select the best, as well as how to keep them at optimum freshness at home. You'll find Valencias, navels and Seville oranges, grapefruits, mandarins, avocados, blood oranges, lemons and other citrus fruits, and can take the mini-train tour (pre-book in the off season) to see even more. At the kiosk, there is additional local produce, as well as farm-fresh fruit and free tastings.
- **Open:** 9.30am–4pm daily
- **Location:** Follow the signs over the bridge at Mildura, eight minutes from the post office on the NSW side

FINLEY

🍎 PRODUCERS

SUN VALLEY AUSTRALIA PTY LTD
211 Murray St, Finley 2713
Phone: 03 5883 2300
Email: svaust@mcmedia.com.au
www.snackfood.com.au
Using natural fruit products, many from the Riverina, Sun Valley manufactures the True Fruit brand of pure fruit bars made with no preservatives, colourings or additives. These are sold nationally in the health food sections of major supermarkets and other shops, as well as at the factory door.
- Open: 8am–4.30pm weekdays

> **MARKETS AND EVENTS**
>
> ✦ **ALBURY–WODONGA WINE AND FOOD FESTIVAL,**
> ALBURY–WODONGA
> 02 6058 2996
> Oct long weekend
>
> ✦ **MIGHTY MURRAY TART TRAWL,**
> ALBURY
> 02 6058 2996
> Mar
>
> ✦ **WORLD'S LONGEST LUNCH,**
> ALBURY–WODONGA
> 02 6058 2996
> Last weekend in Mar

JERILDERIE

🍎 PRODUCERS

BILLABONG PRODUCE
Sergio Rorato
Cnr Oaklands and Cape Roads,
Jerilderie 2716
Phone: 03 5886 1306
Email: rorato@bigpond.com
www.billabongprod.com.au
Billabong's products somehow capture the sunny flavours of this region. There are crushed or pureed tomatoes, chunky dried tomatoes (all made from tomatoes grown by Billabong), as well as marinated eggplant slices, dried capsicums, and garlic, hot chilli, and ready-made pasta sauce. Although the products are widely available, you may buy from the factory too.
- Open: 9am–5pm weekdays

SOUTH COAST AND ILLAWARRA

The lovely coastline of the south coast is home to fishing towns and oyster farms. Inland you won't need to hunt far before finding a great Devonshire tea. Plenty of towns offer homemade breads, cheeses and hydroponic vegetables.

BATEMANS BAY

🍎 PRODUCERS

INNES BOATSHED
Merv and Robin Innes
Clyde St, Batemans Bay 2536
Phone: 02 4472 4052
Those in the know buy fresh fish direct from the fourth-generation Innes family's own boats. The boats go out every day to the Continental Shelf around 20km out. That's where they find the flathead, ling, coral perch, and john dory they bring back. They also catch plenty of royal red prawns and cook them on the boat. So if you want the sweetest ever prawn supper, or the freshest fish for dinner, be at the shore about 4pm when the boats come back. Or buy their super-fresh fish and chips which they have perfected over the past thirty years.
■ **Open:** 9am–4pm Mon–Thurs, 9am–7pm Fri–Sun (longer in summer)

🍴 DINING

ON THE PIER
Alan Imrie, Brendan McClelland, Simon Whiteman (chef)
Old Punt Rd, Batemans Bay 2536
Phone: 02 4472 6405

Licensed
Cuisine: Seafood
Mains average: $$
Local flavours: Seafood, oysters, fruit and vegetables
Extras: Outdoor dining, waterside on the Clyde River.
■ **Open:** Lunch Tues–Sun, dinner Tues–Sat

BEGA

🍎 PRODUCERS

BEGA CHEESE AND BEGA CHEESE HERITAGE CENTRE
Bega Cooperative Society Ltd
18–36 Ridge St, Bega 2550
Phone: 02 6491 7777
Email: bega.admin@begacheese.com.au
www.begacheese.com.au
Over a hundred years old, Bega Cheese Cooperative was formed by a group of local dairy farmers who wanted to control their own industry. Cheddar cheese has been produced in the valley since 1860. The butter factory began production in 1900, and in 1969 a bigger cheese factory was commissioned. Today around forty-five million packets of Bega cheese are sold annually in Australia, and in thirty-nine countries worldwide, although some products are exclusive to the on-site retail centre. The cheese factory and heritage centre combines a kiosk, retail sales and tourist centre. There are viewing rooms of the factory, picnic facilities, and cheese tastings.
■ **Open:** 9am–5pm daily

🏠 STORES

KYDD'S BUTCHERY
Brian Sirl
162 Carp St, Bega 2550
Phone: 02 6492 1162
The meat in Brian Sirl's butchery could hardly be more local. He raises his own Angus beef just a few kilometres away, along with some organic lamb. The other meats come from local farms too and he carries local pork, and home-cured meats. Leg hams are a speciality, and he even makes brawn the old-fashioned way, as well as corned beef, devon, roast pork, frankfurts and saveloys, and has won prizes with his sausages. His marinated steak, using a secret recipe, is very tasty, and can be enjoyed at Grevillea Estate Restaurant.
■ **Open:** 7.30am–5pm weekdays, 7.30am–12pm Sat

BERMAGUI

🏠 STORES

BERMAGUI SEAFOOD SUPPLIES
Marj and Mark Andreatta
Fishermen's Co-op, Bermagui 2546
Phone: 02 6493 4239
Email: seafood@netspeed.com.au
All the fish is fresh from the boats at the wharf behind the shop. 'We buy it as it goes past', say the Andreattas. There is usually flake, flathead, ling, sea bream, ocean perch, trevally, local and farm prawns, oysters, and lobsters caught locally by members of the Fishermen's Co-op.
■ **Open:** 9am–6.30pm daily

🍴 DINING

SALT WATER
Paul Lay
75 Lamont St, Bermagui 2546
Phone: 02 6493 4328
BYO
Cuisine: Modern Australian with a seafood emphasis
Mains average: $$
Specialty: Fresh fish of the day
Local flavours: Fish, seafood, prawns, lobster, mussels, Wapengo oysters
Extras: Right on the jetty, with great views. Paul Lay worked with Ian Hewitson for many years and picked up on his passion for fine food, learning to appreciate good, simple, flavoursome produce.
■ **Open:** Dinner from 6pm Tues–Sun, takeaway fish and chips from 11.30am Tues–Sun

BERRY

🏠 STORES

BERRY WOODFIRED SOURDOUGH BAKERY
Jost and Jelle Hilkemeijer
23 Prince Alfred St, Berry 2535
Phone: 02 4464 1617
Brothers Jost and Jelle Hilkemeijer took over this whitewashed hundred-year-old heritage bakery on the edge of Berry in 2002. Since then they have been turning out traditional handcrafted, wood-fired sourdough bread, made with organic, stone-ground flour from Gunnedah. The bakery incorporates a cafe serving pastries, apple tarts and an enticing selection of foods made on the premises. Look for the organic hummus as well as

babaghanoush, organic African coffee to take away and, of course, that wonderful bread. Recently, locally grown organic fruit and vegetables have been added to the list of products for sale.
- **Open:** 8am-5pm Wed-Sun

DINING

CAVESE TRATTORIA
Gia Frino and Fabio Laudato
3/65 Queen/Albany St, Berry 2535
Phone: 02 4464 3909
Email: cavesetrattoria@bigpond.com
www.cavesetrattoria.bigpondhosting.com
Licensed
Cuisine: Italian
Mains average: $$
Specialty: Cheeses and sausages, pizza, desserts and antipasto products
Local flavours: Cheeses made from local milk
Extras: A small balcony looks out onto the Berry mountains. Homemade cheeses, sausages and desserts for sale (must pre-order).
- **Open:** Lunch Fri-Sun, dinner Tues-Sun

COURTYARD RESTAURANT
Colin Waller
The Hotel Berry,
120 Queen St, Berry 2535
Phone: 02 4464 1011
Email: berrypub@shoal.net.au
www.berryhotel.com.au
Licensed
Cuisine: Modern Australian
Mains average: $$
Specialty: Steak and kidney pie, steaks
Local flavours: Meat, fish, vegetables
Extras: Courtyard with historic 1860s coach-house, used for dining. Four open fires.

- **Open:** Breakfast Fri-Sun, lunch and dinner daily

BODALLA

PRODUCERS

SOUTH COAST CHEESE
Don and Nancy McPhee
22 Potato Point Rd, Bodalla 2545
Phone: 02 4473 5287
Cheese lovers, take note. There are soft and vintage club cheeses available for tasting here, all handmade. The blue combines sharp, vintage cheddar and sharp blue, then there are herb and garlic, sun-dried tomato, basil and garlic, or chilli and paprika, green pepper or chive cheeses.
- **Open:** 7am-4.30pm weekdays, 8am-12pm Sat, 7am-4.30pm, daily (Jan)

CENTRAL TILBA

PRODUCERS

ABC CHEESE FACTORY
Marie Annand, Manager
Bate St, Central Tilba 2546
Phone: 02 4473 7387
Marie.Annand.kig@natfoods.com.au
You can buy international award-winning club cheeses direct from the factory, and watch cheese making in progress in this National Trust village factory. Cheddar was manufactured by the ABC Cheese Cooperative from 1891 until the early 1980s, when the factory was forced to close. Club cheese is a secondary manufacturing process made by combining two or more different varieties of matured cheeses, sometimes with the addition of herbs or other flavourings. Tilba Club

cheeses include vintage, sun-dried tomato, and cracked pepper. The sun-dried tomato cheese was judged best cheese in the world cheese contest in Wisconsin, USA in 1999. On weekdays you may watch cheese manufacturing through viewing windows of the manufacturing room.
- Open: 10am–4.30pm daily

COBARGO
🍷 BREWERY/WINERY

BEGA VALLEY BERRY WINES
Terry Collins
70 Goanna Rd, Cobargo 2550
Phone: 02 6493 6592
Terry Collins's farm brochure begs 'pick me, peel me, eat me' and you would be hard-pressed to disobey this request. The sun-ripened pick your own raspberries, strawberries, boysenberries, youngberries, blueberries, cherries and stone fruit are begging to come home with you. And now there are berry wines, jams and honey too.
- Open: 10am–4pm daily
- Location: 3km south of Cobargo

FALLS CREEK
🍎 PRODUCERS

THE EMU SHOP
Tony and Maryke De Rooy
132 Jervis Bay Rd, Falls Creek 2540
Phone: 02 4447 8505
Email: farm@emushop.com.au
www.emushop.com.au
The small, farm souvenir shop sells, amongst other things, a delicious, hot, gourmet emu pie. Emu meat—fillet, steak, mince and sausages—can also be purchased frozen. There is plenty of parking for coaches, caravans and cars. A farm tour (groups of twenty-five or over, by bookings) includes morning tea and emu-pie tasting. The emu oil is branded Marayong Park and is sold in Coles.
- Open: 9am–5pm Fri–Mon, daily school and public holidays
- Location: Turn left towards Huskisson and Jervis Bay ten minutes south of Nowra; shop is about 1.5km from the highway

GERRINGONG
🍴 DINING

GERRINGONG GOURMET DELI
John and Somprom Svinos
133 Fern St, Gerringong 2534
Phone: 02 4234 1035
BYO
Cuisine: Casual modern Australian
Mains average: $
Specialty: Breakfasts—ricotta pancakes in butterscotch sauce
Local flavours: South Coast cheeses, honey, blueberries
Extras: Also a deli-store with gourmet and grocery lines. Cafe has outdoor dining area.
- Open: Breakfast and lunch daily

GERROA
🍴 DINING

SEAHAVEN CAFE AND GENERAL STORE
Dave Thomson
19 Riverleigh Ave, Gerroa 2534
Phone: 02 4234 3796
Licensed
Cuisine: Modern Australian
Mains average: $$$

Specialty: Lunch—fish and chips, dinner—creative, modern à la carte
Local flavours: Bread, eggs, wines
Extras: Outdoor area with views of Seven Mile Beach.
- **Open:** Breakfast and lunch Wed–Mon, dinner Fri–Sat

GREENWELL POINT
🏠 STORES

DJ'S FISH AND CHIPS
Monica and Jawad Yari
114 Greenwell Point Rd,
Greenwell Point 2540
Phone: 02 4447 1332
These seafood specialists buy the fish straight off the boat, and also have prawns, oysters and fish and chips.
- **Open:** 8.30am–6.30pm Sun–Thurs, 8.30am–7pm Fri–Sat

HUSKISSON
🍴 DINING

THE GUNYAH RESTAURANT AT PAPERBARK CAMP
Jeremy and Irena Hutchings (owners),
Joanne McRae (chef)
Paperbark Camp,
571 Woollamia Rd, Huskisson 2450
Phone: 02 4441 6066
Email: info@paperbarkcamp.com.au
www.paperbarkcamp.com.au
Licensed
Three-course meal: $$$
Specialty: Seafood straight off the boats at Ulladulla, Jervis Bay mussels, Greenwell Point rock oysters.
Local flavours: Mussels, oysters, seafood, locally grown produce where possible.
Extras: Unique setting in a treetop dining room with verandas overlooking the paperbark and spotted gum bushland, or cosy by the open fire in winter. Enjoy a visit from inquisitive possums, and watch for resident sugar gliders. Picnic lunches for guests.
- **Open:** Dinner Tues–Sat

SEAGRASS CAFE
Nathan Fay and Kerrin McKnight
13 Currambene St, Huskisson 2540
Phone: 02 4441 6124
www.seagrass.net.au
Licensed
Cuisine: Modern Australian
Mains average: $$$
Specialty: Fresh flavours with a Thai twist
Local flavours: Seafood, wines
Extras: Outdoor and indoor areas. Grow their own kaffir limes.
- **Open:** Lunch Fri–Sun, dinner Tues–Sun

JASPERS BRUSH
🍴 DINING

SILOS ESTATE WINERY AND RESTAURANT
Alenka and Andrew Knevitt
B640 Princes Hwy, Jaspers Brush 2535
Phone: 02 4448 6160
Email: silos@ozemail.com.au
Licensed/BYO
Cuisine: Seasonal modern Australian
Mains average: $$$
Local flavours: Oysters, homemade breads and Silos Estate wines.
Extras: Stunning views of the vineyard across to the mountains beyond. Outdoor veranda dining.
- **Open:** Lunch Wed–Sun, dinner daily
- **Location:** 6.4km south of Berry

NSW—SOUTH COAST AND ILLAWARRA

KIAMA
🍴 DINING

ESSE RESTAURANT
Simon Everett
Shop 1, 55 Collins St, Kiama 2533
Phone: 02 4232 2811
Licensed/BYO
Cuisine: Seasonal modern Australian
Mains average: $$$
Specialty: Mustard crusted spatchcock with dirty rice and suicide sauce (yes, really!)
Extras: Range of homemade dressings and pestos are for sale. Takeaway available.
- Open: Breakfast and lunch daily, dinner Thurs–Mon

MERIMBULA
🍴 DINING

COME 'N' GET IT RESTAURANT
Christine and Bob Fleming
86 Merimbula Dr, Merimbula, 2548
Phone: 02 6495 1377
BYO
Cuisine: Modern Australian
Mains average: $$
Specialty: Glazed roast duck, tempura prawns with lime and chilli glaze
Local flavours: Seafood—oysters, prawns, bugs, fish
- Open: Dinner Tues–Sat

MOGO
🍴 DINING

WILLOW CAFE
Christine Westrup and Antonio Honorato
Princes Hwy, Mogo 2536
Phone: 02 4474 5445
Email: info@willowcafe.com.au
www.willowcafe.com.au
BYO
Cuisine: Modern international
Mains average: $$
Specialty: Shredded duck with soba noodle salad, fig and honey scones, 'best coffee on south coast'
Local flavours: Narooma smoked warehou fish, Mt Dromedary goat curd, Araluen peaches, organic strawberries, tangelos and rockmelons
Extras: Garden courtyard, with seating under the willows.
- Open: 10am–4pm Wed–Mon, daily during holidays

MOLLYMOOK
🍴 DINING

BANNISTERS RESTAURANT
Andrew Phelan (chef)
Bannisters Point Lodge,
191 Mitchell Parade, Mollymook 2539
Phone: 02 4455 3044
Email: info@bannisterspointlodge.com.au
www.bannisterspointlodge.com.au
Licensed
Cuisine: Modern Australian
Mains average: $$$
Specialty: Southern Thai seafood curry, desserts
Local flavours: Seafood, vegetables
Extras: Every seat in the restaurant has a view of the bushland and Pacific Ocean. Recently awarded NSW Restaurant and Catering Association, 2003, Best Restaurant South Coast Region.
- Open: Lunch weekends, dinner Tues–Sat

MOUNT KEMBLA

🍎 PRODUCERS

REES ORCHARD
Bruce Rees
617 Cordeaux Rd, Cordeaux River,
Mount Kembla 2526
Phone: 02 4271 2047
This third-generation family farm, established in 1886, has been producing tree-fresh fruit for over a century. Located at an elevation of 400 metres the climate is ideal for many different varieties of apples, peaches, pears, plums and persimmons, including some very old varieties, such as twenty-ounce Wine Saps, and Portuguese apples, brought from Wales when the farm began. Pick your own if you like.
- **Open:** 8am–5pm daily

PAMBULA

🍴 DINING

WHEELER'S OYSTERS
Hugh Wheeler
Arthur Kaine Dr, Pambula 2549
Phone: 02 6495 6089 (shop),
02 6495 6330 (factory)
Email: hugh@asitis.net.au
www.wheelersoysters.com.au
Licensed/BYO
Cuisine: Modern Australian seafood
Mains average: $$
Specialty: Oysters
Extras: Sales of the Australian indigenous oyster, the Sydney rock oyster and tours of the factory at 11am (Mon–Sat). Oyster-opening demonstrations using the traditional oyster knife and tastings of cooked and natural oysters. Cooked seafood to take away, but no fish and chips. View over the golf course, inlet and lake from the outdoor dining area.
- **Open:** Lunch daily, dinner Tues–Sat, factory open 7am–5pm daily, shop open 9am–5pm weekdays, 10am–5pm weekends

QUAAMA

🍷 BREWERY/WINERY

FRUIT BALLAD COUNTRY WINES
Jane MacGregor
11 Verona Rd, Quaama 2550
Phone: 02 6493 8382
Email: jane@fruitballad.com.au
www.fruitballad.com.au
The boutique wines here include peach, plum, apricot, apple, pear and berry, as well as unusual ones such as pumpkin. The wines have around the same alcohol content as table wines, and the liqueurs range from 16 to 18 per cent. The rose petal and honey dessert wine is very popular, and so romantic. Just like a bunch of flowers in a bottle, Fruit Ballad reckons.
- **Open:** 10am–5pm Wed–Sun and public holidays
- **Location:** Cnr Princes Hwy and Verona Rd

SUSSEX INLET

🍎 PRODUCERS

KOORAWATHA HYDROPONICS
Gary and Lorraine Muirhead
392 Sussex Inlet Rd, Sussex Inlet 2540
Phone: 02 4441 0282
Email: muirhead@shoal.net.au
www.shoal.net.au/~koorawatha
Although much of the crop goes to market, you should be able to find

Lebanese cucumbers and hydroponic tomatoes when they're in season. The gourmet products such as pasta sauces and green tomato relish are now being made by a commercial kitchen to Koorawatha recipes. Fortunately the bed and breakfast cabin, a romantic couples retreat, survived the 2002 bushfires.
■ **Open:** 8.30am–4.30pm Mon and Wed-Sat

TANJA
🍎 PRODUCERS
TANJA OLIVE OIL
Alan Watt
1426 Bermagui Rd, Tanja 2550
Phone: 02 6494 0015
Email: tanjaolives@bigpond.com
Just imagine spending a day during olive harvest, picking olives that will be pressed into the freshest olive oil you will ever taste, while you dine on a sensational lunch. Then as you leave, you also take some fresh table olives to cure at home. These public days during the harvest are quite rare, but worth attending, and you must book, as numbers are strictly limited. Alan Watt has planted 3000 trees, the oldest now seven years old, on his property and has been producing olives for several years, with his own press imported from Italy. During harvest season (mid April to early June) Tanja offers pick-your-own olives for home pickling, complete with directions, on any day with prior arrangement. Apart from the formal, catered lunches Tanja now offers Pick 'n' Pizza days (using the wood-fired oven) and Pick 'n' Picnic days (self-catered picnics) with barbecues. The full day of harvesting and pressing oil remains the same for all categories. A brochure containing all details is available on request.
■ **Open:** Selected days during harvest (mid April to end May) by appointment
● **Location:** Between Bermagui and Tathra

TATHRA
🍎 PRODUCERS
BEGA DRIED FOODS
Brian Fell
5 Beach St, Tathra 2550
Phone: 02 6494 1995
Email: brian@begadriedfoods.com.au
www.begadriedfoods.com.au
Naturally dried sulphur-free strawberries, mangoes, apples, bananas, fruit salad, pineapples, kiwifruit, pears and tomatoes. These are just some of the unusual items created here. It's mostly local fruit and they are available at health food shops, gourmet delis, or by mail order.
■ **Open:** 9am–5pm weekdays

TILBA TILBA
🍴 DINING
LOVE AT FIRST BITE
James Heaslip
Corkhill Dr (Foxglove Spires Garden), Tilba Tilba 2546
Phone: 02 4473 7055
Email: jamegs@bigpond.com
BYO
Cuisine: Modern Australian
Mains average: $
Specialty: Thai chicken laksa
Local flavours: As much as possible—eggs, tomato, fruit and vegetables, smoked trout
Extras: Outdoor seating and an award-

SOUTH COAST AND ILLAWARRA—NSW

winning garden. Can arrange for lunch in the garden (entrance fee applies) and takeaway picnic hampers.
- **Open:** 9am–4pm daily, shorter hours Tues–Sun during winter but daily during July school holidays

TUROSS HEAD
DINING

PLOYS AT TUROSS
Barbara and Guenther Ploy
2 Trafalgar Rd, Tuross Head 2537
Phone: 02 4473 6143
Email: ploy@ispdr.net.au
BYO
Cuisine: Modern Australian
Mains average: $$
Specialty: Oven-roasted Chinese-style duck wrapped in coriander and spring onion crepes
Local flavours: Beef fillet, vegetables
Extras: Deck has lake and ocean views (as well as ocean views from inside restaurant). Winner Restaurant and Catering Association Awards Best New Restaurant South Coast, 2003
- **Open:** Lunch weekends, dinner Wed–Sat (also Sun in peak season)

ULLADULLA
PRODUCERS

CLYDE RIVER BERRY FARM
Robyn and Ron Russell
River Rd, South Brooman,
via Ulladulla 2539
Phone: 02 4478 1057
Email: bluberry@ispdr.net.au
In 1979, when the Russells planted blueberries, they were the first growers in New South Wales to do so and they became trendsetters. Today they grow twenty high-bush varieties, and dozens of other orchards around the state now grow these delicious berries too. You may pick your own in season, or purchase pre-picked peaches, nectarines, plums, blueberries, raspberries, loganberries, silvanberries, boysenberries, marionberries, lawtonberries and blackberries. Jams and honey are also available.
- **Open:** 10am–7pm daily (Dec–Jan)
- **Location:** Midway between Ulladulla and Batemans Bay, at the end of Sheeptrack Rd

ULLADULLA OYSTERS
Clayton and Leon Harrington
Unit 6, 1–3 Coller Rd, Ulladulla 2539
Phone: 02 4455 5521
Email: ulloysters@shoal.net.au
From the clean, nutrient-rich water of Lake Conjola, surrounded by national park, come these pristine Sydney Rock oysters. They won gold and silver medals at the Sydney Royal Easter Show in 2002, and bronze in 2003. Customers can visit the processing shed and see the different stages of production and taste the oysters, and a range of premium smoked seafood.
- **Open:** 8am–5pm Mon–Sat (daily during Jan)
- **Location:** Follow signs from the Princes Hwy and Kings Pt Dr into the Blackburn Industrial Estate

STORES

FISHERMEN'S WHARF SEAFOOD
John and Josie Young
Ulladulla Wharf, Wason St, Ulladulla 2539
Phone: 02 4455 3906

121

NSW—SOUTH COAST AND ILLAWARRA

Boat-fresh fish as well as crabs and prawns are available at this coastal spot. You can also get takeaway fish and chips at the shop, which is open longer hours in summer.
- **Open:** 8am to 5.30pm Mon–Sat, 8am–5pm Sun

DINING

HARBOURSIDE RESTAURANT
Rob and Lee Manning
Shop 5, 84 Princes Hwy, Ulladulla 2539
Phone: 02 4455 3377
Email: lmanning@shoalhaven.net.au
Licensed/BYO wine
Cuisine: Contemporary Australian
Mains average: $$
Specialty: Fresh, local seafood and beef
Local flavours: Fish, oysters, mussels, prawns, beef, chicken, lamb and whatever else is locally available.
Extras: Waterfront location with dining over two floors, and alfresco terrace. Wine list features South Coast wines.
- **Open:** Lunch and dinner daily

MILLARDS COTTAGE
John and Gwenda Cross
82 Princes Hwy, Ulladulla 2539
Phone: 02 4455 3287
Email: harbour@shoalhaven.net.au
Licensed/BYO wine
Cuisine: Modern Australian
Mains average: $$$
Specialty: Lobster—mornay or natural
Local flavours: Fish, lobster, some local wines, fruit, vegetables, meat
Extras: This restaurant is in Ulladulla's oldest established building, built in 1868. Features views of the harbour and open fires in winter.
- **Open:** Dinner Mon–Sat

TORY'S SEAFOOD RESTAURANT
Joe Puglisi and Lee Layton
30 Wason St, Ulladulla 2539
Phone: 02 4454 0888
Email: torys.seafood@bigpond.com.au
Licensed
Cuisine: Seafood with Italian influence
Mains average: $$
Specialty: King Neptune's platter
Local flavours: Fish and seafood
Extras: Deck overlooking Ulladulla Harbour. Cater for non-seafood eaters, vegetarians and those with special dietary needs. Wheelchair access. Local seafood from a pioneer fishing family's own trawler. A licensed deck below the restaurant is ideal for enjoying seafood with a glass of wine. Takeaway available.
- **Open:** Dinner daily

WALLAGA LAKE

PRODUCERS

HARRY BLACK'S ORCHARD
Harry Black
121 Fairhaven Point Way,
Wallaga Lake 2546
Phone: 02 6493 4548
Email: harryblack@acr.net.au
This amazing orchard grows around sixty types of fruits, twenty of them tropical. The position—north-facing between the ocean, lake and forest—creates a micro-climate that is perfect for raising paw-paws, babaco, white and black sapote, wax jambus, bananas, custard apples, avocados and guavas. There are jaboticabas, wampi (like a lychee) and grummichama, which he says resembles a large cherry, and the four varieties of bananas are doing very well. Wallabies, kangaroos and birds come down to the

car park. There are also one-hour conducted tours for groups. Fee applies.
- **Open:** 9am-4pm daily, by appointment

WANDANDIAN
🍎 PRODUCERS

MARTIN'S ORCHARD
Lorraine and Alwyn Martin
Jerrawangla, Princes Hwy,
Wandandian 2540
Phone: 02 4443 4280
Email: martins@shoalhaven.net.au

This is a good place to break the journey down the highway so you can stock up on peaches and nectarines from trees which the Martins planted thirty years ago, sold direct to the public from the packing shed and at the gate. They have another outlet in town too, in a converted garage on the Princes Hwy.
- **Open:** 8am-6pm Sun-Fri (mid Nov-early Feb)
- **Location:** Five minutes south of Wandandian, 4km south of the Sussex Inlet turn on the highway

RIVERHOUSE HERB FARM
Elke Roland
D2825A Princes Hwy, Wandandian 2540
Phone: 02 4443 4922
Email: elke@riverhouse.com.au
www.riverhouse.com.au

Herbs—culinary, potted and medicinal herbs, 'what grandmother used to grow'—they're all here at this riverside garden. More than twenty years ago, Elke Roland went to Dural and bought some parsley, some chives and a book. Today she has a diploma as a herbalist and this wonderful garden that sells herbs on the Internet, and from the garden.

- **Open:** 9am-5pm Fri-Sun, other days by appointment

WOLLONGONG
🍴 DINING

LORENZO'S DINER
Lorenzo Pagnan
119 Keira St, Wollongong 2500
Phone: 02 4229 5633
Licensed
Cuisine: Modern Australian
Mains average: $$$
Local flavours: Duck, chicken, venison from the Southern Highlands, local beef and veal
Extras: Relaxed ambience and attractive decor with artworks by Pagnan's partner Rebecca Wilford.
- **Open:** Lunch Fri, dinner Tues-Sat

🍷 BREWERY/WINERY

THE FIVE ISLANDS BREWING COMPANY
Michael and Allison Bolt
Eastern Terrace, WIN Entertainment Centre,
Crown St, Wollongong 2500
Phone: 02 4220 2854
Email: fiveislandsbrew@1earth.net
www.fiveislandsbrewery.com
Licensed
Cuisine: Modern Australian
Mains average: $$
Specialty: Asian sampler plate
Local flavours: Food made with own beers
Extras: A huge terrace fifty metres from the sand at this, the largest microbrewery in Australia, which produces ten tap beers.
- **Open:** Lunch daily, dinner Mon-Sat

Darwin

2

Katherine

NORTHERN

TERRITORY

1

Regions:
1 Central Australia
2 Top End

Alice Springs

0　　　　　　　　　　　　　　500
KILOMETRES

WESTERN AUSTRALIA

QUEENSLAND

SOUTH AUSTRALIA

NORTHERN TERRITORY

Comprising a sixth of the country's landmass, this northern outpost stands distant and distinct. Isolation and climate set it apart, after all, Darwin is much nearer to Indonesia than to other Australian capitals, and the city's population reflects a wide range of ethnicity. This flows on, of course, into the foods that are grown and sold here, the hawker-stall ambience of night markets, and the languid air of the place.

Yet the Northern Territory extends almost 1800 kilometres south from Darwin to the South Australian border, one of the world's most inhospitable landscapes, dotted with deserts and dingos, mountains and monoliths, but redeemed by amazing sunsets and night skies ablaze with stars.

This is not a land to be trifled with. Here beef-cattle stations spread for hundreds of square kilometres, and you can play station-hand on some, or dine under more stars than you thought existed at others. There are also bush-tucker tours and camel excursions. You'll bump into eccentric escapees from society, and inevitably encounter the unique 'pull of the Territory' that compels people to return.

Yet you could never call this place barren. Here fruit and vegetables grow and ripen almost overnight. Its bananas, mangoes, oranges, vegetables and herbs are prized by other states 'out of season' and specialty meats include the local wildlife—crocodile, buffalo, camel and venison. Local seafood such as tiger prawns, barramundi, mud crabs and dozens of varieties of fish feature on menus both in Darwin and throughout the country.

WAUCHOPE ICE-CREAM VAN
Patty O'Neill
Stuart Highway, Wauchope
Wacky or what? One hour south of Tennant Creek you'll find the world's most remote ice-cream van. Halfway between Tennant Creek and the Alice.

■ **Open:** Daily

CENTRAL AUSTRALIA

The arid heart of Australia is surprising for the depth and variety of its produce. Native meats and dates and other plants that suit the hot, dry climate thrive here and feature on outback menus.

ALICE SPRINGS

🍎 PRODUCERS

ARIDGOLD FARM
Tim Micklem
Alice Springs 0871
Phone: 08 8956 0926
Email: aridgold@bigpond.com
Date palms have been in Central Australia since the time of the Afghan cameleers who introduced them, no doubt as they spat the pips from the dried dates they carried as food. Dates need plenty of water as well as high temperatures, and Micklem lists over a dozen local date varieties, including one called Aridgold. Tim Micklem planted his first trees twenty-two years ago in 1982 but they mature slowly and the industry itself is still maturing. The hot, dry inland climate of Alice Springs is ideal for growing dates and figs. Figs come on early and fill a gap in the southern markets but are also sold locally. Look for Central Australian Fruits at health food shops Australia-wide, and in supermarkets in the Territory and South Australia.
■ **Open:** By appointment

THE DATE GARDENS
Heather and Mark Bunting
Palm Circuit, Alice Springs 0871
Phone: 08 8953 7558
Email: www.ozemail.com.au/~australiana
These date gardens began in the 1950s with seedlings augmented by shoots from Californian date palms, although the originals were possibly planted by Afghan cameleers over one hundred years ago. The plantation's thirteen varieties of dates are sold locally, and the farm shop also sells date scones, muffins, slices and cakes, as well as sweet tooths' delights—rum and date, chocolate and date, caramel and date and plain date ice creams. Then there are date jams and spreads, fig and date jam, stuffed dates and a small range of chocolate-coated dates. Or you could just settle down for a unique 'Date-vonshire Tea'. There is also a fauna park, licensed restaurant, caravan park, swimming pool and motel units.
■ **Open:** 9am–6pm daily, functions and group bookings at other times by arrangement
● **Location:** South of the Gap, about 5km from Alice Springs

🍴 DINING

BARK HUT TOURISM CENTRE
Tom, Kyron and Leit Starr
Arnhem Hwy, Anaburroo 0822
Phone: 08 8978 8988
Email: barkhutinn@bigpond.com
Licensed
Cuisine: Australian
Mains average: $$

Specialty: Grilled barramundi
Local flavours: Barramundi, buffalo and kangaroo
Extras: Sells locally made beef, crocodile and buffalo jerky. Takeaway available. Animal enclosure with wallabies, wallaroos, emus, and buffalo at certain times of the year. Outdoor area for dining and drinking, caravan park and cabins with a view of Mary River National Park, swimming pool and historic bar. Mary River is home to the famous barramundi fishing spot, Shady Camp ('Don't be fooled', they tell us, 'there's bugger-all shade'), and also boasts the highest concentration of saltwater crocodiles in the world.
■ **Open:** Breakfast, lunch and dinner daily
● **Location:** Between Darwin and Jabiru. About one-and-a-half hours from Darwin, cross the Mary River, then go about 4km further, the Centre is on the left. From Kakadu, it's on the right, 38km past the northern exit.

HANUMAN RESTAURANT
ALICE SPRINGS
Jimmy Shu (owner), Kritkamon Pethsri (chef)
Crowne Plaza Resort,
82 Barrett Dr, Alice Springs 0870
Phone: 08 8953 7188
Email: galangal@bigpond.com
www.hanuman.com
Licensed
Cuisine: Thai-Indian
Mains average: $$
Specialty: Hanuman oysters, Thai-fried whole fish, red duck curry
Local flavours: Thai basil, lettuce
Extras: Exquisitely decorated with Thai artefacts, dining area spills out into the lobby/atrium of the hotel.
■ **Open:** Lunch weekdays, dinner Mon-Sat

OVERLANDER'S STEAKHOUSE
Wayne 'Krafty' Kraft
72 Hartley St, Alice Springs 0870
Phone: 08 8952 2159
Email: sth@overlanders.com.au
www.overlanders.com.au
Licensed
Cuisine: Australian
Mains average: $$
Specialty: Five-course Drover's Blowout
Local flavours: Kangaroo, emu, camel, barramundi, crocodile
Extras: Entertainment during peak tourist seasons.
■ **Open:** Dinner daily

RED OCHRE GRILL
Paula Gordon
The Territory Inn, Todd Mall,
Alice Springs 0870
Phone: 08 8952 9614
Email: asp@auroraresorts.com.au
www.auroraresorts.com.au
Licensed
Cuisine: Australian
Mains average: $$
Specialty: Kangaroo, camel
Local flavours: Kangaroo, camel, bush foods—wattle seed, saltbush
Extras: Courtyard dining.
■ **Open:** Breakfast, lunch and dinner daily

JABIRU
DINING

ESCARPMENT RESTAURANT
Geoff Naumann (Manager)
Gagudju Crocodile Holiday Inn,
Flinders St, Jabiru 0886
Phone: 08 8979 9000
Email: hotel@crocodileholidayinn.com.au
www.gagudju-dreaming.com

NT—CENTRAL AUSTRALIA

Licensed
Cuisine: Australian Indigenous
Mains average: $$$
Specialty: Barramundi, kangaroo
Local flavours: Barramundi, kangaroo, buffalo
Extras: Takeaway pizza and snacks are available. The chef makes relishes for sale from bush fruits that are handpicked by local Aborigines. Set in Kakadu National Park, there is plenty of natural sunlight during the day, and dining happens around the oasis-like pool or guests use barbecues in the courtyard. The hotel is shaped like a crocodile with banquet and conference facilities.
- **Open:** Breakfast, lunch (all-day snacks) and dinner daily
- **Location:** Approximately two-and-a-half hours drive along Arnhem Hwy from Kakadu turn-off

KATHERINE
TOURS

MANYALLALUK—THE DREAMING PLACE
Glenn Dimer
Manyallaluk Tours, PMB 134, Katherine 0852
Phone: 08 8975 4727
Email: manyallaluk@bigpond.com
www.manyallaluk.com
These tours have been organised for the past sixteen years by the local Aboriginal community. Visitors spend the day and get to eat green ants and bush honey on a bush-tucker walk or a bush-medicine tour.
- **Tours:** Mon–Sat according to demand
- **Location:** 110km from Katherine

RENNER SPRINGS
DINING

RENNER SPRINGS DESERT HOTEL MOTEL
John and Judi Blackman
Stuart Hwy (CMB 7), Renner Springs 0862
Phone: 08 8964 4505
Email: rennersprings@bigpond.com
www.rennersprings.com
Licensed
Cuisine: Country style
Mains average: $$
Specialty: Home-baked bread, hamburgers with 'the lot' (very large), corned beef sandwiches on home-baked bread (this has been called the 600-mile sandwich, a massive steak sandwich, made with 'the works' to keep the hungriest traveller satisfied until Darwin)
Local flavours: Beef, lamb, pork, chicken, fish
Extras: The bar area has a welcoming country feel. The covered veranda for dining doubles as a games area, there is a stage for bands and other events.
- **Open:** Breakfast, lunch and dinner daily
- **Location:** 160km north of Tennant Creek and about 820km south of Darwin

TENNANT CREEK
DINING

THE JAJJIKARI CAFE
Mandy Gordon
Nyinkka Nyunyu Art and Cultural Centre, Paterson St, Tennant Creek 0867
Phone: 08 8962 2221
Email: info@nyinkkanyunyu.com.au
www.nyinkkanyunyu.com.au
Licensed

CENTRAL AUSTRALIA—NT

Cuisine: Cafe-style health food
Mains average: $
Specialty: Bush-potato cakes, hamburger with bush spices
Local flavours: Bush tucker, beef, kangaroo, barramundi
Extras: Fully airconditioned restaurant or the option of outdoor seating. Located in a cultural centre with one of the country's best displays of Indigenous culture. Bush-tucker tours may be booked here.
■ Open: 9am–5pm Mon–Sat, 10am–2pm Sun

ULURU–KATA TJUTA NATIONAL PARK
DINING

BOUGH HOUSE RESTAURANT
Ayers Rock Resort, Yulara Dr, Uluru-Kata Tjuta National Park 0872
Phone: 08 8957 7605
Email: info@voyages.com.au
www.voyages.com.au
Licensed
Cuisine: Australian
Buffet: $$$$
Local flavours: Australian game meats—kangaroo, emu, camel, crocodile
■ Open: Breakfast and dinner daily

KUNIYA RESTAURANT
Sails in the Desert Hotel, Ayers Rock Resort, Yulara Dr, Uluru-Kata Tjuta National Park 0872
Phone: 08 8957 7888
Email: info@voyages.com.au
www.voyages.com.au
Licensed
Cuisine: Modern Australian cuisine enhanced with native ingredients

Mains average: $$$$
Specialty: Wagyu beef from South Australia, barramundi from the Northern Territory, crayfish
Local flavours: Bush tomatoes
Extras: The menu features one entree and one main course from each state of Australia, highlighting the regional produce of that state. In addition to traditional dinner seating, there is booth-style seating overlooking the pool. Wonderful at night, with floodlit gum trees.
■ Open: Dinner daily

OUTBACK PIONEER BBQ AND BAR
The Outback Pioneer Hotel, Ayers Rock Resort, Yulara Dr, Uluru-Kata Tjuta National Park 0872
Phone: 08 8957 7600
Email: info@voyages.com.au
www.voyages.com.au
Licensed
Cuisine: Self-cook barbecue
Mains average: $$
Local flavours: Australian game meats—emu sausages and burgers, crocodile skewers, kangaroo, barramundi
Extras: Entertainment and the outdoor-dining experience.
■ Open: Dinner daily
● Location: 15km from the national park's gate

SOUNDS OF SILENCE
Sails in the Desert Hotel, Ayers Rock Resort, Yulara Dr, Uluru-Kata Tjuta National Park 0872
Phone: 08 8957 7888
Email: info@voyages.com.au
www.voyages.com.au
Licensed
Cuisine: Australian
Cost: $$$$ (includes drinks and tour)

Specialty: Outdoor fine dining followed by a 'guided tour of the heavens' with the resident astronomer
Local flavours: Kangaroo, barramundi, emu, crocodile, chicken, homemade chutneys, bush salads
Extras: Before dinner, enjoy canapés and drinks on a private dune-top while watching the sunset over Uluru and Kata Tjuta and listening to the haunting sounds of the didgeridoo and the desert at night. Sounds of Silence received the 1997, 1998 and 1999 Australian Tourism Award for 'Best Tourism Restaurant', which resulted in this exceptional dining experience being entered into the Tourism Hall of Fame.
■ **Open:** Dinner daily

WHITE GUMS RESTAURANT AND ARNGULI FLAME GRILL
Desert Gardens Hotel, Ayers Rock Resort, Yulara Dr, Uluru-Kata Tjuta National Park 0872
Phone: 08 8957 7888
Email: info@voyages.com.au
www.voyages.com.au
Licensed
Cuisine: Modern Australian flame-grill with some bush-food flavours and international influence
Mains average: $$$
Specialty: Vegetarian fare and Top End favourites such as barramundi
Local flavours: Bush tomatoes, wild limes, rosellas, native mint, Davidson plums
Extras: Great sunrise view of Uluru
■ **Open:** Breakfast, lunch and dinner daily

WINKIKU
Sails In The Desert Hotel, Ayers Rock Resort, Yulara Dr, Uluru-Kata Tjuta National Park 0872

Phone: 08 8957 7424
Email: .info@voyages.com.au
www.voyages.com.au.
Licensed
Cuisine: Modern Australian buffet
Mains average: $$$$
Specialty: Largest prawns in the Northern Territory, sushi and sashimi from Japanese master chef
Local flavours: Kangaroo, native herbs, Indigenous chutneys
Extras: Winkiku is Ayers Rock Resort's five-star buffet restaurant, with views over the pool and gardens. There are also private tables for more intimate dining.
■ **Open:** Breakfast and dinner daily, lunch daily for groups by appointments only

WYCLIFFE WELL
🍴 DINING

WYCLIFFE WELL HOLIDAY PARK
Lew Farkas
Stuart Hwy, Wycliffe Well 0862
Phone: 08 8964 1966
Email: lew@wycliffe.com.au
www.wycliffe.com.au
Licensed
Cuisine: Meat and seafood
Mains average: $
Specialty: Chilli or garlic local red-claw crayfish
Extras: This 300-seat restaurant has great views of the starry outback sky and the place is world-renowned as a UFO hot spot as sightings are very common here. Ten acres of lakes produce crabs for the menu and there is the largest range of beers in Australia—300 different labels when fully stocked.
■ **Open:** Breakfast, lunch and dinner daily
● **Location:** 380km north of Alice Springs

TOP END

Many people have been accustomed to thinking of the Northern Territory as primarily desert. These people will soon have to readjust their ideas, as now, the desert is starting to blossom.

Agriculture is becoming a major part of the Territory's economy with the discovery that the Top End can grow almost anything it puts its mind to. Its proximity to South-East Asia is making the area a tempting 'front garden' that is now supplying fresh fruit, such as melons, on a regular basis to several countries. In addition, mangoes and bananas are thriving as well as a large citrus industry.

One great advantage to primary industry in the Northern Territory is the relatively low incidence of plant pests and disease, allowing many things to be grown without chemicals. Excellent dry season conditions allow vegetables and herbs to mature early, coming to the southern markets weeks ahead of other regions.

Yet the oldest primary industry, grazing, is still intact here too. Darwin has a thriving live-cattle export trade to Asian ports, as well as operating export-standard abattoirs and meat-processing works. Buffalo, camel, crocodile and venison are also raised and slaughtered in the Territory.

Abundant seafood adds a final note to the menu in this exciting region. Tiger prawns, barramundi, mud crabs and golden snapper are major items here, but there are dozens of other local fish as well, ideal for export or to supply the local market.

Because of the relatively small population in relation to the wealth of food stuffs produced here, the Northern Territory is ideally placed to contribute to the shortfall in other states, or to export to neighbouring countries.

With space and ideal growing conditions, the Territory may well continue to become a major player in Australia's rural economy.

COBOURG PENINSULA
DINING/TOURS

SEVEN SPIRIT BAY WILDERNESS LODGE
Cobourg Peninsula Wildlife Sanctuary,
Garig Gunak Barlu National Park,
Arnhem Land
Phone: 08 8979 0281
Email: sales@sevenspiritbay.com
www.sevenspiritbay.com
Licensed
Cuisine: Seafood and Cobourg cuisine
Mains average: All meals included in daily tariff
Specialty: Barramundi
Local flavours: Seafood, meat, fruit and vegetables
Extras: Exceptional views over Coral Bay, alfresco dining, tariff includes scenic air transfer from Darwin.
■ **Open:** Breakfast, lunch and dinner daily for guests
● **Location:** 185km north-east of Darwin in Arnhem Land

DARWIN

MARKETS

MINDIL BEACH MARKETS
Tim Robinson
Mindil Beach, Maria Livaris Dr, Darwin 0870
Phone: 08 8981 3454 or 08 8981 3327
Mobile: 0416 278 883
Email: manager@mindilbeachsunset markets.com.au
The market offers a choice of over fifty-five food and drink stalls from thirty or more different countries, plenty of tropical fruits and vegetables, herbs, snacks, Arafura prawns, grilled barramundi, crocodile, buffalo, green pawpaw salads. Winner of a number of tourism awards.
- **Open:** Thurs and Sun evenings during dry season (last Thurs in April–last Thurs in Oct)

DINING

PEE WEE'S AT THE POINT
Simon Matthews and James Wakefield
Alec Fong Lim Dr, East Point Nature Reserve, Darwin 0801
Phone: 08 8981 6868
Email: simrob1@bigpond.com.au
www.peewees.com.au
Licensed
Cuisine: Modern Australian
Mains average: $$$
Specialty: Pee Wee's Taste Plate—a gourmet tour of the chef's entrees
Local flavours: Rosella flowers, carambolas, ocean-caught barramundi and much more—all produce is sourced from local suppliers
Extras: The menu aims to reflect the climate, produce and multiculturalism of the Northern Territory. Unsurpassed views across Fannie Bay to the centre of Darwin. Airconditioned dining inside during the wet season and views of tropical storms over Darwin. During the dry season there is cool, beachfront dining amid the wildlife of the nature reserve. A combination of award-winning, first-class food and service and spectacular city views. (Winner of the Gold Plate Award in 2002 and 2003, and Tucker Seabrook Wine List Award for the NT in 2002).
- **Open:** Dinner daily

CRUSTACEANS ON THE WHARF
Vicki Jenkins
Stokes Hill Wharf, Palmerston, 0831
Phone: 08 8981 8658
Email: crustys@bigpond.com
www.crustys.com.au
Licensed
Cuisine: Seafood
Mains average: $$$
Specialty: Local mud crab
Local flavours: As much as possible
Extras: From May to the end of October, the dining area over the water is open air. From November to April, dining is upstairs and airconditioned, with views over the harbour.
- **Open:** Dinner Mon–Sat

THE HANUMAN RESTAURANT
Jimmy Shu
28 Mitchell St, Darwin 0800
Phone: 08 8941 3500
Email: galangal@bigpond.com
www.hanuman.com.au
Licensed
Cuisine: Thai–Indian
Mains average: $$
Specialty: Hanuman oysters, crispy Thai fried fish

Local flavours: Seafood, Thai herbs
Extras: Alfresco dining area.
- **Open:** Lunch weekdays, dinner daily
- **Location:** Next door to Crowne Plaza Darwin

PALMERSTON
🍷 BREWERY/WINERY

KAKADU MANGO WINERY
Peter Donnolley and Donnolley family
4 Adams Rd, Palmerston 0830
Phone: 08 8931 1166
Mobile: 0421 421 141
Email: kakaduwinery@bigpond.com
If you are lucky enough to have too many mangoes (is that possible?) why not do something really wonderful, like making mango wine? There are several types of wine made with local mangoes at this tropical winery—Xtra dry and demi sec, mango and lime, sweet mango, mango pash, and Kakadu coffee, which is a liqueur blended with mango wine.
- **Open:** 10am–5.30pm daily by appointment (dry season), from 12pm the rest of the year
- **Location:** In the Yarrawonga light business area

was started by the current owner's parents, who were chefs.
- **Open:** 8am–7pm weekdays, 8am–6.30pm Sat, 9am–1pm Sun
- **Location:** Next to ANZ Bank

MARKETS AND EVENTS

✦ **ROYAL DARWIN SHOW,** DARWIN
08 8984 3091
Last weekend in July

✦ **DARWIN SEAFOOD FESTIVAL**
08 8981 4737
End June

✦ **MINDIL BEACH MARKETS,** DARWIN
08 8981 3454
Thurs and Sun evenings during dry season (last Thurs in Apr–last Thurs in Oct)

PARAP
🏠 STORES

PARAP FINE FOODS
Neville Pantazis and Paula Dinoris
40 Parap Rd, Parap 0820
Phone: 08 8981 8597
Email: parapfinefoods@bigpond.com.au
Since 1968, Parap Fine Foods has been specialising in regional food and wine products for all cultures. The business

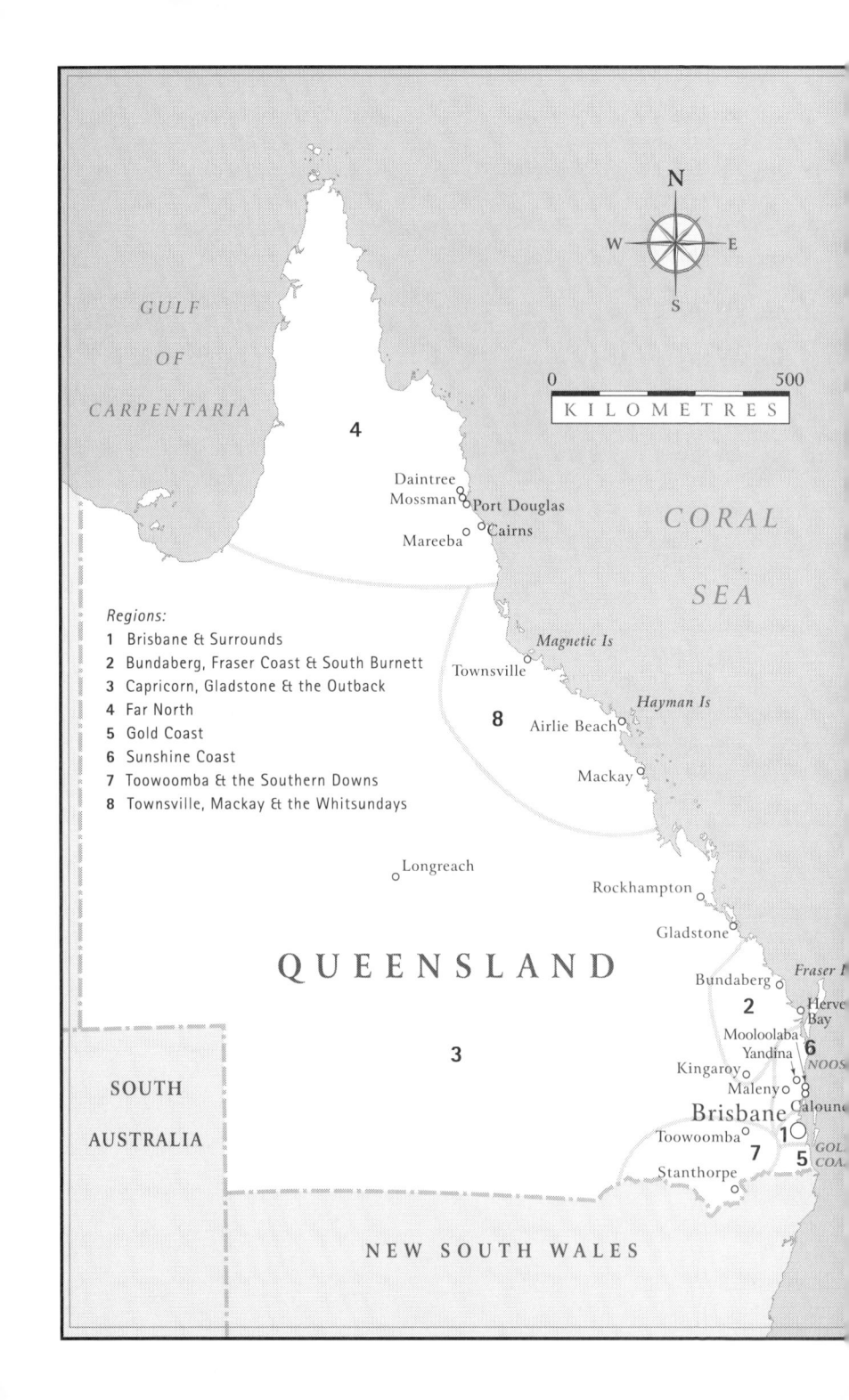

Regions:
1 Brisbane & Surrounds
2 Bundaberg, Fraser Coast & South Burnett
3 Capricorn, Gladstone & the Outback
4 Far North
5 Gold Coast
6 Sunshine Coast
7 Toowoomba & the Southern Downs
8 Townsville, Mackay & the Whitsundays

QUEENSLAND

Almost three times the size of France, Queensland straddles the Tropic of Capricorn and enjoys a widely varied climate and topography. This is reflected, of course, in its produce. In the far north, tea and coffee plantations mirror, although on a very minor scale, the Indian subcontinent, and now, tropical fruits that were once only found in South-East Asia, are being cultivated.

Queensland's 3700 farms produce 200 varieties of fruit and vegetables. Sugar cane is grown throughout the state and the pale green, head-high crop flanks most rural roads, which are crisscrossed by tiny sugar-train tracks. Here you'll also find ginger and macadamias, tea, coffee, fine beef and dairy products, the country's best bread-making flour and even wines.

But this is the party-time state too. Resorts abound. The Great Barrier Reef wraps the coastline for over 2000 kilometres offering limitless water-based activities, and in the far south, the Gold Coast high-rises sparkle like a mythical kingdom.

The richness of the state is reflected in the proportionate lack of food producers that encourage the public to drop in and buy from them. There are fewer roadside outlets, and even less tours and pick-your-own farms than in other states. Agriculture here is solid and few coastal farmers have had to diversify, as have their counterparts in other states.

Inland, the story differs. Many farmers are doing it hard as drought, floods and fires take their inevitable toll. There are some farmstays, some tours, some chances to experience the lifestyle, but these are limited. Brisbane, the capital, sits happily south of the lovely Sunshine Coast, with its own subculture, and north of the Gold Coast and hinterland. No doubt the hotter climate has created Brisbane's relaxed and carefree jackets-off ambience. Old pubs abound, for this is a thirsty climate, and the long coastline's selection of fabulous seafood is in rich and abundant supply at many relaxed venues throughout the city.

Queensland's slogan 'beautiful one day, perfect the next' could just as easily read in food-speak 'bountiful one day, a surfeit the next'.

QLD WEBSITES
Some useful websites include:
www.cuisine.southburnett.com.au
www.australiantropicalfoods.com
www.goldcoastfoodforum.org.au
www.goldcoastwinecountry.com.au

BRISBANE AND SURROUNDS

Bright and bustling Brisbane has a casual ambience. Is it the harsh light, the pub on every corner, the airy, stilted houses known simply as Queenslanders? Or has it more to do with the nonchalant cheek of its inhabitants, laid-back and ready to try anything whether it's a new restaurant, or their luck at the casino.

The city's restaurants are coming of age, winning national awards, becoming a force that allows visitors to experience the state's best produce. Moreton Bay bugs are the local crustacean, of course, named for the bay into which Brisbane's river flows, but there are reef fish and 'barra', the local term for barramundi. Queensland's excellent weather means that many venues are open air, with street and waterside dining commonplace, although never unremarkable.

GATTON

TOURS

LOCKYER DISCOVERY TOURS
Trudy Townson and Rob Bower
PO Box 392, Gatton 4343
Phone: 07 5466 1818
These minibus or coach tours to Bowers Organic Farm demonstrate how organic crops of potatoes, onions, pumpkins, tomatoes, beans are grown. The tours usually begin at 10am and return by 2pm or 3pm.
■ **Open:** For groups only, by appointment

KALLANGUR

PRODUCERS

THE TOMATO PATCH, RON BRAY'S TOMATOES
Kirsten Bray, Chris Brown and Peter Bray
Bray's Rd, Kallangur 4503
Phone: 07 3204 4656 or 07 3204 5346
Email: kirstenbray34@ozemail.com.au
Patient breeding and work has led to these flavoursome tomatoes, which remind all who eat them of what tomatoes ought to taste like. The Brays' contribution (which began with the late Ron Bray) was acknowledged by a Jaguar Award for Primary Produce in 2002. The farm shop sells the tomatoes as well as homemade relish and other seasonal vegetables—herbs, spinach, corn, zucchini and broccoli. The produce is also sold at most of the farmers' markets in Brisbane, and south-eastern Queensland.
■ **Open:** 1pm-4pm, Mon, Wed-Fri; 8am-4pm, Sat; 10am-4pm, Sun

SPRING HILL

PRODUCERS

MAYFIELD CHOCOLATES
Peter Ingall and Danica Antunovich
Shop 8, 101 Wickham Tce, Spring Hill 4000
Phone: 07 3832 1832
Email: peter@mayfieldchocolates.com.au
These chocolates may look like any fine European chocolate from the outside, but one bite and you are into Australian flavours you may never have encountered before. Mayfield Chocolates have

been making these delicious variations on a theme for several years now and tuck things like Kakadu plum, native peppermint, aniseed myrtle, wattle and lemon myrtle into their fillings. Favourites are Bushranger's Dream with Queensland rum and sultana ganache, and Riverina Dreaming with South Australian brandy and apricot. The result is tastebud-blowing. One of the crowning achievements was winning a Gold Medal in Belgium in 1999. Apart from these wild flavours of Australia, other fine Queensland products such as coffee, rum, ginger, macadamias and pineapple pop up regularly in various guises.

- **Open:** 9am–5pm weekdays
- **Location:** Off the Brisbane road opposite the Harbourtown shopping centre

WEST END
DINING

MONDO ORGANICS
Brenda Fawdon, Sonja Drexler, Dominique Rizzo
166 Hardgrave Rd, West End 4101
Phone: 07 3844 1132
Email: eat@mondo-organics.com.au
www.mondo-organics.com.au
Licensed and BYO (wine only)
Cuisine: Modern Australian with an Italian focus
Mains average: $$
Specialty: Organic meals that cater to a range of dietary requirements
Local flavours: As many as possible
Extras: Busy place, this. Every Saturday morning from 7.30am–12pm there is an organic produce market. The cookery school runs throughout the year, there is corporate catering and overseas food and wine tours, and it all happens with organic food as its focus.

- **Open:** Lunch Tues–Fri, dinner Tues–Sat, all-day breakfast weekends, classes Tues night mid Feb–mid Dec

RESTAURANTS

The choice of places to eat in Brisbane and its suburbs is so varied that you would need many weeks to sample even a fraction of what the city has to offer. As in many other Australian capital cities, inner city areas are well catered for, with a wide variety of ethnic eating places. In these parts you'll find food stores, footpath dining, cafes, bars, grills, brasseries and restaurants.

Fortitude Valley is home to Chinatown, and nearby New Farm is stylish and trendy. While food from other countries dominates the dining scene, there is still something for everyone, with menus often making use of the exceptional tropical fruits and seafood for which Queensland is justly famous.

Wander off to some of the upper-income leafy Brisbane suburbs such as Milton, and you will find elegant small eateries in every shopping centre. Paddington and Red Hill feature restaurants in restored Queenslander houses. Of course, Brisbane's many four- and five-star hotels all provide excellent international-standard dining, and often views of the city too. There are excellent views also from South Bank, with many restaurants and cafes, and riverside at Eagle Street Pier, Eagle Terrace.

Other interesting areas include tree-lined Racecourse Road at Hamilton, Dockside at Kangaroo Point, and West End with its many small European and Asian cafes.

QLD—BRISBANE AND SURROUNDS

MARKETS AND EVENTS

✦ **FARMERS' FRESH AND SEAFOOD MARKETS,** BRISBANE
07 3846 4500
6am–12pm first and third Sat of the month

✦ **INDOOROOPILLY FARMERS' MARKET,** INDOOROOPILLY
07 3868 2878
7am–12pm second and fourth Sat of the month

✦ **NORTHEY STREET ORGANIC MARKET,** WINDSOR
07 3289 8042
6am–10.30am Sat

✦ **POWERHOUSE FARMERS' MARKETS,** NEW FARM
07 3868 2878
7am–12pm, second and fourth Sat of the month

✦ **REDCLIFFE FARMERS' FRESH AND SEAFOOD MARKET,** REDCLIFFE
07 3846 4500
3am–3pm third Sun of the month

✦ **BRISBANE FOOD & WINE MONTH,**
brisbanefoodandwine.com.au
July

✦ **BRISBANE MASTERCLASS**
07 3231 3239
July

✦ **CAXTON STREET SEAFOOD AND WINE FESTIVAL,** BRISBANE
07 3369 6969
Early May

✦ **HILTON MASTERCLASS,** BRISBANE
07 3234 2000
Last weekend in July

BUNDABERG, FRASER COAST AND SOUTH BURNETT

Located at the southern end of the Great Barrier Reef, Bundaberg is coral clad and country-bred. The reef offers recreation options and some of the state's best seafood, while the magnificent hinterland provides rugged beauty and crops producing foodstuffs to round out almost any meal.

This fertile area is filled with sugar and mangoes, avocados—and more sugar. Seen from the air, the plantations resemble a pale green corduroy rug thrown carelessly over the countryside. In reality, sugar is a massive industry, one which not only sweetens the economy, but also distils into the famous Bundy Rum.

Mention Kingaroy to any Australian, and the connection is made. Peanuts. Most people may never have seen these nuts growing, but they do know that Kingaroy is the acknowledged peanut capital of the country. Yet there is wine locally too, as well as macadamias, sugar and bacon, and an infant olive oil industry in the Burnett Valley which looks set to expand dramatically in the next few years.

World Heritage-listed Fraser Island, the world's largest sand island, is part of this region that bites into the state around Maryborough. A fishing paradise, the region is equally attractive to fish-lovers, as the local restaurants know exactly what to do with each day's catch.

BUNDABERG

PRODUCERS

BUNDABERG RUM VISITOR CENTRE
Stephanie Hunt
Whittred St, Bundaberg 4670
Phone: 07 4131 2999
Email:
visitor.centre@bundabergrum.com.au
www.bundabergrum.com.au
Book at the visitors centre for tours through this major manufacturer's premises. The package includes a self-guided tour of the museum, a ten-minute video in the theatrette, plus two complimentary drinks from the bar. You may also buy souvenirs and alcohol after the tour. Tours take approximately one-and-a-quarter hours, and run hourly. Fully enclosed footwear is required.
■ **Open:** Hourly tours 10am–3pm weekdays, 10am–2pm weekends and public holidays

BREWERY/WINERY

TROPICAL WINES
Giovanni (John) and Caroline Gianduzzo
78 Mt Perry Rd, Bundaberg, 4670
Phone: 07 4151 5993
Email: tropicalwines@bigpond.com
Exotic fruit wines are not really new because John and Caroline Gianduzzo have been creating them from locally grown fruits such as mangoes, strawberries, mulberries, pineapples, peaches, plums and jaboticaba for about thirty years, using

only Queensland fruit, and a style based on old English cottage wines. The wines have twelve per cent alcohol and retain their fruitiness because they are fermented naturally, just from fruit. Better suited as dessert wines are the strawberry extra, pineapple extra, and mango extra (eighteen per cent alcohol) wines.
- **Open:** 9am–5pm Mon–Sat, 9am–5pm Sun (long weekends only)
- **Location:** 4km from Bundaberg Post Office on the Gin Gin Hwy

FAIRYMEAD HOUSE SUGAR MUSEUM
Botanical Gardens, Thornhill St, Bundaberg 4670
Phone: 07 4153 6786
wendyd@bundaberg.qld.gov.au
This fascinating museum, located in the house once owned by the founder of one of the state's first juice mills, was relocated from beside the mill to its present site, then restored. The display has a video and also documents the history of the sugar industry from 1860 to the current day. Bundaberg is the only place in the world where sugar is grown, milled, refined and distilled all in the same area.
- **Open:** 10am–4pm daily

CHILDERS
PRODUCERS

MAMMINO GOURMET ICE-CREAM
Tina and Anthony Mammino
115 Lucketts Rd, Childers 4660
Phone: 07 4126 2880
Email: mammino@isisol.com.au
www.mammino.com.au
Come here if you love macadamias, the nut trees that were discovered in southern Queensland in 1828 by botanist, Alan Cunningham, and named for Dr John Macadam, the secretary of the Philosophical Institute of Victoria. At this property on top of the hill there are all sorts of macadamia products—flavoured nuts, macadamia flavoured tea and coffee, fudge, beauty products, and really great ice-cream. Little wonder they scooped a Queensland Tourism Award in 2002, and numerous others.
- **Open:** 8am–6pm daily

FRASER ISLAND
DINING

SEABELLE RESTAURANT
Kingfisher Bay Resort, Fraser Island 4655
Phone: 07 4120 3333; Freecall 1800 072 555
Email: reservations@kingfisherbay.com
www.kingfisherbay.com
Licensed
Cuisine: Modern Australian
Mains average: $$$
Specialty: The tastes of Australia
Local flavours: Native meats such as kangaroo, emu and crocodile, Australian bush fruits and herbs, seafood, lamb and beef, free-range eggs, dairy foods, freshly baked bread is served with bush butter
Extras: Rangers take guests on daily bush tucker walks, and there is a Bush Tucker Talk and Taste on Tuesdays and Fridays. Grows herbs and bush foods such as warrigal greens, lemon and aniseed myrtle. There are sour currants, midjim berries, lilly pillies, lemon tea-tree, banksias and paperbark in the resort grounds and bush foods are for sale in the general store.
- **Open:** Dinner daily
- **Location:** Accessed by fast catamaran from Urangan Boat Harbour, Hervey Bay

GOOMERI

🍎 PRODUCERS

SPRING GULLY OLIVES
Joanna and Dan Burnet
8921 Burnett Hwy, Goomeri 4601
Phone: 07 4168 6106
Email: springgullyolives@bigpond.com
www.visitoz.org
The Burnets planted their first olives in 1996 and started selling in 1998. That's one of the best things about olives—they produce quickly. Now there is oil as well as various marinated olives, a tapenade and a unique and tasty relish. Planting olives was 'Plan B' for the couple's large cattle property after cattle prices dropped dramatically in the mid nineties. They decided that olives represented the best long-term investment. Now their product is sold in retail outlets through south-east Queensland and Brisbane, and at festivals and fairs in the region.
- **Open:** 9am–5pm daily
- **Location:** 30km north of Goomeri on the Burnett Hwy

🍴 DINING

THE PUMPKIN PIE COFFEE SHOP
Patsy and Roger Geddes
22 Boonara St, Goomeri 4601
Phone: 07 4168 4477
Email: pgeddes@bigpond.com
Licensed/BYO
Cuisine: Home style
Mains average: $
Specialty: Meat pies (steak, steak and kidney, sweet curry), pumpkin pies, scones, soup (in winter)
Local flavours: Pumpkin, beef
Extras: Next to Rocking Horse Antiques.
Goomeri is the pumpkin capital of Queensland.
- **Open:** 9am–4.30pm daily

HERVEY BAY

🍴 DINING

PIER RESTAURANT AND COCKTAIL BAR
Christine and Gary Dunlop, Carolyn Brummel
573 The Esplanade, Hervey Bay 4655
Phone: 07 4128 9695
Email: chrisdunlop@bigpond.com
www.uranganmotorinn.com.au
Licensed
Cuisine: Queensland seafood
Mains average: $$$
Specialty: Lamb roulade, oysters
Local flavours: Seafood, beef, lamb, pork, fruit and vegetables
Extras: Winner of Best Restaurant Lamb and Mushroom dish (Australian Meat and Livestock Award), coeliac and diabetic friendly menu, alfresco dining at the beachfront.
- **Open:** Breakfast daily, dinner Mon–Sat

MUD CRAB SANDWICHES

On the main road between Bundaberg and Gladstone prepare to break your journey at Miriam Vale—there are signs all through this area for mud crab sandwiches.

KILKIVAN

🍴 DINING

THE LEFT BANK
Rae and Bruce Hurley
10 Bligh St, Kilkivan 4600
Phone: 07 5484 1016
Email: hurley@theleftbank.com.au
www.theleftbank.com.au
BYO
Cuisine: Regional
Mains average: $
Specialty: Beef and red claw crayfish with special dressing, ploughman's lunches
Local flavours: Red claw, beef, cheese, seasonal fruit and vegetables, olives
Extras: Won an award for its beef and red claw dish. There is also a gallery and a B&B. Gourmet regional food hampers available. The Taste the Magic of the South Burnett tour (www.southburnett.com.au) is a food trail tour of the Kingaroy region to an olive farm, redclaw grower and wineries. **Makes for sale:** Kilkivan Collection jams, pickles, dressings, tomato relish, honey, cheesecake, chicken pies.
■ **Open:** Lunch daily, dinner by appointment, tours by demand

KINGAROY

🍎 PRODUCERS

KINGAROY CHEESE
Chris Ganzer
67 William St, Kingaroy 4610
Phone: 07 4162 5990
This cheese company, begun in 2002, is on a roll. It has won eleven gold medals for its cheeses (and eight silver) in major cheese shows. Cheese tasting is available at the cellar door, and tours allow visitors to view the cheese making though a window. Kingaroy Cheese makes soft cheeses at this stage—camembert, triple cream brie, gourmet fetta, Greek fetta, washed rind cow, soft goats cheese, ashed goats cheese, pure cream (very thick), quark, creme fraiche and goats curd.
■ **Open:** 9am–5pm daily, tours 10am and 2pm daily

POTTIQUE LAVENDER FARM
Anne Zalesky-McBride and Lindsay McBride
Kingaroy-Nanango Hwy, Kingaroy 4610
Phone: 07 4162 2781
This twenty-two year old lavender farm, originally selling pottery and antiques, now serves Devonshire teas in a garden setting. There are lavender scones and lavender tea, lavender ice-cream, jam and honey, and Pottique's own range of lavender pottery. The cellar door sells local wines (but not made from lavender, of course!) made especially for Pottique.
■ **Open:** 10am–5pm daily
● **Location:** 10km from Kingaroy

SWICKERS KINGAROY BACON FACTORY PTY LTD
David May
206 Kingaroy-Barkers Creek Rd, Kingaroy 4610
Phone: 07 4164 9500
Email: swickers@bigpond.com
Things have changed a little here in the past couple of years. The butcher's shop sells beef pork and lamb cuts, and sausages made on the premises—so good they snagged third prize in the 2003 Sausage King competition. And you'll still find the wonderful, local South Burnett

pork, but the products now include, under the Hans brand name, hams cooked traditionally using wood-smoking and curing techniques, plus kabana, salami and other smallgoods.
- **Open:** 7.30am–4.30pm weekdays, 8.30am–11.30am Sat

STORES

PROTECO PTY LTD
Graham Hemhold
67 William St, Kingaroy 4610
Phone: 07 4162 5660
Email: proteco@burnett.net.au
Proteco sells a range of products from the winery cellar door. There are wines of course, but also pure, milled, dry-roasted peanuts (naturally, after all, this is Kingaroy) macadamias (from Gympie), cashews and peanut butter. There is a small quantity of locally grown olive oil and table olives, cold-pressed avocado oil, aramas oil (used in aromatherapy), sesame oil, peanut oil (produced near Kingaroy), apricot kernel oil and Kingaroy Cheese.
- **Open:** 9am–5pm daily

THE PEANUT VAN
Rob and Chris Patch
77 Kingaroy St, Kingaroy 4610
Phone: 07 4162 8400
Email: info@peanutvan.com.au
www.peanutvan.com.au
Award-winning and unique, the Peanut Van offers Kingaroy's signature food in a range of flavours (this place pioneered the technique), as well as raw and roasted peanuts. The nuts are packaged the day they are roasted, ensuring freshness. Look for the nine flavours including savoury tomato, as well as chicken, chilli, garlic, or the more traditional curried, hickory smoked, or barbecue flavours, plus Peanut Van peanut paste. Mail order available.
- **Open:** 8.30am to 5pm daily

FARMSTAY

MINMORE FARMSTAY
Graham and Diane Wilson
Minmore, Kingaroy 4610
Phone: 07 4164 3196
BYO
Cuisine: Country
Three-course dinner $
Specialty: Minmore grass-fed roast beef
Local flavours: Beef, wine, fruit, pork, vegetables, peanuts
Extras: Close to Bunya mountains, there are 7000 hectares of bush to explore. All meals in homestead, or by arrangement. Picnic lunches available. A map is sent to each guest before arrival.

DINING

BELLTOWER RESTAURANT
John Crane and Rex Parsons
Cnr Schellbachs and Haydens Roads, Booie, Kingaroy 4610
Phone: 07 4162 7000
Email: brd@big.net.au
www.booie.com
Licensed
Cuisine: Modern Australian up-market cafe
Mains average: $$
Specialty: Smoked duck breast with orange glaze, local roasted peanuts
Local flavours: Peanuts, bunya nuts, duck, most meat, fruit and vegetables, cheese, olives, olive oil

QLD—BUNDABERG, FRASER COAST AND SOUTH BURNETT

Extras: Outdoor area with panoramic rural views. Sells homemade jam and makes own rum, whisky, vodka and gin under the BRD (Booie Range Distillers) label.
- Open: Lunch daily, dinner Thurs–Sat

BURNING BEATS CAFE
Paul Stoddart and Kerry Cotter
194 Kingaroy St, Kingaroy 4610
Phone: 07 4162 3932
Email: info@burningbeats.com.au
www.burningbeats.com.au
BYO
Cuisine: International
Mains average: $
Specialty: Indian and Thai dishes
Local flavours: Organic dairy products, herbs, cheese, meat
Extras: Asian deli, all outdoor dining, bands or entertainment on some nights.
- Open: Lunch Wed–Fri, dinner Wed–Sat

MURGON DAIRY MUSEUM
Murgon Shire Council, 2 Somerville Rd, Murgon 4605
Phone: 07 4168 1499, OR 07 4168 1984 (Murgon Development Bureau)
Email: murgonsc@burnett.net.au
www.murgon.qld.gov.au
This is the only dairy museum in Queensland and it has a very comprehensive collection of milk, butter and cheese-production memorabilia and information outlining the impressive history of the dairy industry in the state. There are also butter churning demonstrations.
- **Open:** 1pm–4pm weekends, or by appointment
- **Location:** About 2km off the Murgon–Gayndah Rd

MOFFATDALE
🍷 BREWERY/WINERY
CLOVELY ESTATE
Jane Parker
Steinhardts Rd, Moffatdale 4605
Phone: 07 4168 4788
Email: sales@clovely.com.au
www.clovely.com.au
Clovely is, at the moment, the largest vineyard in Queensland. The company grows, crushes and makes its wines in one of Murgon's oldest buildings, which was the Murgon butter factory. There are also 320 acres of olive grove or around 37 000 trees. The olives are processed in the same building as the wine and 2003 saw the first yield of olives and olive oil. The cellar door has both wines and olives, there is a B&B on the property and light meals, coffee, local cheeses, and antipasto are available.
- Open: 10am–4pm daily

MURGON
🍎 PRODUCERS
BOTTLE TREE HILL ORGANICS
Wil and Meagan Seiler
1525 Crownthorpe Rd, Murgon 4605
Phone: 07 4168 4669
Email: logboy@burnett.net.au
www.bottletreehill.com
Bottle Tree Hill Organics specialise in free-range pigs and beef and sell their pork through a butcher in Brisbane. Their GABS (Gourmet Australian Bush Sausages), pure pork sausages with Australian bush herbs and no artificial ingredients, are sold through selected delis and farmers' markets and come in

BUNDABERG, FRASER COAST AND SOUTH BURNETT—QLD

MARKETS AND EVENTS

✦ **BLACKBUTT COUNTRY MARKETS,**
BLACKBUTT
07 4163-0090
Second Sat from 6.30am

✦ **MARYBOROUGH HERITAGE MARKETS,** MARYBOROUGH
07 4121 4111
8am to 1.30pm Thurs

✦ **MURGON COUNTRY MARKETS,**
MURGON
07 4168 3864
Second Sun of the month

✦ **SHALOM SUNDAY MARKETS,**
BUNDABERG
07 4159 0190
7am–12pm Sun

✦ **BUNDABERG MULTICULTURAL FOOD AND WINE FESTIVAL,**
BUNDABERG
07 4151 9770
Oct

✦ **CHILDERS MULTICULTURAL FOOD, WINE AND ARTS FESTIVAL,**
CHILDERS
07 4126 3886
Late July

✦ **GAYNDAH ORANGE FESTIVAL,**
GAYNDAH
07 4161 2268
Odd-numbered years in June
(Queen's birthday weekend)

✦ **GOOMERI PUMPKIN FESTIVAL,**
GOOMERI
07 4168 1925
Last Sun in May

✦ **MURGON DAIRY FESTIVAL,**
MURGON
07 4169 5001
Sun of Queen's birthday long weekend in June

✦ **PEANUT FESTIVAL,** KINGAROY
07 4162 3199
Sept

✦ **TARONG MIND WINE AND FOOD IN THE PARK FESTIVAL,** KINGAROY
07 4162 3199
Mar

✦ **THE SHAKIN' GRAPE,** MURGON
07 4168 3864
Third Sat in Oct

---

three different flavours—Desert Cajun spiced with Dorrigo pepper, and lemon myrtle, Outback Blend with native island mint and Dorrigo pepper, and Rainforest with lemon myrtle, Dorrigo pepper, thyme, and lemon peel. Traditional double wood-smoked hams and bacon are available year around.

■ **Open:** By appointment only
● **Location:** 26km north-east of Murgon

TIARO
🏠 STORES

TIARO MEATS AND BACON
Terry and Fran Connors
Mayne St, Tiaro 4650
Phone: 07 4129 2173

This local butcher makes excellent home-smoked hams and smallgoods using pork from Ingham's piggery, slaughtered in Gympie at Nolan's abattoirs. There are two smokehouses and the seventeen-hour process allows the hams to be thoroughly 'cooked-on-the-bone'. Tiaro Meats also manufactures its own smallgoods using beef from its own property, the product is specifically aimed at people who appreciate quality.

- **Open:** 7.30am–6pm weekdays, 7am–12.30pm Sat

YARRAMAN

PRODUCERS

YARRAMAN BACON FACTORY
Darryl and Glennys Woltmanm
Toomey St, Yarraman 4614
Phone: 07 4163 8260

The bacons and hams here are cured following an old-fashioned German recipe used by Darryl Woltmanm's great-grandfather. Woltmanm says this is the only factory left in Queensland doing slow-cured and genuine smoked ham and bacon. Darryl himself has been in the business for close to forty years, and only works with Queensland pork and Australian beef. The shop will host bus tours if given about a month's notice, but health regulations prohibit tours taking place during production. The tours last about an hour and visitors can purchase products direct from the store afterwards.

- **Open:** 8am–5pm weekdays, tours by appointment

CAPRICORN, GLADSTONE AND THE OUTBACK

The Capricorn region, so-named because it straddles the Tropic of Capricorn, is still firmly anchored to the sea and its produce. Industries, sport, shops and restaurants reflect this, with fine fishy offerings everywhere.

It would be impossible to talk about Gladstone without mentioning its seafood. The Gladstone mud crab is legendary, and is served throughout the region. Fishing is a major sport here, but even non-anglers can enjoy someone else's skill at dinner time. Avocado, paw-paw and pineapple feature on menus too, often growing in backyards, and the scent hangs in the air.

BILOELA
FARMSTAY

KROOMBIT PARK
Alan Sandilands
Lochenbar, Valentine Plains Rd,
Biloela 4715
Phone: 07 4992 2186
Email: lochenbar@kroombit.com.au
www.kroombit.com.au
Two-course meal: $
Specialty: Camp-oven cooking on an open fire
Local flavours: Beef
Extras: This award-winning farmstay offers guests a chance to try everything from cattle mustering to camp-oven cooking.
■ **Open:** Year round, phone for details
● **Location:** 35km from Biloela

CALOUNDRA
DINING

THE RED CRAB SEAFOOD BAR
Karen Cronk
Shop 11, Tay Ave, Caloundra 4551
Phone: 07 5491 1009
Takeaway
Cuisine: Seafood
Mains average: $
Specialty: Fish and chips, hamburgers, barra burgers, real crab (when available) sandwiches
Local flavours: Seafood straight from the trawlers
Extras: The Red Crab has been going for thirty-four years and was one of the first seafood outlets—located in the old Fish Board—in Caloundra. The original sign is still outside.
■ **Open:** Lunch and dinner daily
● **Location:** Near the Esplanade

THE BIRTHPLACE OF SURF 'N' TURF
The Capricorn region—is this where the beef and seafood dish 'surf 'n' turf' comes from? This region has great beef and seafood and many local restaurants certainly feature them on the menu. For a good reason. They're fresh and either locally grown or caught.

QLD—CAPRICORN, GLADSTONE AND THE OUTBACK

COOWINGA
🍴 DINING

KOORANA CROCODILE FARM
Lillian (chef) and John Lever
5 Savages Rd, Coowonga, via Rockhampton 4702
Phone: 07 4934 4749
Email: koorana1@iinet.net.au
www.koorana.com.au
Licensed
Cuisine: Australian
Mains average: $
Specialty: Crocodile dishes—pies, ribs, burgers, steak, satay, soup, even Asian noodles with crocodile
Local flavours: Reef fish, Central Queensland beef
Extras: Farm tours (two daily—10.30am and 1pm) lasting about one and a quarter hours. Takeaway crocodile pies available.
- **Open:** Lunch daily, dinner for groups of twenty or more
- **Location:** 50km from Rockhampton off Coowonga Rd

EULO
🍎 PRODUCERS

PALM GROVE DATE FARM AND WINERY
Ian and Nan Pike
Palm Grove, Eulo 4491
Phone: 07 4655 4890
Email: datefarm@bigpond.com
The Pikes have been growing dates for fifteen years, and their outback property, on the banks of the Paroo River, is ideal for this fruit, as well as figs. There are fresh or semi-dried dates, date chews,

> **THE OUTBACK**
> The massive outback region extends to the Northern Territory border; a far-flung, sparsely populated area that takes days to traverse by car. In some places, even this is impossible, as roads deteriorate or are non-existent. Here again, bush tucker and beef take over.

date spread, date wine (made from fermented dates and which comes dry, medium and sweet, with about twelve and a half per cent alcohol), plus date topping and cosmetics. There are tastings and samples, and the fruit products are seasonally available by mail order, although cosmetics are available year round.
- **Open:** 9am–4.30pm daily (Mar–Oct)
- **Location:** Southern side of Eulo along the Adventure Way

🏠 STORES

EULO GENERAL STORE
Garry Berghofer
Leo St, Eulo 4491
Phone: 07 4655 4900
Email: eulostores@bigpond.com
This is an old fashioned country store selling everything from groceries, fuel, camping and fishing supplies, to local produce—honey and honey cosmetic products, date wine (sweet, dry, medium) and date products (date topping, date and chilli spread, date and ginger spread) and fresh figs. You can also get fresh local honey and Yapunyah honey (this is the only place in Australia you can get this type of honey as it's from the blossom of a local tree). There are honey based cosmetics—Paroo Hand cream ('we send loads of this all around Australia and

CAPRICORN, GLADSTONE AND THE OUTBACK—**QLD**

> **QUARANTINE NOTE**
> If visiting from Western Australia, you may need certification to import certain fruits and vegetables into Queensland. All visitors are prohibited from bringing in banana plants and seeds, and may need certification to carry in sugar cane and grapevines. Check with your state agriculture department or the Queensland Department of Primary Industries (Phone: 07 3239 4000).

overseas', they say), revitalising cream, honey ointment, lip balm, body lotion and bath bombs—and furniture polish. Get the idea?
- **Open:** 8am–7pm daily
- **Location:** Adventure Hwy, between Cunnamulla and Thargomindah

GLADSTONE
🍴 DINING

FLINDERS WATERFRONT RESTAURANT
Yvonne Glossop
Cnr Flinders Pde and Oaka Lane, Gladstone 4680
Phone: 07 4972 8322
Email: Flinders2@bigpond.com
Licensed
Cuisine: Modern Australian
Mains average: $$$
Specialty: Mud crab, coral trout, red emperor
Local flavours: Seafood
Extras: Outdoor dining overlooking the water on Gladstone Harbour. Fresh fish filleted daily on the premises. Recipient of numerous tourism and wine awards.
- **Open:** Lunch weekdays, dinner Mon–Sat

KAPERS
Irene and Sascha Dudley, Carol Mitchell
Shop 2, 124B Goondoon St, Gladstone 4680
Phone: 07 4972 7902
Email: kapersby@hotkey.net.au
BYO
Takeaway
Cuisine: Modern Australian
Mains average: $$
Specialty: Cuttlefish in beer batter with chilli-plum sauce, steaks
Local flavours: As much as possible
- **Open:** Dinner Mon–Sat

LONGREACH
🚌 TOURS

BILLABONG BOAT CRUISES
Norm Salsbury and Chris Rumsey
115a Eagle St, Longreach 4730
Phone: 07 4658 1776
Email: info@lotc.com.au
www.lotc.com.au
The camp-oven sunset dinner cruise uses a paddleboat and is most popular in winter when guests enjoy hearty Drover's beef stew (made from local beef) with damper, then dessert and billy tea at the bush port, a campsite on the river under 'a ceiling of stars'. The entertainment features a bush poet and a country music singer.
- **Open:** Dinner Mon–Sat

> **CATCH-YOUR-OWN DINNER**
> Go barramundi fishing at lunchtime? In Rockhampton, you can. Just walk two blocks from the town centre, throw in a line in the Fitzroy River and maybe you'll haul in your dinner.

ROCKHAMPTON

🍴 DINING

ASCOT HOTEL
Will Cordwell
177 Musgrave St, Rockhampton 4701
Phone: 07 4922 4719
Email: will@ascothotel.com.au
www.ascothotel.com.au
Licensed
Cuisine: Stone grill
Mains average: $$
Specialty: Beef and game meats
Local flavours: Beef, crocodile, barramundi, goat, vegetables and salad greens
Extras: Rockhampton is the beef capital of Australia and this restaurant has steak so good, they say, that steak knives are not needed. Lunch and picnic packs can be arranged on request and takeaway is available.
- **Open:** Lunch and dinner daily

EVENTS

✦ **CENTRAL QUEENSLAND MULTICULTURAL FAIR,**
ROCKHAMPTON
07 4927 2055 or 1800 676 701
Aug

✦ **COMEDY AND FOOD FESTIVAL,**
BILOELA
07 4992 9108
Oct

✦ **EMU PARK LIONS CLUB OKTOBERFEST—BEER FESTIVAL,**
EMU PARK
07 4939 6209
Oct

✦ **THE GLADSTONE SEAFOOD FESTIVAL,** GLADSTONE
07 4972 5111
Sept

CAPRICORN TOURISM AND DEVELOPMENT ORGANISATION
'The Spire', Gladstone Rd, Rockhampton 4700
Phone: (07) 4927 2055/ 1800 676 701
Email: infocentre@capricorntourism.com.au
www.capricorntourism.com.au
- **Open:** 9am to 5pm daily

FAR NORTH

Far north Queensland is like another country. The weather is steamy and tropical, strange fruits grow in abundance and brilliant flowers, unseen further south, fill local gardens with colour.

Fertile and lush, this is the place to come for a fix of flavour—from pungent jackfruit to the amazing chocolate pudding fruit, and sublimely delicate reef fish and barramundi.

Everything and anything grows here, from all the components of a civilised afternoon tea (tea, coffee, sugar and cream) to the most exotic of fruits. Herbs and bush tucker, tomatoes with flavour you thought had left the planet forever, and those sugar-sweet little pineapples they simply call 'roughies' are everywhere.

This vibrant region simply bursts with colour and life, perhaps nowhere is it more apparent than in the weekly markets at Cairns or Kuranda.

BIBOOHRA

BREWERY/WINERY

GOLDEN DROP
227 Bilwon Rd, Biboohra 4880
Phone: 07 4093 2750
Email: info@goldendrop.com.au
www.goldendrop.com.au
Fancy tippling some mango wine? This place has been fermenting Kensington red mangoes since early 1999 and now make a range of glowing wines—dry, medium and sweet—as well as a mango liqueur, mango port, mandarincello, limecello and lemoncello.
■ **Open:** 8am–6pm daily
● **Located:** Ten minutes north of Mareeba

CAIRNS

PRODUCERS

ASIAN FOODS AUSTRALIA
Sim Hayward
101–105 Grafton St, Cairns 4870
Phone: 07 4035 2268
Email: sim@asianfoods.com.au
www.asianfoods.com.au
The 100 per cent locally made sauces are of course made using local produce. Varieties include: Sim's Hot Mango Chutney, Spicy Macadamia Sauce, Mango Chilli Sauce, and Hot Mango Pickle. This is the largest Queensland-based Asian food wholesale and retail specialist outside of Brisbane. Supplies both restaurants and stores Australiawide.
■ **Open:** 8am–8pm daily

STORES

CHOCOLATE SENSATIONS
Wallace and Kathy Macdonald, Allison and Louise Green
Reef Hotel Casino Lobby, 35–41 Wharf St, Cairns 4780
Phone: 07 4041 0232
Email: sales@chocsensations.com.au
www.chocolatesensations.com.au
Wallace Macdonald is a pastry chef and chocolatier and was continually asked to

make chocolates for hotels and restaurants. So Wallace and Kathy started the business, the only chocolatier north of Brisbane, and Allison and Louise joined the business in July 2003. Now the team turns out quality handcrafted truffles, pralines and speciality items by combining the finest Belgian chocolate with fresh produce sourced from north Queensland. There are truffles and pralines in thirty flavours, including macadamia nut bark, tropical rocky reef and coconut rough. Local fruits are also dehydrated and dipped in Belgian chocolate. These products are sold to major hotels and cafes from Cairns to Port Douglas.

■ **Open:** 10am–7pm Mon–Sat

NEIL'S ORGANICS

Neil and Norma Robson
Shop 2, 21 Sheridan St, Cairns 4870
Phone: *07 4051 5688*
Email: *caring@neilsorganics.com.au*
www.neilsorganics.com.au

After seven years selling certified-organic and biodynamic fruit and vegetables, meat, poultry, dairy, and grocery lines at the local growers' markets, the Robsons moved into their own premises. They are still selling the same fine foodstuffs, and the shop grew out of the direct sales of their farm produce. Their farm is certified Demeter with the Biodynamic Institute of Australia. For the shop they source as much local produce as possible—including their own of course—and labelling on the shelves and stock states whether something is certified, along with the certifying body, grower and district. Look for certified biodynamic dried mango—only a couple of farmers in Australia are producing this.

■ **Open:** 9am–5.30pm Mon–Thurs, 6am–6.30pm Fri, 8am–12.30pm Sat

DINING

CAFE COCO, YAMAGEN AND KINGSFORDS RESTAURANTS

Cairns International Hotel
17 Abbott St, Cairns 4870
Phone: *07 4031 1300*
Email: *hotel.cih@cairnsinternational.com.au*
www.cairnsinternational.com.au
Licensed
Mains average: $$–$$$
Local flavours: Fruit, vegetables, seafood, beef, cheese, yoghurts, game
Extras: There are three restaurants in this major international-standard hotel ranging from the casual buffet style of Cafe Coco to authentic Japanese fare at Yamagen and up-market modern Australian at Kingsfords. The Lobby Bar cocktail champion makes cocktails with local tropical fruit.

■ **Open:** Breakfast, lunch and dinner daily

RED OCHRE GRILL

Craig Squire, James Fielke
43 Shields St, Cairns 4870
Phone: *07 4051 0100*
Email: *food@redochregrill.com.au*
www.redochregrill.com.au
Licensed/BYO wine
Cuisine: Modern Australian
Mains average: $$$
Specialty: Kangaroo, crocodile, Queensland seafood
Local flavours: Macadamia nuts, herbs, fruits and vegetables, Mungalli cheeses and dairy, buffalo cheeses, seafood, Tablelands pork, veal, chicken and beef, Queensland ostrich and wild boar, farmed crocodile, native foods

Extras: Indoor and covered outdoor dining. Multi-award-winning restaurant with a menu that showcases Australian native foods plus locally grown specialties. Craig Squire founded the Taste of the Tropics Food Trail. Makes and sells packaged bush spices, dukkah, pesto, tapenade, various chilli jams, chutneys, relishes and salsas.
- **Open:** Lunch Mon–Fri, dinner daily

SIROCCO RESTAURANT
Marcus Werner (chef)
Shangri-La Hotel, Pierpont Rd,
Cairns 4870
Phone: 07 4052 7665
Licensed
Cuisine: Modern Australian
Mains average: $$$
Specialty: Use of local produce
Local flavours: Buffalo labna (yoghurt cheese), and yoghurt, all seafood (including farmed and wild barramundi), tropical fruits (such as tamarillo, pawpaw, mango, black sapote and lychees), bushtucker spices and fruits
Extras: Five-star hotel restaurant with waterfront views.
- **Open:** Dinner daily
- **Location:** In the Pier Market Place

TOURS

FOOD TRAIL TOURS— A TASTE OF THE HIGH PLAINS
Warwick James
PO Box 112, Cairns 4870
Phone: 07 4032 0322 (reservations 07 4041 1522)
Email: admin@foodtrailtours.com.au
www.foodtrailtours.com.au
Indulge in a taste of the tropics and take a tropical food and wine day-tour. Pick fruit, visit farms, sip tropical fruit wines and liqueurs, indulge in macadamias and boutique coffee then finish your day with tropical fruit ice-creams. Lunch features barramundi, red-claw crayfish and tropical fruits.
- **Open:** Mon–Sat

RED CLAW
Red claw is a freshwater crayfish native to Queensland. Now commercially farmed, this crustacean is becoming increasingly popular. Freshwater crayfish is traditional in many European cuisines and is becoming more popular in Australia too. In southern states, the closely related yabbies and marron are also farmed, and appear on menus everywhere.

Sweden alone imports 2000 tonnes of freshwater crayfish annually from the United States. Asia is now becoming a market for crays, and this is of great interest to growers because of the proximity of Queensland to South-East Asia, and the high prices they can command there.

Compared with other shellfish, red claw is relatively easy to farm. Almost any land can be used, and the climate is not a major factor, allowing them to be farmed even in tropical and subtropical regions. Like others of its species, red claw is quite amenable to crowding, reproduces easily and grows rapidly. It is easy to harvest, tolerates broad temperature ranges, and even a range of salinity, and is free of disease.

Farming of red claws began in the 1980s and since then the industry has grown quickly. There are over eighty licensed farms in Queensland and the annual production exceeds fifty tonnes.

The slogan of the industry is 'You can't ignore red claw,' and at the rate the industry is growing, it will be impossible to ignore.

TASTE OF THE TROPICS FOOD TRAILS
Nola Craig
PO Box 3065, Cairns 4870
Phone: 07 4040 4415
Email: info@australiantropicalfoods.com.au
www.australiantropicalfoods.com
Five tours have been designed to encourage visitors to experience the region and its diverse range of foods and produce. There are Tastes of: The Tropic Coast, Rainforest, Great Green Way, Northern Tablelands and Mountain Tablelands—each with several wonderful places to enjoy.
- **Open:** 9am–5pm Mon–Fri

CAPE TRIBULATION
PRODUCERS

CAPE TRIB EXOTIC FRUIT FARM AND B&B
Digby and Alison Gotts
Lot 5, Nicole Dr, Cape Tribulation 4873
Phone: 07 4098 0057
Email: digby@capetrib.com.au
www.capetrib.com.au
Experience an organic permaculture orchard growing exotic tropical fruits in the heart of the rainforest. Open for fruit tasting and farm tours, you may taste at least ten different exotic fruits—the selection depends on the season—in a tropical designed outdoor gazebo. There are more than 150 different fruit species here, from all the tropical regions of the world. B&B accommodation is also available (complete with tropical fruit platter for breakfast) and the farm is next to a World Heritage classified rainforest.
- **Open:** Tours at 4pm Tues–Thurs and Sun (bookings recommended)
- **Location:** 70km north of Mossman

DAINTREE
DINING

DAINTREE ECO LODGE AND SPA
Terry and Cathy Maloney
20 Daintree Rd, Daintree 4873
Phone: 07 4098 6100
Email: info@daintree-ecolodge.com.au
www.daintree-ecolodge.com.au
Licensed
Cuisine: Modern Australian
Mains average: $$$
Specialty: Subtle use of bush foods and native ingredients
Local flavours: Beef, fish, seafood, exotic fruits, fruits and vegetables, bush foods
Extras: Located in a lush tropical valley in the World Heritage listed Daintree Rainforest, with indoor and outdoor dining beside lagoon pond. World-renowned day spa. Winner of many awards including 2002–03 Australian Tourism Awards for deluxe accommodation and eco-tourism. Featured as one of the top four spa retreats in the world. Condé Nast (2004) ranked it in the world's top spas—the only Australian property to qualify.
- **Open:** Breakfast, lunch and dinner daily
- **Location:** Turn left into driveway 3km short of Daintree village

DAINTREE TEA HOUSE RESTAURANT
Peter Ryan and Richard Seivers
MS 1880, Mossman–Daintree Rd, Daintree 4873
Phone: 07 4098 6161
Email: daintreeteahouse@bigpond.com
Cuisine: Modern Australian
Mains average: $$
Specialty: Barramundi
Local flavours: All produce sourced from

far north Queensland—seafood, tropical fruits, vegetables
Extras: Indoor and outdoor seating in a rainforest setting. The longest established restaurant in the Daintree area.
- **Open:** Lunch daily, dinner Fri-Sat (and Mon-Wed by reservation only)
- **Location:** 3km south of Daintree

DARADGEE
🏠 STORES

JOHNSTONE RIVER PLANTATION FRUIT SHOP
Marie Dillon
Bruce Hwy, Daradgee 4860
Phone: 07 4063 3867
Email: dillco@bigpond.com
On the banks of the North Johnstone River near to the highway is this shop which sells locally grown tropical fruits and vegetables, including four or five different types of bananas. There are fruit tastings, homemade ice-cream, fruit salads, banana smoothies, dried bananas and preserves as well in a teahouse that was recently opened in a lovely old Queenslander).
- **Open:** 9am-7pm daily
- **Location:** 7km north of Innisfail, next to Johnstone River

INNISFAIL
🍎 PRODUCERS

NUCIFORA TEA
Sybbie and Paula Nucifora
Palmerston Hwy, MS 216, East Palmerston, Innisfail 4860
Phone: 07 4064 5109
Email: nuci@znet.net.au
This 100 per cent locally grown tea, free from pesticides and herbicides, is available from the roadside stall operating on an honesty system. Only the tender young tips of the plant are harvested to provide a top quality tea with a thirst quenching flavour and a delicious aroma. Sales and tastings available.
- **Open:** Daylight hours daily (honesty system), tours by arrangement
- **Location:** 30km west of Innisfail

KURANDA
🍎 PRODUCERS

KURANDA HOME MADE TROPICAL FRUIT ICE CREAM
Betty and Cliff Timmins
Entrance to Kuranda Markets, Kuranda 4881
Phone: 0419 644 933
Email: joy@iig.com.au
www.kuranda-icecream.com.au
Seven years ago, Cliff Timmins felt he was too young to retire, so, at the urging of his wife Betty, they began making ice-creams and selling them next to the Kuranda Markets Arcade in Therwine St from their distinctive red van. The Timminses use fresh local cream and milk (although some of the ice-creams are made with soya milk) and all natural products to create thirty-two different flavours. Some are seasonal, as they use fresh fruit whenever possible, so you can sometimes find exotic offerings such as avocado, black sapote, jackfruit, soursop or mango, and there are always favourites such as the adults only-rated rum and raisin (using real Bundy rum) and chocolate or vanilla.
- **Open:** 9am-4pm daily

🍴 DINING

RAINFORESTATION NATURE PARK
Charles and Pip Woodward
Kennedy Hwy, Kuranda 4872
Phone: 07 4093 9033
Email: res@rainforest.com.au
www.rainforest.com.au
Licensed
Cuisine: Aussie barbecue
Mains average: $$
Specialty: Crocodile, kangaroo
Local flavours: Tropical fruits from the orchard, potatoes
Extras: There are two restaurants, one overlooking a lake, the other surrounded by gardens. It is also possible to have dinners in the wildlife park and visitors can explore the exhibition orchard with over thirty exotic fruit trees.
- **Open:** Lunch daily, dinner by reservation only
- **Located:** Five minutes from Kuranda

MALANDA

🍎 PRODUCERS

TUMBLING WATERS HYDROPONICS
Simon Noons
2 Clarkes Track, Malanda 4885
Phone: 07 4096 6443
Email: hydro1@austarnet.com.au
This is a commercial and retail farm producing fancy lettuces, Asian greens, silver beet and a comprehensive range of potted herbs. You can purchase from the farm gate, wander around the hydroponic farm, or visit the newly opened garden centre. There is also a tea and coffee lounge on the veranda of the beautifully restored Queenslander, with a menu based on an 'eat what you see growing' theme.

- **Open:** 10am to 5pm daily
- **Location:** 5km south of Malanda on the Millaa Millaa-Malanda road

🍴 DINING

MALANDA DAIRY CENTRE
Stewart Lyle
8 James St, Malanda 4885
Phone: 07 4095 1234
Email: mdcesc@tpg.com.au
BYO
Cuisine: Cafe style
Mains average: $
Specialty: Macadamia nut and bocconcini salad, milk shakes made with real fruit
Local flavours: cheese, milk, marinated bamboo shoots, red claw crayfish
Extras: This is an innovative, retro-decor milk-bar style restaurant, showcasing the food and produce of the Tablelands region. There are cheese tasting displays, as well as the 'Taste of the Tablelands' retail section and an interesting tour and interpretive centre for the local dairy industry.
- **Open:** 9am-4.30pm daily, tours 9am-1pm weekdays
- **Location:** Adjacent to the dairy factory

MAREEBA

🍎 PRODUCERS

MAREEBA COFFEE, TICHUM CREEK COFFEE FARM
Mario and Claudia Sorbello
3576 Kennedy Hwy, Davies Creek, Mareeba 4880
Phone: 07 4093 3092
Email: sales@mareebacoffee.com.au
This fledgling coffee plantation came about because the Sorbellos needed to

diversify from tobacco growing. They now also roast coffee and you can visit the coffee seedling nursery and the young coffee plantings. There are free coffee tastings, coffee to drink and sales of roasted and flavoured coffee products. The coffee is sold here and at local supermarkets and souvenir outlets.
- **Open:** 8.30am–4.30pm daily (Easter-end-Sept), Tues–Sun (other months)
- **Location:** Davies Creek, between Mareeba and Kuranda

MJ AND VG JENNINGS
John Jennings
234 Jennings Rd, Mareeba 4880
Phone: 07 4092 2855
Email: jvj@cyberwizards.com.au
Part of a group of eight growers, this farm sells mainly to Port Douglas, Cairns and Brisbane but they will also sell to you if you're got a few days in the area. It's best to call ahead, although they usually have some crays (called red claws up here) in holding tanks. If you want ten kilograms or more, they can ship them to you. Red claw is a freshwater crayfish, native to the waters flowing into the Gulf of Carpentaria. They are relatively easy to grow, but need careful management to achieve good yields. They belong to the same family as marron and yabbies (cherax) and taste much the same as marron.
- **Open:** By prior arrangement, at least three days ahead

NORTH QUEENSLAND GOLD COFFEE PLANTATION
Bruno, Luisa and Maria Maloberti
1140 Dimbula Rd, Mareeba 4880
Phone: 07 4093 2269
Email: sales@nqgoldcoffee.com.au
www.nqgoldcoffee.com.au
The Maloberti family, originally tobacco farmers, were some of the first coffee growers in the state. In the 1970s they established Paddy's Green Coffee Plantation in the prime growing conditions of the Atherton Tablelands, producing only Arabica beans. They are proud of the fact that they have been growing coffee for twenty-five years, and have used no pesticides or sprays for the past nine years, and only organic fertilisers are used. The company roasts their own coffee twice weekly, and the result is a very rare, sweet and aromatic coffee, with no bitter aftertaste. There are light (mocha), medium roast, and espresso roast varieties available, as well as chocolate-coated coffee beans. The coffee is available at souvenir and gift shops throughout north Queensland or by mail order. Complimentary tour with purchases.
- **Open:** 8am–5pm daily, tours on the hour, except 12pm
- **Location:** 10.2km from Mareeba on Dimbulah Rd (Wheelbarrow Way)

BREWERY/WINERY

DE BRUEYS BOUTIQUE WINES
Bob and Elaine de Brueys
Tinaroo Creek Wines Pty Ltd, 189 Fichera Rd, Mareeba 4880
Phone: 07 4092 4515
Email: debrueys@tpg.com.au
This new winery set among mango orchards offers a range of tropical fruit wines (twelve per cent alcohol) and mango or mulberry ports at the cellar door. The De Brueys began with the mangoes, which they have been growing for

twelve years. The wines—currently mango, jaboticaba, mulberry, lychee, and bush cherry—are particularly good with Asian dishes or white meats.
- **Open:** 10am–5pm daily, or by appointment
- **Location:** Off Kennedy Hwy, 2km down Fichera Rd

🏠 STORES

TERMITE FRUIT, VEG AND TAKEAWAY
Janice and Brian Herbohn
3823 Kennedy Hwy, Mareeba 4880
Phone: 07 4093 3197
Email: termite@top.net.au

Way out in the middle of, well, the termite's nests, is this place—a little oasis selling fresh local honey, fruit and vegetables, and roo and croc takeaways. Termite's owners make their own Termite Brand jams and chutneys, dry their own mangoes, bananas and chillies, and the chilli sauce is a specialty. Just what you need for the long journey south.
- **Open:** 8am–6pm daily

THE COFFEE WORKS MAREEBA
Rob and Annie Webber
136 Mason St, Mareeba 4880
Phone: 07 4092 4101
Email: rob@arabicas.com.au
www.arabicas.com.au

'We love coffee', says Rob Webber, and you won't wonder why when you step into his company's airy showroom at Mareeba. The scent of gourmet coffee roasting is almost irresistible and it is all ground from pure Arabica coffee beans. This business has been operating for many years, and blends freshly roasted gourmet coffees, both locally grown Australian ones and international, for a premium result. Australian coffees include Queensland Blue, Black Mountain and Australian Gold. There are also macadamia and macadamia chocolate coffee blends. Of course you'll find a coffee counter here where you can order a coffee, and plenty of information on hand to help you understand the roasting and blending process. The 'roast and post' mail-order service is very popular too.
- **Open:** 9am–4pm daily, tours 10am, 12pm and 2pm

🐄 FARMSTAY

ARRIGA PARK FARMSTAY
Milena and Harold Cek
1720 Dimbulah Rd, Mareeba 4880
Phone: 07 4093 2114
Email: arrigapark@cyberwizards.com.au
www.bnbnq.com.au/arriga
BYO
Cuisine: Farm style
Mains average: $$
Specialty: Farmer's breakfast, black sapote pudding
Local flavours: Fruit and vegetables
Extras: This is a working sugarcane farm and tropical fruit orchard growing mangoes, black sapote and other rare tropical fruit. The farmstay accommodation is in the homestead and there are farm animals, a glorious garden and garden spa. Fruit tasting tours available by appointment only. Home cooked meals and a table for guests under a huge African palm tree. Breakfast, lunch and dinner daily are included in the tariff.
- **Location:** 13km west of Mareeba

MILLAA MILLAA

🍎 PRODUCERS

MUNGALLI CREEK DAIRY AND OUT OF THE WHEY CAFE
Robert and Danny Watson, Sally McPhee
Mungalli Creek Farmhouse, 256 Brooks Rd, Millaa Millaa 4886
Phone: 07 4097 2232
Email: mungalli@bigpond.com

This biodynamic dairy farm is the supplier of the largest range of biodynamic, organic and farm fresh products in Far North Queensland, and the only farmhouse cheese-making operation in north Queensland. The range includes milk, butter, cream, havarti, fetta, blue vein cheese, yoghurt with indigenous fruit flavours and quark with indigenous herb flavours. Visitors can enjoy tastings in the factory and meals in the appropriately named cafe, which specialises in Devonshire teas and cheese platters. The products are also available at stores throughout North Queensland.
- **Open:** 10am–4pm daily
- **Location:** 10km south of Millaa Millaa

MISSION BEACH

🍷 BREWERY/WINERY

PARADISE ESTATE WINES
Mary and Tony Lankester
6-3 Dewar St, Mission Beach 4852
Phone: 07 4088 6888
Email: info@paradisewines.com.au
www.paradisewines.com.au

Opened in 2004, this is Australia's only winery to specialise in banana wine and they also blend different fruits with it to create a tropical experience. There are five wines available, each dry and sweet, creating ten wines. There's Paradise Gold (banana and French oak), Tropical White (banana, coconut, pineapple and mangosteen), Tropical Red (banana, mango, pitaya or dragon fruit, and vanilla French oak), Rainforest White (banana, bush limes, mango and persimmon), and Rainforest Red (banana, Australian native Davidson plum, strawberry, blueberry and French oak).
- **Open:** 10.30am–5pm Wed–Sun, or by appointment
- **Location:** North Mission Beach

MORESBY

🍎 PRODUCERS

MARY'S PASTA PRODUCTS
Bert Pagano
21 Moresby Rd, Moresby 4871
Phone: 07 4063 2463
Email: maryspasta@comnorth.com.au

This place was once a family-owned grocery store, but in the 1960s, when supermarket chains opened in town and signalled the end for general stores, Bert Pagano's parents needed to change their business. So his mother, Mary, started making ravioli and selling them in the shop. As people began tasting and buying it regularly, each passing year saw less groceries being sold and pasta sales increasing. No wonder, as Bert says, 'Many of our customers tell us that Mary's ravioli are the best they have ever eaten, anywhere in the world'. Flavoured pastas include spinach, chilli, pepper, lemon myrtle, and kaffir lime.
- **Open:** 7am–6pm Mon–Sat
- **Location:** 15km south of Innisfail

MOSSMAN

🍎 PRODUCERS

DAINTREE TEA COMPANY
Nicholas family
Cape Tribulation Rd (PMB 15),
Mossman 4873
Phone: 07 4098 9139
Email: daintreetea@austarnet.com.au
www.daintreetea.com.au
Halfway between Cape Tribulation and the ferry across the Daintree River, many travellers are surprised to see ordered rows of tea bushes growing. The Daintree Tea Plantation began in 1978, but has only been in production since 1991. Only black teas are grown and processed on the premises. The tea is sold in most north Queensland tourist shops, by mail order, or from the property.
- **Open:** By appointment only
- **Location:** 5km from Hutchinson Creek bridge across the Daintree

SHANNONVALE TROPICAL FRUIT WINE COMPANY
Tony and Trudie Woodall
Shannonvale Rd, Mossman 4873
Phone: 07 4098 4000
Email: ttwoodall@yahoo.com.au
www.shannonvalewine.com.au
This five-acre orchard grows seventy different fruit trees from which the Woodalls obtain most of the fruit they need for wine making. The wine tasting area is under a vine-covered pergola with great views. They apply current commercial wine making techniques to organically raised tropical fruit and currently have twelve different wines available for tasting and sales including carambola (star fruit), grapefruit, purple star apple, passionfruit, mulberry, jaboticaba, mango, orange, black sapote and ginger. They say that they 'welcome organised or disorganised tours of any number of people'.
- **Open:** 10am–4.30pm Tues–Sun
- **Location:** Fifteen minutes north of Port Douglas and ten minutes south of Mossman

SOMERSET HORTICULTURAL FARM
Bruno and Deirdre Scomazzon
Scomazzon's Rd, Mossman 4873
Phone: 07 4098 3446 or 07 4098 2446
Email: deir@ledanet.com.au
The Scomazzons' open-sided fruit stall may look, at first glance, like any other fruit stall as you whiz past on the Captain Cook Hwy heading for Daintree. It's not. Slow down and check out the huge range of tropical fruit—much of it grown on the property beside the stall. You will find rowlinnias (like a massive custard apple, yet tastier), red Thai pawpaw, abiu, Malay apples, canistel, black sapote, jackfruit, soursops and of course bananas, pineapples, mandarins and avocados.
- **Open:** 7.30am–6pm Mon–Sat, 10am–4pm Sun
- **Location:** About 6km north of Mossman

🏠 STORES

YUM YUMS
Aafke McPhie
20 Front St, Mossman 4873
Phone: 07 4098 1490
Plenty of fresh exotic fruits, vegetables, organics, health and Asian foods here. Yum Yums is famous for its homemade fruit smoothies and soft-serve yoghurt ice-cream.
- **Open:** 8am–5.15pm weekdays, 8am–12pm Sat

FAR NORTH—QLD

> **SUGAR CANE JUICE**
> While in this area, watch out for places that sell sugar cane juice such as Somerset Horticultural Farm at Mossman as well as some of the local markets. It's a uniquely Queensland taste sensation.

DINING

TREEHOUSE RESTAURANT
P&O Resorts, Silky Oaks Lodge and Healing Waters Spa, Finlayvale Rd, Mossman 4873
Phone: 07 4098 1666
Email: silky.oaks@poresorts.com
www.poresorts.com
Licensed
Cuisine: Modern Australian
Mains average: $$
Specialty: Seafood
Local flavours: Seafood, fruit
Extras: Open-air restaurant that sits high above the Mossman River. Luxury accommodation and spa.
- **Open:** Breakfast, lunch (best for non-guests), dinner, morning and afternoon teas (for guests) daily
- **Location:** First left on Syndicate Rd after Mossman, then left on Finlayvale Rd and travel to the end of the road

MOSSMAN GORGE

TOURS

KUKU YALANJI DREAMTIME WALKS
Emma Burchill, Manager
Bamanga Bubu Ngadimunku Inc,
Mossman Gorge Rd, Mossman Gorge
Phone: 07 4098 2595
Email: tours@yalanji.com.au
www.yalanji.com.au

At the edge of the Mossman Gorge Park, rangers will take you on a fascinating journey that touches on Aboriginal medicine, foods and lore. You will see fruits that are 'booyang'—poisonous—and others that are delicious, such as white apples and wild pears. See how the leaves of the soap tree lather up to clean the dirtiest hands, and discover the branches that smell like heat-rub. Pandanus centres can be boiled like cabbage, and part of the umbrella tree will alleviate toothache. Walks take about an hour and a half, and there is an art and craft shop at the centre too.
- **Open:** 10am, 12pm, 2pm, weekdays

PALM COVE

DINING

CANECUTTERS RESTAURANT
Matthew Lock (executive chef)
Novotel Palm Cove Resort, Coral Coast Dr, Palm Cove 4879
Phone: 07 4059 1234
Email: res@novotel-pcr.com.au
www.novotel-pcr.com.au/
Licensed
Cuisine: International
Mains average: $$$
Specialty: Themed buffet dinners
Local flavours: Seafood, meat, fruit and vegetables
Extras: Dine on the terrace overlooking the pool. The Saturday seafood buffet is very popular.
- **Open:** Breakfast, lunch and dinner daily

THE REEF HOUSE RESTAURANT
Philip Mitchell (executive chef)
The Sebel Reef House and Spa, 99 Williams Esplanade, Palm Cove 4879

QLD—FAR NORTH

Phone: 07 4055 3633
Email: info@reefhouse.com.au
www.reefhouse.com.au
Licensed
Cuisine: Modern Australian
Mains average: $$$
Specialty: Seared barramundi
Local flavours: Seafood, locally grown vegetables, salad greens
Extras: Acclaimed restaurant with outdoor deck dining.
■ Open: Breakfast, lunch and dinner daily

PORT DOUGLAS

STORES

MOCKA'S PIES
Anne and Peter Lloyd
9 Warner St, Port Douglas 4871
Phone: 07 4099 5295
Email: thelloyds@dodo.com.au
Famous in Port Douglas for thirty-something years, these pies include plain steak, as well as bacon, mushroom, pea, onion, kidney, or chicken, all made on site. 'We always put good quality chunky meat in', they say, 'and they are all homemade, even the pastry'. The vegetarian pie is popular too, as the Lloyds find that even confirmed meat-eaters often prefer lighter fare. The original Mocka started making pies at the Central Hotel, then moved to the shop across the road in 1973. Anne Lloyd started working for him, then finally bought him out when he retired. Sensibly, she has kept to the same recipes, although the shop moved from Macrossen Street a few years ago.
■ Open: 9am–3pm daily (unless sold out which, Anne says, can quite often happen by early afternoon in the middle of the season)

DINING

FIORELLI'S RESTAURANT AND BAR
Grant and Joanne Williams
Shop 16, Marina Mirage, Port Douglas 4871
Phone: 07 4099 5201
Email: info@fiorellisrestaurant.com.au
www.fiorellisrestaurant.com.au
Licensed
Cuisine: Tropical produce
Mains average: $$$
Specialty: Mother of All Seafood Platters
Local flavours: Kangaroo, crocodile, seasonal tropical fruits
Extras: Situated on the boardwalk at the marina, with views of the boating harbour.
■ Open: Breakfast, lunch and dinner daily

NAUTILUS RESTAURANT
Grahame Wearne, Erwin Salaun (chef)
17 Murphy St, Port Douglas 4871
Phone: 07 4099 5330
Email: nautrest@austarnet.com.au
www.nautilus-restaurant.com.au
Licensed
Cuisine: Modern Australian
Mains average: $$$$

SUGAR INDUSTRY MUSEUM
Bruce Hwy, Mourilyan 4858
Phone: 07 4063 2656
Email: asim@znet.net.au
www.sugarmuseum.org.au
Located about 70km south of Cairns, the Australian Sugar Industry Museum has displays of the sugar industry, including processing, video updates on the current industry, and a gallery. The retail shop sells every type of sugar product from molasses to coffee sugar, and the cafe uses local produce.
■ **Open:** 9am–5pm, weekdays; 8am–3pm, Sat; 9am–3pm, Sun (off season: Dec–April, 9am–3pm Sat, 9am–12pm Sun)

FAR NORTH–QLD

Specialty: Whole coral trout, mud crab dishes
Local flavours: Local seafood, fruit and vegetables, Queensland Avon beef
Extras: All dining is outdoors in a rainforest setting, either the Palm Courtyard or Forest Deck.
■ **Open:** Dinner daily

ON THE INLET

Mark Stephens (general manager),
Thomas Clever (chef)
3 Inlet St, Port Douglas 4871
Phone: 07 4099 5255
Email: ontheinlet@bigpond.com
www.ontheinlet.com.au
Licensed
Cuisine: Seafood
Mains average: $$
Specialty: Whole coral trout (for one or two), whole baby barramundi, live local mud crabs and crayfish
Local flavours: Mud crabs, coral trout, barramundi, yabbies, prawns, daily fresh fish
Extras: On the Inlet has its own seafood wholesaler, Port Douglas Seafoods. The restaurant is located on Dickson Inlet and built on suspended decking over the water and has a retractable roof and views across the Coral Sea to the sugar town of Mossman and Daintree.
■ **Open:** Lunch and dinner daily
● **Location:** Next to the Marina Mirage on Wharf Street

RAINFOREST HABITAT

Doug Ryan
Port Douglas Rd, Port Douglas 4871
Phone: 07 4099 3235
Email: info@rainforesthabitat.com.au
www.rainforesthabitat.com.au
Licensed
Cuisine: Australian

Set price meal, drink included: $$$
Specialty: Breakfast buffet
Local flavours: Everything is local
Extras: Dine with the birds that wander through the dining area at this truly amazing rainforest recreation.
■ **Open:** Breakfast and lunch daily, dinner functions Fri–Sat (by appointment July–Oct)
● **Location:** Corner Captain Cook Hwy

SASSI CUCINA

Tony Sassi and Sato
Cnr Wharf and Macrossan Streets,
Port Douglas 4871
Phone: 07 4099 6100
Email: dine@sassi.com.au
www.sassi.com.au
Licensed
Cuisine: Modern Italian seafood and Japanese
Mains average: $$$
Specialty: Seafood, including sushi from the sushi bar
Local flavours: Seafood, herbs, salad greens, fruit and vegetables
Extras: Choice of airconditioned or alfresco with views across the park to the sea.
■ **Open:** Lunch and dinner daily

2 FISH SEAFOOD RESTAURANT

Jeff Gale (owner/manager) and Sheldon Wearne (owner/chef)
7/20 Wharf St, Port Douglas 4871
Phone: 07 4099 6350
Email: dine@2fishrestaurant.com.au
www.2fishrestaurant.com.au
Licensed
Cuisine: Contemporary Australian seafood
Mains average: $$$
Specialty: Fresh, locally caught seafood
Local flavours: Reef fish fillets
■ **Open:** Lunch and dinner daily

SILKWOOD EAST
🍷 BREWERY/WINERY

MURDERING POINT WINERY
Berryman family
161 Murdering Point Rd, Silkwood East 4856
Phone: 07 4065 2327
Email: info@murderingpointwinery.com.au
www.murderingpointwinery.com.au
This family-owned boutique winery opened in November 2003, after four years of planning and development. The Berryman family have been cultivating the lush, fertile soils of Silkwood in the heart of the tropical rainforest belt since 1952. Over this time they have produced first quality sugar cane, pineapples, pawpaws, passionfruit and now nine tropical fruit wines created from mango, passionfruit, pineapple and the lesser known native fruits lemon aspen and Davidson plum.
- **Open:** 10am–6pm daily or by appointment

SMITHFIELD
🍴 DINING

THE COFFEE WORKS CAFE
Karlo Beringer
Smithfield Shopping Centre (next to Woolworths), Smithfield 4878
Phone: 07 4038 2838
This cafe is the ideal showcase for tablelands-grown local coffee—nine varieties, fresh from the roaster. There is a regional-flavoured menu, and there are local products for sale, as well as local teas.
- **Open:** 8am–5.30pm Mon–Sat (8am–8pm Thurs)
- **Location:** 12km north of Cairns

TARZALI
🍎 PRODUCERS

TARZALI LAKES FISHING PARK
Peter C Whiddett
Lot 3, Millaa-Malanda Rd, Tarzali 4885
Phone: 07 4097 2713
Email: tarzalilakes@ledanet.com.au
www.ledanet.com.au/~tarzalilakes
Tarzali Lakes is an aquaculture farm growing and selling North Queensland jade perch to the local market with over 50 000 perch in seven large ponds on forty acres. You can catch and release, catch and cook on the barbecues or buy to take away. Fish can be purchased fresh daily from the farm gate, and orders in lots of 20kg can be shipped to your door.
- **Open:** 10am–6pm Thurs-Tues, Wed by appointment only
- **Location:** Halfway between Millaa Millaa and Malanda

TOLGA
🏠 STORES

THE BIG PEANUT
BJ and VL Coombes
Kennedy Hwy, Tolga 4882
Phone: 07 4095 4246
As you'd expect, this shop sells peanuts in all flavours, but also local fruit, jams and honey, seasonal vegetables and great north Queensland coffee.
- **Open:** 7.30am–5.30pm daily
- **Location:** 2km north of Tolga

THE HUMPY
Peter and Giovana Griffiths
Kennedy Hwy, Tolga 4882
Phone: 07 4095 4102

This new look fruit and vegetable shop stocks as much local produce, straight from the growers, as they can get hold of. Of course the fresh goods are seasonal, but there's always local Tablelands pepper, biscuits, honey, jams, coffee and tea.
- **Open:** 8am–6pm daily
- **Location:** North of Tolga

DINING

THE HOMESTEAD RESTAURANT PLANTATION AND LODGE
Toni Dahl
Beantree Rd, Tolga 4882
Phone: 07 4095 4266
Email: homestead@qldnet.com.au
Licensed
Cuisine: Modern Australian
Mains average: $$
Specialty: Seafood dishes
Local flavours: As much as possible, including fruit and meat
Extras: A tropical fruit display orchard with 123 exotic fruit trees and a 5000-tree avocado plantation is the setting for this large restaurant with barbecue area, platypus viewing (yes, really) and accommodation available. There are homemade jams, chutneys and dried fruits for sale.
- **Open:** 10am–3pm daily, lunch daily, dinner Wed–Mon

WALKAMIN

BREWERY/WINERY

MOUNT UNCLE DISTILLERY
Mark Watkins
1819 Chewko Rd, Walkamin 4872
Phone: 07 4086 8008
Email: info@mtuncle.com
www.mtuncle.com

Mt Uncle Distillery produces a wide range of 100 per cent natural and preservative-free liqueurs and spirits, including what they call 'elixir de musa'. This is a banana liqueur with a well-rounded, light exotic flavour. There are also macadamia, avocado and sucrier banana products. The distillery produces limoncello, limecello, coffee liqueur, mulberry liqueur, mango liqueur and introduced banana brandy in 2004. Stay tuned for banana rum. All liqueurs are medal-winners.
- **Open:** 10am–5pm daily
- **Location:** Off Kennedy Hwy

YUNGABURRA

BREWERY/WINERY

WINEWORKS DOWNUNDER
Lake Eacham Roadhouse, cnr Gillies Hwy and Lake Barrine Rd, Yungaburra 4872
Phone: 07 4096 2117
Email: admin@wineworksdownunder.com
www.wineworksdownunder.com
This historic winery with a beautiful old Queenslander house is the only grape winery in North Queensland. These are the oldest vines in Queensland, and there is a grape wine for sale, but also delicious fruit wines made from local fruits—lychee, grapefruit, water cherry, black sapote, as well as Snakebite (a chilli and lime eye-opener) and some intriguingly named grape blends. All available here, or at the Cairns night markets, Kuranda Heritage markets, and Willows markets, Townsville.
- **Open:** 10am–5pm daily
- **Location:** Second road on right after Wondella is Flaggy Creek Rd

MARKETS AND EVENTS

✦ ATHERTON MARKET, ATHERTON
07 4096 5382
From 7am first Sat

✦ COOKTOWN–ROSSVILLE MARKET,
COOKTOWN
07 4060 3907
9am–12pm Sat, fortnightly

✦ COOKTOWN MARKET, COOKTOWN
07 4060 3937
8am–12pm Sat (weather permitting)

✦ GORDONVALE MARKET,
GORDONVALE
07 4056 1940
7.30am–12pm, first Sat (except Jan)

✦ HERBERTON MARKET, HERBERTON
07 4096 2408
7.30am–12pm third Sun

✦ KURANDA MARKETS, KURANDA
07 4093 8772
7am–3pm Wed–Fri and Sun

✦ MAREEBA MARKETS, MAREEBA
07 4092 3418
8am–12pm second Sat

✦ PORT DOUGLAS MARKETS,
PORT DOUGLAS
07 4098 5159
8.30am–2pm Sun

✦ RAVENSHOE MARKET, RAVENSHOE
07 4097 6044
8am–12pm fourth Sun

✦ RUSTY'S MARKETS, CAIRNS
07 4051 5100
7am–9.30pm Fri, 6am–2pm weekends

✦ YUNGABURRA MARKET,
YUNGABURRA
07 4095 2111
Fourth Sat

✦ ATHERTON FOOD FESTIVAL,
ATHERTON
07 4091 0700
First weekend in Oct

✦ ATHERTON MAIZE FESTIVAL,
ATHERTON
07 4091 0700
Sept

**✦ CARNIVALE—LONGEST LUNCH
AND HARVEST OF THE CORAL SEA,**
PORT DOUGLAS
07 4099 5775
Late May

**✦ MISSION BEACH BANANA
FESTIVAL,** MISSION BEACH
07 4068 7099
July

**✦ PACIFIC TOYOTA PALM COVE
FESTIVAL,** PALM COVE
07 4031 1321
Sept

✦ SALAMI FESTIVAL, YUNGABURRA
07 4095 3330
Second weekend in Sept

GOLD COAST

A short drive to the south of Brisbane, the Gold Coast's Hollywood looks lure many at least for a while. Apparently brash and glitzy, the soft underbelly is its proximity to a fascinating and unspoilt hinterland on one side and that glorious ocean, flanked by ruler-straight beaches on the other.

Those towering five-star and international hotels also have excellent restaurants and many are justly proud of serving Queensland produce to diners.

Make sure you visit the hinterland and drop in to the unique cafes at Springbrook, Binemail, Burra, Beechmont and Tamborine Mountain, many of them joyously celebrating fresh local produce.

CURRUMBIN
STORES

SUPERBEE HONEYWORLD
Hillary and Raewyn Wright
35 Tomewin St, Currumbin 4223
Phone: 07 5598 4548
Email: superbee@superbee.com.au
www.superbee.com.au
A great family spot with lots of honey tastings, honey and honey products to buy, and the children can learn just how clever these tiny insects really are, especially by watching the Live Bee Show.
■ **Open:** 8am–5.30pm daily
● **Location:** Opposite Currumbin Wildlife Sanctuary

MAIN BEACH
DINING

SHUCK ON TEDDER
Scott and Anna Budgeon (owners),
Alex Chipizubov (chef)
20 Tedder Ave (cnr Woodroff),
Main Beach 4217
Phone: 07 5528 4286
Licensed
Cuisine: Seafood
Mains average: $$$
Specialty: Oysters, seafood, steaks
Local flavours: Locally caught fish and seafood
Extras: Winner Best Seafood Restaurant on the Gold Coast 2003, Restaurant and Caterers Award, good spot for people watching (aka celebrity spotting!).
■ **Open:** Lunch and dinner daily

VIE BAR
Steve Szabo
Palazzo Versace, 94 Seaworld Dr,
Main Beach 4217
Phone: 07 5527 0311
Email: vie@palazzoversace.com
Licensed
Cuisine: Modern Australian
Mains average: $$$
Local flavours: Seafood
Extras: Restaurant overlooks Broadwater.
■ **Open:** Lunch and dinner daily

KINGSCLIFF ROADSIDE STALLS
Watch out for roadside fruit and vegie stalls in season along the road through the Tweed Valley to Kingscliff.

QLD—GOLD COAST

MT COTTON
🍴 DINING

RESTAURANT LURLEENS
Terry Morris (owner),
Erik van Alphen (chef)
Sirromet Wines, 850 Mount Cotton Rd,
Mt Cotton 4165
Phone: 07 3206 2999
Email: wines@sirromet.com
www.mountcottonestate.com
Licensed
Cuisine: Modern Queensland
Mains average: $$$
Specialty: Slow pan-roasted New England white rabbit au basilic
Local flavours: Organic vegetables from the winery gardens, herbs, fruit, eggs from free-range chickens and other local produce such as Moreton Bay bugs
Extras: Alfresco dining available on the balcony overlooking Moreton Bay and North Stradbroke Island.
■ **Open:** Breakfast, lunch, morning and afternoon tea daily, dinner Thurs–Sat
● **Location:** Old Cleveland Rd exit, Pacific Motorway

MT TAMBORINE
🍴 DINING

MT TAMBORINE COFFEE PLANTATION AND FINGERPRINT GALLERY
Aemelia Smith
64 Alpine Tce, Mt Tamborine 4272
Phone: 07 5545 3856
Email: Bakerp007@aol.com
www.fingerprint_gallery.com.au
Cuisine: Light lunches
Mains average: $
Specialty: Soups and scones

Local flavours: Organic vegetables, natural organic coffee
Extras: Set amongst 900 coffee trees, the verandas offer great bushland and plantation views. The coffee grown here is said to have the lowest caffeine content (one per cent) of any coffee in the world. You can buy the coffee and there is also a mail order service and an art-and-craft gallery.
■ **Open:** 10am–4pm daily

MUDGEERABA
🍎 PRODUCERS

CHARELLA FARMSTEAD GOAT DAIRY
Parsons Family
151 Berrigans Rd, Mudgeeraba 4213
Phone: 07 5525 3525
Email: drcharlesparsons@hotmail.com
The Charella range of goats cheese, has had astounding success. Charella cheese has been served on Qantas First Class cheese platters and is used by many top Gold Coast restaurants and hotels, as well as those in Brisbane and the Sunshine Coast. But don't go looking for it in the shops. 'We sell direct to the public', say the Parsons, 'strictly, direct'. It was this firm commitment that set Charles Parsons looking for the best possible way to meet his customers. Farmers' markets were just becoming popular a couple of years ago and so he waded in to begin the Mudgeeraba markets in early 2001. He and the four other small cheese (two goat, three cow) producers needed a place to sell their cheeses, milk and yoghurt. The result of these markets was so successful that he has taken on the running of other markets in the area.
■ **Open:** By appointment

GOLD COAST—QLD

NORTH TAMBORINE
🍷 BREWERY/WINERY

TAMBORINE MOUNTAIN DISTILLERY
Michael and Alla Ward
87-91 Beacon Rd, North Tamborine 4272
Phone: 07 5545 3452
Email: info@tamborinemountaindistillery.com
www.tamborinemountaindistillery.com
This privately owned distillery makes liqueurs, and spirits such as schnapps, gin, vodka, rum and whisky and offers tastings as well as sales. Internationally award-winning (rated second only to Benedictine Dom in one contest), it is the smallest distillery in the world, yet in the International Wine and Spirit Competition in London, it won three awards. You should also know that liqueur fruit is made here, and the liqueur chocolates are filled with forty per cent alcohol. The products are only sold from the property, and Alla Ward handpaints every bottle.
■ **Open:** 10am–3pm, Wed–Sun

TAMBORINE
🍴 DINING

VERANDAH RESTAURANT
David and Janette Bladin
Albert River Wines, 117 Mundoolun Connection Rd, Tamborine 4270
Phone: 07 5543 6622
Email: info@albertriverwines.com.au
www.albertriverwines.com
Licensed
Cuisine: Modern Australian
Mains average: $$$
Local flavours: Vegetables
Extras: Extensive views of the hinterland. The restaurant overlooks the courtyard where a jazz band plays every Sunday from 12pm to 3pm. The Vineyard Chapel on site is used for weddings and functions.
■ **Open:** Lunch Fri–Sun
● **Location:** Fifteen minutes from Canungra

UPPER COOMERA
🍴 DINING

THUMM ESTATE WINES
Robert and Janet Thumm
87 Kriedeman Rd, Upper Coomera 4209
Phone: 07 5573 6990
Email: winemaker@thummestate.com
www.thummestate.com
Licensed
Cuisine: Cafe style
Mains average: $
Extras: Local deli selling cheeses, alfresco dining overlooking the majestic Wongawallen valley. Makes Divine Super Grape spread—a world first.
■ **Open:** 9.30am–5pm, daily
● **Location:** Take exit 57, Tamborine-Oxenford Rd, eight minutes off the Hwy

---

GOLD COAST FOOD AND WINE TRAILS
Gordon Hay, PO Box 5042, Qld 9729
Phone: 07 5582 8771
Email: haveitall@goldcoast.qld.gov.au
www.goldcoast.qld.gov.au/tourism
The Gold Coast City Council has recently developed three food trails: Brisbane to the Gold Coast, Gold Coast to Beaudesert, and New South Wales to the Gold Coast, crammed with interesting information on the best food and wine-related places to visit as you travel around.

MARKETS AND EVENTS

◆ **HINTERLAND COUNTRY MARKETS,** TAMBORINE MOUNTAIN
07 5545 3200
Second Sun of month

◆ **MAIN BEACH FARMERS' MARKET,** MAIN BEACH
07 5584 6217
First and third Sat

◆ **MARINA MIRAGE FARMERS' MARKETS,** MARINA MIRAGE
07 5525 3525
7am–12pm first and third Sat

◆ **MUDGEERABA FARMERS' MARKET,** MUDGEERABA
07 5525 3525
First and third Sat 6am–11am Oct–Apr
7am–12pm May to Sept

◆ **PARKLANDS FARMERS' MARKETS,** PARKLANDS SHOWGROUND
07 5525 3525
7am–12pm
Second and fourth Sat

◆ **REDLANDS FARMERS' MARKET,** MOUNT COTTON
07 3821 4460
8am–1pm last Sat

◆ **SOUTHPORT FARMERS' MARKET,** SOUTHPORT
07 5584 6217
Second and fourth Sat

◆ **TAMBORINE MOUNTAIN COUNTRY MARKETS,** MOUNT TAMBORINE
0417 618 379
8am–3pm second Sun

◆ **BEENLEIGH CANE FESTIVAL,** BEENLEIGH
07 3807 6316
May

◆ **GOLD COAST SIGNATURE DISH COMPETITION,** MAIN BEACH
December

◆ **TAMBORINE SHOW,** MT TAMBORINE
tambourineshow@hotmail.com
September

FOOD INFORMATION
For web information on this exciting area, check these sites:
www.goldcoast.qld.gov.au
www.goldcoastfoodforum.org.au

SUNSHINE COAST

Rich in beauty, produce and ambience, the Sunshine Coast offers each visitor a subtly different cocktail of enjoyment. The superb coastal fringe, elegantly draped with holiday accommodation options, suits some. Others flee to the hinterland, delving into tiny villages, the haunt of artists and artisans.

Markets, especially the famous Saturday Eumundi Markets, bring it all together and visitors can be sure of finding the best local wares and produce.

This is a region of gardens and farms that sell direct to markets in Brisbane and beyond, growing the state's pineapples and macadamias, vegetables and herbs. Yet just when you despair of finding somewhere you can get to know the produce, up pops the Big Pineapple.

BELLI PARK
PRODUCERS

KENILWORTH ORGANIC OLIVES
Joel and Shirin Donaldson
1870 Eumundi-Kenilworth Rd,
Belli Park 4562
Phone: 07 5447 0047
Email: olive1@iprimus.com.au
Three years ago the Donaldsons bought this farm from a conventional farmer and have been turning it organic ever since. With a passion for olive oil and the growing of olives, their aim was to create an extra special olive oil. They are careful to press the olives from their 1200-plus olive trees at exactly the right time to capture the flavours they need. They then blend the result with specially selected olive oils to create an organic, cold-pressed extra virgin olive oil. Look for it in antique green bottles at selected retail outlets in the Sunshine Coast corridor.
- **Open:** By appointment
- **Location:** 19km inland from Eumundi, off the Bruce Hwy

CABOOLTURE
PRODUCERS

HAVEN GLAZE
Cindy Elliott
5 Auster Court, Caboolture 4510
Phone: 1800 551 528 / 07 5495 2045
Email: info@gingerpeople.com
www.gingerpeople.com
This ginger processing business uses plenty of local ginger. You can taste and buy at the factory door. There's crystallised ginger, either bulk or in retail packs, and wonderful ginger chews.
- **Open:** 8am–4pm weekdays

CALOUNDRA
DINING

TANJA'S BEACH PAVILION
Gerard Schoendorfer
8 Levuka Ave, Kings Beach,
Caloundra 4551
Phone: 07 5499 6600
Email: euronatural@ozemail.com.au
Licensed
Cuisine: Organic modern Australian

Two-course set menu: $$
Specialty: Seafood platters for one or two
Local flavours: As much as possible, including vegetables, fruit, meat, seafood
Extras: Homemade ice-cream. This organic restaurant is part of a chain of five restaurants. All bread is organic and baked in the restaurant's bakery. Only organic meat is used, plus free-range or organic chickens and eggs, and all cakes are free of artificial additives. Takeaway is available from the cafe.
- Open: Breakfast, lunch and dinner daily, à la carte dinner Wed-Sun

CHATSWORTH
DINING

HIGHBURY MOUNTAIN—ON FARM CAFE
Chris and Peter Clifford
MS 483, 1 Irvine Rd, Chatsworth 4570
Phone: 07 5482 4222
Email: highbury5@yahoo.com
www.freewebs.com/highbury
BYO
Cuisine: Light meals
Mains average: $
Specialty: Devonshire teas with rosella jam, steak, coffee
Local flavours: Beef, chicken, seafood, vegetables
Extras: Rosella (fruit, not birds) plantation, gift shop, self-contained cottage with lovely views over the Mary Valley, Highbury homestead with three bedrooms. Gift shop selling homemade and local products.
- Open: Lunch, morning and afternoon teas daily, dinner Fri-Sat (bookings for large groups)
- Location: 1.5km from the Bruce Hwy, five minutes north of Gympie

CONONDALE
PRODUCERS

THE STOCK EXCHANGE AND VILLAGE ORGANIC FARM
Julie Shelton and Pat Forsman
14 Crystal Waters, Kilcoy Lane, Conondale 4552
Phone: 07 5494 4699
Email: tvof@bigpond.com
www.crystalwaters.org.au
This couple's passion is for 'producing nutritious food for the local community'. And, since 1999, their range of organic dairy products (including milk, butter, cream and ice-cream), beef, poultry and pork products have been doing just that. They only butcher and process animals that have been born and raised on the property. They also supply a large range of organic wholefoods through a retail outlet so that customers can get most of their food needs from the one source. While the Village Organic Farm products are currently only available to residents of Crystal Waters, they intend to seek accreditation as a registered dairy, cheesery and butchery so that their products can be offered to all of their customers.
- Open: Retail outlet: 8am-11am, Wed and Sat. Phone to arrange a farm visit
- Location: 2km Maleny side of Conondale, left into Aherns Rd, then 7km further to Kilcoy Lane

CRYSTAL WATERS
PRODUCERS

LINDEGGER ORCHARD
Max and Trudi Lindegger
59/65 Kilcoy Lane, Crystal Waters 4552
Phone: 07 5494 4741

Email: office@ecologicalsolutions.com.au
www.ecologicalsolutions.com.au/crystal-waters
Max Lindegger, one of the founders of the Crystal Waters Eco Village, also runs this certified organic orchard. He specialises in growing kaffir lime trees for their leaves, and also sells pecan nuts at the Crystal Waters market and from the property. This village is worth seeing as it is a twenty-first century model for low impact, sustainable living.
- **Open:** By appointment

EUMUNDI
PRODUCERS

AUSTRALIAN NOUGAT COMPANY
Scott Schirmer
4 Tallgum Ave, Eumundi, 4562
Phone: 07 5442 7617
Email: ausnougat@email.tc
www.ausnougat.com.au
How to make macadamias even better. Macadamia Bliss, a French-style soft nougat, is sold in major department stores, confectionery shops, duty free, and specialty stores, but it all begins here.
- **Open:** 8am–4pm weekdays, 9am–1pm Sat
- **Location:** 3km east of Eumundi on Eumundi–Noosa Rd

STORES

EUMUNDI MARKETS
Peter Homan, Organiser
Historical Association Grounds, Main St, Eumundi 4562
Phone: 07 5442 7106
Email: manager@eumundimarket.com.au
www.eumundimarket.com.au

Famous throughout the country, these markets with around 500 stallholders, are run by the Historical Association. Their motto is 'make it, bake it, sell it, grow it'.
- **Open:** 8am–1pm Wed, 6.30am–2pm Sat

GYMPIE
PRODUCERS

THE NUT FARM
John and Lesley Groves
335 Marys Creek Rd, Gympie 4570
Phone: 07 5483 4746
Email: johnandlesley@thenutfarm.com.au
www.thenutfarm.com.au
After six years in this business, the Groveses are very busy. They grow macadamias, avocados, and also have a small Limousin cattle stud. They have developed their own line of macadamia products and have nuts in various flavours that include chocolate-coated nuts, as well as some that are roasted in macadamia honey gathered from hives on their farm. There are also dry-roasted, salted, and natural macadamias and the Nut Farm's own cold-pressed macadamia oil, pressed in Kingaroy. More? Yes—natural macadamia honey, macadamia paste, as well as the Nutters skincare range. All these are available from the farm, or by mail order throughout Australia. We said they were busy.
- **Open:** Phone first

KENILWORTH
PRODUCERS

COOLABINE FARMSTEAD GOATS CHEESES
Dee Dunham

Coolabine Rd, Kenilworth 4574
Phone: 07 5446 0616
coolabine@coolabinefarmstead.com
www.coolabinefarmstead.com
The cheese from this chemical free farm in the lovely Obi Obi valley won the prize as maker of the best non-bovine cheese at the Sydney Royal Easter Show in 2003. The dairy also won four gold and a championship at the Brisbane Show last year. Their eighty-strong herd of Nubian and Saanen goats are 'like pets', says Dee. They are milked twice daily and even treated with homeopathic remedies when necessary. If you want to learn how to make goat's cheese, Coolabine runs one-day workshops from time to time.
- **Open:** 10.30am–4pm Wed–Sat, or by appointment
- **Location:** See website for help to find the farm

STORES

KENILWORTH COUNTRY FOODS
Henry Gosling
45 Charles St, Kenilworth 4574
Phone: 07 5446 0144
Email: kcf@coastnet.net.au
This is the home of Malling Red and Malling Roma cheeses, developed in Australia by Danish cheese-maker Peter Hansen at the Malling factory in the 1920s, which won a gold medal in Brisbane the first year he entered them. When the Kraft factory in Kenilworth closed a few years ago, the ex-employees bought the factory so that they could stay employed and also continue the business. It's a good thing they did, as their speciality cheeses and yoghurts, hand-crafted by traditional methods and using milk from local dairy herds, are very good. Sold throughout Queensland, mostly in the south-east corner and Sunshine Coast, some find their way to capital cities in other states. The range includes hard cheeses, ricotta, fetta flavoured and processed cheeses and traditional set-in-the-tub yoghurts. Look for Kenilworth Classic Mild, Classic Vintage and Classic Matured.
- **Open:** 9am–4pm, weekdays; 10am–3pm, weekends
- **Location:** Twenty minutes from Hwy 1, turn off at Eumundi, signposted at Kenilworth

MALENY

STORES

MAPLE STREET CO-OP
Alan Harrington (Manager)
37 Maple St, Maleny 4552
Phone: 07 5494 2088
Email: maplest.coop@serv.net.au
This place, in an old building in the main street has been operating for a dozen years as an organic co-op. There is a wide range of locally grown organic fruit and vegetables and a host of other products which are, wherever possible, organic. There are no GMO products, and plenty of bulk food, books, and other things.
- **Open:** 9am–5.30pm weekdays, 9am–4pm weekends

DINING

COLIN JAMES FINE FOODS
Colin and Jean Cunningham
37 Maple St, Maleny 4552
Phone: 07 5494 2860
Email: finefoods@bigpond.com

Licensed
Cuisine: Gourmet light meals
Mains average: $
Specialty: Gourmet sandwiches on homemade bread, antipasto platters, cheese platters
Local flavours: As much as possible
Extras: Maleny is a centre for many other activities—walks, galleries, views. The attached store has around 150 varieties of imported and Australian cheeses at what is a surprise packet of a place in a small country town. But even though the cheeses are good, the gold award ice-cream, which is made on the premises using local milk, seems to be even more popular. There are thirty flavours, making use of local produce such as macadamias and mangoes, and they have scooped up the Champion Dairy Dessert award at the Brisbane Show for the last two years. The shop also stocks fine deli lines, smallgoods, meats, pâtés and fish.
■ **Open:** 9am–5pm, weekdays; 9am–4pm, weekends

TERRACE SEAFOOD RESTAURANT
Craig, Pauline and Wynn Mitchell
Mary Cairncross Corner, cnr Mountview and Landsborough-Maleny Roads, Maleny 4552
Phone: 07 5494 3700
Email: cm46@ozemail.com.au
www.terraceseafood.com.au
Licensed/BYO wine
Cuisine: Seafood
Mains average: $$$
Specialty: Seafood chowder, Terrace for two (seafood platter)
Local flavours: Seafood
Extras: Six private cottages behind the restaurant with packages available. Courtesy bus to the coast. Winner of many awards, and inducted into the Queensland Tourism Hall of Fame in 2002.
■ **Open:** Lunch and dinner daily

MOOLOOLABA
🍎 PRODUCERS
SUNCREAM HOMESTYLE ICE-CREAM
Gerry and Jodie Nelson
207 Brisbane Rd, Mooloolaba 4557
Phone: 07 5444 7706
Email: suncream@powerup.com.au
Jodie Nelson's homemade ice-creams (think ginger ice-cream, and honey roasted macadamia) are made from sweet, fresh Jersey milk from the nearby Blackall Range, and the gelato-style sorbets are made from local tropical fruits. Nevertheless, it's her ice-cream cakes that sell best. The couple's experience in restaurants made it a natural progression to begin this business with the ice-cream parlour attached to an ice-cream kitchen. A mobile ice-cream parlour also appears at local fetes and festivals.
■ **Open:** 9am–8.30pm, daily (closed Sun am, and closes earlier on Tues)
● **Location:** Follow Brisbane Rd towards Kawana. Be in left-hand lane through the Bundilla shopping area and at the last set of lights, take the service road on the left and go to end of the road, Suncream is next to The Bridge Seafoods.

🍴 DINING
FISHERIES ON THE SPIT
Sandy Wood
21 Parkyn Pde, Mooloolaba 4557
Phone: 07 5444 1165
Email: marketing@debrettseafood.com.au

www.debrettseafood.com.au
Licensed/BYO
Cuisine: Seafood
Mains average: $
Specialty: Grilled broadbill, seared tuna
Local flavours: Local seafood
Extras: Outdoor dining under umbrellas, as part of the multi-award-winning De Brett Seafood complex, which won a Jaguar Gourmet Traveller Award for Primary Industries, 2001. Also sells homemade condiments, smoked tuna and marlin.
- **Open:** 7.30am–8pm daily

NAMBOUR
🍎 PRODUCERS

SHIPARD'S HERB FARM
Ricky and Isabell Shipard, Angela Stewart
139 Windsor Rd, Nambour 4560
Phone: 07 5441 1101
www.herbsarespecial.com.au
The Shipards have a massive range of culinary herbs and spices, some medicinal, others rare edibles. They have an astounding collection of more than 900 exotic plants and seeds, which could be the largest in the country. As you would expect, the owners are extremely knowledgeable about their specialty and Isabell's recent book on herbs, *How can I use Herbs in my Daily Life?* a practical guide to growing and using herbs, has proved very popular. Come here for seeds and plants, natural licorice snack packs, and other herbal products.
- **Open:** By appointment
- **Location:** Third on the right past Sunshine Coast TAFE College, Windsor Rd

NOOSA
🍴 DINING

RIVA WATERFRONT RESTAURANT
Vic and Karen McInman (owners),
Matthew Swannell (chef)
Noosa Wharf, Quamby Place, Noosa 4567
Phone: 07 5449 2440
Email: welcome@riva-noosa.com
www.riva-noosa.com
Licensed
Cuisine: Modern European
Mains average: $$$
Specialty: Tapas tasting plate for two
Local flavours: Organic greens, Queensland barramundi, pork, duck
Extras: Deck dining over the water, views, modern decor.
- **Open:** Lunch Fri–Sun, dinner daily

NOOSA HEADS
🍴 DINING

BERARDO'S RESTAURANT AND BAR
Jim Berardo
52 Hastings St, Noosa Heads 4567
Phone: 07 5447 5666
Email: info@berardos.com.au
www.berardos.com.au
Licensed/some BYO nights
Cuisine: Modern Australian
Mains average: $$$
Specialty: Sashimi
Local flavours: All organic fruit and vegetables, cheese, meat, seafood
Extras: Piano bar, weddings and functions. Berardo's On the Beach has a beach ambience and takeaway available.
- **Open:** Dinner daily

SUNSHINE COAST—QLD

CATO'S RESTAURANT AND BAR
Sheraton Noosa Resort and Spa,
16 Hastings St, Noosa Heads 4567
Phone: *07 5449 4754*
www.noosaeguide.com/catos
Licensed/BYO (Sun–Thurs)
Cuisine: Australian seafood
Buffet: $$$$
Specialty: Caesar Salad Cato's Style with seared scallops or tandoori chicken
Local flavours: Seafood, prawns, crabs, salad greens
Extras: Overlooks busy Hastings St.
■ **Open:** Breakfast, lunch and dinner daily

NOOSA SOUND
🍴 DINING

RICKY RICARDO'S
Matthew Golinski (chef)
Noosa Wharf, Quamby Place,
Noosa Sound 4567
Phone: *07 5447 2455*
Email: *info@rickyricardos.com*
www.rickyricardos.com
Licensed
Cuisine: Mediterranean
Mains average: $$$
Specialty: Tapas
Local flavours: Everything is local, including seafood, fruit, vegetables, cheese
Extras: Outdoor dining, and waterside position.
■ **Open:** Lunch and dinner daily

NOOSAVILLE
🍴 DINING

SILVER BISTRO
Wolfgang Groh
251 Gympie Tce, Noosaville 4566
Phone: *07 5449 9577*
Email: *silverbistro@bigpond.com*
Licensed/BYO wine
Cuisine: Modern Australian
Mains average: $$$
Specialty: Whole crispy reef fish
Local flavours: Seafood
Extras: Located on the river, there is outdoor seating to enjoy the views. Takeaway fish and chips available.
■ **Open:** Dinner daily

PALMVIEW
🍎 PRODUCERS

STRAWBERRY FIELDS
Von and Maurie Carmichael
133 Laxton Rd, Palmview 4553
Phone: *07 5494 5146*
Email: *david@strawberryfields.com.au*
www.strawberryfields.com.au
If you are passing Palmview, stop in for the sweetest, juiciest strawberries you've ever tasted, grown right here on the farm. Maurie Carmichael's main aim is to 'grow strawberries that taste nice and sweet'. And as an ex-president of the Queensland Strawberry Growers Association, you can be sure he knows how. The Chandler strawberries he plants are one of the sweetest varieties on the market, and his twelve hectares of berries are allowed to ripen on the bush to achieve ultimate flavour. Best of all, you may come and pick your own.
■ **Open:** 8am–5pm daily (June–Nov)
● **Location:** Located opposite the Ettamogah Pub

TANAWHA
🍎 PRODUCERS

SUPERBEE HONEY FACTORY
David and Coral Bell
Tanawha Tourist Dr, Tanawha 4556
Phone: 07 5445 3544
Email: superbee@superbee.com.au
www.superbee.com.au
Superbee's honey, supplied by beekeepers from around the state, is processed as little as possible because, say the Bells, when honey is heated over 35°C (hive temperature) the natural goodness starts to break down. Throughout the year they rotate approximately twenty-eight varieties on their tasting bar, and have a large variety of floral honeys, as well as a live beekeeping demonstration. Established nearly twenty years ago, Superbee has recently developed a wholesale and exporting market. Royal jelly and other bee products are also available here. The products are sold in various shops throughout Australia under the Superbee and Buderim Honey labels.
- **Open:** 9am -5pm daily
- **Location:** One hour north of Brisbane on the Bruce Hwy

WOOLOOGA
🍎 PRODUCERS

GYMPIE FARM CHEESE
Camille Mortaud
1615 Wide Bay Hwy, Woolooga 4570
Phone: 07 5484 7223
Email: camille@spiderweb.com.au
Glorious goats cheese, French farm-style. While you can only visit this farm, which creates magnificent goats cheeses, by appointment, you can buy them at the Brisbane and Eumundi markets. French-born Camille Mortaud was so nostalgic for chevre after he moved to Australia that he decided to make his own. Now he also makes a soft, semi-matured cows milk cheese and a European-style cultured butter using local Jersey cream.
- **Open:** By appointment only

WOOMBYE
🍎 PRODUCERS

THE BIG PINEAPPLE–SUNSHINE PLANTATION
Peter Auld
Nambour Connection Rd, Woombye 4559
Phone: 07 5442 1333
Email: info@bigpineapple.com.au
www.bigpineapple.com.au
This is the only pineapple plantation open to the public in Australia and a revelation to all of us who have no clear idea of how pineapples actually grow. The Big Pineapple grows many other fresh tropical fruits as well, and you can see all of these if you take the sugarcane train through the plantation. Join the Macadamia and Rainforest Tour for a rainforest and macadamia orchard trip, finishing up at the southern hemisphere's largest macadamia factory. After that you can glide on a canal through an illustrated history of horticulture from earliest times through to the future of hydroponics. There are two restaurants, Sunshines and Plantations, serving full meals with many dishes featuring pineapples and macadamias, of course. You can also shop for fresh and processed tropical fruits, or enjoy an animal nursery.
The Big Pineapple is a major supplier of

pineapples to Golden Circle and has been operating since 1971 showcasing 'the king of fruits'. The latest products are a range of seven gourmet jams and marmalades—featuring pineapple, of course.
- **Open:** 10am–5pm daily during tourist season (May–Oct), 9am–5pm weekdays, 10am–5pm weekends and public holidays (at other times)

YANDINA

🍎 PRODUCERS

NUTWORKS MACADAMIA PROCESSORS
Jim Atkinson (General Manager)
112 Pioneer Rd, Yandina 4561
Phone: 07 5472 7777
Email: info@nutworks.com.au
www.nutworks.com.au

Macadamias are the stars here. You can buy them dipped in chocolate or flavoured in other ways. There is an information centre where you can see a video about macadamias, enjoy some free tastings, then pop into the Hardnut Cafe downstairs for a snack and stock up with goodies at the retail shop to take home. The factory processes from March to December, but you can view the factory year round.
- **Open:** 9am–5pm daily
- **Location:** Yandina exit off Bruce Hwy, opposite the Ginger Factory

🛒 STORES

BUDERIM GINGER LTD
Craig Todd, Retail Marketing
50 Pioneer Rd, Yandina 4561
Phone: 07 5446 7100
Email: buderimg@buderimginger.com
www.buderimginger.com

Set up in 1941 as a farmers' cooperative to take up export opportunities when the war ended, Buderim Ginger Ltd is Australia's only ginger processor. The company produces a wide range of ginger products (including jams, crystallised ginger, toppings, and stir-fry mixes) and confectionery (such as sweet preserved ginger). They also bulk process products for other food manufacturers. Sixty per cent of production goes directly to export, the rest to local markets. Today 400 000 visitors visit each year, wandering through gardens full of different ginger varieties, watching production, and buying products in the shop. There is also an annual Ginger Flower Festival in January.
- **Open:** 9am–5pm daily
- **Location:** Travel north on the Bruce Hwy, seventy-five minutes from Brisbane, then take the Yandina exit

🍴 DINING

PICNICS AT THE ROCKS
Wendy and Neville Cutting (owners),
Kim Walker (head chef)
Coolum–Yandina Rd, Yandina 4561
Phone: 07 5446 8191
Email: PicnicsattheRocks@bigpond.com
www.therocks-yandina.com
Licensed
Cuisine: Modern Australian
Mains average: $$
Specialty: Seafood, Picnics' tea-smoked Atlantic salmon, lime tart with coconut snaps
Local flavours: Buderim ginger, macadamia nut ice-cream, Laguna Bay snapper
Extras: Waterfront setting on the banks of the Maroochy River, with alfresco

QLD—SUNSHINE COAST

dining on a deck under sails. Makes Kitchen Door packaged products for sale.
- **Open:** Lunch daily, dinner Fri–Sat, sunset Happy Hours Thurs–Sat (3pm–6pm)
- **Location:** From Bruce Hwy, take the Yandina–Coolum exit, head towards Coolum and the restaurant is 2km from the exit on the right. From the Sunshine Motorway, exit at the Coolum–Yandina roundabout, travel towards Yandina and the restaurant is 15km along the Yandina–Coolum Rd on the left

🍴 DINING/CLASSES

THE SPIRIT HOUSE—RESTAURANT AND COOKING SCHOOL

Helen and Peter Brierty
20 Ninderry Rd, Yandina 4561
Phone: 07 5446 8994 (restaurant);
07 5446 8977 (office)
Email: admin@spirithouse.com.au
www.spirithouse.com.au
Licensed
Cuisine: Contemporary Asian with a Thai emphasis
Mains average: $$$
Specialty: Crispy whole fish with tamarind chilli sauce
Local flavours: All fruit, vegetables and herbs sourced locally, also ginger, galangal, lemongrass, chilli, lime
Extras: Restaurant seating is in a series of private tropical garden courtyards nestled around bamboo and papyrus fringed lily pond. Five acres of tropical gardens. Grows Asian salad greens, Thai basil, lemongrass, Vietnamese mint and snake beans in the kitchen garden and hydroponic farm. The cooking school (hands on, lunch included) is booked out months ahead and the cooking school recipe book was published by New Holland in late 2004. See website for the Spirit House range of curry pastes, frozen curry meals, Thai fish cakes, spring rolls, plus Thai sauces and relishes.
- **Open:** Lunch daily, dinner Wed–Sat, cooking school Wed–Sun
- **Location:** Take the Yandina exit off the Bruce Hwy, next roundabout take the Coulson Rd exit, turn into School Rd, then right into Ninderry Rd. Spirit House is 100m on the right.

MARKETS AND EVENTS

✦ **EUMUNDI MARKETS,** EUMUNDI
07 5442 7106
6am–1pm Wed and Sat

✦ **NOOSA FARMERS' MARKETS,** NOOSAVILLE
7am–12pm second and fourth Sun

✦ **POMONA ECO MARKETS,** POMONA
07 5448 2365
8am–1pm Fri

✦ **POMONA ORGANIC MARKETS,** POMONA
07 5448 2365
8am–12pm third Sun

✦ **SLOW FOOD NOOSA,** POMONA
07 5448 2365
Jan

TOOWOOMBA AND THE SOUTHERN DOWNS

The Toowoomba region has always ridden on the sheep's back, and there still remain a few mansion-like homesteads that reflect the rich history and energy of the early pastoralists in this area. Today, vegetables such as potatoes and broccoli are grown in the area, as well as most of Queensland's beetroot crop.

This large region, which is now even dabbling in wine-making, adjoins the Southern Downs. It may be called the Southern Downs, but visitors experience a few 'ups' as well, as they climb onto the Granite Belt tablelands. There is talk today of 'borderless tourism' and in this region it makes enormous sense as the Granite Belt merges effortlessly with the far northern tablelands of New South Wales. The higher altitude has always been ideal for apple and pear growing, but Italian immigration earlier this century also set the stage for today's superb wine industry.

Many cool-climate fruits and berries thrive here and, in season, dozens of orchards and farms uncover their roadside signs or set up stalls selling sun-ripened tree-fresh fruit. A leisurely drive down the Fruit Run is one of the best ways to spend a summer day, yielding sweet bargains that will last for days.

BALLANDEAN

DINING

VINEYARD CAFE AND COTTAGES
Janine (chef) and Peter Cumming
New England Hwy, Ballandean 4382
Phone: 07 4684 1270
Email: info@vineyard-cottages.com.au
www.vineyard-cottages.com.au
Licensed
Cuisine: Modern Australian
Mains average: $$$
Specialty: Desserts
Local flavours: Beef, lamb, pork, trout, fruit and vegetables, herbs, figs
Extras: Views and extensive gardens producing spectacular vases of David Austen roses for the restaurant, which is located in a historic church. There are also self-contained cottages with packages that include meals in the cafe.
■ **Open:** Breakfast daily (for guests), lunch weekends, dinner Fri–Sat (daily for guests)

BUNYA MOUNTAINS

DINING

BUNYA FOREST GALLERY AND TEAROOM
Kay Joyce
14 Bunya Ave, Bunya Mountains 4405
Phone: 07 4668 3020
Email: info@bunyaforest.com
www.bunyaforest.com.au
Cuisine: Bush food

QLD—TOOWOOMBA AND THE SOUTHERN DOWNS

Mains average: $
Specialty: Bunya nuts in everything possible—scones, focaccia, salad, with a salsa
Local flavours: All bush food ingredients
Extras: Grows native mint, lemon myrtle, lilly pillies and bunya nuts. Sells a range of homemade preserves, salad oils and condiments. There is also a gallery, cool temperate rainforest views and veranda dining. You can go off discovering bush foods if you wish and Kay Joyce will design walks to suit requests.
■ **Open:** 10am–5pm Tues–Sun, daily in school holidays

CROWS NEST
PRODUCERS

BUNNYCONNELLEN OLIVE GROVE AND VINEYARD
Peter and Janie Simmonds
Swain Rd, M/S 357, Crows Nest 4355
Phone: 07 4697 9555
Email: bunnyconnellen@bigpond.com.au
www.theoliveoilcompany.com.au
'Total madness', say the Simmondses when asked why they began this business six years ago, saying they are 'still reeling from enormity of the decision!' Despite this, their extra virgin olive oils, Sinolea pressed olive oil, infused oil, and table olives are doing very well. Maybe it's because the product is subtly different in that the table olives are prepared in olive oil using a unique process that gives them a richness not experienced with olives in brine. If you love the idea of living in an olive grove, you can—short-term at least—by checking into the self catering B&B, 'The Studio', which is ideal for couples and overlooks the vines and olive grove.
■ **Open:** By appointment

● **Location:** 12km north-west of Crows Nest, off New England Hwy

GLEN APLIN
PRODUCERS

THE CHERRY PATCH
Lex and Elizabeth Ferris
New England Hwy, Glen Aplin 4381
Phone: 07 4683 4296
Email: cherries@halenet.com.au
This place is a vision of loveliness during cherry blossom season. The small shop is set amongst cherry trees and crammed full of cherry products such as liqueurs, jams, chutneys, sauces and spreads, as well as fresh cherries in season and dried cherries at any time. Lex Ferris has been growing this fruit for ten years, since he switched from a job with the electricity board. His background was in wine, though, so perhaps he was simply returning to his first area of interest. Cherry season is from the end of October to mid December.
■ **Open:** 8.30am–4.30pm daily
● **Location:** 7km south of Stanthorpe, next to Cominos Winery

MOUNT STIRLING OLIVES
Jim Miller and Vivienne Quinn
Collins Rd, Glen Aplin 4381
Phone: 07 4683 4270
Mob: 0412 677 584
Email: jimmiller@quinnassoc.com.au
www.mtstirlingolives.com
Mount Stirling Olives is a major commercial producer in the area, and also crushes for other growers. Although the grove was established eleven years ago, in what was an abandoned stone-fruit orchard, the cellar door has only been open since

1999. In a nice bit of circularity, they also sell honey from the bees that propagate the olive trees. There are eight varieties of olives here—some for oil, others to pickle, and there is also tapenade and dukkah sold from the farm gate. If you like the place a lot and want to stay, there is farmstay accommodation too.
- **Open:** 9.30am–4pm weekends and public holidays, or by appointment
- **Location:** 8km south of Stanthorpe

GREENMOUNT
PRODUCERS
KIALLA PURE FOODS
G and S McNally
342 Greenmount–Etonvale Rd, Greenmount 4359
Phone: 07 4697 0300
Email: kiallafoods@bigpond.com
www.kiallafoods.com.au
This is the largest 100 per cent certified-organic cereal grain manufacturer in Australia, supplying products to the domestic and export market. You can hardly get more back-to-basics than this. Flour here is milled from organic wheat grown on the property, and from other local organic growers. You can also buy fresh flour from the mill.
- **Open:** 8am–5pm weekdays
- **Location:** Twenty minutes on the Warwick side of Toowoomba

JONDARYAN
DINING
JONDARYAN WOOLSHED
David Totenhofer,
Evanslea Rd, Jondaryan 4403
Phone: 07 4692 2229
Email: info@jondaryanwoolshed.com
www.jondaryanwoolshed.com
Licensed/BYO
Cuisine: Traditional Australian
Mains average: $
Specialty: Billy tea and damper, corned meat and damper, roast Jondaryan lamb
Local flavours: Damper made from locally grown and milled wheat, lamb from the merino-dohne flock on the property
Extras: Outdoor museum, administered by the Jondaryan Shire Council, and dining room set up in the historic homestead, one of many buildings on the site. The woolshed, which was built in 1859–1861 and opened to the public in 1972, was taken over by the Shire in late 2002. It is now used for demonstrations of pioneer life and to host heritage events. They can also cater for weddings, corporate functions and coach groups.
- **Open:** Lunch weekends or by appointment, tours—Wed–Sun
- **Location:** 3km south of the Warrego Hwy from Jondaryan

STANTHORPE
PRODUCERS
THE BRAMBLE PATCH BERRY GARDENS
Don and Patsy Stirling
381 Townsends Rd, Stanthorpe 4380
Phone: 07 4683 4205, shop
Berry liqueurs, jams, chutneys, vinegars, coulis and other products line the walls of the cafe-bar-showroom in this delightful place. The extensive range of delicious gourmet berry products (try the ice-cream!) are made by hand without the use of preservatives. Basically, it's all

QLD—TOOWOOMBA AND THE SOUTHERN DOWNS

just pure fruit—raspberries, strawberries, boysenberries, blackberries, tayberries, silvanberries, manonberries—and a lot of hard work. There are over sixty different items that are only available here or by mail order, although the Stirlings' vinegars and coulis are used in local restaurants, as well as in Brisbane and interstate. The only berries they buy are the ones that will not grow in this warmer climate, such as blackcurrants.

■ **Open:** 10am–4pm daily
● **Location:** 15km south of Stanthorpe, 7km north of Ballandean, Townsends Rd is the last turn-off

🏠 STORES

HAWKER BROS BUTCHERS
Mal Newley
54 Maryland St, Stanthorpe 4380
Phone: 07 4681 2150

Mal Newley first started making leaner sausages a few years ago now, when his business partner had a heart bypass. The pair experimented and the trials paid off because they won the pork section for the Darling Downs area in the 1998 Sausage King competition. In 1999 they won in the gourmet style section in the national titles in Sydney. Of course they use local pork from Stanthorpe—in fact all meat sold in the shop is local. Now they have also perfected a lean beef sausage and do a great Italian sausage, plus other gourmet style ones such as pork and apple, or beef and burgundy.

■ **Open:** 6.30am–5.30pm weekdays, 6am–12.30pm Sat

VINCENZO'S AT THE BIG APPLE
Vince Catanzaro
Cnr New England Hwy and Maryland Rd, Thulimba, Stanthorpe 4380
Phone: 07 4681 3004
Email: vincec@vccatanzaro.com.au

Vince Catanzaro is redefining the name 'Big Apple'. This new facility is the northern gateway to the Granite Belt and showcases fresh, quality local fruit, wines, fruit liqueurs and gelati blended with local fruit. It is an ideal pit-stop for the traveller as there's a deli and a cafe featuring ready-to-eat Italian home-style meals and good coffee. There is catering for functions and groups of up to one hundred guests. Fully licensed with tastings and sales of local wines, The Big Apple is also a tourist information centre.

■ **Open:** 9am–6pm daily
● **Location:** North of Stanthorpe

THE FRUIT RUN—FROM SOUTH OF WARWICK TO NORTH OF STANTHORPE

The main road swoops off the main highway south of Warwick, following the Old New England Highway, and reconnects north of Stanthorpe. It is clearly signposted. Further south, watch out for other roadside businesses such as The Fruit and Nut Shop, for local mushrooms and tomatoes, and the Cherry Patch. In all of this area, hand-lettered signs pop out as soon as orchardists have some freshly picked fruit, so be ready to brake!

TEXAS

🍎 PRODUCERS

WILGA VALE VENISON
Joan White
Wilga Vale, Texas 4385
Phone: 07 4653 1179

TOOWOMBA AND THE SOUTHERN DOWNS—QLD

Email: whitejoan@bigpond.com
www.farmersonlinemarket.com
A wide range of venison cuts and venison sausages, cryovac-packed and frozen, are available at this deer farm. Several types of deer are run too—fallow, red, rusa and chital deer—on Wilga Vale, which has been operating since 1982 as a farming enterprise. Look for these products, and those of several other local producers, under the Border Highland Fresh brand in local shops.
- **Open:** Phone first

THULIMBAH

🍎 PRODUCERS

HI VALUE FRUIT AND BERRY GARDENS
David and Roslyn Sutton
10 Halloran Dr, Thulimbah 4376
Phone: 07 4685 2464
Email: hivalue@flexi.net.au
www.hivalue.com.au
The Suttons have come up with some great solutions for recycling fruit that has been deemed worthless because of its size, colour and shape, yet is still tasty and fresh. They make eight or so apple juices, made by apple variety, eight fruit liqueurs, traditional flat apple cider from Hi Value farm apples, apple cider and berry vinegars, all of which are sold under the (appropriately named) Hi Value brand. All this and more are available at the farm, as well as fresh fruit in season, jams, sauces and chutneys. Otherwise, you'll find the juices in Queensland and NSW specialty gourmet delis and fruit outlets.
- **Open:** 10am–5pm daily
- **Location:** 13km north of Stanthorpe on New England Hwy

🏠 STORES

THE STANTHORPE APPLE SHED
Memo Mattiazzi
Maryland Rd, M/S 1981, Thulimbah 4376
Phone: 07 4683 2207
Email: matto@halenet.com.au
Memo Mattiazzi has been a major grower and supplier of apples in this region for quite a while. Before the wines took off, the whole area around Stanthorpe was known best for its apples and stone fruits. In season (early January to April) you will be able to choose from bins of up to ten different varieties including pink ladies, fujis, and gala apples. At other times, the apples come from controlled-atmosphere storage.
- **Open:** 8am–5.30pm daily, look out for the signs to see when they're open
- **Location:** Halfway between the old and new road, 15km north of Stanthorpe

STANTHORPE WINE CENTRE
William Higgins
Granite Belt Dr, Thulimbah 4376
Phone: 07 4683 2011
Email: stanwine@halenet.com.au
www.summitestate.com.au
Although this is called a wine centre, it's actually much more, selling local marinades, flavoured oils and souvenirs. You will be able to enjoy a cup of great coffee, as well as light meals and snacks made using local produce. There's a vineyard talk, as well as bush poetry, and the centre also caters for coaches. Lunches by appointment for tour groups.
- **Open:** 9am–5pm daily

QLD—TOOWOOMBA AND THE SOUTHERN DOWNS

TOOWOOMBA

🍎 PRODUCERS

STAHMANN FARMS
Cnr McDougall St and Industrial Ave, Toowoomba, 4350
Phone: 07 4699 9400
Email: mailorder@stahmann.com.au
www.stahmannfarms.com.au
Although this office mainly handles the large mail-order business for this major pecan nut farm, you can drop in and simply buy pecans, puddings, fudge and a number of other products at factory prices. Now with a bunch of exciting flavours, it's worth a visit. The farm itself is far away in Moree, New South Wales, or you can find the products in supermarkets, or by mail order on the web as well.
■ Open: 8am–5pm weekdays

🍴 DINING

GIP'S RESTAURANT
Jon and Julianne McCorley
Clifford House, 120 Russell St, Toowoomba 4350
Phone: 07 4638 3588
Email: jonjules@optusnet.com.au
www.gipsrestaurant.com.au
Licensed/BYO wine
Cuisine: Modern Australian
Mains average: $$$
Local flavours: Duck, lamb, chicken, beef, organic eggs, vegetables, organic olives
Extras: Enclosed garden area with historic Clifford house as a backdrop. Takeaway available.
■ Open: Breakfast Sun–Fri, lunch daily, dinner Mon–Sat

KARINGAL CAFE
Naomi Luscombe and Irfan Tokmak
35 Bell St, Toowoomba 4350
Phone: 07 4637 8000
Email: reservation@karingalcafe.com.au
www.karingalcafe.com.au
Licensed
Cuisine: Modern Australian
Mains average: $$
Specialty: Lamb rump stuffed with olive, prosciutto and sun-dried tomatoes, Mediterranean grilled octopus salad
Extras: Alfresco dining for thirty to forty people in a private, airconditioned function room. Private dining room for groups of up to fifteen people. Named Best Cafe Restaurant in Australia, by the Restaurant and Catering Association, 2002 and Best Cafe Restaurant for Fine Dining, Australian Darling Downs 2003.
■ Open: Breakfast, lunch and dinner daily

MANGO
Mango is the connecting link amongst north Queensland regions, as these fragrantly scented marvels of the fruit world are grown better here than anywhere else. But there is sugar cane too and salt, melons, bananas, prime beef on the inland properties, and always the magnificent seafood, especially muddies (mud crabs) from the reef.

TOOWOMBA AND THE SOUTHERN DOWNS—QLD

MARKETS AND EVENTS

✦ **GLENGALLAN SEASONAL FARMERS MARKETS,** ALLORA
07 4667 3866
First Sun of each season

✦ **MARKET AT THE ABBEY,** WARWICK
07 4661 3122
8am–12pm first Sun

✦ **STANTHORPE IN SEASON FARMERS' MARKET,** STANTHORPE
07 4684 1226
7am–12pm third Sun, Nov–Apr

✦ **STANTHORPE MARKET IN THE MOUNTAINS,** STANTHORPE
07 4661 3122
9am–1pm second Sun

✦ **STANTHORPE SHOWGROUND MARKETS,** STANTHORPE
07 4661 3122
Fourth Sun

✦ **WARWICK BAND CENTRE MARKETS,** WARWICK
07 4661 3122
First and third Sun

✦ **WARWICK TOWN HALL CARPARK MARKETS,** WARWICK
07 4661 3122
Second and fourth Sun

✦ **APPLE AND GRAPE HARVEST FESTIVAL,** STANTHORPE
07 4681 4111
Feb–Mar, in even-numbered years

✦ **BRASS MONKEY SEASON,** SOUTHERN DOWNS
07 4661 3122
Jun–Aug

✦ **GOURMET IN GUNDY,** GOONDIWINDI
07 4671 3264
Second Sun in Sept

✦ **GRANITE BELT SPRING WINE FESTIVAL,** STANTHORPE
07 4683 4311
Mid Oct

✦ **WINE AND FOOD AFFAIR,** STANTHORPE
07 4681 0411
Feb in odd-numbered years

TOWNSVILLE, MACKAY AND THE WHITSUNDAYS

Queensland's coastal areas are like magnificent opals, brilliant yet with hidden fire. So rich is the Burdekin that, in addition to being the sugar-producing capital of Australia, it is known as the state's fruit and salad bowl.

The epicentre of the Great Barrier Reef is its playtime islands. While some reflect a 'lifestyles of the rich and famous' theme, many are family-friendly, others remote and romantic. While the style of dining varies of course according to guest's needs and bank accounts, the magnificently fresh local produce is always first-class.

Inland, the land becomes harsher and home to beef cattle. Here too, guests are welcome at farmstays, where they can experience the 'real Queensland'.

AIRLIE BEACH
DINING

THE CLIPPER RESTAURANT AND BAR
Greg Waites
Coral Sea Resort, 25 Oceanview Ave,
Airlie Beach 4802
Phone: 07 4946 6458
Email: res@coralsearesort.com
www.coralsearesort.com
Licensed
Cuisine: Modern Australian
Mains average: $$$
Specialty: Seafood
Local flavours: Seafood, Rockhampton beef, tropical fruit

Extras: Absolute ocean frontage, outdoor dining beside the pool.
■ **Open:** Breakfast, lunch and dinner daily

AYR
DINING

PEPPERS ON QUEENS
Kerry Meizner (owner-chef) and Yvette Phipps
The Country Ayr Motel,
197-199 Queen St, Ayr 4807
Phone: 07 4783 1700
Email: sales@countryayr.com.au
www.countryayr.com.au
Licensed
Cuisine: Modern Australian
Mains average: $$$
Specialty: Moreton Bay bugs
Local flavours: Prawns, cuttlefish, barramundi, mangoes
Extras: Outdoor dining area near the pool. Takeaway available.
■ **Open:** Dinner Mon-Sat

BOWEN FARMSTAY
Bowen Visitors Information Centre,
Bowen 4805
Phone: 07 4786 4222
Email: info@bowentourism.com.au
www.bowentourism.com.au
Farmstay is often available at local cattle properties such as Bogie River, Strathmore, and Johnny Cake Stations. Call for information.

FLAGGY ROCK
🍎 PRODUCERS
FLAGGY ROCK EXOTIC FRUIT GARDEN
Hans and Janice Plessing
Bruce Hwy, Flaggy Rock 4741
Phone: 07 4950 2156
Email: hansandjanice@bigpond.com
Set in a tropical fruit orchard with a landscaped picnic area, this is definitely worth a stop. Especially if you can take time to really enjoy the homemade ice-cream, couldn't-be-fresher squeezed juices and home-cooked meals.
- **Open:** 8.30am–4.30pm daily (closed Feb)
- **Location:** North of Rockhampton about two-thirds of the way to Mackay

HAMILTON ISLAND
🍴 DINING
OUTRIGGER RESTAURANT
Vivienne Perhard
Hamilton Island Resort,
Hamilton Island 4803
Phone: 07 4946 8582
Email: FBAdmin@hamiltonisland.com.au
www.hamiltonisland.com.au
Licensed
Cuisine: Modern Australian
Mains average: $$$$
Local flavours: Barramundi
Extras: This restaurant features a Polynesian style ambience and is located on the shores of Catseye Bay with romantic views. There are several restaurants and cafes on the island, and some will even cook guests' catches of reef fish. The Great Barrier Feast is celebrated every Queens Birthday weekend in June.
- **Open:** Breakfast, lunch and dinner daily

HAYMAN ISLAND
🍴 DINING
LA FONTAINE
Marcus Dudley (chef)
Hayman Island Resort, Hayman Island 4801
Phone: 07 4940 1902
Email: dining@hayman.com.au
www.hayman.com.au
Licensed
Cuisine: Modern French
Mains average: $$$$
Specialty: Fillet of John Dory with sweet potato and leek compote, vanilla and mussel cream
Local flavours: Blue swimmer crab, Pacific oysters, John Dory, tuna, tropical fruit.
Extras: Multi-award winning Hayman Island Resort has been voted one of the ten best resorts in the world.
- **Open:** Dinner selected evenings

MACKAY
🏠 STORES
MACKAY FISH MARKET
Graham, David and Brian Caracciolo
2 River St, Mackay 4740
Phone: 07 4957 6497
Email: sales@mackayreeffish.com.au
www.mackayreeffish.com
It would be hard to find fresher local seafood and fish than the bounty that is on offer here, which is unloaded directly off the boats. There is a full range of green and cooked prawns, bugs, crabs, fish, calamari, octopus and crays, also game meats, venison, crocodile meat, frozen fruit and vegetables, as well as cooked seafood products.
- **Open:** 8.30am–5.30pm daily

QLD—TOWNSVILLE, MACKAY AND THE WHITSUNDAYS

🍴 DINING

LIGHTHOUSE SEAFOOD RESTAURANT
Rick Shelley
1 Mulberin Dr, Mackay Marina,
Mackay 4740
Phone: 07 4955 5022
Email: lighthouse@atthemarina.biz
Licensed
Cuisine: Seafood
Mains average: $$
Specialty: Seafood platter for two
Local flavours: Seafood direct from the local trawlers moored in the marina
Extras: Located on the beautiful Mackay Marina with a large deck for dining.
■ Open: Breakfast, lunch and dinner daily

MUDDIES RESTAURANT
Kimon, Tichsia and Tino Zapantis
Illawong Beach Resort, 73 Illawong St,
Mackay 4740
Phone: 07 4957 8427
Email: illawong@illawong-beach.com.au
www.illawong-beach.com.au
Licensed
Cuisine: Seafood
Mains average: $$
Specialty: Seafood platters, fresh mud crab
Local flavours: Everything is local
Extras: On Illawong Beach
■ Open: Breakfast and dinner daily, lunch weekends

🚌 TOURS

MACKAY DISCOVERY TOURS
Peter and Bronwyn Dawes
Andergrove Van Park, Beaconsfield Rd,
Mackay 4740
Phone: 07 4942 4922
Email: andvanpk@mackay.net.au
www.andergrovepark.com.au

> **SEAFOOD OUTLETS—MACKAY**
> Mackay has a host of great seafood outlets serving local reef fish and other seafood. River Street has several to choose from, and the combination of water views and fresh seafood is delightful. Rosslyn Bay Fishermen's Co-op (Phone: 07 4933 6105) has plenty of reef fish, bugs and mud crabs straight off the boats.

These seasonal sugar cane day tours let you meet the locals and see a working farm. Lunch is at General Gordon Heritage Hotel in Homebush.
■ Open: On demand in season (June–Oct)

MAGNETIC ISLAND
🍴 DINING

MAGNETIC ISLAND TROPICAL RESORT
Emma Bainbridge (owner),
Andrew Cooper (chef)
56 Yates St, Nelly Bay,
Magnetic Island 4819
Phone: 07 4778 5955
bookings@magneticislandresort.com
www.magneticislandresort.com
Licensed
Cuisine: North Queensland seafood and steaks
Mains average: $$
Specialty: Barramundi topped with scallops and prawns in a brandy sauce
Local flavours: Barramundi
Extras: Tropical palm gardens with National Park backdrop. Winner of Best Tourism Restaurant North Queensland several years running.
■ Open: Breakfast and dinner daily, lunch for groups only

MAGNETIC MANGO

Gary Cutler
63–83 Apjohn St, Horseshoe Bay,
Magnetic Island 4819
Phone: 07 4778 5018
BYO
Cuisine: Homestyle cooking
Mains average: $
Specialty: King prawns, avocado and fresh mango, mango smoothies.
Local flavours: Mangoes, seafood
Extras: Generous use of mangoes from Cutler's 400 trees (they also grow coconuts, citrus, pineapples, macadamias, bananas, and herbs). The restaurant is under the mango trees. Everything on the ever-changing menu, including all the bread and cakes, is made fresh every day and is chemical free. They also sell homemade mango jam, three kinds of mango chutney, and honey from their own bees as well as fresh fruit in season.
- **Open:** 10am–4.30pm daily, dinner Sat–Mon

MUTARNEE

STORES

FROSTY MANGO

Alf Poefinger
Bruce Hwy, Mutarnee 4816
Phone: 07 4770 8184
Email: frostymango@bigpond.com
www.frostymango.com

Queensland can get pretty hot, and it's then that some frosty mango makes great sense. You'll find ice-cold juices and ice-creams here as well as snacks, and they're good—as you would expect them to be because most of the food and beverages are based on tropical fruit grown in the shop's own orchards. To show how much they are appreciated, Frosty Mango was awarded the Restaurant and Caterer's 2002 Award for Top Cafe in North Queensland.
- **Open:** 8am–6pm daily
- **Location:** 65km north of Townsville

CATCH YOUR OWN FISHERMAN'S BASKET—CARDWELL

In Cardwell, you can catch your own fisherman's basket of prawns or mud crabs from the foreshore. Or go reef fishing for sweetlip, coral trout or cod. In the mangroves look for mangrove jack, barramundi or trevally.

SOUTH TOWNSVILLE

DINING

MICHEL'S RESTAURANT

Michel Flores and Jason Makara and Marco Gulisano
7 Palmer St, South Townsville 4810
Phone: 07 4724 1460
Email:
michelsrestaurant@bigpond.com.au
Licensed/BYO
Cuisine: Modern Australian
Mains average: $$
Specialty: Twice-cooked duck
Local flavours: Fruit and vegetables, seafood, meat
Extras: Alfresco dinning is available with terrific views of the river and Townsville.
- **Open:** Lunch Tues–Fri, dinner Tues–Sun

QLD—TOWNSVILLE, MACKAY AND THE WHITSUNDAYS

SCIROCCO CAFE BAR AND GRILL
Tanya Hudson
61 Palmer St, South Townsville 4810
Phone: 07 4724 4508
Email: scirocco@ozemail.com.au
www.sciroccocafe.com
Licensed
Cuisine: Mediterranean-Asian
Mains average: $$
Specialty: Typhoon barramundi steamed in a banana leaf
Local flavours: Barramundi, green pawpaw
Extras: Close to waterfront, outdoor dining on the deck. Makes a range of products for sale including Cajun spice, chilli jam and peanut chilli jam.
- **Open:** Brunch Sun, lunch Tues–Fri, dinner Tues–Sat

TULLY VISITOR AND HERITAGE CENTRE
Bruce Hwy, Tully 4854
Phone: 07 4068 2288
Email: tourism@znet.net.au
Here you can see the whole process of sugar processing on tours that last from one-and-a-half hours to two hours. Enclosed shoes and shirts with sleeves must be worn. Pay at the information centre.
Tours: 10am, 11am and 2pm in season (June–Nov)

TOWNSVILLE
🍴 DINING

SEAGULLS RESTAURANT
Barry Toohey (owner),
Marcel Ricchetti (chef)
Seagulls Resort, 74 The Esplanade, Townsville 4810
Phone: 07 4721 3111
Email: resort@seagulls.com.au
www.seagulls.com.au
Licensed
Mains average: $$
Specialty: Garlic prawns
Local flavours: Seafood, including fish and prawns, fruit and vegetables
Extras: Indoor, airconditioned dining or alfresco dining overlooking tropical gardens and pool. Restaurant also has a children's colouring-in menu. Four-star accommodation, winner of the 2003 Queensland Tourism Awards for Deluxe Accommodation. Located on the seafront and set in three acres of tropical landscaped gardens.
- **Open:** Breakfast, lunch and dinner daily

MARKETS AND EVENTS

✦ **AUSTRALIAN ITALIAN FESTIVAL,**
INGHAM
07 4776 5288
Second weekend of May
(Mothers Day weekend)

✦ **GUMLU CAPSICUM FESTIVAL,**
GUMLU
07 4784 8126
Sun after June long weekend

✦ **TASTING MACKAY,** MACKAY
07 4953 5353
Mid July

SOUTH AUSTRALIA

Established without convict labour, South Australia appears different, in some respects, to the other states. Certainly the large German migration last century affected both the land-use and culture of the state, and greatly influenced the food and wine industries. The Barossa thrives because of this. The German accent is less apparent now, but still evident in the wide variety of deli items, breads and smallgoods, and of course the wines produced in the area.

The Adelaide Hills, Clare Valley and Fleurieu Peninsula are also rich with wine and food producers. In all these areas, farmhouse cheeses, berry producers, yabby farmers and apiarists abound.

Kangaroo Island is also rich with produce and is a case-in-point for how diversification, due to falling prices in traditional industries, can be turned to advantage by gutsy and energetic farmers.

Yet much of the land in South Australia is wild and inhospitable to small holdings. In the north and west, large stations run cattle and sheep, or raise wheat. The peninsula coastlines mine the wealth of seafood, oysters, fish, tuna and abalone in the southern ocean.

TRIVIA: SOUTH AUSTRALIA'S UNUSUAL FOODS

Balfours frog cakes: These unusual beasties are really just patty cakes, cut and iced to resemble frogs and dating back to the 1920s when they were served in city tearooms. Still made with the same recipe, they were originally green (of course) but now come in pink and white, as well as yellow for Easter, and red for Christmas. Sightings have even been made of piles of frogs as birthday and wedding cakes! Now declared a BankSA Heritage Icon.

Pie floaters: A real South Australian tradition, particularly late at night from pie carts located around the city. The deal is that you put a meat pie upside down in mushy green pea soup and top it with tomato sauce. Delicious, say the locals. Look for the carts at North Terrace, outside SkyCity, and Franklin Street, outside the General Post Office. They are carrying on the tradition begun in 1864 when J Gibbs opened a pie stall on the corner of Rundle and King William streets. Now declared a Bank SA Heritage Icon.

SOUTH AUSTRALIA

Regions:
1. Adelaide & Surrounds
2. Barossa
3. Clare Valley & Yorke Peninsula
4. Eyre Peninsula, Flinders Ranges & the Outback
5. Kangaroo Island
6. Limestone Coast
7. Murray & Riverland

ADELAIDE AND SURROUNDS

Adelaide has to be one of Australia's more lovely cities; small and perfectly planned, yet with a character all of its own. It has been a formidable force in the food area, and has led the country more than once in adventurous dining. Most interestingly, this was the state that legalised the consumption of kangaroo and emu meat long before the other states. Its proactive promotion and use of bush foods has no doubt inspired the rest of the country to take notice of these wonderful indigenous resources too.

Dining in Adelaide is eclectic. Some of Australia's finest dining experiences can be had here, but the cafe culture is alive and well too, with everything from breakfast bars to pavement cafes taking advantage of Adelaide's superb sunshine. Late night bars and bistros are also a feature.

The Central Markets are an absolute must for anyone even slightly interested in food. Serious food lovers will find it hard to tear themselves away as this is one of the world's best regional food markets.

And for a truly wacky food experience, certainly no visit to Adelaide would be complete without at least one pie floater.

SA WEBSITES
www.adelaide.southaustralia.com

ADELAIDE
🍎 PRODUCERS

CHARLESWORTH NUTS
Mark and Brett Charlesworth
29–31 Township Rd, Marion 5043
Phone: 08 8296 8366
Email: purchasing@charlesworthnuts.com.au
www.charlesworthnuts.com.au
There are eight stores in Adelaide, as well as an outlet at the Central Markets (where it all began in 1934), plus the factory shop, where you can buy the wonderful confectionery and nuts from this seventy-year-old, third generation South Australian company. Much of its product is made with Australian produce, including dried fruit from the Riverland, and nuts from the Riverland and elsewhere in Australia.
■ **Open:** Factory shop—8.30am–5pm weekdays, retail shops—8.30am–5.30pm weekdays

FERGUSON AUSTRALIA PTY LTD
Debra and Andrew Ferguson
Cnr Days and Regency Roads, Regency Park 5010
Phone: 08 8346 8764
Email: admin@fergusonaustralia.com
www.fergusonaustralia.com
This company has been in operation for over twenty-five years, specialising in producing quality gourmet lobster products—medallions, cooked picked lobster meat, lobster oil, and other lines—using fresh lobster caught by the company's own commercial fishing vessels. Gourmet

SA—ADELAIDE AND SURROUNDS

food lines have evolved in the last four years and are Australian and world firsts. The lobster oil, made from the shell of the lobster and used as an accompaniment to many dishes, is a knockout. All these products are sold in gourmet delis and fine food stores around Australia. Tanks are at 48 Kohinoor Rd, Kingscote, and are open to the public, but times need to be pre-arranged by phoning 08 8346 8764.

- **Open:** 9am–5pm weekdays
- **Location:** Part of Regency TAFE complex

ROBERN MENZ (MFG) PTY LTD
Phil Sims
71 Glynburn Rd, Glynde 5070
Phone: 08 8365 4700
Email: rmenz@robernmenz.com.au
www.robernmenz.com.au,
www.medlow.com

Since 1850, four generations have contributed to this business being what it is today—a major confectioner and glacé fruit processor. The factory door shop at the manufacturing site has a wide range of local fruits and fruit-based confectionery and sells firsts, seconds and large bulk packs, all at discounted prices, and the products are also widely available. Also at Kiosk 64, Central Market Arcade.

- **Open:** 9am–5pm weekdays, 9am–12pm Sat

SUNTRALIS FOODS
Rod Kleeman
U7/61 O'Sullivan Beach Rd, Lonsdale 5160
Phone: 08 8384 7511

Rod Kleeman began Suntralis eleven years ago and it now packages over 350 lines of nuts, dried fruits and confectionery at the Adelaide premises, using as much local South Australian fruits and produce as possible. The products are only available from the factory shop, where you can also get bread-making flour, beans and other cooking ingredients.

- **Open:** 9am–5pm weekdays, 9am–1pm Sat

MARKETS

ADELAIDE CENTRAL MARKET
Adelaide City Council Customer Centre
Gouger and Grote Streets, Adelaide 5000
Phone: 08 8203 7203
Email: centralmarket@uunet.com.au
www.adelaide.sa.gov.au/centralmarket

These unique markets began in 1869 as a way for the local market gardeners to sell their produce. Today traders selling fruit and vegetables, bakery products, seafood, meat, poultry, nuts and confectionery, gourmet foods, cheeses and continental smallgoods make up the mix. Cafes throughout the markets ensure people have the energy to keep going in this very large site. Car parking is readily available, located above the market.

- **Open:** 7am–5.30pm Tues, 9am to 5.30pm Thurs, 7am–9pm Fri, 7am–3pm Sat

STORES

GULF SEAFOODS PTY LTD
Malcolm Stremberger
57 Duthy St, Malvern 5061
Phone: 08 8271 4225

Fresh seafood, poultry, game, gourmet foods, and takeaways are the line-up here. Operating for twenty-five years, Gulf Seafoods bases its success on very fresh local produce and the best quality.

- **Open:** 9am–6pm weekdays, 9am–4pm Sat

ADELAIDE AND SURROUNDS—SA

HAIGH'S CHOCOLATES VISITORS CENTRE
154 Greenhill Rd, Parkside 5063
Phone: 08 8372 7070
Email: admin@haighs.com.au
www.haighs.com.au
Haigh's dates back to the early 1900s in Adelaide. Alf Haigh began his career as a confectioner making boiled sweets in Jamestown, and Beehive Corner was his first shop, on the corner of Rundle Mall. The company, run by fourth-generation Haighs, still uses original recipes and makes its own chocolate from imported, raw cocoa beans. There is a new range of bush food flavours—quandong, lemon myrtle, macadamia, wattle seed crunch—and part of the proceeds will be donated to the Botanic Gardens of Adelaide to help promote the use of native Australian plants. Haigh's is also home to the Easter (chocolate) bilby, and there are free, twenty-minute guided tours (bookings 08 8372 7077) of the factory. There are six stores in Adelaide, and five in Melbourne. Haigh's won the chocolate Championship at the Sydney Royal Easter Show in 2003.
- **Open:** 8.30am–5.30pm weekdays, 9.30am–4.45pm Sat
- **Guided tours:** 1.30pm and 2.30pm Mon–Sat

MAROUDAS OLIVES
Margarita Flabouris
18 Cawthorne St, Thebarton 5069
Phone: 08 8354 0322
Email: maroudasolives@hotmail.com.au
Maroudas Olives has been growing and producing some of the finest quality organic kalamata olives and Organic EVOO in Australia for over thirty years, continuing a family tradition that goes back generations. They have won many awards for oils and olives, no doubt because they say they have a 'passionate belief in growing and producing the highest quality organic olives and olive oil'. There are cellar door sales and the products are also sold at various gourmet delis, organic gourmet shops and providores throughout Australia.
- **Open:** 9am–5pm weekdays

NATIONAL WINE CENTRE OF AUSTRALIA
Leigh Craig (Manager)
Cnr Botanic and Hackney Roads,
Adelaide 5000
Phone: 08 8222 9222
Email: nwc.info@adelaide.edu.au
www.wineaustralia.com.au
This major centre, established just three years ago as a government initiative, profiles sixty-two wine regions throughout Australia. It also includes de Castella's restaurant, featuring fresh local produce.
- **Open:** 10am–5pm daily
- **Location:** Adjacent to the Botanical Gardens

NEWSLINK DISCOVER SOUTH AUSTRALIA
Zoe Groom
Qantas domestic terminal, Adelaide airport, Adelaide 5000
Phone: 08 8234 366
Email: discoversa@newslink.com.au
This is a great place to visit if you are flying interstate. You can stock up on South Australian gourmet food including Willunga almonds, South Australian Company Store products and wines while waiting for your boarding call.
- **Open:** 5am–9pm daily

SA—ADELAIDE AND SURROUNDS

🚌 TOURS

A TASTE OF SOUTH AUSTRALIA WINE TOURS
Anne Kennedy
Adelaide
Mobile: 0419 861 588
Email: info@tastesa.com.au
www.tastesa.com.au
Taste of South Australia can tailor special food and wine tours for individuals or groups with travel provided in limousines or coaches.
■ **Open:** Phone for details

ADELAIDE'S TOP FOOD AND WINE TOURS
Tony and Helen Rensberg-Phillips
Adelaide
Phone: 08 8263 0265
Email: ocars@senet.com.au
www.food-fun-wine.com.au/tours.php
This company organises food and wine lifestyle tours such as Market Adventures (at the Adelaide Cental Market), Grazing on Gouger, Dawn Market Tour and Connoisseur's Choice—all of them great ways to catch up with South Australian food and dining. Tour group sizes are kept small and personalised for the greatest benefit. Check out the website—some tours can be booked online.
■ **Open:** By appointment

SUSIE'S BOUTIQUE TOURS
Burnside 5066
Mobile: 0417 841 008
Email: susie@susiestours.com.au
www.susiestours.com.au
Gourmet tours of McLaren Vale, Barossa and Central Markets (Tues–Sat).
■ **Open:** 9am–5.30pm daily

TAUONDI CULTURAL TOURS
Mike Gray
1 Lipson St, Port Adelaide 5015
Phone: 08 8341 2777
Email: tcagency@tauondi.sa.edu.au
These forty-minute, guided Tappa Mai (bush food) tours of the Adelaide Botanic Gardens are run by accredited Aboriginal guides. Tickets from SA Visitor and Travel Centre or the botanic gardens kiosk, which is where the tour begins.
● **Tours:** 10.30am Wed–Fri and Sun

🍷 BREWERY

COOPERS BREWERY LIMITED
461 South Rd, Regency Park 5010
Phone: 08 8440 1800
Email: coopers@coopers.com.au
www.coopers.com.au
Begun in Adelaide in 1862, Coopers is now Australia's oldest remaining family-owned brewery. Now a BankSA Heritage Icon, it produces 400 000 hectolitres every year. The new brew-house is two and a half times the size of the previous one, and tours are planned to recommence in early 2005.
■ **Open:** Tour times vary, phone for details

ADELAIDE AND SURROUNDS—SA

🍴 DINING

MAGILL ESTATE RESTAURANT
Dianna Battistella
78 Penfold Rd, Magill 5072
Phone: 08 8301 5551
Email:
magillestaterestaurant@penfolds.com.au
www.penfolds.com.au
Licensed
Cuisine: Modern Australian with a Mediterranean influence
Mains average: $$$$
Local flavours: As much as possible
Extras: Panoramic views over the pond and vineyards as well as city views.
■ **Open:** Lunch Wed–Fri (Dec only), dinner Tues–Sat

ADELAIDE HILLS AND FLEURIEU PENINSULA

Hardly anything compares to a leisurely drive through the Adelaide Hills on a perfect late-summer or autumn day. Apart from the views across the city, there are surprises at every turn, as you discover orchards or market gardens, bursting with produce for sale. You can dine for a week on just what you'll find in this area, or, you can take a hamper and cosy down at some scenic picnic spot with a bottle of the region's wine.

The Fleurieu Peninsula to the south is equally delightful; a fertile wineland, that partners almonds and honey, berries, farmhouse cheeses, venison and trout. If there were doors on the region, they would be inscribed 'Gourmet Food Hall'. As there are not, we are all free to wander in, sample the produce, buy some and come away with a basketful of goodies and the resolve to return.

BALHANNAH

🍎 PRODUCERS

THE OLDE APPLE SHED
GV and JA Morriss
55 Main Rd (also called Onkaparinga Valley Rd), Balhannah 5242
Mobile: 0413 380 251
Email: jessie@adam.com.au
www.visitadelaidehills.com.au/oldeappleshed

Specialising in wax-free apples and pears straight from the grower, these orchardists also stock regional and local produce and make homemade apple pastries and condiments. With more than ten varieties of apples and three varieties of pears, as well as cherries, this minimal-spray, pest-integrated orchard, established in 1968, is a wonderful place to visit.
■ **Open:** 10am–5pm daily
● **Location:** Approximately 5km from the freeway, 200 metres on the left past Mitre 10, and about 300 metres from Balhannah town centre

🏠 STORES

ORGANIC PASTA SHOP
Melissa Pavia
6/24 Bridge St, Balhannah 5242
Phone: 08 8388 4566
Email: orgpasta@picknowl.com.au

The fresh, organic (certified by NASAA) pasta comes in a range of flavours as well as varieties, such as lasagne, ravioli and Asian noodles. The pasta is unique because only spring water is used in the recipes. Melissa had been making fresh pasta at home because she was unable to buy fresh, organic pasta. Realising there was a niche market, she opened this shop

199

three years ago. Now there is a range of dried pasta including bush flavours, fusion flavours and the wheat alternative, spelt.

- **Open:** 10am–5pm weekdays, 10am–1pm Sat

ADELAIDE RESTAURANTS

How can you describe Adelaide's dining? You could start by considering the diversity of its popular dining areas: Gouger Street, close to the Central Market, is cosmopolitan with some sensational restaurants; Henley Beach Road, Torrensville, is the place to go for casual, family-style dining, as well as Greek restaurants; Hindley Street West is basically a coffee-shop strip; King William Road is home to the classier shops of Adelaide with good coffee and gourmet food; O'Connell Street offers dozens of affordable and interesting spots to eat; and Rundle Street, the heart of Adelaide's Fringe Festival, has a twenty-four hour buzz.

BIRDWOOD
STORES

BIRDWOOD WINE AND CHEESE CENTRE
Trevor and Sue Manning, Glen and Lyn Venning
Shop 3, 22 Shannon St, Birdwood 5234
Phone: 08 8568 5067
Email: winecentre@bigpond.com.au
This centre promotes South Australian produce with tastings of wines and cheeses from around the state. The customers choose from around forty wines, (changed fortnightly) and the samples are served with cheese and other SA products, such as olives, conserves or fruit. You can 'experience the flavours of South Australia' in this one-stop shop. The owners have an interest in cheese from their dairying backgrounds, and set up this centre as another means of drawing tourists to Birdwood, and to showcase the products of small businesses in the area.

- **Open:** 10am–6pm Wed–Sat, 12pm–6pm Sun, other times by appointment

BRIDGEWATER
DINING

PETALUMA'S BRIDGEWATER MILL
Le Tu Thai (chef)
Mount Barker Rd, Bridgewater 5155
Phone: 08 8339 3422
Email: bridgewatermill@petaluma.com.au
www.bridgewatermill.com.au
Licensed
Cuisine: Modern Australian
Mains average: $$$
Local flavours: Adelaide Hills produce used where possible
Extras: Outside deck overlooking historic waterwheel and stream.

- **Open:** Lunch Thurs–Mon, dinner for functions or groups

CRAFERS
DINING

MOUNT LOFTY HOUSE
Kent Aughey
74 Mt Lofty Summit Rd, Crafers 5152
Phone: 08 8339 6777
Email: relax@mtloftyhouse.com.au
www.mtloftyhouse.com.au

Licensed
Cuisine: Modern country house
Mains average: $$
Specialty: Bread and butter puddings, summer berry puddings
Local flavours: Bread, herbs, venison, duck, beef
Extras: Kitchen garden grows vegetables and herbs for the restaurant. One-day and weekend cooking and health retreats. Tiered lawns, private gardens, heated pool, tennis court, volleyball court and views of the Piccadilly Valley from all areas of the house.
■ **Open:** Breakfast and dinner daily, lunch Sun

CURRENCY CREEK
DINING

CURRENCY CREEK ESTATE WINES
Shaw family, Michael Duckow (chef)
Winery Rd, Currency Creek 5214
Phone: 08 8555 4069
Email: enquiries@currencycreekwines.com.au
www.currencycreekwines.com.au
Licensed
Cuisine: Modern regional
Mains average: $$
Specialty: Matching food and wine
Local flavours: Produce from Fleurieu Peninsula
Extras: Rural setting with views over the vineyard and creek.
■ **Open:** Lunch daily, dinner Tue-Sat
● **Location:** 10km north of Goolwa

GOODWOOD
PRODUCERS

LIMONCELLO AUSTRALIA PTY LTD
Libero de Luca
97 Goodwood Rd, Goodwood 5034
Phone: 08 8357 7744
Email: libero@ambralimoncello.com.au
www.ambralimoncello.com.au
There could hardly be a better way to make use of plump, specially grown citrus from South Australia's Riverland. Limoncello, a traditional Italian digestif, is made according to a centuries-old formula and results in a fragrant and delicious liquor which is low alcohol and has no artificial flavours, preservatives or colourings. Once you have acquired a taste for limoncello, move on to Ambra Agrumello (orange), Ambra Fragolino (strawberry), Cream of Lemon Liqueur and Ambra Chocolatino, an orange and chocolate liqueur. A twist on the liqueur is Baba Al Limoncello, small sponge cakes preserved in Limoncello syrup.
■ **Open:** 9am–5pm weekdays

GOOLWA
DINING

SIGNAL POINT CAFE
Carol and Greg Butler
The Wharf, Goolwa 5214
Phone: 08 8555 1722
Email: signalpointcafe@bigpond.com
Mains average: $
Local flavours: Alexandrina milk, cheese, cream
Extras: Outdoor seating and view over Goolwa Wharf precinct and River Murray.
■ **Open:** 8am–5pm daily

SA—ADELAIDE AND SURROUNDS

> **THE COCKLE SHUFFLE**
> November thru March do the 'cockle shuffle' on Fleurieu Peninsula beaches, checking the sands for edible cockles. Contact Goolwa Visitor Information Centre on 08 8555 1144 for information (open 9am–5pm daily).

GUMERACHA
🍴 DINING

VINEYARD BALCONY
Chain of Ponds Wines, Adelaide-Mannum Rd, Gumeracha 5233
Phone: 08 8389 1415
Email: admin@chainofponds.com.au
www.chainofpondswines.com.au
Licensed
Cuisine: Italian slow food
Mains average: $$
Specialty: Seasonal, regional platters
Local flavours: Hare, baby goat, figs, baby greens, homemade salami sausage, estate olives, olive oil
Extras: Outdoor tables and umbrellas, petanque strip, B&B in 1880s cottage, views of pine forests, grounds and garden.
■ **Open:** Lunch weekends, dinner by arrangement

HAHNDORF
🍎 PRODUCERS

BEERENBERG FARM
Grant and Carol Paech
Mount Barker Rd, Hahndorf 5245
Phone: 08 8388 7272
Email: admin@beerenberg.com.au
www.beerenberg.com.au
If you want a real Australian success story, this is it. The Paech family have been on this land since 1840, but only started making jams in 1969, as a means of diversifying. Strawberry jam was first, and next came chutneys, sauces, marinades, dessert sauces, pickled onions, all originally from Carol's kitchen using traditional recipes for a good old-fashioned taste. It worked, and now these jams, pickles and condiments are in gourmet food shops, supermarkets, hotels and airlines. If you'd rather use your own recipes, you may pick-your-own strawberries here from October to May.
■ **Open:** 9am–5pm daily
● **Location:** Off the South Eastern Fwy at Hahndorf

HAHNDORF VENISON
John Delaine
River Rd, Hahndorf 5345
Phone: 08 8388 7347
Email: venison@optusnet.com.au
www.hahndorfvenison.com.au
The Delaines sell their venison vacuum-packed for hygiene and convenience. There are all cuts from their farmed fallow or red deer, as well as smoked venison, sausages (the herbs used in these are grown on the farm), and smallgoods. The company also produces and differentiates 'elk'. Venison cuts include viande, a premium grade venison, and vevette, a 180–200 gram venison steak.
■ **Open:** By appointment

🏠 STORES

MATTISSE
Matti Malchi
Shop 14-15, 23 Main St, Hahndorf 5245
Phone: 08 8388 1258
Email: mattisse@mattissebakery.com
http://mattissebakery.com
For the last two years, baker Matti Malchi has been trapping the wild yeasts from the air over the Adelaide Hills and letting them work wonders with his sourdough artisan breads. His hand-crafted loaves are made in the traditional manner from only natural ingredients—unbleached flour, purified water, salt and wild yeast cultures—and baked in a hearth oven. Sold here and in selected delis and shops around Adelaide.
■ **Open:** 8am-5pm Tues-Sat, 9am-5pm Sun

🍴 DINING

GERMAN ARMS HOTEL
Peter and Monica Colotti
69 Main St, Hahndorf 5245
Phone: 08 8388 7013
Email: germanarms@bigpond.com
Licensed
Cuisine: German Australian
Mains average: $
Specialty: Steaks, German desserts
Local flavours: Beerenberg products
Extras: Outdoor dining area.
■ **Open:** Lunch and dinner daily

HAHNDORF INN
Jason and Olivia Duffield, and family
35 Main St, Hahndorf 5245
Phone: 08 8388 7063
Email: hahndorfinn@adam.com.au
www.hahndorfinn.com.au
Licensed/BYO

Cuisine: Steaks
Mains average: $
Local flavours: South Australia's finest steaks, smallgoods, cheese, chutney, mustard
Extras: German beers on tap. Outdoor eating area overlooking a leafy laneway and historic Hahndorf's main street. Next to the Hahndorf Inn Showcase Store, which features SA-made products.
■ **Open:** Breakfast, lunch and dinner daily

MUGGLETONS GENERAL STORE AND RESTAURANT
Nicole Barrie
Shop 2/38 Main St, Hahndorf 5245
Phone: 08 8388 7555
Email: muggletons@pinnacle.net.au
www.muggletons.com
Licensed
Cuisine: Traditional Australian
Mains average: $$
Specialty: Scones, bangers and mash
Local flavours: Sausages made especially for the restaurant, fruit, vegetables, jams, sauces, pickles and mustards made on site
Extras: Old-style country store attached.
■ **Open:** 9am to 5pm daily (breakfast, lunch and afternoon teas)

HEATHFIELD

🍎 PRODUCERS

NIRVANA ORGANIC PRODUCE
Deb Cantrell and Quentin Jones
184 Longford Rd, Heathfield 5153
Phone: 08 8339 2519
Email: nirvanafarm@chariot.net.au
If you are craving hot roasted chestnuts then you have to come to this place in chestnut season. Specialising in quality, fresh berries and nuts, gourmet preserves,

farm tours and educational courses in organic farming methods, they have far too many products to list here. 'Diversity is the key to our farming methods and also our business', they say. All the jams, jellies, vinegars and cordials are made from fruit grown on the property. Farm tours allow you to see and sample whatever is in season, and short courses and workshops on orchard management and techniques are held from time to time.

■ **Open:** 9am–6pm daily during berry and nut harvests, closed Sun and Wed morning at other times
● **Location:** 3.3km from the Stirling roundabout freeway exit

HINDMARSH ISLAND
DINING

RANKINE'S LANDING TAVERN
Peter and Sue Rankine
The Marina, Hindmarsh Island 5214
Phone: 08 8555 1122
Licensed/BYO
Cuisine: Modern Australian
Mains average: $$
Specialty: Prawn caesar salad, fish of the day, fish and chips, lower deck burgers
Local flavours: Veal, lamb, seafood, fruit and vegetables, cheese, wines
Extras: A peaceful waterfront location with views over the marina and lower Murray.
■ **Open:** Breakfast on weekends, lunch and dinner daily
● **Location:** Turn right off Randell Rd

HOUGHTON
PRODUCERS

WILLABRAND
Willa Wauchope
Glen Ewin, Lower Hermitage Rd, Houghton 5131
Phone: 08 8380 5508
Email: willa@willabrand.com
www.willabrand.com
Willa Wauchope began making fig products at the historic Glen Ewin jam factory in 1997 and hit the big time with his wonderful Willabrand figs. If you love chocolate, and you love figs, just make sure you try these chocolate-coated figs. Your tastebuds will never be the same again. Sold Australia-wide at good gourmet shops, some health food shops and fruit shops. You can pick your own fresh figs too and visitors are welcome.
■ **Open:** 9am–5pm weekdays, 10am–4pm weekends
● **Location:** 80 metres along Lower Hermitage Rd, first driveway on left

INMAN VALLEY
PRODUCERS

GALLOWAY YABBIES
Carol and Jim Schofield
261 Main Rd, Inman Valley 5211
Phone: 08 8558 8215
Email: galloway21@bigpond.com
Fourteen years ago, the Schofields decided to diversify from dairying. 'A hobby that just grew', is how they describe their business. Now the yabbies (live, pickled, and turned into pâté) and the tours and fishing keep them more than busy. The product is sold here, but you could also

find it on your plate at the Adelaide Hilton, Salopian Inn or Red Ochre Restaurant.
- **Open:** 11am–4pm Fri–Mon (11am–4pm daily during school holidays), or by appointment
- **Location:** 16km from Victor Harbor on the Yankalilla Rd

RIPE CHERRY TRAIL
Karen Shepherd, Adelaide Hills Visitor Information Centre
Phone: 08 8389 8800
Email: tourism@adelaidehills.com.au
Twelve participating orchards in December and January with self-drive map.

KANMANTOO
STORES

KANMANTOO BACON AND QUALITY MEATS
Robert and Elain McInnes
Mines Rd, Kanmantoo 5252
Phone: 08 8538 5097
Email: kanmantoo.bacon@senet.com.au
If you are craving some old-style bacon, naturally wood-smoked, this is the place to go. The McInneses use local pork and traditional methods, and you'll find the results of their skills widely available in South Australia, and also at selected outlets in Sydney.
- **Open:** 8am–4.30pm weekdays, weekends by appointment

KENTON VALLEY
PRODUCERS

NETHERHILL STRAWBERRY FARM
Doug and Margaret Reid
Netherhill Rd, Kenton Valley 5233
Phone: 08 8389 1046
Email: netherhl@chariot.net.au
www.netherhill.com.au
Although the main market for these magnificent strawberries is Brisbane, there is a farm shop here where export quality seconds and jam fruit are sold. There are light lunches for bus groups and the jams and pickles come from the farm kitchen. The list is long. Start with the jams—strawberry, plum, apricot, cherry, blackberry, fig and a range of marmalades. Before your mouth waters too much, get yourself down there (or to the Barossa farmers' market) to see for yourself.
- **Open:** 9am–5.30pm daily (Nov–May)
- **Location:** 1km along Netherhill Rd, which runs off the Lobethal–Gumeracha Rd, 3km from Gumeracha

LANGHORNE CREEK
PRODUCERS

NEWMANS HORSERADISH
BJ Meakins Pty Ltd
Lake Plains Rd, Langhorne Creek 5255
Phone: 08 8537 3086
Email: horseradish@olis.net.au
Home-grown horseradish is not something you find every day. Although this company has been producing Newman's Original red-label horseradish for fifty years, it still processes what it grows. Other horseradish

products are also made here, as well as crushed garlic, chilli-garlic, mustards and crushed ginger. All products are available throughout Australia.
- **Open:** 9am–4.30pm weekdays
- **Location:** 4km south of Langhorne Creek

LENSWOOD

TOURS

OTHERWOOD FARM TOURS
RD and CF Brockhoff
Swamp Rd, Lenswood 5240
Phone: 08 8389 8418
If you fancy a drive though a working orchard on a coach tour, contact these people. They also grow and sell apples, kiwifruit, lemons, limes, plums and avocados for market, as well as other fruit for their own use (and for jam making) and these are sold to visitors and at local markets.
- **Open:** By appointment
- **Location:** On the left between Oakwood and Collins Hill roads

LOBETHAL

PRODUCERS

UDDER DELIGHTS
Sheree Dunford
15/1 Adelaide-Lobethal Rd, Lobethal 5241
Mobile: 0413 000 790
Email: udderdel@adam.com.au
www.adelaidehillsfood.com.au/udderdelights
Since 1999, Udder Delights (now isn't that a great name?) has been in the business of creating truly artisan goats cheeses with delicate flavours and beautiful presentation. It's a classic tale of turning adversity into a positive. The Dunford family ran a goat dairy and after experiencing difficulties securing a milking contract decided to build a goats cheese factory themselves. The cheeses, made with milk from dairies at One Tree Hill and Owen, have been awarded nine national gold awards and a national trophy. Look for Fromage Blanc, a delicate fresh goat curd, Brancoleite, a multi national award-winning chevre, marinated Brancoleite, Cabraleite, a fetta style goats cheese, camembert, and Auveshe, matured triangles of goats cheese, hand rolled in ash, and finished with white mould.
- **Open:** By appointment
- **Location:** In the Adelaide Hills Business and Tourism Centre

STORES

LOBETHAL BAKERY
Helmut Trinkle
80 Main St, Lobethal 5241
Phone: 08 8389 6496
Where else but South Australia would German baking be regarded as good Australian regional food? This family bakery has been specialising in turning out top-class breads such as pretzels and sourdough rye, for thirty years. There are also shops in Woodside (phone: 08 8389 7584) and Stirling (phone: 08 8370 1999).
- **Open:** 8.30am–5.30pm weekdays, 8.30am–2pm Sat

LYNDOCH LAVENDER FARM
Jill and Evan Allanson
Cnr Hoffnungsthal and Tweedies Gully Roads, Lyndoch 5351
Phone: 08 8524 4538
Email: llf@lyndochlavenderfarm.com.au
www.lyndochlavenderfarm.com.au

At the farm shop cafe there are light meals including a delicious platter of local cheese, pickles and meats and a choice of superb Allanson shiraz or riesling, available by the glass or bottle. The morning and afternoon tea menu includes unusual lavender treats such as scones, cake, biscuits, ice-cream and lavender conserves, which are all available for tasting. All food lines—jams, chutneys, mustards, vinegar, cordial, nougat and lavender chocolate—are made at the farm, many from original recipes. There is also a comprehensive range of skin care and home care items.
- **Open:** 10am–4.30pm daily (Sept–Feb), 10am–4.30pm weekends (Mar–Aug)
- **Location:** 3.5km from Lyndoch on the Lyndoch–Williamstown Rd, turn left into Hoffnungsthal Rd, after 2.5km turn right into Tweedies Gully Rd

McLAREN FLAT

PRODUCERS

DAVID MEDLOW CHOCOLATES
David and Sharon Medlow-Smith
Kangarilla Rd, McLaren Flat 5171
Phone: 08 8323 8818
Email: sharon@pecktins.com
www.pecktins.com
Acknowledged as leaders in pectin confectionery (that's gels and jellies) Medlow pioneered the Australian use of wines (including muscat and vintage port) in confectionery. Located on twenty-one acres of garden and lake they retail a full range of toothsome treats.
- **Open:** 10am–4.30pm daily
- **Location:** At McLarens on the Lake

STORES

McLAREN VALE OLIVE GROVES
David Lloyd
Lot 34, Warner Rd, McLaren Flat 5171
Phone: 08 8323 8792
Email: mail@olivegroves.com.au
www.olivegroves.com.au
We're told that you'll find 'arguably the best pickled kalamata olives in Australia' here. That and olive oil made from the fruit of verdale trees that are over 120 years old. You can get the oils here or by mail order, although, it's worth visiting this lovely area and also checking out the pesto and wine at the old worker's cottage that has been turned into a cellar door. You may sample and purchase olives, olive oils, fabulous cosmetics and many local gourmet foods. The cottage is surrounded by a large undercover paved area where an olive picker's platter is served. Tours through the factory show how oil used to be processed.
- **Open:** 10am–5pm daily
- **Location:** 3km along Tourist Dr 60, turn left at sign

THE ALMOND AND OLIVE TRAIN
Tonny Beagley and Colin Church
Main Rd, McLaren Flat 5171
Phone: 08 8323 8112
Email: info@thealmondtrain.com.au
www.thealmondtrain.com.au
This converted railway carriage now sells a range of local produce, including olives and almonds, olive oils, sauces and crafts. There are a couple of dozen different flavoured sweet and savoury almonds, most of them unique to this shop. The concept works well as a tourist attraction plus a way to promote the almond industry. Expect exotic options such as barbecue,

salt and vinegar, crunchy cheese, garlic, and smoked almonds, but also sweet temptations like toffee, honey or maple.
- Open: 10am–4.30pm daily
- Location: Next to Hardy's Winery

🍴 DINING

THE BARN
Phillip Taylor, Daryl Moyle, John Grosvenor
Cnr Chalk Hill and Main Roads,
McLaren Flat 5171
Phone: 08 8323 8618
Email: thebarn@senet.com.au
www.thebarnrestaurant.com.au
Licensed (BYO for special wines)
Cuisine: Modern Australian
Mains average: $$
Specialty: Emphasis on fresh, local ingredients
Local flavours: Veal, venison, turkey, kangaroo, cheeses, olives, almonds, fruit and vegetables
Extras: Courtyard dining under vines, a huge walk-in wine room with over 400 different wines, cigar lounge, art on display.
- Open: Lunch daily, dinner Wed–Sat (or by arrangement)

WOODSTOCK COTERIE
Scott Collett and Kay Cazzolato (chef)
Woodstock Winery, Douglas Gully Rd,
McLaren Flat 5171
Phone: 08 8383 0156
Email: woodstock@woodstockwine.com.au
www.woodstockwine.com.au
Licensed/BYO
Cuisine: Seasonal regional modern Australian
Mains average: $$
Specialty: Regional platter, turkey muscat pâté

Local flavours: As many as possible—cherries, pears, cheese, meats
Extras: Paved courtyard, extensive lawns and native gardens, vineyard views, barbecue area, takeaway platters available.
- Open: Lunch daily, breakfast and dinner (by booking)

McLAREN VALE

🍎 PRODUCERS

CORIOLE VINEYARDS
Mark and Paul Lloyd
Chaffeys Rd, McLaren Vale 5171
Phone: 08 8323 8305
Email: contact@coriole.com.au
www.coriole.com
Begun in 1967 by Mark and Paul's parents, to further Dr Hugh Lloyd's interest in viticulture, Coriole Vineyards now also produces a limited amount of very fine extra virgin olive oil and vinegars, in addition to premium wines. The vinegars include red wine vinegar and an aged sweet vinegar, both produced in very small amounts. The company now also owns Woodside Cheese Wrights in the Adelaide Hills.
- Open: 10am–5pm weekdays, 11am–5pm weekends and public holidays

PICK YOUR OWN FRUIT AND KIMBER WINES
Peter and Bridget Kimber
Chalk Hill Rd, McLaren Vale 5171
Phone: 08 8323 8642
Email: kimber@comstech.com
www.kimberwines.com
Twenty-two years ago, an oversupply of wine grapes and a poor grape price encouraged Peter Kimber to try other means of using his land in the lovely McLaren Vale area. Now he opens his

2000 plus pesticide-free fruit tree orchard to the public during picking season (mid December to the end of February) and allows pick-your-own visitors to do just that among his apple, apricot, peach, plum, prune, nectarine and peacharine trees. The cellar door facility is nearby and a fruity addition there is a stunning peach liqueur.

- **Open:** 8am–6pm daily during picking season, phone for availability at other times
- **Location:** 3.3km along Chalk Hill Rd from the main McLaren Vale road

WINE COAST AVOCADO FARM
Nicky Robinson
Lot 2, Stump Hill Rd, McLaren Vale 5171
Phone: 08 8323 8869

The frost-free environment of McLaren Vale means that cool-climate avocados ripen much later than in Queensland. Because Nicky Robinson raises several varieties (including Hass and the late-season Reed, as well as a few earlier ones) her pesticide-free avocados are available for nine to ten months of the year.

- **Open:** Daylight hours daily (closed June–July)
- **Location:** Turn left off the main McLaren Vale road and look for the signs

STORES

ALDINGA TURKEY KITCHEN
Luke Sutherland
Cnr Foggo and Kangarilla Roads, McLaren Vale 5171
Phone: 08 8323 8077
Email: turkman@aldingaturkeys.com.au

At this unique outlet you are able to taste free turkey samples and actually see the turkeys being cooked in the full-scale kitchen. Turkey mince, chops, and other cuts, as well as sausages and breast rolls are for sale. The Turkey Gallery has a collection of artefacts and every visitor to the complex receives a free recipe book. Aldinga raises the turkeys, and they have a flock of around 50,000 at any one time. In the snack bar you can try turkey pies and turkey bacon muffins.

- **Open:** 10am–4.30pm Tues–Sat, 11am–4pm Sun

DINING

MARKET 190
Wayne (chef) and Naomi Angove, Craig and Christine Blacker
190 Main Rd, McLaren Vale 5171
Phone: 08 8323 8558
Email: wayne@market190.com
www.market190.com.au
Licensed/BYO
Cuisine: Modern Australian
Mains average: $$
Specialty: Lemon curd tarts, coffee
Local flavours: Olives, almonds, organic fruit and vegetables, seafood
Extras: Undercover courtyard, table and chairs at the front under market umbrellas, children's area, including a cubbyhouse near the front tables. Makes for sale jams, topping and pâté and sells cheese and other produce.

- **Open:** 8am–5pm daily (winter), 8am–6pm daily (summer)

SALOPIAN INN
Pip Forrester
Cnr McMurtrie and Willunga Roads, McLaren Vale 5171
Phone: 08 8323 8769
Email: salopian@bigpond.com
Licensed

Cuisine: Regional modern Australian
Mains average: $$$
Specialty: Seasonally changing menus
Local flavours: As much Fleurieu Peninsula produce as possible
Extras: Outdoor dining with views over the vineyards to the ranges, art collections, private dining room, open fires in winter. Sells homemade dukkah, olive oil and kasundi.
- **Open:** Lunch Thurs–Tues, dinner Fri–Sat

MEADOWS

🍎 PRODUCERS

BD FARM PARIS CREEK PTY LTD
Ulli and Helmut Spranz
Paris Creek Rd, Meadows 5201
Phone: 08 8388 3339
Email: ulli@bdfarmpariscreek.com.au
www.bdfarmpariscreek.com.au
In 1988, realising that there were no real biodynamic yoghurts on the market, the Spranzes set about changing that by establishing a biodynamic dairy farm with processing plant. Their organic/biodynamic yoghurts, including fruit yoghurts, quark (cottage cheese), milk, butter and cream are produced under strict biodynamic principles using milk from their own farm and those neighbouring farms that have converted to biodynamic principles under their guidelines. No chemicals are used on the farms and no artificial ingredients are used in production. No hormones, genetic modification, homogenisation, or thickeners are allowed either. Sold in South Australia as well as interstate at health food stores and selected supermarkets. Also exported to South-East Asia.
- **Open:** Phone for appointment

- **Location:** 8km from Meadows, or 10km from Strathalbyn, at Paris Creek

MOUNT BARKER

🍎 PRODUCERS

SPRINGS SMOKED SEAFOODS
Richard Harris
3 Enterprise Court, Mount Barker 5251
Phone: 08 8398 1000
Email: mferguson@springssalmon.com.au
www.springssalmon.com.au
In 1990, Richard Harris decided to use the same cottage, kiln-smoking techniques developed by his grandfather in Suffolk because there was no fresh, smoked seafood being produced in Australia at that time. He migrated to Australia from the United Kingdom to set the business up and now produces smoked salmon, grown in clean Tasmanian waters, ocean trout, mackerel, and kippers, using traditional kilns and imported English oak. These products are widely available.
- **Open:** 8am–5pm weekdays
- **Location:** 1km from Mount Barker High School off Strathalbyn Rd

🛒 STORES

FEAST! FINE FOODS
Richard and Elizabeth Gunner
Mount Barker Plaza, Mount Barker 5252
Phone: 08 8391 1139
Mobile: 0408 864 420
Email: rgunner@ozemail.com.au
www.cooronangusbeef.com.au
Feast! Fine Foods is owned by the family that produces MSA-standard Coorong Angus Beef, which is grain finished and raised without the use of hormones. This

is also the exclusive retail outlet for the branded beef and other family-produced specialty meat products such as Coorong Coast Veal, and locally sourced products such as Pure Suffolk Lamb, Emerald Hills Beef and Greenslades Chicken, plus the usual range of cuts and other products. Owner, Richard Gunner, was the winner of Young Leader of the Year 2002.
- Open: 8am–5.30pm Mon–Sat

DINING

BULLOCKS WOOD OVEN EATERY
John and Deanna Bullock
46 Gawler St, Mount Barker 5251
Phone: 08 8391 3331
BYO
Mains average: $
Specialty: Thin and crispy wood-oven pizza, cooked in ninety seconds
Local flavours: Olives, olive oil, cheeses (fetta and haloumi), fruit and vegetables, organic breads
Extras: Hundred-year-old restored shop with outdoor dining. The purpose-built oven was inspired by Russell Jeavons of Russell's Pizzas, McLaren Vale.
- Open: Breakfast and lunch daily, dinner Fri–Sat

MOUNT COMPASS

PRODUCERS

THE BLUEBERRY PATCH
Grant and Merry Gartrell
Nangkita Rd, Mount Compass 5210
Phone: 08 8556 9100
Email: blueberrypatch@ozemail.com.au
The climate, soil and water in this region suit the Gartrells' blue-blooded berries perfectly and they have been growing and selling their premium-quality fresh blueberries (and some raspberries) for twenty years. They also offer blueberry jams, blueberry ice-cream, pick your own and pre-picked farm sales. There are picnic tables in a pleasant garden setting, so what better place to enjoy your ice-cream? Blueberry, of course.
- Open: 9am–5pm daily (mid Dec–mid Feb), other times by appointment
• Location: 5.5km along Nangkita Rd from Mount Compass

TOOPERANG TROUT FARM
Bill Walker
Clelands Gully Rd, Mount Compass 5210
Phone: 08 8556 9048
You can make a day of it and fish here, or you can buy fresh or smoked trout, trout products or honey at the farm shop. At the tearooms you can also find light refreshments and trout meals. It's a fun family day, catching trout to take home, or buying it if you are unlucky. Tooperang trout have been bred and reared in fresh spring water on this property, in the scenic, glacier-formed Tooperang valley for the past nineteen years.
- Open: 10am–5pm weekends and public holidays
• Location: Second turn left, past township

MOUNT JAGGED

PRODUCERS

ALEXANDRINA CHEESE CO
Krystyna McCaul
Sneyd Rd, Mount Jagged 5210
Phone: 08 8554 9666
Email: alexandracc@ozemail.com.au
Located on a family farm at Mt Jagged, this is a wonderfully welcome stopping point on

the way to Victor Harbor. Since 2001 they have begun producing cloth-bound, rinded cheddars, edam and gouda cheeses as well as dairy products, which include a national award-winning double cream made with milk from the farm's own Jersey herd. You can buy the products in the cheesery, or stay longer at the snack bar for some good coffee, a cheese platter, or cakes.

- **Open:** 10am–4.30pm weekends and public holidays, 10am–4.30pm Wed–Sun (during school holidays)
- **Location:** 8km south of Mount Compass

MYLOR

🍎 PRODUCERS

AUSSIE FRUIT BOY
Bruce Maxwell
42 Boyle Swamp Rd, Mylor 5153
Phone: 08 8388 5334
Email: bm@picknowl.com.au

If you love cherries, figs, persimmons, peaches, and nashi pears, this orchard will be heaven for you. There is a small-time operation making jams and savoury preserves with these lovely ingredients.

- **Open:** By appointment

NAIRNE

🍴 DINING

THE ALBERT MILL RESTAURANT AND PIZZERIA
Amanda Ahlburg (owner),
Donna Parkes (chef)
4 Junction St, Nairne 5252
Phone: 08 8388 6152
Email: albertmill@optusnet.com.au
www.visitadelaidehills.com.au/albertmillbnb
Licensed/BYO

Cuisine: Modern Australian
Mains average: $$
Specialty: Local produce and homemade dishes
Local flavours: Onkaparinga Valley meats, Mount Barker strawberries, Stirling fruit and vegetables
Extras: Located in an historic three-storey flour mill, with beautiful gardens and outdoor dining area.

- **Open:** Lunch and dinner Wed–Sun

POORAKA

🍎 PRODUCERS

AGON BERRY FARM
Rod Lewis
Main Rd, Pooraka 5095
Phone: 08 8556 8428

Forty years growing berries has to mean something, and in Agon Berry Farm's case it equates to the production of the finest strawberries, and other berries including raspberries, blackberries, blueberries, youngberries, boysenberries, tayberries, lawtonberries, silvan and brambleberries. Rod Lewis grows and packages the berries then wholesales and retails them. Look for the distinctive red van on the side of the road at his property—he has sold strawberries from it year round for the last five years. There are other berries from mid November to the end of April (and yes, you may pick your own), as well as home-made jams, and sauces, sweetcorn, swedes and turnips in season. Council regulations stipulate that farmers may only sell what they grow, so you know the berries have only travelled a few metres to get to you.

- **Open:** 9am to 5pm daily (Nov–end of Apr), tour groups welcome—bookings essential

● **Location:** 1km on the Adelaide side of Mount Compass, about halfway between Adelaide and Victor Harbor

PORT ELLIOT
🍴 DINING

FLYING FISH
Lea Steimanis and Frieda De Leeuw
1 The Foreshore, Port Elliot 5215
Phone: 08 8554 3504
Email: flying.fish@bigpond.com
Licensed/BYO
Cuisine: Modern Australian
Mains average: $$$
Specialty: Fresh-caught fish
Local flavours: Fish and seafood
Extras: Outdoor dining overlooking Horseshoe Bay.
■ **Open:** Lunch daily, dinner Fri-Sat

PORT WILLUNGA
🍴 DINING

STAR OF GREECE CAFE
John Garcia and Zanny Twopeny-Garcia
Esplanade, Port Willunga 5173
Phone: 08 8557 7420
Email: starofgreece@bigpond.com
Licensed/BYO
Cuisine: Mediterranean-style regional
Mains average: $$$
Specialty: Squid, yellow fin, whiting
Local flavours: Fish and seafood, veal, olives, cheese, cream, almonds
Extras: Great view with outdoor seating to enjoy it from.
■ **Open:** Lunch daily, dinner Fri-Sat

SELLICKS BEACH
🍴 DINING

VICTORY HOTEL
Doug Govan
Main South Rd, Sellicks Beach 5174
Phone: 08 8556 3083
Email: vichot@senet.com.au
Licensed/BYO
Cuisine: Modern Australian
Mains average: $$
Local flavours: Whiting, venison
Extras: Outdoor area, views, a wine cellar with over 400 wines, two golden retrievers to keep children amused.
■ **Open:** Lunch and dinner daily

STIRLING
🏠 STORES

THE ORGANIC MARKET
Grahame Murray and Bronwyn Griffiths
5 Druids Ave, Stirling 5152
Phone: 08 8339 4835
Email: organics@ozemail.com.au
www.visitadelaidehills.com.au/organic-market
This certified retailer of organic fruit and vegetables also has a good range of cheeses, breads and much more. There's a vegetarian cafe (phone: 08 8339 7131) attached to this well-established shop and it mainly uses organic, and therefore often regional and local produce, as well as catering for vegans, and coeliac or other special diets. The market also sells bulk foods and showcases Adelaide Hills produces as well as other South Australian regional produce from small producers.
■ **Open:** 8.30am–5.30pm daily
● **Location:** Behind State Bank

STRATHALBYN

🍎 PRODUCERS

TALINGA GROVE
Helen and Daryl Morgan
Talinga Rd, Strathalbyn 5255
Phone: 08 8536 3911
Email: hmorgan1@hotkey.net.au
www.talinga.com.au
Talinga Grove's award-winning, quality extra virgin olive oil and infused extra virgin olive oils, as well as dukkah produced on the farm from local Langhorne Creek almonds, are available for tastings and sales at Bremerton Lodge Winery cellar door, Langhorne Creek, as well as other gourmet outlets.
■ Open: by appointment

🍴 DINING

GOURMET CONNECTION
Jack and Gill Roach
30 High St, Strathalbyn 5255
Phone: 08 8536 4222
Email: groach1@hotkey.net.au
Licensed
Cuisine: Modern Australian
Mains average: $$
Specialty: Caesar salad
Local flavours: Woodside cheeses, smallgoods, fruit and vegetables
Extras: Continental deli selling a wide range of local produce, including cheese and olive oil, attached, outdoor dining, courtyard and function rooms. The rear courtyard has a wood fired pizza oven, which is fired up for specialty pizza nights and also private functions.
■ Open: Breakfast and lunch Wed–Sun, dinner Thurs–Sat (private groups by appointment at other times)

VERDUN

🍎 PRODUCERS

TUMBEELA NATIVE BUSHFOODS
Warren and Eva Jones
Beaumont Rd, Verdun 5245
Phone: 08 8388 7360
Email: tumbeela@ozemail.com.au
The Joneses planted their first trees in 1995 and Tumbeela, which pioneered much of the early South Australian plantings of native shrubs, now mainly grows *Backhousia citriodora* (Lemon Myrtle) and *Tasmania lanceolata* (Mountain Pepper). Warren and Eva have about 800 trees on their Verdun property, including stands of riberry bushes. Riberries can be eaten fresh or used in puddings, pies or chutneys for a big, apple and clove-like flavour boost.
■ Open: By appointment

🍴 DINING

STANLEY BRIDGE TAVERN
Con and Julie Katsaros
Onkaparinga Valley Rd, Verdun 5245
Phone: 08 8388 7249
Licensed
Cuisine: Modern Australian
Mains average: $$
Specialty: Mezze platters, cured Atlantic salmon
Local flavours: Venison, duck
Extras: Outdoor beer garden.
■ Open: Lunch and dinner daily

VICTOR HARBOR

🍎 PRODUCERS

WAHROONGA DAIRY GOAT FARM
Margaret Watkins
RMB 80, Back Valley Rd, Victor Harbor 5211
Phone: 08 8554 5285
Email: wahroonga@chariot.net.au
www.wahroonga-farm.com
What began as a simple exercise to use up excess goats milk has developed into a gourmet delight—goats cheese made by the French method, resulting in small rounds (140g chevrotin). Sold locally under the Wahroonga Farm Produce brand, and used in restaurants.
- **Open:** By appointment
- **Location:** 12km west of Victor Harbor

🍴 DINING

VICTOR HARBOR WINERY
Neville and Adrienne Scott
Cnr Adelaide and Mont Rosa Roads, Victor Harbor 5211
Phone: 08 8554 6504
Licensed
Cuisine: Regional
Mains average: $
Specialty: platters
Local flavours: Cheese, venison, rabbit, field mushrooms in season
Extras: Situated in hills and surrounded by vineyards, there are eating areas inside and out with great views. A restaurant is due to be completed late 2004.
- **Open:** Lunch weekends, dinner Mon once a month (themed)

WILLUNGA

🍎 PRODUCERS

GLAETZER'S BLUEBERRY HILL
Karen and Chris Glaetzer
Pages Flat Rd, Willunga 5172
Phone: 08 8556 1204
You'll find only certified-organic fruits and berries used at this lovely place. Over twenty years ago the Glaetzers wanted a home-based business. They got it, along with a host of strawberries, blueberries, raspberries, and other soft berries. Millions of scratched fingers later they still make berry jams, while a friend bakes the blueberry pies. They now grow organic apples too, and all the fruit is also available weekly at Willunga farmers' markets.
- **Open:** 9.30am–6.30pm Thurs–Mon in season (late Dec–early Mar), 9.30am–5.30pm weekends and public holidays at other times

🍴 DINING

RUSSELL'S PIZZA
Russell Jeavons and Katrine Kylka
13 High St, Willunga 5172
Phone: 08 8556 2571
Licensed/BYO
Mains average: $
Specialty: Pizzas, desserts
Local flavours: Cheese, fruits, flour, meats, olives, oil,
Extras: Outdoor areas, dancing, great ambience. Makes and sells Willunga almond dukkah, which is roasted in the wood-fired brick oven.
- **Open:** Dinner Fri only

WOODSIE

STORES

MELBA'S CHOCOLATES
Joy Foristal
Heritage Park, 22 Henry St,
Woodside 5244
Phone: 08 8389 7868
Email: melbachoc@ozemail.com.au
Willy Wonka, eat your heart out! Melba's chocolate factory is a working tourism factory in a heritage-listed complex, forty minutes from Adelaide. You could be excused for thinking that this place has functioned for a century, but the fact is it was once a cheese and butter factory, and was only revived as a chocolate factory in 1990. All the old Australian favourites are here, including inch licorice, traffic lights and cow pats.
- **Open:** 10am–4.30pm daily
- **Location:** Follow the brown tourist signs

MARKETS AND EVENTS

✦ **BRICKWORKS MARKET,**
TORRENSVILLE
08 8352 4822
9am–5pm Fri–Sun and public holidays

✦ **TASTING AUSTRALIA,** ADELAIDE
08 9388 8877
Oct in odd-numbered years

✦ **BATTUNGA COUNTRY GROWERS' MARKET,** MACCLESFIELD
08 8388 9792
10am–3pm Sun

✦ **GOOLWA WHARF MARKET,**
GOOLWA
08 8554 3476
9am–3.30pm first and third Sun, Easter Sun, every Sun in Jan

✦ **HEART OF THE HILLS MARKET,**
LOBETHAL
08 8389 5615
10am–4pm weekends and public holidays

✦ **PORT ELLIOT MARKET,**
PORT ELLIOT
08 8554 3476
First and third Sat plus Easter Sat and every Sat in Jan

✦ **VICTOR HARBOR COUNTRY MARKET,** VICTOR HARBOR
08 8556 8222
Second and fourth Sun and every Sun in Jan

✦ **WILLUNGA FARMERS' MARKET,**
WILLUNGA
08 8556 4297
8am–12.30pm Sat

✦ **OI OI OISTER DAYS,**
McLAREN VALE
08 8323 8689
First Sun May–Nov

BAROSSA

Entire books have been written to try to distil into words the magic of this area. Largely they have failed, because the Barossa weaves history with sensuality; tangles the tangible with myth and vision. The roots of the place are solid and sensible, full of Germanic hard work and integrity. Yet somehow the light and air of this new country that offered those northern settlers freedom and a future, worked subtly on them after their arrival in the mid 1800s. They continued to make their bread and sausages and wine, and display their pickles, yet the new-found peace gave them room to also relax and enjoy the fruits of their labours.

For that is what you sense here. An enjoyment of the creation of people's hands and the good land. This place buzzes year long with festivals and music, punctuated of course with popping corks. Food and wine intermingle perhaps better in the Barossa than anywhere else in the country. All the necessary adjuncts—dried fruit, cheeses, processed meats, pantry products and chicken—are here.

Basically, the Barossa is South Australia's picnic hamper. Just don't forget to bring the corkscrew.

ANGASTON
🍎 PRODUCERS

ANGAS PARK FRUIT COMPANY PTY LTD, CHIQUITA SOUTH PACIFIC LTD
3 Murray St, Angaston 5353
Phone: 08 8561 0830
Email: janice.martin@chiquita.com.au

If you want to taste the Barossa, this is as good a place as any to begin. This company was established in 1911 for the express purpose of processing fruit grown in the area and has become the largest processor and packer of Australian dried fruit. The range includes dried and glacé fruit such as apricots, peaches, pears, figs, apples, oranges, pineapples, quinces, and kiwifruit, nuts and confectionery, which are beautifully presented in the retail shop. While here, you may even see how it's done through the factory viewing window. These products are available locally as well as all over Australia, and are exported to more than twenty countries. Also on the Sturt Hwy at Berri.

■ **Open:** 9am–5pm Mon–Sat, 10am–5pm Sun and public holidays

BAROSSA FARMERS' MARKET
Vintners Sheds, cnr Nuriootpa, Stockwell and Angaston Roads, Angaston 5353
Phone: 08 8563 3603
Mobile: 0402 026 882
Email: enquiries@foodbarossa.com
www.foodbarossa.com

These markets only commenced in 2002, but have already made waves and received plenty of publicity. The major players are well represented—but so are smaller producers. It's a safe bet to assume that anyone who's anyone in food in the Barossa will be there, either buying or (more likely) with a stall selling their wares. The rules are pretty straightforward—to sell here you must grow or make the product and it must be grown or produced in the Barossa. And even the

breakfast is in the sure hands of Barossan winemakers and chefs. A must-do.
- Open: 7.30am-11.30am, Sat

BAROSSA VALLEY CHEESE CO
Frances McClurg and Victoria Glaetzer
67b Murray St, Angaston 5353
Phone: 08 8564 3636
Email: bvcc@internode.on.net

Working as a winemaker in France for three years, Victoria had an opportunity to taste some amazing cheeses. When she returned to the Barossa Valley, also known for its fine produce, she decided to indulge her dream of creating specialist Barossan cheese. After several years of experimenting in their kitchen, Victoria and her mother, Frances McClurg, decided to open a shop in the heart of Angaston. Each week they now hand-make between twenty and forty kilograms of mainly soft, fresh cheeses including camembert, brie and washed rind styles, as well as a soft, tangy vache curd and a creamy soft fetta. Appropriately yellow walled, the cheese shop sells tasting plates and there is a window so you can see the cheeses quietly maturing.
- Open: 10am-5pm Wed-Fri, 10am-4pm Sat

FARM FOLLIES AND HUTTON VALE WINES
Jan and John Angas
Hutton Vale, Stone Jar Rd, Angaston 5353
Phone: 08 8564 8270
Email: enquiries@huttonvale.com
www.huttonvale.com

Hutton Vale is home to the fifth, sixth and seventh generations of the Angas family, who run a mixed farm. Farm Follies is a foundation member of Food Barossa and began over a decade ago, to make the most of seasonal produce which ties in well with their support of the Slow Food movement. Their range of artisan chutneys and pickle products includes Oops Onion and Mint, Spiced Quince Honey, Chilli Tease, Kashmir Kasundi, Hutton Vale red gum honey, and various other condiments.
- Open: By appointment

STORES

ANGASTON COTTAGE INDUSTRIES
Gillian Gravestock
38 Murray St, Angaston 5353
Phone: 08 8564 2596

Gillian Gravestock has been here for thirty-two years, so she must have seen a lot of talent displayed in this shop. It all began when a group of local women noticed that too much local fruit was going to waste and started this place as an opportunity for people to make money working at home. So there are homemade cakes, biscuits, jam, sweets, marmalade, chutney, and sauces, as well as local honey, free-range eggs and garden produce, all subject to seasonal variation.
- Open: 10am-4pm weekdays, 9.30am-12pm Sat, 11am-4pm public holidays

SCHULZ BUTCHERS
Peter Barratt (operator)
Franz Knoll (owner)
42 Murray St, Angaston 5353
Phone: 08 8564 2145
Email: barossaf@arcom.com.au

Opened in 1939, this is the sort of butcher's shop you would hope to find in the Barossa. Famous for its smokehouse, which gives a distinct regional flavour to cold-smoked bacon (smoked for two days) and a

range of German smallgoods such as mettwurst, and black and white puddings.
- Open: 8am–5.30pm weekdays, 8am–11.30am Sat

SOUTH AUSTRALIA COMPANY STORE AND KITCHEN

Leandra Davoren
27 Valley Rd, Angaston 5353
Phone: 08 8564 3788 (store);
08 8564 2725 (cafe)
Email: sa_store@marianiaust.com.au
www.sacompanystore.com.au
Licensed
Cuisine: Cafe-style regional
Mains average: $$
Specialty: Fresh Barossa and South Australian produce, 'amazing cakes!'
Local flavours: Seasonal fruits and vegetables, milk-fed lamb, corn-fed chicken, fresh fish
Extras: Makes SA Company Stores' own brand of condiments, sauces and jams for sale. Views over Yalumba vineyards, outdoor veranda and deck dining, and lawns for children to play on. The store is a refurbished army barracks, with special SA sections for giftware, and the largest range of SA gourmet foods in one location.
- **Open:** Store—10am–5pm daily, Cafe—lunch and afternoon tea daily, Happy Hour on Fri nights
- **Location:** Turn right before Yalumba

DINING

ANGASTON GOURMET FOODS CAFE
Aaron Penley and Graham Butler
36 Murray St, Angaston 5353
Phone: 08 8564 2429
Mains average: $
Specialty: Thirty-six different types of baguette

Local flavours: As many as possible including meat, breads
Extras: Indoor-outdoor dining, and all food used is Barossa produce. You'll find a superb showcase of local brands and produce here, as well as continental and Australian lines. The owners are, in their own words, 'passionate about food, wine tourism and life!' So each purchase comes with a generous free addition of enthusiasm for the region's bounty. Look for local cheeses, bread, meats, pickles, jams, chutneys and much more.
- **Open:** Cafe—lunch daily, store—8am–6pm daily

BARR-VINUM
Sandor Palmai (chef) and Bob McLean
10 Washington St, Angaston 5353
Phone: 08 8564 3688
Email: barrvinum@bigpond.com
Licensed/BYO
Cuisine: Provincial, classic yet modern
Mains average: $$$
Local flavours: Bread, olive oil, chicken, rabbit, vegetables, fruit
Extras: Grows Bob McLean's own Barr-Vinum fat tail lamb. Formerly a station master's residence, the building is welcoming and the gardens provide the perfect setting for a pre-dinner drink. The wine room is well worth a visit too.
- **Open:** Lunch Tues–Sun, dinner Tues–Sat (and Sun on long weekends), morning and afternoon tea daily

SALTERS AT SALTRAM
Richard Bate
Saltram Winery, Nuriootpa Rd, Angaston 5353
Phone: 08 8561 0200
Email: saltersbarossa@ozemail.com.au
www.saltramwines.com.au

SA—BAROSSA

Licensed/BYO
Cuisine: Regional modern Australian
Mains average: $$$
Specialty: Coconut mussels
Local flavours: Fruit, vegetables, meat, smallgoods
Extras: Outdoor dining on the veranda of this heritage listed, 140-year-old homestead. Winery with cellar door in the same building.
■ Open: Lunch daily, dinner Thurs–Sat

VINTNERS BAR AND GRILL
Peter Clarke
Cnr Stockwell and Nuriootpa Roads,
Angaston 5353
Phone: 08 8564 2488
Email: enquiries@vintners.com.au
www.vintners.com.au
Licensed/BYO
Cuisine: Modern Australian
Mains average: $$
Specialty: Smoked salmon pizza
Local flavours: As much as possible, including meat, fruit and vegetables
Extras: Outdoor dining overlooking gardens and vineyards.
■ Open: Lunch Wed–Sun, dinner Mon–Sat

LYNDOCH
DINING

CAFE Y
Yaldara Estate
Kas Martin (executive chef)
Yaldara Estate, Hermann Thumm Dr,
Lyndoch 5351
Phone: 08 8524 0200
Email: kasm@yaldara.com.au
www.yaldara.com.au
Licensed
Cuisine: Innovative regional

Mains average: $$
Local flavours: As much as possible
Extras: Magnificent chateau housing cellar door and wine sales. Cafe Y is located overlooking the North Para River, and has a modern, casual dining area. Also larder–produce store and cheese room. Sells homemade pâté, relish, dukkah, biscuits and truffles. Winery tours run daily.
■ Open: 9.30am–5pm daily

CHATEAU BARROSA
Hermann Thumm, John Sharman (chef)
Hermann Thumm Dr, Lyndoch 5351
Phone: 08 8524 5055
Email: dpitt@chateaubarrosa.com.au
www.chateaubarrosa.com.au
Licensed/BYO
Cuisine: Barossa Valley regional
Mains average: $$
Specialty: Many dishes feature grape syrups and grape nectars made at the winery
Local flavours: Grape nectar and syrup (like honey, a natural sweetener)
Extras: Outdoor area with swimming pool and spa, views to the Barossa Ranges, twelve hectare landscaped rose garden. Sells Chateau Barrosa grape nectar and syrup. The spelling of Barrosa isn't a mistake—the owners are playing with the word Barossa of course.
■ Open: Breakfast, lunch and dinner daily
● Location: 1km from Lyndoch on the Gawler Hwy

NURIOOTPA
STORES

LINKES CENTRAL MEAT STORE
Graham and Lola Linke
27 Murray St, Nuriootpa 5355
Phone: 08 8562 1143

This business is seventy-five years old and Graham Linke is a third-generation butcher still producing those fine, German-heritage smokehouse products—kassler rib, jaegerbraten, bacon, and many other smallgoods—as well as prime cuts of locally raised meats.
- **Open:** 8.30am–5.30pm Mon–Thurs, 7am–5.30pm Fri, 8.30am–12pm Sat

DINING

MAGGIE BEER'S FARM SHOP
Maggie Beer
Pheasant Farm Rd, Nuriootpa 5355
Phone: 08 8562 4477
Email: farmshop@maggiebeer.com.au
www.maggiebeer.com.au
Licensed
Cuisine: Regional
Mains average: $$
Specialty: Pheasant Farm pâté, verjuice jellies, game, offal
Local flavours: Vegetables, chicken, cheese, ham
Extras: Grows pheasant and partridges and sells about twenty products including jams, verjuice, sauces, olive oil, pastes, preserved lemons and pâté. Deck and grassed area next to a beautiful dam featuring turtles, ducks, fish and native birds. Retail outlet for Maggie Beer products and cellar door.
- **Open:** Shop—10.30am to 5pm daily, lunch available daily
- **Location:** Off Seppeltsfield–Samuel Rd

ROSEDALE

TOURS

BAROSSA DAIMLER TOURS
Libby and John Baldwin
Main Rd, Rosedale 5350
Phone: 08 8524 9047
Email: baldwin@barossadaimlertours.com.au
www.barossadaimlertours.com.au
More than just a tour, this is an award-winning food and wine experience. There are individual wine and food tours, as well as history, culture and heritage, tailored according to tastes. Winner of Barossa's Best Tour in regional awards in 1996, 2000 and 2002.
- **Open:** Daily, by appointment

ROWLAND FLAT

DINING

MIRANDA WINES
Luigi Miranda
Barossa Valley Way, Rowland Flat 5352
Phone: 08 8524 4537
www.mirandawines.com.au
Licensed (Miranda wines)
Cuisine: Italian
Mains average: $
Specialty: Pumpkin ravioli, gnocchi and blue vein cheese
Extras: Vineyard views and outdoor area, cellar door atmosphere and view of barrel room.
- **Open:** Lunch daily

TANUNDA

PRODUCERS

BAROSSA VALLEY PRODUCE
Steve Zimmermann
Basedow Rd, Tanunda 5352
Phone: 08 8563 2477
If you're looking for fine horseradish, dill cucumbers, beetroot spread, or pickled

SA—BAROSSA

onions, you'll find them here. Steve Zimmermann's grandparents made this sort of food and he has been carrying on the tradition, producing some highly flavoursome regional produce with a definite Barossan-German accent for the past twenty years or so. Sold under the Zimmys brand, you'll find them at local Foodland supermarkets.

- **Open:** By appointment
- **Location:** 2km from Tanunda

TANUNDA'S NICE ICE
Tony and Chris Bowen
Shop 8, Kavel Arcade, 46 Murray St, Tanunda 5352
Phone: 08 8563 3601
Email: icecreammantony@telstra.com

Just one more, and there will be one hundred ice-cream flavours to choose from at this very nice ice-cream shop. They are all made on the premises, some are fat free, and many are made with local produce. All use natural ingredients, and add no artificial flavours or ingredients. Check out the local flavours in this line-up: apricot, lemon, plum, peach, quandong, quince, rhubarb, plus unfermented wine grape juices such as shiraz, alicante bouchet, cabernet and chardonnay.

- **Open:** 8.30am–5.30pm daily

STORES

APEX BAKERY
Fechner Family
Elizabeth St, Tanunda 5352
Phone: 08 8563 2483
Email: apexbakery@bigpond.com

For eighty years this bakery has been producing some of the Barossa's finest bread. The recipes haven't changed and the original wood-fired ovens are still in use. Look for original-recipe salt sticks and caraway pretzels, along with more mainstream offerings, which are available from the bakery shop or other outlets around the area.

- **Open:** 9am–5.30pm weekdays, 8.30am–12pm Sat

FOOD BAROSSA
Allison Bockman
6/109 Murray St, Tanunda 5352
Phone: 08 8563 3603
Email: hampers@foodbarossa.com
www.foodbarossa.com

This great range of hampers with a 'fork in the cork' logo, can be ordered here, but be aware that this is not a shopfront. A showcase of hampers is available for viewing, purchasing or ordering at the Barossa Wine and Tourism Association Visitor's Centre in Murray St.

- **Open:** 9am–5pm weekdays

DINING

BAROSSA VINES CELLAR DOOR
Grant Burge Wines
Krondorf Rd, Tanunda 5352
Phone: 08 8563 7675
Licensed
Cuisine: Local produce
Mains average: $
Specialty: Regional platters

BAROSSA WINE AND TOURISM ASSOCIATION
66–68 Murray St, Tanunda 5352
Phone: 1300 852 982
Email: info@barossa-region.org
www.barossa-region.org
- **Open:** 9am–5pm weekdays, 10am–4pm weekends

Extras: Views of the surrounding area.
- Open: 10am–5pm daily

LA BUONA VITA
Juergen and Venetta Leib
89 Murray St, Tanunda 5352
Phone: 08 8563 2527
Email: labuonavita@ozemail.com.au
www.labuonavitarestaurant.com
Licensed
Cuisine: Italian Australian
Mains average: $
Specialty: Focaccias
Local flavours: Salami, peperoni, vegetables, homemade bread, biscuits, pies
- Open: Lunch and dinner daily

KRONDORF ROAD CAFE
Richard and Ingrid Glastonbury
Kabminye Wines, Krondorf Rd, Krondorf, Tanunda 5352
Phone: 08 8563 0889
Email: wine@kabminye.com
Licensed (Kabminye Wines)
Cuisine: Traditional Barossa Valley
Mains average: $
Specialty: Original Silesian and Wendish dishes
Local flavours: Linke's smallgoods, Wiech's egg noodles, Lyndoch Bakery bread, Balleycroft quark, Zimmys dill gherkins and pickled onions, Rosie's free range eggs, Lowke's carrots
Extras: Makes and sells apricot jam, fig preserve, plum and apricot mus (a traditional Silesian jam recipe with minimal sugar), olives and olive oil. Extensive 270-degree views of the Barossa Ranges, and outdoor veranda seating. 'Heaven in a Paddock', as one visitor wrote.
- Open: Lunch daily
- Location: At the Kabminye Wines cellar door

1918 BISTRO AND GRILL
Alan (chef) and Jennie Lennon
94 Murray St, Tanunda 5352
Phone: 08 8563 0405
Licensed/BYO
Cuisine: Modern Australian
Mains average: $$
Local flavours: Venison
Extras: Shady garden in summer.
- Open: Lunch and dinner daily

PETER LEHMANN WINES
Para Rd, Tanunda 5352
Phone: 08 8563 2100
Email: cellardoor@peterlehmannwines.com.au
www.peterlehmannwines.com.au
Licensed (Peter Lehmann wines)
Barossa produce platters $$
Local flavours: Smoked meats, matured cheddar cheese, olives, dill cucumber, wood-fired oven bread, pear chutney, beetroot relish
Extras: Platters may be enjoyed on the veranda or in the grounds.
- Open: Lunch daily

TOURS

ROCKFORD–PS MARION
Robert O'Callaghan and Pam O'Donnell
Krondorf Rd, Tanunda 5352
Phone: 08 8563 2720
Email: pam@rockfordwines.com.au
Fancy a steam-powered dinner? You need to be quick, as these book out many months ahead. The century-old paddle-steamer PS Marion is moored at Mannum, near the ferries. Rockford wines and some of Adelaide's best chefs combine to present outstanding wine and food weekends spent cruising the River Murray. Rockford has a stated aim of

SA—BAROSSA

raising funds this way to keep the old girl (built in 1898) afloat for another hundred years. Passengers' aims often include full-time relaxation, as they drift along the Murray while being luxuriously wined and dined.

■ **Open:** Cruises run three times a year and the rest of the time the vessel is available for private charter

TRURO
🍎 PRODUCERS

BAROSSA OLIVES
Martin and Auriel Wright
Moorundie Street (Sturt Highway),
Truro 5356
Phone: 08 8564 0141
Email: barossaolives@internode.on.net

This retail outlet selling olives and olive oil right in the centre of town is the ideal place to see where the fast-growing South Australian olive industry is heading. Producers include Domenics, Rio Vista, Hickorys Run, Olive Vale, Marne River, Talinga Grove, and Rubric. Associated products for sale include olive wood artefacts from Bethlehem, olive leaf extract, oil cans, bottles, pitters, tea towels, aprons, and olive trees, including bonsais. Also in Adelaide at Shop 4, 45 Jetty Rd, Glenelg.

■ **Open:** 10am-5pm daily

WILLIAMSTOWN
🍎 PRODUCERS

STEVENS QUALITY RASPBERRIES
Cheryl and Grant Stevens
Williamstown 5351
Phone: 08 8524 6407
Mobile: 0428 130 694
Email: stevo59@ihug.com.au

Growing raspberries for twenty years, making jam for seven, the Stevenses primarily began this business to utilise the land and make a living. They have succeeded deliciously. Now they sell their raspberries in kilogram containers, mainly to restaurants, but their luscious jam and toppings can be found throughout the Barossa region.

■ **Open:** By appointment

MARKETS AND EVENTS

✦ **BAROSSA FARMERS' MARKET,**
ANGASTON
08 8563 3603
7.30am–11.30am Sat

CLARE VALLEY AND YORKE PENINSULA

They used to call this the Clare Valley, but perhaps its history merits a loftier title. There is a Silesian heritage, like the adjoining Barossa, but the early Irish settlers dubbed it with a name to remind them of home. Here you will find small vineyards, boutique winemakers, charming picnic spots and honey, deer, pantry products, olive oils, quandongs and bacon.

While the Clare Valley's rolling hills are lush and green in winter, blossoming in spring, not far away the land flattens and dries, becoming home to kangaroos and emu. Near here farmers raise ready-salted lamb. While this may seem like a typical Aussie leg-pull, the fact is the lambs feed on saltbush and the flavour permeates their flesh in the same way as the French come by pre-sel or the Welsh their herb-infused lamb.

Classic country, this may be, but surprises like this make you realise that this region too has its own distinct and delightful character.

Email:getcloser@clarevalley.com.au
www.clarevalley.com.au
www.yorkepeninsula.com.au

AUBURN

STORES

GLOVER'S OF AUBURN
Chris Elmitt and Janeece Madigan
17 Main North Rd, Auburn 5451
Phone: 08 8849 2022
Email: chris.elmitt@bigpond.com
A wonderful place to catch up with plenty of regional and South Australian produce. It's all here—jams, condiments, mustards, sauces, olive oil products, lavender, dried fruit and nuts, soaps, mettwurst, pickles and honey.
- **Open:** 10am–5.30pm Thurs–Mon
- **Location:** Cnr Main North Rd and St Vincents St, opposite the Rising Sun Hotel

DINING

MOUNT HORROCKS
Stephane Toole, Jo Barrington Case (chef)
The Old Railway Station, Curling St, Auburn 5451
Phone: 08 8849 2243
Email: sales@mounthorrocks.com
www.mounthorrocks.com
Licensed/BYO
Cuisine: Fresh, light dishes
Mains average: $
Local flavours: As much as possible
Extras: Veranda dining overlooking the vines.
- **Open:** Lunch weekends and public holidays

BLYTH

🍎 PRODUCERS

MEDIKA GALLERY
Ian Roberts
16 Moore St, Blyth 5462
Phone: 08 8844 5175
Email: medika@chariot.net.au
www.medikagallery.com.au

Twenty years ago, Ian Roberts simply wanted a studio to paint in. He got that and much more—a lifetime's supply of native peaches, or quandongs. The gallery showcases art and craft but also sells gallery garden-grown quandong jam and chutney as well as Blyth Plains olive oil. Across the road 130 quandong trees have been planted for the community to use. These have been especially selected as grub-free, and capable of producing good-sized and quality fruit. These hardy trees need no pruning, watering, or spraying, and fruit is harvested simply by shaking the tree.

■ **Open:** 10am–5pm weekdays, 2pm–5pm weekends

BURRA

🍷 BREWERY/WINERY

THOROGOOD'S FARMHOUSE CIDER CELLAR
Susan and Tony Thorogood
John Barker St, Burra 5417
Phone: 08 8892 2669
Email: cider@thorogoods.com.au
www.thorogoods.com.au

Thorogoods is a small country cidery that is a cider drinker's paradise. The ciders are alcoholic, full bodied, clean, refreshing and delicate, not the product of a chemist's laboratory but made from apples, apples and more apples. Look for hand crafted apple wines, apple champagne, apple liqueurs and even apple beers. They have good names too—Gold Dust, the original classic medium dry apple wine; Summer Lightning, a sweet apple cider fermented and matured in old American oak; Misty Morning, a fruity and mellow sparkling; Sweet Panic, an apple dessert wine; and Billy B's, a fourteen per cent alcohol apple beer. The orchard, established in 1990 has 800 trees of more than seventy varieties. The owners hope to produce 30 000 litres of cider annually. Sold at the cellar door, Clare Wine Store and Baily & Baily Liquor Stores in Adelaide.

■ **Open:** 12pm–4.30pm daily
● **Location:** First on right when entering Burra

🏠 STORES

McPHEE'S MEATS
Gary McPhee
2 Market St, Burra 5417
Phone: 08 8892 2009

If you have never tasted saltbush merino then you haven't really tasted country-style meat. At least that's what Gary McPhee says when discussing the prime cuts he has in both his Burra and Clare butcher shops. The McPhees deal direct with station owners for this seasonal product with a unique flavour, a result of grazing on natural grasses and saltbush, and reminiscent of the French pre-sel lamb. They have a cured bush bacon, and smoke bacon and ham using unique and older-style techniques. There is another store in Main North Road, Clare and McPhee's also delivers to Adelaide homes and office blocks.

■ **Open:** 8am–5.30pm weekdays, 8am–12pm Sat

CLARE

🍎 PRODUCERS

VALLEY OF ARMAGH
Don Hiller and Robyn Hill
St Georges Tce, Armagh, Clare 5453
Phone: 08 8842 1237
Email: oliveoil@rbe.net.au
This six-year-old olive oil grove and vineyards has tastings at the cellar door. They say the move here was a C-change. Does the C stand for challenge, you wonder? Certainly the result has been the production of high quality extra virgin olive oil, and the beginnings of Glendalough Wines.
- **Open:** 10.30am–4pm daily
- **Location:** 2.4km west of Clare off the Clare–Blyth Rd

🍴 DINING

LONDON HILL
Philip Scarles and Amanda Waldron
Eldredge Vineyards, Spring Gully Road
Clare 5453
Phone: 08 8842 3086
Email: bluechip@capri.net.au
Licensed (for Eldredge Wines)
Cuisine: Fresh, regional
Mains average: $$
Specialty: Free-range, grain-fed poultry
Local flavours: Poultry, Burra saltbush lamb, venison, olive oil, fruits, nuts, vegetables, honey, eggs—anything in season
Extras: Makes wine jellies, jams, chutneys, mustards and dukkah for sale under the London Hill label. Alfresco dining area attached to stone cellar door, cottage, panoramic waterside views of vines and bushland, landscaped lawns and gardens. Takeaway available.
- **Open:** Lunch weekends and public hols

NEAGLES ROCK VINEYARD RESTAURANT
Jane Willson and Anne-Marie Mitchell
Main North Rd, Clare 5453
Phone: 08 8843 4020
Email: nrv@neaglesrock.com
www.neaglesrock.com
Licensed/BYO
Cuisine: Fresh seasonal
Mains average: $$
Specialty: Confit of duck, octopus
Local flavours: Yabbies in season, lamb, buffalo, tapenade, olive oil
Extras: Outdoor area set in the vineyard with garden and trees as a backdrop, a full-size, gravel petanque area for customers' use (boules are available). Guest chef dinners, and aged red wine tastings and dinners.
- **Open:** Lunch daily, dinner Fri–Sat (midweek for groups or functions)
- **Location:** 600m south of caravan park

CRADOCK

🍴 DINING

CRADOCK HOTEL
Julie Tadeo (chef) and David Frost
Main St, Cradock 5432
Phone: 08 8648 4212
Email: cradockhotel@bigpond.com
Licensed/BYO
Cuisine: Australian
Mains average: $$
Specialty: Saltbush mutton, saltbush mutton sausages, kangaroo baguettes
Local flavours: Saltbush mutton, bush tomato chutney, pistachios, quandongs, fish from Spencer Gulf, Clare valley wine and southern Flinders Ranges wines
Extras: A small hotel built in 1881 in a tiny outback town that is now almost

deserted. Bush food is used in dishes during peak tourist times. Takeaway and catering available.
■ **Open:** Breakfast daily (house guests), lunch and dinner daily (closed four weeks in mid-summer)

MINLATON
🍴 DINING

HARVEST CORNER INFORMATION AND CRAFT
Janet Cameron (Tourism Administration, coordinator)
29 Main St, Minlaton 5575
Phone: 08 8853 2600
Email: harvestcorner@netyp.com.au
This craft co-op and gallery has tea rooms and sells homemade produce—jams, pickles, chutneys, chocolates, and breads. They say that you can 'taste the years of experience cooked into every mouthful'. It's made from local produce too.
■ **Open:** 10am–5.30pm weekdays, 10am–4.30pm Sat

MINTARO
🍴 DINING

REILLY'S WINERY AND RESTAURANT
Justin Ardill, Andre van der Veken,
Cnr Burra and Hill Streets, Mintaro 5415
Phone: 08 8843 9013
Email: reillys@chariot.net.au
Licensed
Cuisine: Northern Italian influenced
Mains average: $$
Specialty: Garlic prawns, pastas
Local flavours: Fruit and vegetables, meat
Extras: Outdoor dining, overlooking the vineyard. Operates as a cellar outlet.

■ **Open:** Lunch daily, dinner Mon, Wed, Fri–Sat

PENWORTHAM
🍎 PRODUCERS

PENNA LANE WINES AND PRODUCTS
Ray and Lynette Klavins
Penna Lane, Penwortham 5453
Phone: 08 8843 4364
Email: klavins@rbe.net.au
www.pennalanewines.com.au
This new winery already has award-winning wines. You can try them along with a range of estate-grown and made organic jams, pickles, chutneys, sauces and relishes (in packaging hand-painted by Lynette) or turn up around lunchtime and order a platter to share, which will feature these products plus good homemade bread, or soup in winter.
■ **Open:** 11am–5pm Thurs–Sun and public holidays
● **Location:** Ten minutes south of Clare

🍴 DINING

COLEMANS AT KILIKANOON
Susanne Coleman
Penna Lane, Skillogalee Valley,
Penwortham 5453
Phone: 08 8843 4377
Email: admin@kilikanoon.com
Licensed/BYO
Cuisine: Country
Mains average: $$
Specialty: Homemade chicken liver pâté, gourmet burgers
Local flavours: Lamb
Extras: Lovely outdoor area, gardens and view across Skillogalee valley.
■ **Open:** Lunch Fri–Sun

PORT AUGUSTA
🍴 DINING

AUSTRALIAN ARID LANDS BOTANIC GARDEN
Stuart Hwy, Port Augusta 5700
Phone: 08 8641 1049
Email: aridlands@portaugusta.sa.gov.au
www.australian-aridlands-botanic-garden.org
Licensed
Cuisine: Light meals
Mains average: $
Specialty: An Australian native food dish is always on the menu
Local flavours: Quandong (native peach)
Extras: Grows indigenous foodstuffs in the gardens. The visitors centre houses an up-market cafe with conference or meeting room facilities. Overlooks Australia's largest Eremophila Garden and the beautiful Flinders Ranges.
■ **Open:** Brunch, morning and afternoon tea, lunch daily
● **Location:** 1.4km from the start of the Stuart Hwy via Port Augusta

SEVENHILL
🍴 DINING

SKILLOGALEE RESTAURANT
Diana (chef) and Dave Palmer (winemaker)
Trevarrick Rd, Sevenhill 5453
Phone: 08 8843 4311
Email: skilly@chariot.net.au
www.skillogalee.com
Licensed (own wines)
Cuisine: Eclectic Modern Australian
Mains average: $$
Specialty: Good food with the best quality ingredients
Local flavours: Lamb, chicken, herbs, fruit, honey, wine, eggs, olive oil
Extras: Veranda and terrace under olive trees in cottage garden setting amongst the vines. Gourmet picnic baskets, platters and soups for B&B guests.
■ **Open:** Morning and afternoon teas and lunch daily, dinner by arrangement
● **Location:** 3km from main North Rd

THORN PARK COUNTRY HOUSE
David Hay (chef) and Michael Speers
College Rd, Sevenhill 5453
Phone: 08 8843 4304
Email: stay@thornpark.com.au
www.thornpark.com.au
Licensed/BYO
Cuisine: Imaginative country cuisine
Set-price dinner: $$$$
Specialty: Saltbush merino, grain-fed beef, fusion food
Local flavours: As many as possible
Extras: The homestead is surrounded by beautiful gardens and views. There are regular residential cooking schools and winemaker weekends. This is a country house 'hotel'—a gourmet retreat—and was awarded the Jaguar Award for Gastronomic Travel in 2001.
■ **Open:** Breakfast daily (for guests), dinner by arrangement
● **Location:** Opposite Sevenhill Cellars

TWO WELLS
🍎 PRODUCERS

COUNTRY FRESH EGGS
Anne and Dion Andary
Hart Rd, Two Wells 5501
Phone: 08 8520 2001
Email: dayseggs@ozemail.com.au

SA—CLARE VALLEY AND YORKE PENINSULA

This egg distributor is happy to make farm-gate sales direct from the grading floor. The Andarys have been doing this for fifteen years now. The eggs you buy are barn laid, which means the birds are free to range within certain boundaries.
- Open: 8.30am–5pm daily
- Location: Off the main Wakefield Hwy, turn left after Two Wells

VIRGINIA
PRODUCERS

PRIMO ESTATE WINES
Joe and Dina Grilli
Old Port Wakefield Rd, Virginia 5120
Phone: 08 8380 9442
Email: info@primoestate.com.au
www.primoestate.com.au
Although specialising in red and white table wines, sparkling red wine and fortifieds, Primo's extra virgin olive oil, sold under the Joseph label and introduced about ten years ago, was Joe Grilli's idea and now sells Australiawide and overseas. Foothills extra virgin olive oil comes from century-old trees in Adelaide's earliest olive plantings, and now there are Joseph La Casetta aged vinegars, made from white grapes, to add to the collection.
- Open: 10am–4pm weekdays

WIRRABARA
STORES

WIRRABARA OLD BAKERY
Dennis Wheatley
Wirrabara 5481
Phone: 08 8668 4225
A passer-by rolled up to the Old Bakery and said, 'My friend says I must try one of these pies.' Turns out that the visitor and his friend were from Oslo, Norway, so it seems that word is getting around about these outback pies and pastries. Apart from the traditional beef pie, there are some real Aussie flavours like buffalo, kangaroo, and rabbit stew. The bunnies come from a local breeder 'just up the road', says Wheatley. This is not just a bakery, but a licensed tearoom seating twenty indoors and up to fifteen outside. The quandong pies with Golden North ice-cream are a must. Enjoy a glass of local Flinders wine with this very Australian meal. Most of the ingredients are local. Do say hello to Val.
- Open: 8am–6pm daily

WIRRABARA FOREST
PRODUCERS

O'REILLY'S ORCHARD
Jackie and David O'Reilly
Watts Gully Rd, Wirrabara Forest 5481
Phone: 08 8668 4245
Email: oreillyorchard@centralonline.com.au
Certified organic fruit and vegetables, and naturally grown stone fruit is the business of this family-run property, which is committed to sustainable integration and self-sufficiency. Farm gate sales and deliveries to local areas. Farm tours with seasonal tastings can be negotiated by appointment.
- Open: Daylight hours daily
- Location: 10km west of Stone Hut and 12km south-west of Wirrabara

EYRE PENINSULA, FLINDERS RANGES AND THE OUTBACK

A casual glance at a map might cause you to believe that this area is sparse and uninteresting. This most remote of all South Australia's peninsulas is where the towns are spread further apart, the topography flattens, and you can bet the temperature will soar in summer.

Yet for anyone who loves the sea and its bounty, this is a rich place indeed. This is the home of many foods that citysiders only see on glossy menus—Coffin Bay oysters, shucked as you watch, tuna so fine you'll be lucky to get to it before it's exported, and whiting and shark that turns into possibly the best fish and chips ever. And all this is only further enhanced by the salty tang in the air.

Totally different and wildly beautiful, the Flinders Ranges and Outback offer so much, and there is a whole pantry of native foods tucked away here. Take a tour with Indigenous people, stay on a station, or simply wander the area by yourself. You will be amazed at what you will find.

MARKETS AND EVENTS

+ AUSTRALIAN OYSTER FESTIVAL, CEDUNA
08 8625 2780
Oct long weekend

A BIGHT OF OYSTERS

CEDUNA OYSTER GROWERS: Oysters available April–December

+ GREAT BIGHT OYSTERS
08 8625 2900

+ ASTRID OYSTERS
08 8625 3554

+ CEDUNA CLEARWATER OYSTERS
08 8625 2933

+ SOUTHERN RIGHT OYSTERS
08 8625 3430

+ WEST EYRE SHELLFISH
08 8625 2822

CEDUNA

STORES

CEDUNA BAKERY COFFEE LOUNGE
Jason Greatbaten
Shop 2, 35 Poynton St, Ceduna 5690
Phone: 08 8625 2611
For the past five years the local bakery has got in on the act during the annual October Oysterfest and bakes superb oyster pies during the weekend the festival is on. Make sure you try one.
■ **Open:** 7am–5.30pm weekdays, 7am–12.30pm Sat

SA—EYRE PENINSULA, FLINDERS RANGES AND THE OUTBACK

🍴 DINING

CEDUNA OYSTER BAR
Hoffrichter family
Eyre Highway, Ceduna 5690
Phone: 08 8626 9086
Licensed
Specialty: Oysters (fresh, frozen, open, closed, cooked)
Local flavours: Oysters, of course!
Extras: Eat in or takeaway. A great place for travellers on the highway to drop in.
- **Open:** 9.30am–6pm Mon–Sat, 1pm–6pm Sun
- **Location:** Western end of Ceduna, 300m north of fruit fly inspection point

COFFIN BAY
🍴 DINING

THE OYSTERBEDS
Damien and Emma Mannix
61 Esplanade, Coffin Bay 5607
Phone: 08 8685 4000
Licensed
Cuisine: Seafood
Mains average: $$

TASTE OF THE OUTBACK FESTIVAL
Iga Warta, Copley 5732
Phone: 08 8648 3737
Email: enquiries@igawarta.com
www.igawarta.com
The Iga Warta cultural tourism centre runs tours all year round. However, special events such as Taste of the Outback and a music and cultural festival, are held annually (in 2005 they expect the festival to be in mid May). Iga Warta has been awarded a Jaguar Award for Gastronomic Travel.

Specialty: Seafood platter
Local flavours: Seafood—local oysters, SA whiting, garfish, yabbies, scallops
Extras: Right on the waterfront with magnificent views of the bay, alfresco dining.
- **Open:** Lunch and dinner Tues–Sun

COPLEY
🍴 DINING

QUANDONG CAFE AND BUSH BAKERY
Fiona Gosse
Railway Tce, Copley 5732
Phone: 08 8675 2683
Mains average: $
Specialty: Pies and pasties, quandong pies
Local flavours: Quandongs, kangaroo
Extras: Shady courtyard dining. Makes quandong jam and sauces for sale. Takeaway available.
- **Open:** 9am–4pm daily (closed Nov–Feb)

COWELL
🏠 STORES

COWELL MEAT SERVICE
David and Natalie Wiseman
9 Main St, Cowell 5602
Phone: 08 8629 2051
Email: cowellmeat@bigpond.com
David Wiseman followed his father's lead in butchery and has been running this place for the past twenty years. 'People come from all over Australia to try some fair dinkum country killed meat.' It's tender and tasty, he promises, and fresh from paddock to plate.
- **Open:** 8am–5pm weekdays

HAWKER
🍴 DINING

CAPTAIN STARLIGHT'S, PADDY DODGER'S BISTRO
Mark Reynolds (chef)
Wilpena Pound Resort, Flinders Range via Hawker, 5434
Phone: 08 8648 0004, Freecall 1800 805 802
Email: admin@wilpenapound.com.au
www.wilpenapound.com.au
Licensed
Cuisine: Hearty country
Mains average: $$
Specialty: Local quandong pie
Local flavours: Indigenous Australian combined with local produce
Extras: Kangaroos on the lawn in front of the restaurant.
- **Open:** Breakfast, lunch and dinner daily

INNAMINCKA
🍴 DINING

COOPER CREEK HOMESTAY
Julie and Geoff Matthews
Cnr Mitchell and Stuart Streets, Innamincka 5731
Phone: 08 8675 9591
Email: fourmatthews@bigpond.com
Cuisine: Homestyle
Mains average: $$
Specialty: Outdoor eating in cooler months—campfire and camp oven meals
Local flavours: Sauces from Paroo Food Ventures
Extras: Eat with the family, and enjoy camp oven dinners under the stars.
- **Open:** Breakfast, lunch and dinner daily (by arrangement)
● **Location:** 300 metres from the township

MELROSE
🍴 DINING

NORTH STAR INN HOTEL
David Rosenzweig
41 Nott St, Melrose 5483
Phone: 08 8666 2110
www.nomadic-enterprise.com.au
Licensed
Cuisine: Australian country
Mains average: $
Specialty: The Whuflungdunga farm feast
Extras: The original inn was built in 1853 and has been renovated and restored but retains the style and charm of an old Australian hotel. Whenever possible the owners will purchase a whole beast from a local farmer who runs a few head of cattle. This superb beef (33B brand) couldn't be more organic. The Whuflungdunga is David Rosenzweig's own creation. It uses gas and wood in a contraption that is a cross between an oven and a barbecue, used here for mutton, beef, fish, chicken and occasionally kangaroo. Takeaway available.
- **Open:** Lunch and dinner daily, bookings essential

ORROROO
🛒 STORES

DEW'S MEATS PTY LTD
Taryn Ackland
East Extension, Orroroo 5431
Phone: 08 8658 1063
Email: taryna@ozemail.com.au
South Australia was the first state to legalise kangaroo meat, and interestingly this remains one of the few wild-caught meats available in this country for

SA—EYRE PENINSULA, FLINDERS RANGES AND THE OUTBACK

human consumption. Dew's meats have been selling this very lean and healthy meat since 1977, and have a full range of cuts, including steak, roasts and tails, as well as marinated and minced meats.
- **Open:** 8am–4pm weekdays
- **Location:** Check the Ororoo information board, or ask in town for directions

PARACHILNA
DINING

PRAIRIE HOTEL
Jane and Ross Fargher
Cnr High St and West Tce, Parachilna 5730
Phone: 08 8648 4895
Email: info@prairiehotel.com.au
www.prairiehotel.com.au
Licensed/BYO
Cuisine: Modern Australian using indigenous ingredients
Mains average: $
Specialty: Feral antipasto, feral mixed grill
Local flavours: Kangaroo (hunted locally), quandong pies
Extras: Sells homemade bush tomato chutney and emu liver pâté and takeaway is available. There are spectacular sunsets from the veranda of the hotel.
- **Open:** Breakfast, lunch and dinner daily
- **Location:** Western side of the Flinders Ranges, between Hawker and Leigh Creek

PORT AUGUSTA
PRODUCERS

NECTARBROOK DISCOVERY PLANTATION
Graham and Iris Herde
28 Jervois St, Nectarbrook,
Port Augusta 5700
Phone: 08 8634 7077
Email: gherde@centralonline.com.au
www.nectarbrook.com
Quandong *santalum acuminatum* is a dry-land Australian native tree that produces a uniquely flavoured fruit. The Herdes process the fruit into what they call 'the world's best quandong jam' and sell it under their Aridland label. They have been running this experimental orchard for ten years, and it has been producing for six years. In 2003 they added a B&B. Now, in addition to the jam, they sell quandong and sandalwood seedling plants through Arid Lands Botanic Gardens and other SA locations, including Adelaide Central Market's Gourmet To Go.
- **Open:** By appointment
- **Location:** Off the national highway, 1.3km south of Port Augusta

PORT LINCOLN
DINING

MOORINGS
Grand Tasman Hotel,
94 Tasman Tce, Port Lincoln 5606
Phone: 08 8682 2133
Email: gthotel@chariot.net.au
www.grandtasmanhotel.com
Licensed
Cuisine: Modern Australian seafood
Mains average: $$
Local flavours: Seafood
Extras: Overlooking Boston Bay.
- **Open:** Breakfast and dinner daily, lunch on weekdays

TOURS

YACHTAWAY CRUISING HOLIDAYS
Lindsay Alford
The Marina, Port Lincoln 5606
Phone: 08 8682 5585
Mobile: 0428 822 166
Visit a working, southern bluefin tuna farm and watch harvesting and feeding.
- **Open:** Phone for details

SMOKY BAY

TOURS

SMOKY BAY OYSTER TOURS AND HOLMES OYSTERS
Colleen Holmes
6-8 Sandy Creek Dr, Aquaculture Park, Smoky Bay 5680
Phone: 08 8625 7077
Email: holmesoysters@bigpond.com
Tours of the oyster growing facility as well as tastings and fresh oysters for sale. See how an oyster farm operates.
- **Open:** 3pm-4pm weekdays, other times by appointment

WARNERTOWN

PRODUCERS

GULF BUFFALO
Bob and Christine Cook
Abattoir Rd, Warnertown 5540
Phone: 08 8634 3043
Mobile: 0419 866 050
Email: gulfbuffalo@bigpond.com
For centuries Asia has recognised the value of the water buffalo, both as a working beast and a food source. Australia is just beginning to discover that the feral buffaloes of the north can be used in a new industry. Bob and Christine Hook farm them in the southern Flinders Ranges and sell the light, lean meat to be used for a range of products such as mettwurst, peperoni, jerky and beef sticks, as well as prime eye fillets, scotch and porterhouse buffalo steaks. All meats are chemical-free. Observe them grazing, or, if you have kids, Bob will find a calf for them to sit on.
- **Open:** 9am-5pm most weekends, groups by appointment
- **Location:** 16km from Port Pirie

SOUTH AUSTRALIAN MARICULTURE

The highly prized greenlip abalone may not be large, but it is beautiful. And unlike its much larger cousin, the more common abalone caught in the deep waters of the Southern Ocean, they are accessible to us all.

These delicious shellfish have been farmed since 1993 in the pristine waters off Boston Point, Port Lincoln, in what is now the largest abalone growing area in Australia. It takes three years to raise the abalone to their market size of seventy to eighty milli-metres, but even then they are young and succulently tender. The annual nursery production is currently around 3.2 million juveniles a year.

Graded according to size during growing and harvest, the abalone can then be packed on site immediately after harvest for optimum freshness. They are sold live in the shell, frozen in the shell, as frozen meat, or vacuum packed and pressure cooked.

Even the shells are particularly attractive; an ideal serving 'dish' for the meat, which can be arranged on ice or serving dishes for a spectacular effect. For details contact Mr Daryl Evans of the SA Abalone Growers Association on 08 8682 5485 or dlevans@bigpond.net.au.

KANGAROO ISLAND

At one time, Kangaroo Island was just a sleepy farming place, one that few outsiders bothered much about. Then disaster hit—at least as far as grain, wool and lamb prices were concerned—and farmers, many of whom had been raised on the island, faced a sad dilemma. Would they leave, or would they stay and change direction?

Those who stayed suddenly found themselves thrust into new industries. The land had to be treated differently, tourism became involved and new markets were cultivated.

Now the island has changed direction and olives, marron, venison, olive oil and cheese from farmlands here fill shelves in gourmet shops across the country. The name Kangaroo Island no longer evokes a sleepy 'Where's that?' but, instead, animated responses.

Not content just to diversify and survive, the island industries have surpassed the standards of many more-established colleagues on the mainland.

CYGNET RIVER

PRODUCERS

ISLAND PURE
Susan and Craig Berlin
Gum Creek Rd, Cygnet River 5223
Phone: 08 8553 9110
When sheep farming took a downturn, many people on Kangaroo Island looked at alternative farming options. This couple's choice has been our good luck as it

tourki@kin.net.au
www.tourkangarooisland.com.au

resulted in the development of this dairy and its high quality products—sheeps milk yoghurts and cheeses. Island Pure pot-cultured yoghurt is a thick and ideal substitute for cream, high in calcium and with double the vitamin B of other yoghurt. Cheesemaker Susan Berlin also makes haloumi, a gourmet fetta, Greek-style kefalotiri, Spanish-style manchego and ricotta. The products are distributed Australiawide in gourmet specialty shops, health food stores and cheese shops.
■ **Open:** 1pm–5pm daily
● **Location:** 12km west of Kingscote, 1.5km along Gum Creek Rd, off the main Kingscote–Cygnet River Rd

FLINDERS CHASE

DINING

THE CHASE CAFE
Joe and Linda Tippett, Rod and Joy Cowin
Flinders Chase Visitor Centre,
South Coast Rd, Flinders Chase 5223
Phone: 08 8559 7339
Email: kishoes@internode.on.net
Licensed
Cuisine: Light lunches
Mains average: $
Specialty: Local cheese with cranberry, taste of Kangaroo Island platter
Local flavours: Whiting, cheese, fruit

Extras: Outdoor dining area overlooks natural bushland. Takeaway is available and the visitors centre with information and souvenirs is next door.
- Open: 9am–5pm daily
- Location: At the national park entrance

KANGAROO—A LOW-FAT ALTERNATIVE

Viewpoints differ on the consumption of kangaroo meat. South Australia led the way in making it legal, but only in the past few years has the rest of Australia allowed it.

No Australian wants to see wildlife decimated, but kangaroos are in dangerously high numbers (for the farmlands at least) and carefully controlled culling is necessary. This is always carried out under the supervision of the National Parks and Wildlife authorities.

All kangaroo meat is subjected to the same strict hygiene, food handling and inspection laws as other meat, so it is quite safe to eat. Kangaroo meat is an extremely lean, low-fat meat (less than two per cent fat) and is ideal for cooking in a number of ways. Because it oxidises quickly, it should be used as soon as the packaging is opened. The meat should be cooked quickly at a high heat to no more than medium-rare then well rested. Ideal barbecued, pan-fried or roasted.

EXPLORERS' RESTAURANT

Allan Mellor and Jennifer Bowers
Kangaroo Island Wilderness Resort,
1 South Coast Rd, Flinders Chase 5223
Phone: 08 8559 7275
Email: reservations@austdreaming.com.au
www.austdreaming.com.au
Licensed
Cuisine: Modern Australian

Mains average: $$
Specialty: Local produce, seafood and freshwater crustacean (marron)
Local flavours: As much as possible, including crayfish, octopus (pickled in own special recipe), King George whiting, marron, fruit and vegetables, sheeps milk, cheese and yoghurts, farmhouse cheeses including brie and camembert, free-range eggs, honey
Extras: A selection of locally-produced Island wines is available. The 'West End Bar' looks out onto a forest of trees and ferns, which are lit up at night. During the day many species of birds, some rare, may be observed. Residential cooking schools also planned.
- Open: Breakfast, lunch and dinner daily
- Location: At the entrance to Flinders Chase National Park on Kangaroo Island

KINGSCOTE
🍎 PRODUCERS

EMU RIDGE EUCALYPTUS
Larry and Bev Turner
Willsons Rd, Kingscote 5223
Phone: 08 8553 8228
Email: emuridge@kin.on.net

The Turners decided to look for other ways to make an income when plummeting wool prices made farming on the island difficult. Remembering that in the early days on the island, eucalyptus distilling was a thriving business, and as the eucalypts were still thriving, they hit on a plan. Now they farm native flora and fauna rather than sheep, and specialise in eucalyptus oil distilled using a steam engine that allows the venture to be self-sufficient. They also produce honey and honey chocolates, native currant jam and chocolates, and many other

products, which are sold in health food shops, mainly in South Australia.
- **Open:** 9am-2pm daily
- **Location:** Fifteen minutes from Kingscote at MacGillivray

ISLAND SEAFOOD

Bevan Golding
22 Telegraph Rd, Kingscote 5223
Phone: 08 8553 2728
Island Seafood are local fish buyers who wholesale and retail to the public, as well as supplying hotels and restaurants on Kangaroo Island. The main fish caught locally is King George whiting (no surprise as Kangaroo Island is home to this fish), garfish, snapper, snook, local oysters (unavailable to the public here), crayfish, and southern rock lobsters.
- **Open:** 9am-5.30pm Mon-Sat
- **Location:** On the main road into Kingscote, next to Caltex

---

LIGURIAN BEE

Kangaroo Island's Ligurian bee has been declared a BankSA Heritage Icon—the island is the only place in the world with a pure surviving strain of these midget bees. As the island is free of bee diseases no bees, second-hand beehives or any other equipment may be introduced.

CLIFFORD'S HONEY FARM

Jenny and Dave Clifford
Elsegood Rd, Kingscote 5223
Phone: 08 8553 8295
The Cliffords' quiet-natured bees still do sting, says Dave Clifford, and he's had his share from over 200 hives of his Ligurian bees. Originally sheep farmers who kept bees as a hobby, the Cliffords set up this farm five years ago, when the honey business blossomed. Their honey ice-cream has become something of a local institution and in addition to the honey they also sell local crafts, including beeswax candles made by their daughters. Their honey is sold at outlets on the island, as well as through Australia on a Plate, Sydney, and some Adelaide shops.
- **Open:** 9am to 5pm daily
- **Location:** Between Moores and Barretts Roads

🏠 STORES

KINGSCOTE IGA FRIENDLY GROCER (GRIFFITHS)

Dauncey St, Kingscote 5223
Phone: 08 8553 2047
Not a large store, but here you will find Kangaroo Island olive oil, Clifford's honey and products from other island producers.
- **Open:** 9am-5pm weekdays, 9am-12pm Sat

🍴 DINING

OZONE SEAFRONT HOTEL

Chris and Debbie Schumann (managers)
The Foreshore, Kingscote 5223
Phone: Freecall 1800 083 133, 08 8553 2011
Email: bookings@ozonehotel.com
www.ozonehotel.com
Licensed
Cuisine: Bistro
Mains average: $
Specialty: King George whiting
Local flavours: Cheese, seafood, honey, lamb, bread
Extras: Extensive ocean views from bistro dining area
- **Open:** Breakfast, lunch and dinner daily

KANGAROO ISLAND—SA

RESTAURANT PIZZA CAFE BELLA
Philip Ford, Debra Dexter, Chan Thornley, Rachel Bowers
54 Dauncey St, Kingscote 5223
Phone: 08 8553 0400
Email: fourdex@bigpond.com
Licensed
Cuisine: Modern Australian
Mains average: $$
Specialty: Pizza, King George whiting
Local flavours: American River oysters, King George whiting, Ordways chicken, seafood, sheeps milk dairy produce, olive oil, fruit and vegetables
Extras: Alfresco dining
■ **Open:** Lunch and dinner daily, pizza all day

TOURS

ADVENTURE CHARTERS OF KANGAROO ISLAND
Craig and Janet Wickham
Kingscote 5223
Phone: 08 8553 9119
Email: wildlife@kin.on.net
www.adventurecharters.com.au
Tours can be tailored to include some of the best food producers on this island—and the gourmet picnic lunch is one of the best.
■ **Open:** On demand

PARNDANA
DINING

KAIWARRA FOOD BARN AND COTTAGES
Uzi Shein
South Coast Rd, Parndana 5223
Phone: 08 8559 6115
Licensed
Cuisine: Kangaroo Island produce

Mains average: $
Specialty: Kangaroo Island marron sauteed in garlic butter, shearer's plate, Devonshire teas
Local flavours: Marron, King George whiting, Island Pure sheeps cheese, calamari, wine
Extras: Outdoor area with children's playground. The walls of the barn are covered with information on a wide range of local products and history.
■ **Open:** Breakfast and lunch daily
● **Location:** Opposite Seal Bay turn-off

PENNESHAW
PRODUCERS

HOG BAY APIARY
Jim and Betty McAdam
South Terrace, Penneshaw 5222
Phone: 08 8553 1237
Email: hogbay@kin.on.net
www.kigateway.kin.on.net/hogbay/hogbay1.htm
It started as a hobby (does this sound familiar?) and ten years later the gentle pastime of beekeeping keeps the McAdams more than busy. Kangaroo Island is the oldest bee sanctuary in the world and the only place where the pure Ligurian honeybee is known to exist in its natural state. Honey is produced by the bees from the many flowering nectar sources in these pristine environmental conditions. There are no bee diseases, so no need for chemical treatment. Varieties include sugar gum, eucalypt, wildflower and cup gum, available as conventional or creamed honey.
■ **Open:** By appointment
● **Location:** Log cabin on right at the end of the road (one house before the corner) with rosemary hedge

SA—KANGAROO ISLAND

🍴 DINING

PENGUIN STOP CAFE
Marion Chambers (chef)
Middle Tce, Penneshaw 5222
Phone: 08 8553 1211
Email: marion@penguinstopcafe.com
www.penguinstopcafe.com
Licensed/BYO
Cuisine: Seasonal, regional
Mains average: $$
Specialty: A fresh approach to local and wild products
Local flavours: Cheese, fish, salad, wines, wild foods (samphire, boxthorn berries, cranberries, muntries)
Extras: Wilderness Gourmet Catering operates from this restaurant. Outdoor seating, sandpit for children, penguins live under the building.
■ **Open:** Lunch Thurs–Mon, dinner Thurs–Sat

OLD MULBERRY TREE
Located at Reeves Point, Kangaroo Island, this is South Australia's first fruit tree. Planted in 1836, it still produces mulberries, which are made into jams for sale.

STOKES BAY

🍴 DINING

ROCKPOOL CAFE
Jenny Morris
Stokes Bay–North Coast Rd,
Stokes Bay 5223
Phone: 08 8559 2277
Email: jenny@kiexperience.com
www.kiexperience.com
Licensed
Mains average: $$
Specialty: Seafood platters at dinner
Local flavours: Marron, crayfish, fish, sheeps cheese
Extras: Located right beside the sea.
■ **Open:** Lunch and dinner daily Sept–May, closed other months

🚌 TOURS

KANGAROO ISLAND COAST TO COAST TOUR
Kangaroo Island Sealink
Phone: 13 13 01
Email: bookings@sealink.com.au
www.sealink.com.au
This two-day tour introduces visitors to the local produce and cottage industries of Kangaroo Island. Visit a traditional eucalyptus oil distillery, learn about the minute Ligurian bee at Clifford's Honey, watch shearing and enjoy an Aussie pub lunch at Parndana. Later you might visit a sea lion colony before sampling fresh local cheeses at Island Pure.
■ **Open:** Tours daily, phone for details

LIMESTONE COAST

This area doesn't make a fuss about itself. There are no oompah-bands or showy fiestas here. The south-east just quietly gets on with doing what it does best—producing fine local food that so perfectly partners their great regional wines.

Kingston is the crayfish centre of this region and its succulent flavour is eagerly awaited each year until its season, which lasts from October to April. Other times, the waters are thick with fish such as snapper, bream, trevally and mullet. To take care of the rest of the meal, what better foods than fresh-from-the-orchard nuts, berries and fruits?

Email: tourism@seedb.org.au
www.thelimestonecoast.com

AVENUE
🍎 PRODUCERS

AVENUE EMUS
Neville and Julie Thomas
Thomas Rd, Avenue 5273
Phone: 08 8766 0085
Email: ave_emus@seol.net.au
The Thomases sell emu products and goat meat from their farm gate, and also have farm tours available. The most popular product is their emu mettwurst (salami, if you are not from South Australia!), which is noticeably less fatty than other mettwursts. Tastings are available. Then there is goat meat (much like a sweeter, very lean lamb) that is packaged into a range of cuts, ready to cook, cryovac-packed and frozen, but tastings are only available at the Penola farmers' markets. The mettwurst was developed to make use of the cuts of emu meat that are not popular with restaurants.
■ **Open:** 10am-4pm weekends, or by appointment
● **Location:** North of Avenue, 7km along Minnie Crow Rd, then turn right onto Thomas Rd (gravel road) and travel a further 5km

COONAWARRA
🍎 PRODUCERS

THE COONAWARRA LAVENDER ESTATE
Julie and Peter Kidman
Coonawarra 5277
Mobile: 0419 247 149
coonawarralavender@yahoo.com.au
www.coonawarralavender.com.au
While you may make an appointment to visit this property, the fragrant products manufactured here—lavender jelly, mustard, vinegar, honey, tea, herbs, shortbreads, and chocolate—are available by mail order as well. The Kidmans began this venture, they say, with a 'vision for the diversity of lavender'. They've certainly illustrated this with an extensive range of lavender-based skin-care products.
■ **Open:** By appointment

SA—LIMESTONE COAST

🍴 DINING

REDFINGERS CAFE BAR AND GRILL
Andrew Henry and Jodie Cahill (chef)
Memorial Dr, Coonawarra 5277
Phone: 08 8736 300
Email: redfingers@ozemail.com.au
Licensed
Mains average: $$
Specialty: Coonawarra Ale pies, vanilla slice, curries, Coona dogs
Local flavours: Beef, lamb, kranskies, eggs, herbs, vegetables, fish, olives, olive oil, relishes, jams, yoghurt
Extras: Courtyard dining area shaded by hundred-year-old peppercorn trees. Takeaway available.
- **Open:** Breakfast, lunch and dinner Wed–Mon

UPSTAIRS AT HOLLICK
Ian Hollick (owner), Sean Emery (chef), Ian Perry (chef)
Ravenswood Lane, Coonawarra 5263
Phone: 08 8737 2752
Email: kate@ hollick.com
www.hollick.com
Licensed
Cuisine: Modern Australian
Mains average: $$
Specialty: Duck dishes
Local flavours: Lamb and beef, other produce seasonally
Extras: This is the only restaurant attached to a winery in Coonawarra and it has outdoor decking and views over the Coonawarra vineyards. Takeaway available.
- **Open:** Lunch daily, dinner Fri–Sat, other times by appointment

KALANGADOO

🍎 PRODUCERS

KALANGADOO ORGANIC
Chris and Michelle McColl
Old Kalangadoo Rd, Kalangadoo 5278
Phone: 08 8737 2028
Email: mccoll@penola.limestonecoast.net
What do you do when you love growing apples but are disillusioned with the traditional marketing chain? If you are like the McColls you decide to grow your own and do it your way. Which means you start an organic apple orchard producing apple juice, sun-dried apples, and fresh apples that are sold at the roadside stall, farmers' markets and selected retail outlets and restaurants in the region. These are seasonal organic apples and each variety is available for only a two to three week period, as they are picked at optimum maturity and sold to the consumer within three days of picking. Any fruit not sold within three days is juiced or dried. There are also potted apple trees—heritage and modern varieties.
- **Open:** 8am–dusk most days (late Feb–mid May)
● **Location:** 13km north of Kalangadoo

BIG LOBSTER

The locals call him Larry and the tourists line up for photos with him. This seventeen-metre, four-tonne, tourist attraction plus cafe celebrates the importance of the lobster fishing industry in the area. Local produce is displayed and served in the cafe. You can find Larry on the Princes Highway, at Kingston SE.

LIMESTONE COAST—SA

KINGSTON

🍴 DINING

LACEPEDE SEAFOOD
S and S Hyland
Marine Pde, Kingston 5275
Phone: 08 8767 2549
Email: shopwise@bigpond.com
This place serves crayfish but it is only open during cray season. Which makes it doubly important to plan ahead to be here to buy lobster and lobster pâté to take away.
- **Open:** 9am–6pm Oct–Apr (cray season)

MENINGIE

🍎 PRODUCERS

THE COORONG FISHERMAN
Glen and Tracy Hill
Lot 10, Yumali Rd, Meningie 5264
Phone: 08 8575 1489
Email: hillfish@lm.net.au
Fishing and fish processing is the name of the game. You can order fish but it is strictly by arrangement, so please phone at least a day in advance.
- **Open:** By appointment

🍴 DINING

THE CHEESE FACTORY MENINGIE'S MUSEUM RESTAURANT
Peter and Maureen Van Heusden
3 Feibig Rd, Meningie 5264
Phone: 08 8575 1914
Email: mpvh@bigpond.com.au
www.coorongcheese.modnet.com.au
Licensed
Cuisine: Steaks
Mains average: $
Local flavours: Coorong mullet

Extras: There is a community museum next door, and an outdoor area on the shores of Lake Albert, with an abundance of bird life.
- **Open:** Lunch and dinner Tues–Sun

MILLICENT

🍎 PRODUCERS

LIMESTONE COAST TROUT
Darren Winter, Carlien Lavers, Henk Versluis
91 Lossie Rd, Millicent 5280
Phone: 08 8733 1407
Mobile: 0407 791 669
Email: darrenwinter@vistara.com.au
Grown in spring water, this fresh and smoked rainbow trout is available here or at the Sydney and Melbourne fish markets, as well as at local butchers and restaurants around south-eastern SA.
- **Open:** 8am–4pm weekdays (summer and school holidays), phone for other times
- **Location:** Follow the heavy vehicle bypass road and you will see the sign and factory to the south

GREEN FARMHOUSE
Annette Green
Millicent 5280
Phone: 08 8735 2043
Email: gfh@seol.net.au
www.greenfarmhouse.com
Like many other farmers, Annette Green decided to diversify and look at other money-making avenues. She decided to develop a range of products that would introduce bush foods to the general public. Her Corroboree Dust, Boobialla seasoning and Coolamon Herbs are available in many places including local farmers' markets.
- **Open:** By appointment
- **Location:** 26km north of Millicent on Hwy 1, then turn right towards Furner

243

MOUNT GAMBIER

🍎 PRODUCERS

KENTISH AND SONS PTY LTD
Kentish Family
Kentish Rd, Mount Gambier 5290
Phone: 08 8739 8230
Email: malcolm@kentish.biz
www.kentish.biz
Once you could only buy 'potatoes' in Australia. Recently we have become more particular and this family knows all about it. They specialise in packaging red potatoes into various sized packs for the consumer market, including speciality packs of flow-wrapped punnets and net bags. Established in 1963, Kentish's aim is that the potatoes should be grown with as little chemical assistance as possible.
- **Open:** 7.30am–5pm weekdays
- **Location:** North on Penola Rd to 15km post, then east on Kentish Rd, 6km to packing shed

THE CHEESE MASTER SPECIALITY CHEESES
Ian Sims
4 Pyne Close, Mount Gambier 5290
Phone: 08 8725 1176
Ian Sims is the only person in the southeast of the state who is working with cheddar cheese, an amazing fact for a region that has such rich dairy country. His club cheese uses Victorian cheddar, but he says this is okay, as it is actually made from local milk, which is sold interstate. Ian shreds the cheese, incorporates powdered flavours, then presses the cheese again into a block, giving a smooth texture. There is no heat treatment, and flavours include pepper, garlic, bacon, chilli, sage and tomato apricot, ginger, date and walnut. A local person, Sims has been in the cheese game all his life. He was working in a Victorian dairy when it suddenly closed down, so he bought some of the equipment and decided to go it alone. Cheese Master Speciality Cheeses are sold at Yoey's Traditional Fine Foods as well as She's Apples in Mount Gambier. Sims also supplies local hotels, bakeries and delis.
- **Open:** 8am–3pm weekdays

🏠 STORES

YOEY'S TRADITIONAL FINE FOODS
Bernie Kain
32 James St, Mount Gambier 5290
Phone: 08 8725 7710
You can still find Mil-lel and Yahl cheeses here, even though most of the local milk is now sold to Victoria and cheese is not made in the region any more. There is Dutch licorice and continental meat here too.
- **Open:** 8.30am–5.30pm weekdays, 8.30am–12.30pm Sat

🍴 DINING

SAGE AND MUNTRIES CAFE
Graeme and Leigh Armstrong
78 Commercial St, Mount Gambier 5290
Phone: 08 8724 8400
Mobile: 0417 872 144
Email: sagemunt@bigpond.net.au
Licensed/BYO
Cuisine: Indigenous Australian
Mains average: $$
Specialty: Use of indigenous fruits and herbs
Local flavours: Varies seasonally—asparagus, raspberries, loganberries, venison, buffalo, crayfish, rabbit, fish
- **Open:** Breakfast, lunch and dinner Mon–Sat

YABBIES

Yabbies are freshwater crustaceans native to Eastern Australia, and are farmed in all the states of southern Australia, except Tasmania. Western Australia is the major producer.

In their natural habitat, yabbies thrive in creeks and billabongs, relying on the murky waters to hide them from bird predators. Because yabbies are sensitive to chemicals, growers raise them in an environ-ment free of pesticides, antibiotics and chemicals, often in ponds or dams that replicate their natural habitat.

When buying live yabbies, make sure they are lustrous in colour, and that their claws are intact. Place them in a bucket with a little unsalted water (too much and they will drown) cover with a damp cloth and store them in a cool place, such as the laundry.

Like most crustaceans, yabbies may be stunned before cooking by placing them in the freezer for a short time, then plunged into boiling water. They need only to be simmered for about five minutes or until they turn a bright pinkish-red. Once cooked, the yabby can be foil-wrapped and refrigerated for up to two days, or frozen for up to three months.

The flesh is exceptionally sweet and succulent and is ideal for use in soups and bisques, seafood platters, in salads or barbecued.

NARACOORTE

🍎 PRODUCERS

THE VEG SHED—KESTER'S APPLES
Robert and Joan Kester
Unit 4/44 Robertson St, Naracoorte 5271
Phone: 08 8762 2111
Email: kesters.apples@bigpond.com

This fruit and vegetable retail venture is in addition to the Kesters' apple orchard, and provides a fresh, quality product to the public. The apples come in fresh three times a week and are first grade, with a long shelf life.
- **Open:** 8am-6pm Mon-Fri, 8.30am-12.30pm Sat

🍎 PRODUCERS

YELLAND YABBIES
Ken and Elizabeth Yelland
Naracoorte 5271
Mobile: 0428 621 613

Yabbies are the specialty here, as you'd suspect—live, cooked, pickled and any other way the Yellands can think of. 'It's supposed to be our retirement', they say, repeating something many others in this book have said, but adding, 'We love it!' One of the reasons is that *cherax destructor* (the proper name for yabbies) is so incredibly popular. 'Everyone loves to eat them', say the Yellands. With wineries throughout this region, their yabbies are in great demand for platters, and are snapped up at various fairs and markets in the region too.
- **Open:** By appointment

PADTHAWAY

🍴 DINING

PADTHAWAY HOMESTEAD
Janice Fort
Riddoch Hwy, Padthaway 5271
Phone: 08 8765 5555
Email: jan@padthawayhomestead.com.au
www.padthawayhomestead.com.au
Licensed
Cuisine: Modern Australian

Three-course degustation menu (with wines): $$$$
Specialty: Homemade gnocchi with duck confit and lime beurre blanc, ravioli of Robe crayfish, daily home-baked breads, including cabernet sauvignon bread
Local flavours: Limestone Coast organic lamb, Naracoorte yabbies, Robe crayfish, Coorong mullet, all fruit and vegetables sourced from the region
Extras: Rose gardens and native flower gardens situated in the middle of Padthaway Estate's vineyard. One of the oldest cottages in the Tatiara Region.
■ **Open:** Breakfast daily (guests), lunch (on request only), dinner daily (for guests and bookings)
● **Location:** Fifty minutes north of Coonawarra

PENOLA

🍎 PRODUCERS

PETTICOAT LANE HERB GARDEN
Jenny Hinze
Davidson Cottage, Petticoat Lane,
Penola 5277
Mobile: 0428 816 795
Email: hinzepen@penola.limestonecoast.net
This is a very smart community idea, and one that should be in every town. Or maybe each town simply needs a Jenny Hinze, whose passion for herbs goes way back. The garden works on an honesty system, and people may come at any time to cut as many herbs as they need, then pay in the place provided. Even bags and scissors are supplied. Since August 2003, Hinze has also operated the herb garden shop. Here she sells pestos, relishes, jelly, dressings, dried herbs, packaged teas and culinary blends. Out-of-date herbs are turned into paper, and seeds appear on cards so the receiver may grow their own herbs. As a member of the regional food group, she also stocks some produce from other members.
■ **Open:** 10am-4pm daily (Sept-May), 10am-4pm Mon-Fri (winter)
● **Location:** In the National Trust precinct of Petticoat Lane

🏠 STORES

DJ AND CA MEEK BUTCHERS
David and Chris Meek
66 Church St, Penola 5277
Phone: 08 8737 2330
For twenty-nine years Meek Butchers has been meeting the meat needs of the locals. Using cuts from locally grown animals, and doing interesting and innovative things to them such as making red-gum smoked hams and bacon, sweet chilli kransky or gourmet sausages, they have won a faithful following of customers.
■ **Open:** 8am-5pm weekdays

🍴 DINING

DIVINE CAFE AND GOURMET DELI
Suzie Chant
39 Church St, Penola 5277
Phone: 08 8737 2122
Email: susie@penola.limestonecoast.net
Cuisine: Regional modern Australian
Mains average: $
Local flavours: Salmon, smoked salmon, barramundi (farmed) Coorong mullet, boarfish, venison, wagyu beef, wagyu sausages, free-range eggs, local bacon, berries, cherries, honey, local potato wedges

Extras: Outdoor tables at the front of the shop and takeaway is available. There is also B&B accommodation.
- **Open:** 9am–5pm daily

PIPERS OF PENOLA
Tim Foster
58 Riddoch St, Penola 5277
Phone: 08 8737 3999
Email: pipers@pipersofpenola.com.au
www.pipersofpenola.com.au
Licensed/BYO
Cuisine: Modern Australian
Mains average: $$
Specialty: Spectacular desserts, innovative cuisine
Local flavours: All local meat, seasonal fruit, herbs and Limestone Coast seafood
Extras: The restaurant is housed in a restored 1908 Methodist church with original flooring and ceiling. Diners look out through glass doors onto a courtyard. Function room with a natural stone wall sourced from a local vineyard.
- **Open:** Lunch weekends, dinner daily

ROBE
PRODUCERS

MAHALIA COFFEE
Mahalia and Paul Layzell
Cnr Robe and Flint Streets, Robe 5276
Phone: 08 8768 2778
Email: mahalia@mahaliacoffee.com.au
OR mahalia@seol.net.au
www.mahaliacoffee.com.au
Mahalia roasts and blends coffee of the world, and makes some jams. The owners also run training courses in espresso making, and visitors may view the coffee roasting process.
- **Open:** 10am–4pm weekdays

TINTINARA
PRODUCERS

BOUNDARY PARK OLIVES
Jeff and Kathie Bridge
Tintinara 5266
Phone: 08 8756 5018
Fax: 08 8756 5018
Email: boundpk@lm.net.au
In a pristine area of the Limestone Coast, Boundary Park is pressing extra virgin olive oil from Frantoio olives. This is an organically maintained grove and was begun because the Bridges had an interest in the Mediterranean lifestyle and diet, and because they figured that a small olive grove could be managed by one person. First released in very limited quantities in 2002 to a local outlet, the 2003 harvest saw the first commercial crop with an extremely positive response to this Tuscan-style olive oil with 'a hint of the mallee!' Sold in many local outlets and at the Central Markets in Adelaide.
- **Open:** By appointment

WOLSELEY
PRODUCERS

TEATRICK LAVENDER ESTATE
Liz Ballinger
Custon Rd, Wolseley 5269
Phone: 08 8753 2268
Email: teatrick@lm.net.au
www.teatricklavender.com.au
Lavender has an olde worlde feel to it and you can slip back in time here if you are part of a group as you will be treated to lavender Devonshire tea—lavender tea, lemonade scones and lavender conserve. The Lavender Shoppe (see, we said it had

SA—LIMESTONE COAST

an olde ring to it) has lavender products and plants, and foodwise there are lavender biscuits, chocolates, tea, honey, conserve and fruit cake. Inspirational, as you can see first-hand how gardens can still grow with limited watering, and the companion planting of lavender.

- **Open:** 10am–4pm Wed–Sun Oct–Mar (groups by appointment)
- **Location:** 4.6km south of Wolseley

MARKETS AND EVENTS

✦ **LIMESTONE COAST FARMERS MARKET**
08 8737 2855
10am–1pm, approximately monthly in various local towns: ring the information line above for assistance

✦ **ROBE VILLAGE FAIR,** ROBE
08 8723 1644
Nov

MURRAY AND RIVERLAND

While most would name the Barossa, or even the New South Wales Riverina, as the largest wine producer in Australia, actually, this region gets the honour. The table-flat irrigation lands watered by the mighty Murray make for ideal conditions to raise not only wine grapes, but also all types of citrus fruits, table grapes, peaches and apricots. These are further processed into juice or dried fruits and supply the rest of the country, as well as being sold for export.

Murray cod, once threatened by the introduced and out-of-control European carp, is now returning and is a sublime local delicacy on many menus. Long distance travellers find the irrigated areas an unexpected oasis, a cool respite from the inland harshness where bush tucker and emus are grown.

A river runs through this land, and the land shows its gratitude by producing gold.

BERRI

STORES

BERRI DIRECT, BERRI LTD
Old Sturt Hwy, Berri 5343
Phone: 08 8582 3428
Email: samfielke@berriltd.com.au
The good news is that you can again buy direct from the factory shop. There is a comprehensive range of Berri Ltd products and Riverland Gourmet Products and everyone benefits, as the excess stock of two brands is cleared, allowing savings for customers. On the Berri side of things, look for juices, cordials, mineral water and fruit drinks. Under the Riverland Gourmet brand you'll find dried fruit, nuts, olives, olive oil, jams, chutneys, marinated fruit, chocolate fruit and condiments.
- **Open:** 9am–5pm daily
- **Location:** 3km from Berri on the Renmark side of Berri

THE BIG ORANGE
Riuskills Inc,
Sturt Hwy, Berri 5343
Phone: 08 8582 4255
This fifteen-metre high fibreglass building houses a cafe and souvenir shop, and has been a landmark in this flat irrigated countryside for more than twenty years. It celebrates a district, which is known as a large citrus growing area. There are special citrus juices for sale and citrus foods and other local produce such as Red Ochre, Viva olives, Kangara Natural juices and Vitor Citrus.
- **Open:** 9.30am–4.30pm daily
- **Location:** Near the Monash bypass, between Renmark and Berri

DINING

MALLEE FOWL RESTAURANT
Howard and Caryl Michael
Sturt Hwy, Berri 5343
Phone: 08 8582 2096
Email: malleefowlrest@bigpond.com
Licensed
Mains average: $$
Specialty: Steaks, roo fillet
Local flavours: Barramundi
Extras: Awarded Best Themed Restaurant

in Australia by the Restaurant and Catering Association (Australia) 2002. Set amongst mallee trees on twenty-four acre property, 'The Mallee Fowl Nest' is an outdoor entertainment area featuring bush-style cooking, including a camp oven and a bush cook.
- **Open:** Lunch and dinner Wed–Sat
- **Location:** 1km from The Big Orange

LOXTON

🍎 PRODUCERS

THE SOUTH AUSTRALIAN OLIVE CORPORATION PTY LTD
Mark Troy, Managing Director
Starcevich Rd, Loxton 5333
Phone: 08 8584 5811
Email: saolcorp@riverland.net.au
This company presses olives for oil between May and August, but packages olives and oil all year. It is here that Viva's verdale, kalamata, Spanish queen and manzanillo olives, known throughout South Australia and nationally, are processed. The company presses olives for local growers, making this the largest olive producer in Australia. The fruit is sold in plastic packs, jars or in bulk, and there are also interesting olive marinade combinations such as lemon and garlic, and chilli and spice.
- **Open:** 8am–5pm weekdays, factory open until 3.30pm
- **Location:** 1km on the eastern side of Loxton adjacent to the sporting complex

RENMARK

🏠 STORES

NUTS ABOUT FRUIT
Jenny Frahn
35a Renmark Ave, Renmark 5341
Phone: 08 8586 4090
Email: nutsaboutfruit@riverland.net.au
If you are nuts about the wonderful fruit in South Australia's Riverland area, this must be a stop. Fruit from the local dried fruit factory makes an interesting gift or souvenir and the gift packs and baskets are hand packed on the premises, as required, to guarantee freshness. Not content to simply feature a full range of dried fruit, there are also nuts, confectionery, Havenhand chocolate fruits, honey, jams, Viva olives, biscuits, Buzz honey and Red Ochre sauces, all beautifully presented.
- **Open:** 9.30am–5.30pm weekdays, 9am–12pm Sat, 10am–1pm most public holidays, other times by appointment

WAIKERIE

🍎 PRODUCERS

HAVENHAND CHOCOLATES
Dean and Janette Grosse
22 Peake Terrace, Waikerie 5330
Phone: 08 8541 2134
Email: dean@havenhandchocolates.com
www.havenhandchocolates.com
Many of the fillings in these handmade chocolates are made using dried fruit base mixes made from locally grown fruit—apricots, raisins sultanas, and oranges. There are milk, dark and white

chocolates to suit every chocoholic. Sold locally and elsewhere in South Australia.
- **Open:** 10am–5pm daily

RIVERLAND SMALLGOODS
Brian and Sharon Burnett
8 Peake Tce, Waikerie 5330
Phone: 08 8541 2114
Email: beburnett@austarnet.com.au

A family recipe that has been handed down through four generations is the basis for the specialty of this business—traditional German mettwurst. 'We knew our mettwurst recipe was something special,' say the Burnetts, 'and so we decided to market it outside our local butcher shop.' All of their products are made to heritage recipes, and all smoked products are naturally smoked, the baked products are roasted traditionally. The salamis, csabai, peperoni, kabanas and several other products are widely available.
- **Open:** 8am–5.30pm weekdays, 8am–12pm Sat

DINING

MALLYONS ON THE MURRAY
Nick and Rita Builder
PMB 51, Weston Flat, Highway 64, Waikerie 5330
Phone: 08 8543 2263
Email: weston@riverland.net.au
Licensed (organic wines)
Cuisine: Light meals
Mains average: $
Specialty: Pancakes, warm chicken salad, Murray River mud cake
Local flavours: Organic produce from own orchard
Extras: Spectacular views of the Murray River and outdoor dining. The residence was a hotel built in 1860 for the paddle steamer industry and later for Cobb and Co. The gardens of this solar-powered property feature a herb garden, bulb garden, orchid house and an extensive rose garden. Frankie the duck, Jesse the Border Collie and Remo the Kelpie greet visitors. The owners are members of WWOOF, a cultural exchange program, so often host people from other countries. There is also a bush cafe and gallery with a large range of locally created art and craft in the restored stone barn, circa 1841. Farm walks are encouraged and camping on the river flat is allowed.
- **Open:** Lunch and teas Thurs–Mon, closed July
- **Location:** 18km east of Morgan, 25km from Waikerie on the north shore of the River Murray

---

SA WEBSITES
www.foodbarossa.com
www.adelaidehillsfood.com.au
www.safoodonline.com

BASS STRAIT

King Island
○ Currie
2

Flinders Island
Whitemark ○
2
Furneaux Group

0 — 100
KILOMETRES

Fleurieu Group

Wynyard ○   Burnie   Pipers Brook ○
   Devonport ○
         Launceston ○
5
   ○ Deloraine
3
   + *Cradle Mountain*

TASMANIA

○ Coles Bay

4

Richmond ○
Hobart ○ ○ Cambridge
1

○ Port Arthur

SOUTHERN OCEAN

TASMAN SEA

N
W — E
S

Regions:
1 Hobart & Surrounds
2 Bass Strait Islands
3 The North
4 The South
5 The North-West

TASMANIA

Tasmania tends to be the leader with things: truffles, saffron, wakame, wasabi—even, years ago now, Atlantic salmon in the antipodes. There are also some fantastic spin-offs of indigenous foods (pepperberry liqueur?) and who would have thought of 'cloud juice' (aka rain)?

They have a saying in Tasmania—'small enough to know, and big enough to share'—that just about sums up the state which sometimes inexplicably gets dropped off Australian logos.

Tasmania claims green grass, green wildernesses and a clean green environment that produces pure foods of the highest standard. No idle boast this, because as an island-state, Tasmania is able to defend its primary industries against many pests and diseases. Expect careful and stringent quarantine inspection for plants, animals, fruit and vegetables at all entry points.

Once simply dubbed the Apple Isle, Tasmania is now known for much more—everything from seafood to confectionery, beer to poppy seeds. A garland of pick-your-own orchards and berry farms loops around Hobart, the capital, itself a food-lover's paradise. The food from this island is not just good, it is gourmet-plus. How about this for a costly little shopping list: abalone, saffron, oysters, Atlantic salmon, yosterberries, game—and those truffles!

As well as the established mainstream beef, dairying and sheep industries, this fertile land supports many exotic animals such as deer, emu, quail, wallaby, possum and Cape Barren geese. Crops are equally diverse and include onions and potatoes, peas and other fresh vegetables, grains, hops, buckwheat, wasabi, poppy seeds, honey, amaranth, quinoa, bush foods and flowers. Orchards of apples (Tasmania is still Australia's largest apple-exporting state, with 1.5 million trees) and other cool-weather fruits, as well as a thriving wine industry, complete the smorgasbord on this amazing island.

HOBART AND SURROUNDS

Hobart's deepwater port rivals Sydney's and is the destination of that very famous yacht race each year. But that's where the similarities cease. Hobart is small, tucked at the base of mighty Mount Wellington, which is sometimes snow-capped in winter.

Hobart is also old. Australia's second oldest capital, established in 1804, its sandstone warehouses remind the visitor that this is a maritime city, established as much for defence as settlement. A stroll around Sullivan's Cove will take you back in time, but the city itself will bring you back to the present era.

For this is a place on the move with cafe strips, smart restaurants opening almost weekly it seems, and a deep pride in local produce that shines on menus throughout the city.

Harbourside, you can pick up fish-to-go from the fishing punts or dine in the best surroundings on perfectly prepared seafood. This city was born with the water on its doorstep and continues the fine tradition of dealing with its bounty with care and skill.

HOBART
🍎 PRODUCERS

CREATIVE CHICKEN
Peter Walker and Rodney Berry
1 Innes St, Glenorchy 7010
Phone: 03 6273 5567
Email: CreativeChickens@aol.com

Whole yolla (pronounced oola) or mutton birds, are traditional fare of Tasmanian Aborigines. They are harvested on Bass Strait islands around March and April each year, and are only available for about a month, in season, at this factory outlet that, as its name implies, deals mainly with chicken.
■ **Open:** 8am–6pm weekdays, 8am–4pm Sat

DIAMOND STILL SPRING WATER
John Gall and Ian Jackson
93 Brooker Ave, Hobart 7001
Phone: 03 6234 5577
Email: info@eski-ice.com.au
www.eski-ice.com.au

Clouds formed in the path of the roaring forties eventually turn into water that is then gathered from a spring, near Beaconsfield in north-eastern Tasmania. If you would like to take home some of Australia's purest water, this place has made a business of supplying it in everything from 350 millilitre to eleven litre bottles. It is also a pioneer in making disposable ice tray packs.
■ **Open:** 8.30am–5pm weekdays

ISLAND PRODUCE TASMANIA
Catherine Mayhea and Scott McKibben
16 Degraves St, South Hobart 7004
Phone: 03 6223 3233
Email: info@islandproduce.com.au
www.islandproduce.com.au

Among the restored ruins of the Cascades Female Factory (the female equivalent of Port Arthur), despite its grim beginnings, there is now a very

HOBART AND SURROUNDS—TAS

sweet ending. This is where thirteen varieties of gift-boxed fudge, four flavours of handmade truffles, and five flavours of sauce (ideal for ice-cream, drinks or as dipping sauce) are created using superb Tasmanian produce wherever possible. Tours of the Female Factory Historic Site (phone 03 6223 1559) let you catch up with both the past and present use of the area.

- **Open:** As part of tours
- **Location:** Turn off Cascade Rd at the sign for Hobart tip

JOANNA'S JAMS
Joanna Dean
13 Avon Rd, Cascades 7004
Phone: 03 6224 2654
Email: joannasjams@ozemail.com

'It's like my grandmother used to make', is a comment Joanna often hears from people sampling her jam at Salamanca Markets. Another frequent comment is, 'It's real jam!' There's a reason for this. Joanna makes jam the way her mother and grandmother did—using Tasmanian raspberries, and other berries such as strawberries, blackberries and tayberries from Sheffield Berry Farm and Cygnet. Find her and her jams at the markets each weekend, or she will sell from her premises.

- **Open:** By appointment

THE STRUDEL COMPANY
Amanda and Bruce Howell
10a Patriarch Dr, Huntingfield 7055
Phone: 03 6229 8331
Email: tasstrudel@iprimus.com.au

The Howells have a passion for food, and when they started this business nine years ago, it was a natural step for them to use predominantly Tasmanian products. This means they have a range of vegetarian, vegan and Tasmanian Atlantic salmon strudels, plus an extensive range of cakes and individual desserts, appropriately labelled Devilish Desserts. The company supplied the Olympic Games with vegan strudel and smoked Atlantic salmon strudel, and also supply the TT line with individual desserts and vegetable strudels. Look for them at places around Hobart and Tasmania, as well as Salamanca Markets.

- **Open:** 8am–4pm weekdays
- **Location:** At Kingston, behind the fork in the road Mitre 10 hardware store

🍷 BREWERY/WINERY

LARK DISTILLERY PTY LTD
Lyn and Bill Lark
14 Davey St, Hobart 7000
Phone: 03 6231 9088
Email: info@larkdistillery.com.au
www.larkdistillery.com.au

In 1992, when the Larks obtained Tasmania's first distillery licence since 1839, they may not have realised where it would end up. Today their single malt whisky rates highly amongst top Scottish malts, they export to Denmark and the United Kingdom, and their labels are available in many parts of Australia, as well as in over seventy bars and nightclubs in Victoria. But from a regional point of view, their unique Tasmanian Bush Liqueur (distilled from native pepperberry), Pepperberry Gin and Vodka, Tasmanian Apple Schnapps and Tasmanian Apple Liqueur are most notable. Several years ago, Lyn Lark, perhaps Australia's only female distiller, was trying to make an Aussie version

TAS—HOBART AND SURROUNDS

of gin. Wondering if Tasmanian native mountain pepperberries collected in the mountain heathlands might have a similar result, she went one better and created a fabulous Tasmanian bush liqueur.
- **Open:** 9am–6pm daily (closed Sun in winter)
- **Location:** Next door to the Hobart travel and information centre and adjacent to Constitution Dock and Mawson Place

TAVERNER'S PRODUCTS
David and Ruth Thomas
40 Negara Cres, Goodwood 7010
Phone: 03 6273 2466
Email: taverners@trump.net.au
www.taverners.au.bz
Ginger beer, mead, fruit wine, beer—' a microbrewery in the broadest sense', says David Thomas. Fourteen years ago, the Thomases recognised that the by-product of an island full of blossoming fruit trees is honey. And fortunately for us, they dreamed up a great use for it. Taverner's range includes sparkling meads and nectar made from different varieties of Tasmanian honeys, and they have added ginger beer, ales, and other wines and liqueurs. The labels (Taverners, Gillespies, Hazards, and Thornlea) are sold in several states.

HOBART CITY AND SUBURBS
The waterfront area has emerged in recent times as Hobart's premier restaurant location. The area around Elizabeth Pier, Victoria Dock, Hunter Street and Salamanca Place offers a large choice of dining options. Other small clusters on Elizabeth Street, North Hobart and the village areas of Sandy Bay are worth visiting.

- **Open:** 9am–5pm weekdays, weekends by appointment
- **Location:** Just off Bowen Bridge Rd

STORES

BAYSIDE MEATS
Gary and Noeleen Cooper
628 Sandy Bay Rd, Sandy Bay 7005
Phone: 03 6225 1482
Email: baysidemeats@netspace.net.au
This is the place to come if you want some fine Tasmanian wallaby loin fillets, sausages, prosciutto, or kangaroo. There is Tasmania's own fallow venison, mutton bird, and Rannoch quail (which is sold whole, butterflied, smoked, or in breasts) plus quail eggs. Then there is emu, ostrich, local beef and pork from Taylor Bros Elderslie (guaranteed hormone-free), lamb from Longford and Nichols chicken. Bayside Meats also stocks goat, wild hare and rabbit that are shot locally, spatchcock, goose in season, pheasant, turkey and gourmet sausages, made on the premises. Dried Tasmanian wakame is available as flakes or as a condiment, powdered and fresh salted.
- **Open:** 6am–6pm weekdays, 6am–1pm Sat

LESLEY BLACK'S
David Schnitzer
149c Collins St, Hobart 7000
Phone: 03 6231 2100
Email: lesleyblacks@bigpond.com
Begun as a hobby ten years ago, this unusual range of chutneys and pickles, which includes hot eggplant, tomato and red pepper, banana, lemon, apricot, autumn apple, wholegrain mustard, and country relish has really taken off. Based wherever possible on local produce, they

HOBART AND SURROUNDS—TAS

are a yummy addition to almost anything. The product is readily available, sold throughout Tasmania in all major hotels, both state casinos, restaurants, cafes and delis.

■ **Open:** 10am–4pm weekdays, other times by appointment

LIPSCOMBE LARDER

John and Rena Fiotakis
527 Sandy Bay Rd, Sandy Bay 7005
Phone: 03 6225 1135
Email: lipscombelarder@bigpond.com.au
www.lipscombelarder.com

If you have only a short time in Tasmania and want an overview of the state's fabulous food, this is a good place to begin. Both local and imported gourmet produce, including fine Tasmanian cheeses and a large range of Tasmanian wines are stocked here. The larder bakes bread and makes gourmet takeaway food including cakes, desserts, ready-to-heat gourmet meals and stocks. Lipscombe Larder began eleven years ago and uses the freshest Tasmanian produce. The patisserie items are sold in the retail store, in other gourmet retail stores throughout Tasmania and in restaurants.

■ **Open:** 7am–7pm daily
● **Location:** About 3km from city centre along Sandy Bay Rd

MURE'S FISH CENTRE

Will and Judy Mure
Victoria Dock, Hobart 7000
Phone: 03 6231 2121
Email: fishing@mures.com.au
www.mures.com.au

The Mure family has been dealing with fish for over forty years and knows virtually all there is to know about Tasmanian fish. The business has grown to include seafood restaurants and a factory, as well as the original fishmongering. The range of Mure's products, still mainly only available in Tasmania, is enormous—pâtés, such as smoked salmon or smoked trout and trevally; preserves; salmon, prawn or scallop roulades; salmon, crayfish or seafood terrines; plain or smoked salmon burgers; fish soups; cold-smoked salmon or trevally; boronia-cured salmon; gravalax; and hot-smoked, whole Atlantic salmon or rainbow trout.

■ **Open:** 7am–7pm daily

SALAMANCA MARKETS

Salamanca Place, Hobart 7000
Phone: 03 6223 2700
Email: salamancassm@hotkey.net.au
www.hobartcity.com.au

If you don't have time to travel around the state, the quickest and easiest way to see what is in season is to drop in here. Depending on the time of year, you will find apples and pears, berries, potatoes, apricots, plums, cabbages or herbs plus wines, confectionery, jam, honey, pâté and cheese.

■ **Open:** 8am–3pm Sat

FISH PUNTS

Another interesting 'market' is found at the fish punts moored at Constitution Dock. These have been part of Hobart's waterfront activities for some time, daily supplying fresh fish caught off the coast of Tasmania. Check here to see what is fresh and seasonal.

TAS—HOBART AND SURROUNDS

STINGRAY SEAFOODS
Jodie Way
308 Elizabeth St, North Hobart 7000
Phone: 03 6234 5977
Email: joway@ozemail.com.au
Fresh local seafood, processed, attractively arranged, and sold on site, is the recipe here. Jodie Way's father was a cray fisherman and she would go fishing with him from an early age, so she knows the industry from the inside. Weather permitting, she aims to have 'the freshest and largest variety of seafood in Hobart', and it is sold here as well as to restaurants, hotels and other seafood retailers.
■ **Open:** 9am–6pm weekdays, 9am–2pm Sat

DINING

LOWER DECK
Gary Shepherd
Mure's Fish Centre, Victoria Dock, Hobart 7000
Phone: 03 6231 2121
Email: lowerdeck@mures.com.au
www.mures.com.au
Licensed
Cuisine: Seafood
Mains average: $
Specialty: Fisherman's basket, grilled blue eye
Local flavours: Fish and seafood, fruit, vegetables, berries, honey, cheese
Extras: Outdoor dining, views of Constitution Dock.
■ **Open:** Lunch and dinner daily

UPPER DECK
Will and Judy Mure
Mure's Fish Centre, Victoria Dock, Hobart 7000
Phone: 03 6231 1999
Email: upperdeck@mures.com.au
www.mures.com.au
Licensed
Cuisine: Seafood
Mains average: $$$
Specialty: Char-grilled blue eye with honey-soy marinade
Local flavours: Fish and seafood, fruit, vegetables, berries, honey, cheese
Extras: Views over Constitution Dock and Mt Wellington.
■ **Open:** Lunch and dinner daily

TOURS

CASCADE BREWERY TOURS
Richard Gerathy
Carlton and United Breweries,
140 Cascade Rd, South Hobart 7004
Phone: 03 6221 8300
Peter Degraves, an architect and draughtsman, arrived in Hobart in 1824 and led a colourful life that landed him in the town clink for a period. While there he planned a brewery, and on

---

MARKETS AND EVENTS

✦ **SALAMANCA MARKETS,** HOBART
03 6223 2700
8am–3pm Sat

✦ **TASTE OF TASMANIA,** HOBART
03 6238 2711
Dec–Jan

release built the Cascade Brewery at the foot of Mount Wellington. A year later he was producing beer, ales and porter for the colonists using pure water from Tasmania's wilderness for brewing. Subtitled 'Australia's oldest brewery', Cascade features the Tasmanian Devil on its logo. Companion company Cascade Beverages began as a cider factory in 1883, was devastated by bushfires in 1967, but today produces carbonated cordials and fruit juices made from local fruit. Tours take about two hours, but remember this is a work site so you must wear flat, covered shoes (no thongs) and be able to climb stairs.

- **Tours:** 9.30am and 1pm, weekdays. Extra tours at peak times.
- **Location:** Turn right at Southern Outlet (marked A6 City on signage) then left at the next lights into Macquarie St, following signs, and 2km to brewery parking

TASMANIAN ATLANTIC SALMON

A triumph of modern marketing and production, Tasmanian Atlantic salmon is now a standard inclusion on almost any menu in Australia. Yet it was not always so. The industry is surprisingly new given its widespread acceptance and recognition. Begun in 1986-87, the industry has leapt ahead, with these fine fish fetching premium prices in Japan. About half the annual production is exported, with one of its selling points being its freedom from disease. Of course the clean waters surrounding Tasmania only enhance the appeal of this fish.

Pre-dating the salmon industry by about a century is rainbow trout, now being sea-raised, although originally introduced to streams as a recreational prey. Freshwater farms occasionally transfer the young fish to sea cages to allow them to grow out adequately. Another eighties industry, this too is gaining momentum.

As an adjunct to the salmon industry, the blue mussel spats that settle on the salmon cages are gathered as the cages are cleaned and then cultivated on hanging ropes, allowing them to mature into amazingly flavoursome and grit-free delicacies.

BASS STRAIT ISLANDS

KING ISLAND

Picture an elegant green platter edged with lacy white, and filled with everything you could want to eat; all of it top quality, all of it fresh today; set on a brilliant blue cloth.

Imagine this and you have King Island, a tiny dot in Bass Strait that on the map appears to swing Tasmania like a pendant from Australia's south-eastern tip. This island, just sixty-four kilometres long and twenty-four kilometres wide, has perhaps one of the richest and most diverse food selections in the world: many varieties of fish and seafood that includes abalone, scallops, prawns, crayfish, huge king-sized oysters and rock lobsters, breads, pies, biscuits and cakes, a dozen or more types of cheese, cream so rich you almost need a knife for it, milk, butter and yoghurt. Then there's honey produced by bees, drunk from feasting on white clover, and some of the world's best beef, free-range pork, game birds, wild wallaby and smallgoods; even some local wine.

The island was named in 1801 after King, an early governor of the infant New South Wales colony, yet, interestingly, so much of the produce is also king-sized. The kelp is gigantic, the bulls are enormous. Even the crabs, crayfish and oysters, although delicate in flavour, are huge, too big for many Asian markets, we are told, where small is often more desirable.

Take your bathers to the Reef and your sunscreen to Perth, but remember to pack a shopping bag for King Island.

CURRIE, KING ISLAND

🍎 PRODUCERS

KING ISLAND PRODUCE
Peta and Dennis Klumpp
Currie 7256
Phone: 03 6463 1147
Email: peta@kip.com.au
www.kip.com.au

If there was a prize for the most unusual food product in this book, this one would have to be a finalist. The Klumpps decided a couple of years ago to see if they could do something with the storm-cast bull kelp that festoons the shores on this remote island. They obtained a licence from the Tasmanian Department of Primary Industries to collect it and set about experimenting, coming up with a kelp chutney, kelp lemon spread, hot kelp pickles and hot and spicy kelp sauce. Although the products are sold in many retail outlets on King Island, including the KI airport shop, you might want to meet this highly innovative couple.

■ **Open:** By appointment

🏠 STORES

KING ISLAND AIRPORT KIOSK
Dot McEwan
Airport Terminal, Currie, King Island 7256
Phone: 03 6462 1010

This may only be a small island airport with a tiny kiosk and shop, but like so much of King Island it cannot help but make a statement about fine food. It is a great place to purchase a cross-section of the island's produce as you leave for the mainland, and preserve the taste of King Island for weeks to come—cheeses, chutneys, cryovac-packaged meat, salami, pickles, herbal teas and honey in season.

■ **Open:** 7.30am–6pm daily (when planes are due or leaving)

KING ISLAND BAKERY
Audrey Hamer
5 Main St, Currie, King Island 7256
Phone: 03 6462 1337
Email: kibakery@kingisland.net.au

This is the only bakery on the island, but it is an excellent one. Not only will you find great meat pies filled with the island's own beef, but seafood such as King Island crayfish and cheese and asparagus as well. Get up early and you will find the local fishermen stocking up with bread and baked goods to take out with them on their boats. The pies export to Victoria and Hobart, but if you can handle thirty-six, they will mail order them. Don't despair if you become addicted to the shortbread made at the bakery. It comes in plain, chocolate-chip and walnut and is now stocked in specialty shops in other states and also exported to Japan, China, Singapore and Fiji.

■ **Open:** 6am–5pm weekdays, 7.30am–4pm weekends

🍴 DINING

BOOMERANG BY THE SEA
John Finlay
Boomerang Motel, Golf Club Rd, King Island 7256
Phone: 03 6462 1288
www.bythesea.com.au
Licensed

Cuisine: Modern Australian
Mains average: $$
Specialty: Local produce
Local flavours: Seafood
Extras: Views of Bass Strait and the golf course.

■ **Open:** Breakfast daily, lunch Wed–Sun and dinner Sat

LOORANA, KING ISLAND

🍎 PRODUCERS

KING ISLAND DAIRIES
National Foods Ltd
869 North Rd, Loorana, King Island 7256
Phone: 03 6462 1348
www.kidairy.com.au

King Island Endeavour Blue won the coveted Grand Champion Cheese for the second time in four years at the Australian Grand Dairy Awards in 2003. A hundred years before that, when a simple butter factory began in 1903, nobody could have envisioned the later developments at this site. Cape Wickham double brie, Surprise Bay cheddar, Phoques Cove camembert, Bass Strait Blue, and Lighthouse blue brie—these evocative names fit this stormy, windswept island perfectly. King Island Dairies began in 1987 as a specialty cream and cheese producer and went on to become a multi-award-winning dairy. Cheddar and brie were the first cheeses produced, and it was not until 1992 that a separate blue cheese factory was built.

■ **Open:** 8am–5pm daily, tours by arrangement
● **Location:** 11km from Currie

TAS—BASS STRAIT ISLANDS

FLINDERS ISLAND
This small island is packed full of surprises. Mainly agricultural, yet with a variety of activities, this is a must-visit dot in Bass Strait. For tourism information contact Lou Mason, Latitude 40, 116 Dutchman Rd, Lackrana Tasmania 7255
Phone: 03 6359 6526, or check the web at www.focusontas.net.au/regions/flinders.

WHITEMARK, FLINDERS ISLAND

STORES

A TASTE OF FLINDERS ISLAND
Diana and Gary Sykes
3 Walker St, Whitemark,
Flinders Island 7255
Phone: 03 6359 2005
For a real taste of the island, this butcher specialises in Flinders' lamb and wallaby, with a limited amount of local beef. The wallaby is of course pasture-fed and tastes like lamb. Shot in the wild, the cuts include fillets, topside, whole wallaby to bake, rolled roast, mince, and sausages. Try the wallaby and bacon sausages, and do make sure you also taste the lamb—pre-salted by nature as they graze on Flinders' salty island grass.
■ **Open:** 8.30am–5pm weekdays, 9.30am–11.30am Sat (Sat in summer only)

FLINDERS ISLAND BAKERY
Mark and Darnielle Fenn
Lagoon Rd, Whitemark, Flinders Island 7255
Phone: 03 6359 2105
A chef by trade, Fenn makes wonderful pies using local beef and lamb and is known for his bee-sting cakes. The bakery makes all the bread for the Furneaux Islands and now operates as a bakery-cafe.
■ **Open:** 7am–5pm weekdays, 9am–2.30pm weekends (summer only)

DINING

FLINDERS INTERSTATE HOTEL
Peter Aitken
Patrick St, Whitemark,
Flinders Island 7255
Phone: 03 6359 2114
Email: interstatehotel@trump.net.au
Licensed
Cuisine: Hotel bistro
Mains average: $
Specialty: Marinated wallaby, mutton bird
Local flavours: Local fish (flathead, flake, garfish, scallops in season), lamb, mutton bird
Extras: This 1912 heritage hotel building is a favourite with the locals, but drop in and you're sure to find an old-timer ready to have a yarn. The Aitken family have run the hotel for around sixty years. Mutton bird enthusiasts particularly love the tomato sauce, sold on request.
■ **Open:** Breakfast daily, lunch weekdays and dinner Sat

THE NORTH

Truly a gateway into the state for anyone making the often bumpy crossing from what the locals laughingly call 'the mainland', this region nurtures the fertile grape-growing banks of the Tamar River. Further south and east is a microcosm of all the state has to offer, a coastal playground, heritage buildings, farming country and lovely landscapes.

And if this region has much of Tasmania's wine growing land, then it also has the cheese to go with it. Heidi and Pyengana cheeses are known and respected worldwide and are able foils for the produce of the Tamar winelands.

COLES BAY

PRODUCERS/TOURS

FREYCINET MARINE FARM
Andrea Cole
Flacks Rd, Coles Bay 7215
Phone: 03 6257 0140
Email: oystergrower@hotmail.com
Andrea Cole has been farming oysters in this area for twenty-three years. She started because she was frustrated at not being able to get fresh oysters in this coastal area, so she decided to try to grow them, and did this so successfully they are now branded as Freycinet Oysters. Both Pacific and native oysters are raised in Moulting Bay then finished in the appropriately named Oyster Bay. There are also mussels and scallops. Tours of the oyster farm—an onshore talk-and-taste, where Andrea discusses the method, then wades out to bring back oysters to open and taste—are great for conference groups. Bookings essential. Boat tours of up to twelve passengers enjoy both the onshore component, then a two-hour trip that demonstrates how the boat works the oyster lines, followed by tastings. The peaceful, self-contained Shuckers Cottage on the property has a great incentive—six free oysters per person per day.
■ **Open:** 8am–8pm daily, tours 10am daily
● **Location:** Off the Tasman Hwy on the way to Coles Bay (C302), 18km from junction

DINING

THE EDGE RESTAURANT
Ray Johnston
Edge of the Bay Resort, 2308 Main Rd, Coles Bay 7215
Phone: 03 6257 0102
www.edgeofthebay.com.au
Licensed
Cuisine: Modern Australian
Mains average: $$
Specialty: Seafood
Local flavours: Seafood, scallops, farmed oysters, vegetables, cheese—the best of local produce
Extras: Great views, boutique resort accommodation.
■ **Open:** Dinner daily

MADGE MALLOYS
Barbara and Ian Barrett
3 Garnet Ave, Coles Bay 7215
Phone: 03 6257 0265
Email: madge@eftel.com
Licensed
Cuisine: Seafood
Mains average: $$
Specialty: Cajun flathead on grilled vegetables and mango cream sauce
Local flavours: Oysters, scallops, fish from Coles Bay, farmed trout
Extras: Ian is the fisherman as well as the chef, so local fresh catch of the day is tops. There is outdoor seating on front veranda, weather permitting, and window seats have a water view.
■ **Open:** Dinner Tues–Sat

DELORAINE

🍎 PRODUCERS

EELS AUSTRALIS
John and Sophie Ranicar
95 Montana Rd, Deloraine 7304
Phone: 03 6362 2539
Email: eelsaust@bigpond.com
John is the son of smokehouse pioneers Sue and Piers Ranicar, who owned and operated Tasmanian Smokehouse until 1994, the first commercial smokehouse in the state. Eels Australis produces and sells live, raw and smoked eels for the local, domestic and overseas markets. The glass eels, or elvers, are collected from local lakes and rivers then fattened in the farm. Salmon and trout are sourced locally. The smoking process is done in mechanical kilns with the eels suspended over slow-burning Tasmanian hardwood smoke. Trout, salmon, chicken, and cheese are also smoked here, and there are tours of the indoor recirculation eel farm. You can see them in tanks being fed, touch them, learn about their life cycle and about recirculation fish farming.
■ **Open:** 8am–5pm daily
● **Location:** 6km from Deloraine, left onto Mole Creek Rd

TRUFFLES
Tasmanian Truffle Enterprises
Tim and Adele Terry
844 Mole Creek Rd, Deloraine 7304
Phone: 03 6363 6194
www.tastruffles.com.au
The first Perigord black truffle grown in Australia was harvested at the Askrigg Trufferie in Tasmania in June 1999. This climaxed several years of hoping and trusting that truffles were successfully growing under the trees that had been specially planted to become their hosts. The Terrys were delighted to announce that the 2004 harvest was the best ever, with chefs delighted with the quality.

🍴 DINING

ARCOONA HERITAGE GUESTHOUSE
Patrick and Olwen Waters
13 East Barrack St, Deloraine 7304
Phone: 03 6362 3443
Email: arcoona@vision.net.au
www.arcoona.com
Licensed/BYO
Cuisine: Modern Australian
Mains average: $$
Specialty: The freshest local ingredients
Local flavours: Beef, poultry, seafood, smoked salmon, cheese, chocolate, bread
Extras: Award-winning and beautifully restored heritage house, B&B and function

THE NORTH—TAS

centre, set in extensive landscaped grounds.
- **Open:** Breakfast daily (for guests), dinner by reservation

DILSTON
🍴 DINING
A ROSTELLA EXPERIENCE
Ronald Szypura
135 Rostella Rd, Dilston 7252
Phone: 03 6328 1409
Email: rostella1@yahoo.com
www.rostella.com.au
Licensed
Cuisine: European
Mains average: $$
Specialty: Rostella antipasto
Local flavours: Varies seasonally, including home-grown greens and fruit
Extras: Live jazz with Sunday lunch.
A stylish restaurant that concentrates on local produce, located in an old shearing shed.
- **Open:** Lunch Wed–Sun, dinner Wed–Sat
- **Location:** 15km north of Launceston

HILLWOOD
🍎 PRODUCERS
HILLWOOD STRAWBERRY FARM, FRUIT WINE AND CHEESE CENTRE
Alan and Kate Focken
105 Hillwood Rd, Hillwood 7252
Phone: 03 6394 8180
Email: kfocken@tassie.net.au
Free wine and cheese tastings, pick your own berries, strawberries or raspberries and cream, and other Tasmanian products—do you need any more incentives to make a stop here? Over the thirty years it has been operating, Hillwood has learnt what brings people in. Of course you might find Hillwood fruit wines in other places, even interstate, but it won't be the same as seeing them here. All the spray-free fruit for the wines, jams and vinegars comes from this farm, and the Fockens' enthusiasm for this lovely part of Tasmania is worth tapping into as well. Alan Focken says it takes a lifetime to build up something like this, but he also admits he has loved every minute of it.
- **Open:** 8am–5.30pm daily
- **Location:** About fifteen minutes from Launceston, off the East Tamar Hwy (A8), look out for the signs

> **QUARANTINE NOTE**
> Only fruit, vegetables, plants and animals with certification are allowed to enter Tasmania.

LAUNCESTON
🍎 PRODUCERS
WESTHAVEN DAIRY
Lorraine and Geoff Mance
89 Talbot Rd, Launceston 7249
Phone: 03 6343 1559
Email: westhaven@intas.net.au
www.westhavendairy.com
French style chevre and fetta cheese and a range of probiotic cows milk yoghurts are the deal here. Geoff Mance has been milking goats for thirty years, and manufacturing cheese for nine, so pay attention chevre lovers! The fresh cheeses are rolled in different coatings, and you can find them in Tasmania, Melbourne, and Sydney. And in case you are interested,

TAS—THE NORTH

Westhaven also makes YoPet, a drinking yoghurt for animals.
- **Open:** 9am–5pm daily (factory sales only)
- **Location:** South Launceston, part of the PFD Food Distributors complex in Talbot Rd

PACIFIC OYSTERS

Tasmanian Pacific oysters were the first Australian oysters imported live into the United States. This major industry now produces about 3.5 million dozen oysters annually. Oysters are rich in protein, vitamins and minerals and low in fat, kilojoules and cholesterol, a fact which, when coupled with their delightful taste and aroma, make them an almost perfect food.

For centuries oysters have been credited as an aphrodisiac. It's no wonder—true or not, many people have fallen for their charms.

Hatcheries at leases around Tasmania produce these minor miracles with great care and expertise and restaurants throughout the state serve them proudly.

If buying them on the half shell, look for a plump white to creamy coloured meat, with a fresh sea smell. Unopened oysters should be kept chilled until opening, but must be covered and refrigerated after opening and eaten within two or three days.

There is always controversy whether to rinse oysters after opening. Some purists argue that the natural liquor found in the freshly opened oyster is part of the allure. Others argue for washing away grit and shell residue.

As with so many food rites, it hardly matters. The main thing is to experience these delectable specialties of the sea.

🍷 BREWERY/WINERY

J BOAG & SON BREWING LTD
39 William St, Launceston 7250
Phone: 03 6332 6300
Email: info@boags.com.au
www.boags.com.au

Everywhere you go in Tasmania you will see the distinctive Boag sign. Established in 1881, this company set about brewing what they modestly now call 'the best beer in the world' on the banks of the River Esk in Launceston, using Tasmanian hops and barley. In addition, the company makes Boags Strongarm Bitter and Boags Draught, the latter only available in Tasmania. Each weekday, there are one-and-a-half hour conducted tours that include a full circuit of the brewery from the brewhouse to the packaging line, information on the brewing process and Boag's fascinating history. Bookings essential, and fully enclosed footwear must be worn.
- **Open:** 8.30am–5pm weekdays

🏠 STORES

GOURLAY'S SWEET SHOP
Michael and Anita Wood
12 The Quadrant, Launceston 7250
Phone: 03 6331 4053 (shop),
03 6331 4730 (factory)

This business was established in 1896, and remains a family business, turning out all those wonderful old-fashioned goodies you thought you'd never see again such as hard boiled sweets, acid drops, bullseyes, coconut ice, and handmade chocolates and fudges. You can buy from the shop or the factory at 147 Patterson St.
- **Open:** Shop—9am–5.30pm weekdays, 9.30am–2pm Sat, factory—9am–4.30pm Sat, 10.30am–3.30pm Sun (Jan–Apr)

THE NORTH—TAS

THE MILL PROVIDORE AND GALLERY
Alice and Paul Bradbury
Ritchies Mill, 2 Bridge Rd, Launceston 7250
Phone: 03 6331 0777
Email: info@themill.net.au
www.themill.net.au
Opened in 2002, right at the entrance to the Cataract Gorge, this food outlet, Ritchies Mill, is located in Launceston's original flour mill and includes a deli stocking a wide range of Tasmanian cheeses, cooking equipment, smallgoods and olive oils. A handmade 'chocolate cupboard' features mouth-watering treats, and elsewhere are terrines and rillettes made by Calstock guest house's owner and chef, Remi Bancal.
▪ Open: 9am–6.30pm daily

PETUNA SEAFOODS PTY LTD
Peter and Una Rockliff
Wellington St, Launceston 7250
Phone: 03 6344 1876
Email: info@petuna.com
www.petuna.com
The Petuna group of companies operates deep-sea trawling, seafood processing, aquaculture, wholesaling and exporting ventures and has been instrumental in promoting Tasmanian seafood within Australia and overseas. This is Tasmania's largest seafood operator, founded and owned by the Rockliff family who have been involved in Tasmania's fishing industry since 1950. Specialising in fresh fish from Petuna's fishing boats and sea farms, as well as freshly shucked oysters and many other varieties of shellfish and smoked fish. Tasmanian fish available here include fresh, frozen or smoked ocean trout, Atlantic salmon, Petuna saltwater Charr, pink ling, blue eye trevalla, orange roughy, white cod, blue grenadier, coral perch and many more wild fish species. There is also an outlet in East Devonport (phone: 03 6427 9033).
▪ Open: 9am–5.30pm weekdays, 9am–12pm Sat

WURSTHAUS AT OLIVERS
Annie Myers
15 The Quadrant, Launceston 7250
Phone: 03 6331 9171
Email: lmyers@bigpond.net.au
This is the Launceston franchise, established in 2002, of the very popular Hobart Wursthaus (phone: 03 6224 0644). This is a gourmet deli specialising in Tasmanian fresh meats, smallgoods, wines, cheeses and a large range of other Tasmanian products. Perhaps best known for its gourmet sausages, all fresh sausages, meats and smallgoods are made and prepared in the factory in Cambridge, near Hobart, and there's also Tasmanian smoked trout and salmon, cheeses, chutneys, sauces and jams. These come together well in the 'build your own lunch' facility, where customers prepare their own filled roll or salad.
▪ Open: 8.30am–6pm weekdays, 8.30am–3.30pm Sat
● Location: Between Brisbane and St John Streets

DINING

FEE AND ME
Peter Crowe and Fiona Hoskin (chef)
190 Charles St (cnr Charles and Frederick streets), Launceston 7250
Phone: 03 6331 3195
Email: info@feeandme.com.au
www.feeandme.com.au
Licensed
Cuisine: Regional

267

TAS—THE NORTH

Mains average: $$
Specialty: Crayfish and abalone, delicious desserts
Local flavours: Everything is local
Extras: Great views of Prince's Park and surrounding churches at night. Multi-award-winning restaurant.
- Open: Dinner Sat

RED PEPPERS CAFE
Kaye McNaney
Shop 1, Centreway Arcade,
82 Brisbane St, Launceston 7250
Phone: 03 6334 9449
Mobile: 0427 812 914
Email: kayemcnaney@hotmail.com
BYO
Mains average: $
Specialty: Gluten-free, vegan and organic meals, snacks and treats
Local flavours: As much local produce as possible plus organic herbal teas
Extras: Outdoor tables and takeaway available.
- Open: Breakfast and lunch Sat

STILLWATER RIVER CAFE, RESTAURANT AND WINE BAR
Don Cameron (chef)
2 Bridge Rd, Launceston 7250
Phone: 03 6331 4153
Email: stillwater@microtech.com.au
www.stillwater.net.au
Licensed
Cuisine: Modern Australian
Mains average: $$$
Specialty: Local Tasmanian seafood and game with 'a few Asian twists'
Local flavours: Almost everything is from Tasmania
Extras: There are over 400 wines on the wine list, which was awarded top marks in the Tucker Seabrook awards. Outdoor courtyard overlooks the Tamar River and yacht club. A casual cafe by day, fine dining at night. Head chef Don Cameron and his wife Imelda grow beef cattle and have raised venison and rabbits for restaurants in the past. There is an extensive selection of Tasmanian and international gourmet goodies available upstairs at The Mill Providore and Gallery.
- Open: Breakfast, lunch and dinner daily
- Location: At the bottom of Paterson St across from Penny Royal in Ritchies Mill (an 1830s flour mill)

CLASSES/DINING

DRYSDALE INSTITUTE OF TAFE
Narelle Wynwood, Paul Herbig
Cnr Charles and Frankland Streets,
Launceston 7250
Phone: 03 6336 2930, restaurant bookings: 03 6336 2902
Email: Narelle.Wynwood@tafe.tas.edu.au
or Paul.Herbig@tafe.tas.edu.au
www.tafe.tas.edu.au/Drysdale
During the year, short cooking classes and produce demonstration classes are available here. The training restaurant, Drysdale, is open to the public.
- Open: Classes—phone for details, restaurant—breakfast Fri, lunch Thurs–Fri, dinner Wed

LEBRINA
BREWERY/WINERY

BROOK EDEN VINEYARD
Sheila Bezemer
167 Adams Rd, Lebrina 7254
Phone: 03 6395 6244
Email: jojobowen@bigpond.com

Although it is a winery, established in 1993 and now producing fine table wines, Brook Eden also promotes farm-fresh and dried produce and cheese, such as the local Ashgrove Farm Cheese. There are free tastings and you may buy packs of old English cheeses such as Cheshire, tasty Lancashire and Red Leicester. It's no coincidence, of course, that these cheeses superbly complement Brook Eden's own wines.
- **Open:** 10am–5pm daily
- **Location:** On the wine route, well signposted

PERTH

PRODUCERS

TASMANIAN HONEY COMPANY
Julian Wolfhagen
25a Main Rd, Perth 7300
Phone: 03 6398 2666
Email: tashoney@microtech.com.au
www.tashoney.com.au

For twenty-five years, Julian Wolfhagen has been collecting the unique leatherwood honey drawn from the filigree of lacy flowers that, as he expresses it, 'settle like a mantle' across Tasmania's wilderness each summer. His unique cool-technique extraction and packaging process ensures that the delicate floral essences and wild vitality of the soft-candied honey are not destroyed, and now he is working towards organic certification. His leatherwood honey has a purity level as high as ninety-nine per cent compared to the international standard of seventy per cent, and is exported to the United Kingdom, Germany, Korea, Japan, and the United States. Wolfhagen also produces honey fruit spreads, blending meadow honey with fruits such as apricot, apple, strawberry, orange and pineapple, as well as ginger. His honeys (meadow, leatherwood, Christmas bush and tea-tree) are packaged in beautifully decorated cans.
- **Open:** 8.30am–5pm weekdays, 10am–4pm weekends (Dec–Apr)
- **Location:** 16km south of Launceston on the main Hobart highway

PIPERS BROOK

DINING

THE WINERY CAFE
Pipers Brook Vineyard
1216 Pipers Brook Rd, Pipers Brook 7254
Phone: 03 6382 7527
Email: cellardoor@pipersbrook.com
www.pipersbrook.com
Licensed
Mains average: $$
Specialty: Regional vineyard platters
Local flavours: Ocean trout, smoked Rannoch quail, Black Forest game terrine, local vegetables and salad greens, Heidi gruyere cheese, King Island Lighthouse Blue, fresh tomato tarts, freshly glazed local ham
Extras: Uses chutneys and rosemary jelly from Tasmanian makers, and these are also for sale at the cellar door. There is courtyard dining next to the cellar door, which is a delightful place to be on a sunny day. This is Tasmania's largest vineyard and there are twice-daily guided tours.
- **Open:** Lunch, morning and afternoon tea, daily
- **Location:** On route C818 (off B81 and B82), follow wine route signs from Launceston

PIPERS RIVER
🍴 DINING

BAY OF FIRES WINERY CELLAR DOOR
The Hardy Wine Company
40 Baxters Rd, Pipers River 7572
Phone: 03 6382 7622
Email: bayoffires@hardywines.com.au
www.hardywines.com.au
Licensed
Mains average: $$
Specialty: Cheese plates, illy coffee, T2 teas, biscuits and cakes
Local flavours: Tasmanian cheeses
Extras: Cafe-style ambience.
- Open: 10am–5pm daily

PYENGANA
🍎 PRODUCERS

ANCHOR FARM
Terrence Rattray
325 Anchor Rd, Pyengana 7216
Phone: 03 6373 6270
Email: mail@anchororganics.com.au
www.anchororganics.com.au
Terrence Rattray says simply that he grows 'the best potatoes'. Plenty of others seem to think so too. Each year, Rattray digs more than 300 tonnes of specialist organic potatoes as well as growing broccoli, carrots, onions, and pumpkin alongside forests at Pyengana in northeast Tasmania. The Dutch cream potatoes are unique to this business.
- Open: 8am–5pm daily, phone first
- Location: Just over 3km from Pyengana

PYENGANA DAIRY COMPANY
Jon Healey
St Colomba Falls Rd, Pyengana 7216
Phone: 03 6373 6157
Email: pyengana@mail.com
Jon Healey's great-great-grandfather was the local cheese-maker here in 1895. The factory closed down, and the tradition was almost lost until Healey decided to revive the practice, using milk from the farm's Friesian dairy herd, creating traditional stirred-curd cheddar, matured in cloth, and aged from one week to eighteen months. Pyengana Dairy is now one of Australia's largest producers of cloth-bound cheddar cheese and has won many awards. Although this cheese is mature at six months, many people choose to reverently age it for much longer. From January to June each year the Healeys produce a French-style washed rind soft cheese called George. Look for it here, or in cheese rooms and fine delis throughout Australia.
- Open: 9am–5pm daily (10am–4pm daily in winter)
- Location: 2km off the St Helens–Launceston Rd

🍴 DINING

PUB IN THE PADDOCK
Anne Mary Free
250 St Colombia Falls Rd, Pyengana 7216
Phone: 03 6373 6121
Email: pubinthepaddock@bigpond.com
www.pubinthepaddock.com.au
Licensed
Cuisine: Good pub tucker
Mains average: $
Specialty: Homemade kangaroo patties, chicken strips, secret recipe Pub sauce that complements the steak

Extras: There are views of the paddock and its valley from the dining area and this pub has been popular since the 1880s. Takeaway available.
- **Open:** Lunch and dinner daily

ROBIGANA
🍎 PRODUCERS

LYNTON FRUIT AND VEGIE FARM
64 Deviot Rd, Robigana 7276
Phone: 03 6394 4680
Email: lyntonfarm@bigpond.com.au
Ideally located on the wine route, this enterprising place knows that people will also want to stock up on local fruit and vegetables as they taste wines. It specialises in growing spray-free tomatoes, capsicums, and cucumbers and there are also homemade scones with cream and freshly made raspberry jam. The special Berry Delight made year round with raspberries, blackberries, strawberries, silvan berries and blueberries is, in their words, 'just berry berry delightful'. Can't get to the farm? Look for the produce in the greengrocer's shop in Legana Grove shopping centre, Legana.
- **Open:** 9am–5.30pm weekdays, 9am–5pm weekends
- **Location:** On the wine route around the scenic Tamar River, follow the blue tourist signs

ROCHERLEA
🍎 PRODUCERS

LENAH GAME MEATS
John Kelly and Katrina McKay
315 Georgetown Rd, Rocherlea 7248
Phone: 03 6326 7696
Email: lenah@bigpond.net.au
John Kelly says that 'wallaby is the veal of kangaroo'. It has a finer texture, sweeter flavour and lighter colour than kangaroo, and he believes it is nature's finest meat. Lenah Game Meats (lenah is the local Aboriginal word for wallaby) harvest wallabies that are less than three years of age and which have been raised grazing on lush pastures. He also sells possum, as well as venison, to the retail and restaurant trade, and wholesales a wide range of game meats. The company believes that the careful harvest of native meats could lead to a more sustainable meat industry.
- **Open:** 8am–6pm weekdays

ROSEVEARS
🍴 DINING

DANIEL ALPS AT STRATHLYNN
Daniel Alps, chef
Pipers Brook Wine, 95 Rosevears Dr, Rosevears 7277
Phone: 03 6330 2388
Email: strathlynn@pbv.com.au
Licensed
Cuisine: Regional modern Australian
Mains average: $$
Local flavours: Mostly organic and local, including fruit, vegetables, meat
Extras: Outdoor dining on a deck with views over the vineyard. Regular cookery classes with Daniel Alps or guest chefs.
- **Open:** Lunch daily, dinner by arrangement

ST MATTHIAS VINEYARD
Moorilla Estate, 113 Rosevears Dr, Rosevears 7277
Phone: 03 6330 1700
Email: stmatthias@moorilla.com.au

TAS—THE NORTH

www.moorilla.com.au
Licensed
Cuisine: Modern Australian
Mains average: $
Specialty: Gourmet barbecues
Local flavours: All Tasmanian fresh produce and premium wines
Extras: Gourmet platters to eat outdoors and stunning views of the upper reaches of the Tamar. Functions and wedding receptions are welcomed by appointment. On Sunday there are gourmet barbecues and local jazz artists on the deck.
- **Open:** 10am–5pm daily, functions by appointment
- **Location:** 2km from Legana on C733

ROSS
STORES

THE ROSS VILLAGE BAKERY
Chris and Dot Lloyd-Bostock
15 Church St, Ross 7209
Phone: 03 6381 5246
Email: rossbakery@vision.net.au
www.rossbakery.net.au
In this 150-year-old bakery, the Lloyd-Bostocks have been baking sourdough breads on the bricks of the original oven, and producing handmade gourmet pies and traditional pastries and cakes for the past nine years. The sourdoughs—a French levain, a rye, and a cornbread—are made with 'proper sourdough starters' begun nine years ago. Then there are the seafood pies (scallops and fish), fisherman's pasties (scallops, fish in a curry sauce), lamb mint and pea pies, and Eccles cakes. Just to keep them busy, they also run a B&B and guesthouse alongside the bakery.
- **Open:** 9am–5pm daily

SCAMANDER
PRODUCERS

EUREKA FARM
Ann Calder and Denis Buchanan
89 Upper Scamander Rd, Scamander 7215
Phone: 03 6372 5500
Email: dlbuchanan@vision.net.au
BYO
Mains average: $
Specialty: Summer puddings
Local flavours: Everything is made on the premises, including natural fruit ice-cream which won the award for Tasmania's best ice-cream in 2003
Extras: Outdoor terrace with seats under the trees and a log fire for cold days. The orchard grows raspberries, strawberries, brambleberries, stone fruits and apples and you can buy fresh fruit from the produce store near the cafe. Sells a huge range of award-winning jams, sauces and pickles, which are made from own fruit. Takeaway available.
- **Open:** 8am to 6pm daily (Oct–June)
- **Location:** 1km along Upper Scamander Rd, 3km south of Scamander

ST HELENS
PRODUCERS

SALTY SEAS
Anita Paulsen
16 Esplanade, St Helens 7216
Phone: 03 6376 1252
Winner of a Jaguar Award for Excellence in 2002 in the primary produce category, this business, which commenced in 1998 as a youth employment project, now sources and processes a huge range of seafood products for restaurants in

Sydney and Melbourne. There are wild, dive-harvested clams, periwinkles, Angasi oysters, wild mussels, giant Pacific oysters, cultured Pacific oysters, scalefish (including mowong and wrasse), Tasmanian stripey trumpeter (which you can pick live from the tank to have filleted to order), garfish and flathead. Live fish go direct to Chinese restaurants in Sydney and Melbourne each week, as well as giant oysters (up to thirty centimetres in length). 'The secret ingredient', says owner Anita Paulsen, 'is Tasmania's pure waters'. She buys everything straight from the boats or local divers, and the public can buy here too. She will even cook crays to order over summer.
- Open: 9am–4pm weekdays (year round), 9am–4pm daily (Jan–Feb)

DINING

THE CAPTAINS CATCH
Roderick Alan and Irena Faulkner
Marina Pde, St Helens 7216
Phone: 03 6376 1170
BYO
Cuisine: Local seafood
Mains average: $
Specialty: Blue-eye trevalla (blue-eye cod) and chips, local fresh crayfish
Local flavours: Blue-eye, crayfish, oysters
Extras: The fish and seafood are collected fresh from the boats as they come in and oysters are harvested daily and shucked to order. There is also a live crayfish tank with crays cooked fresh each morning. Takeaway available.
- Open: Lunch daily
- Location: Only premises on St Helens Wharf

SWANSEA
DINING

KABUKI BY THE SEA
Terry Lanning and Mitsuo Nakanishi
Rocky Hills, Tasman Hwy, Swansea 7190
Phone: 03 6257 8588
Email: rockyhills@vision.net.au
www.kabukibythesea.com.au
Licensed
Cuisine: Japanese
Mains average: $$
Local flavours: Wallaby, abalone, oysters, smoked Atlantic salmon
Extras: Perched on sea cliffs overlooking Great Oyster Bay and the Hazards Range, near Swansea on the east coast, with accommodation for those who can't tear themselves away. Takeaway available.
- Open: Lunch, morning and afternoon tea daily, dinner during the tourist season
- Location: 12km south of Swansea

THE LEFT BANK COFFEE AND FOODBAR
Subi Mead and Helen Bain
7 Maria St, Swansea 7109
Phone: 03 6257 8896
Email: theleftbank@tassie.net.au
BYO
Cuisine: Modern Australian
Mains average: $
Specialty: Lemon tart, all-day breakfast or lunch
Local flavours: As much Tasmanian produce as possible
Extras: Outdoor courtyard area within garden setting, good coffee.
- Open: 9am–5pm Wed–Mon (closed Wed May–July, closed Aug)
- Location: Look for the red door in the main road

WESTBURY

🏠 STORES

ANDY'S BAKERY AND CAFE
Andy Oliver
45 Bass Hwy, Westbury 7303
Phone: 03 6393 1846
Mobile: 0417 583 673
Email: andyscom@bigpond.com
www.andystasmania.com
Andy's travels in thirty-six countries over thirty years have given him plenty of ideas for baking, and the large range of breads, including sourdoughs, plus a yoghurt bread made with culture, shows he paid attention to what he saw. The result is award-winning products, and what some say is the best-ever ciabatta. The open door policy (Andy's never closes) and homemade ice-cream using Tasmanian berries ensure a faithful following of happy customers.
- **Open:** All day every day

YORKTOWN

🍎 PRODUCERS

YORKTOWN ORGANICS
Clare and Bruce Jackson
120 Bowen Rd, Yorktown 7270
Phone: 03 6383 4624
Email: yorganics@dodo.com
Look for the Scarecrow label at Salamanca Markets, or Grand Central Station and Ye Olde Greengrocer in Launceston. Or come here and see how good freshly harvested berries and vegetables can be. This lovely NASAA-certified organic property grows strawberries, raspberries, brambleberries, salad greens, cherry tomatoes (in a large hothouse), cucumbers, capsicums, squash and many other vegetables. This is intensive organic farming at its peak.
- **Open:** 9am–5pm weekdays, 9.30am–12pm weekends
- **Location:** 8km from Beaconsfield towards Tamar River mouth

MARKETS AND EVENTS

◆ **DELORAINE SHOWGROUNDS MARKET,** DELORAINE
03 6369 5321
9am–1pm first Sat (Jan–May)

◆ **FARMERS' MARKET,** DELORAINE
03 6369 5321
9am–1pm third Sat (Jan–May)

◆ **FESTIVALE,** LAUNCESTON
03 6323 3152
Feb–Mar

◆ **GREAT ABALONE BAKE-OFF,** BINALONG BAY
03 6376 8335
First Sat in Feb

THE SOUTH

Hobart reigns supreme in this most southern region, yet the city relies on the rich surrounding agricultural lands to provide so much of its needs. Most Tasmanians would not agree that the 'apple isle' tag is truly descriptive today, but the Huon region still produces much of Tasmania's apple crop.

Even here, though, there has been growth and change, and now a berry trail meanders through the hills, there is cider and more wines, honeys, meats, herbs and stone fruit. Even saffron, that gold-standard of condiments, is grown, and there are jams and ice-cream, seafoods, wakame and muttonbird.

Somehow, all the foods you could ever need have been stocked here in one corner of one island in an out-of-the-way latitude. A sort of cold larder for Australia.

BERRIEDALE
DINING

MOORILLA ESTATE WINERY RESTAURANT
655 Main Rd, Berriedale 7011
Phone: 03 6249 2949
Email: wine@moorilla.com.au
www.moorilla.com.au
Licensed
Cuisine: Modern Australian
Mains average: $$
Specialty: Food and wine matching
Local flavours: Fish, mushrooms, apples, beef

Extras: Museum of antiquities, cellar door, outside dining with vineyard and water views.
- **Open:** Lunch daily

BIRCHS BAY
PRODUCERS

GRANDVEWE CHEESES
Diane Rae and Alan Irish
59 Devlyns Rd, Birchs Bay 7162
Phone: 03 6267 4749
Email: diane_alan@bigpond.com
www.grandview.au.com
Grandvewe claims to be Australia's only organic sheep dairy, and there is also a cellar door with three hectares under vines. You can see this certified organic farm in action, including the milking of one hundred East Friesland sheep twice daily. Cheeses include pecorino, Blue-By-Ewe and Primavera, and you may buy them direct from the cheesery door.
- **Open:** 10am–5pm daily
- **Location:** 2km south of Woodbridge look for the tourist signs to the Farmhouse Cheesery

DINING

THE PENGUIN CAFE ON BRUNY
Mary Richards and Martin Watson
710 Adventure Bay Rd, Adventure Bay, Bruny Island 7150
Phone: 03 6293 1352
Email: penguincafe@iprimus.com.au
www.dharamsalanet.com/penguincafe
Occasional licence/BYO

TAS—THE SOUTH

Cuisine: Contemporary Australian
Mains average: $
Specialty: Salmon burgers, cottage pies, apple crumbles, salads
Local flavours: Tasmanian cheese, organic berries, salads, fruit juice
Extras: International theme nights, garden area with views of Storm Bay and Tasman Peninsula. Gourmet dinner or picnic hampers by order.
■ Open: Breakfast and lunch daily, dinner Sat

THE HOTHOUSE CAFE
Stuart and Liz Bennett
Morella Island Retreat, 46 Adventure Bay Rd, Bruny Island 7150
Phone: 03 6293 1131
Email: fun@morella-island.com.au
www.morella-island.com.au
Licensed
Cuisine: Contemporary country
Mains average: $
Specialty: Wallaby curry
Local flavours: Oysters, salmon, fish, berries
Extras: Grows own oysters. The cafe is really a large greenhouse in the middle of a huge garden, with wide views of the area. There are five self-contained holiday units.
■ Open: Breakfast and lunch daily
● Location: Morella is a thirty-minute drive south of Hobart on the main Bruny Rd

CAMBRIDGE

🍎 PRODUCERS

BARILLA BAY TASMANIA
David Forrest
1388 Tasman Hwy, Cambridge 7170
Phone: 03 6248 5458
Email: sales@barillabay.com.au
www.barillabay.com.au
In 1980 the Forrest family was looking for an ideal oyster-farming location and chose Barilla Bay because of its cold, pristine waters, tidal water replacement and proximity to the airport. Their oyster hatchery pioneered the single-seed oyster technique which, by attaching the oyster to a single grain of sand, creates a more evenly shaped oyster. Today they supply live or split and packed oysters to local and mainland markets, as well as exporting to the United States. You can watch your oysters being opened, or buy them live. Barilla will even pack them appropriately for you if you are flying out, and can deliver to the airport. Oyster meat and native Angasi oysters are available occasionally. There is also a recently opened restaurant with a menu emphasising local seafood.
■ Open: 7.30am–6pm weekdays, 7am–7.30pm weekends, restaurant—breakfast Wed–Sun, lunch and dinner daily
● Location: Twenty minutes south-east of Hobart, five minutes from the airport, just beyond the airport turn-off

ISLAND OLIVE GROVE
Wendy Roberts
222 Denholmes St, Cambridge 7170
Phone: 03 6248 5432
Mobile: 0409 232 250
Email: wendy@islandolivegrove.com
www.islandolivegrove.com
You wouldn't be able to see Wendy Roberts if she wore all the ribbons she has won, which include one for Grand Champion in the National Fine Food Awards, and she is justly proud of these achievements. Island Olive Grove is a ten-year old olive farm producing superb

olive products, including antipasto, tapenades and olive oils, which are sold in gourmet outlets throughout Australia and as far away as Harvey Nichols in London. Roberts also makes Tasmania's only pink verjuice. You can buy the products directly (by appointment) or order via the website. They expect to open a major visitors centre at the end of 2004.

- Open: By appointment

BREWERY/WINERY

TASMANIA DISTILLERY
Patrick Maguire
1/14 Lamb Place, Cambridge 7170
Phone: 03 6248 5399
Email: tasdistillery@iprimus.com.au
www.tasdistillery.com.au

Australia's most historic whisky distillery has recently moved some 5km from Hobart down the Richmond Road to Cambridge, but still incorporates the distillery and the Steam Packet Preserving Co. The old site, in what used to be Hobart's red-light district, is still open to the public for free, self-guided tours. The new premises include a large open area where visitors are able to observe the making of whisky and ask questions about the process. There is also a museum display and coffee shop. The attractive gift shop continues to sell a wide range of goods. Look for liqueured jams, such as marmalade and whisky, hot whisky mustard, liqueured fruits, leatherwood honey and tea-tree honey. All foods are made without artificial colourings, flavourings or preservatives.

- Open: 9am–5pm weekdays
- Location: Five minutes from Hobart on the Richmond Rd

DINING

COAL VALLEY VINEYARD
Gillian Christian and Todd Goebel (owners) Justin Harris (chef)
257 Richmond Rd, Cambridge 7170
Phone: 03 6248 5367
Email: coalvalleywine@bigpond.com
www.coalvalley.com.au
Licensed/BYO
Cuisine: Modern Australian
Mains average: $$
Specialty: Duck confit; ocean trout from the west coast of Tasmania
Local flavours: Ocean trout, Rannoch quail, Nichols free-range chicken (from north-west Tasmania)
Extras: Breathtaking water views across the vines can be enjoyed whether dining inside or sipping wine on the terrace.

- Open: Lunch daily, dinner by appointment (closed June)

MEADOWBANK ESTATE VINEYARD AND RESTAURANT
Gerald Ellis
699 Richmond Rd, Cambridge 7170
Phone: 03 6248 4484
Email: bookings@meadowbankwines.com.au
www.meadowbankwines.com.au
Licensed
Mains average: $$
Specialty: Daily, fresh market fish
Local flavours: As much as possible, including Barilla Bay oysters, Tasmanian salmon (cured at Meadowbank), Eleni's yoghurt cheese, Tasmanian asparagus, Huon Valley mushrooms (shiitake, oyster, honey browns), Doo Town venison, South Arm pink-eye potatoes, Thorpe Farm goats cheese, King Island cheese, Rannoch quail and quail eggs, Ashbolt

olive oil, Wursthaus meats, Spring Bay scallops
Extras: Takeaway picnics available. There is an outdoor terrace with market umbrellas, a boules court, art gallery, meeting and function facilities, disabled access, playground, and vineyard views. Sells homemade aioli in the shop and there are plans for more Meadowbank food merchandise.
■ **Open:** 10am–5pm daily, lunch daily
● **Location:** Turn left at Cambridge primary school and Meadowbank is 4km further

CLAREMONT
TOURS

CADBURY FACTORY TOURS
Cadbury Rd, Claremont 7011
Phone: 03 6249 0333; Freecall: 1800 627 367
To anyone growing up in Australia, the name Cadbury evokes memories of wonderful chocolate. Many do not realise, though, that it is a Tasmanian operation, firmly anchored to the lush banks of the Derwent River since 1921. That glass and a half of full-cream milk we all heard about from an early age, actually came from Tasmanian cows. You can visit or buy from Cadbury on a tour, and because this is a working factory, make sure you wear enclosed shoes, no jewellery, suitable clothes and are able to climb stairs. Leave the cameras behind too. There is a short introductory movie and then the fifty-minute tour takes visitors through the production area of the factory to learn the process of chocolate making. Included in the tour is a chocolate sample bag and access to the staff wholesale chocolate shop where you can shop till you drop. You may also arrive on a river cruise (you'll need your cameras for this) that leaves Wrest Point Jetty at 10am weekdays to coincide with a special 11.15am tour. For details phone 03 6234 9294.
■ **Open:** Tours at 9am, 9.30am, 10am and 1pm weekdays (also 8am at peak times, according to demand)

CYGNET
PRODUCERS

STEENHOLDT'S ORGANIC PRODUCTS
Christopher and Paula Steenholdt
Petcheys Bay, Cygnet 7109
Phone: 03 6295 0141
There are over one hundred varieties of apples and pears at this NASAA certified organic orchard. These are the highly flavoured, best-lasting types, and are sold fresh at Salamanca Markets each week or from the farm gate by appointment.
■ **Open:** By appointment

DINING

RED VELVET LOUNGE AND FIREBIRD BAKEHOUSE
Juniper Shaw and David Roberts
24 Mary St, Cygnet 7112
Phone: 03 6295 0466
Email: redvelvet@southcom.com.au
BYO
Cuisine: Mainly vegetarian
Mains average: $
Specialty: Vegetarian dishes
Local flavours: Tofu, goats cheese, organic vegetables, berries, organic wood-fired bread baked on premises
Extras: Breads are baked in 1912 wood-fired oven. Grows own herbs and seasonal

TASMANIA—WHAT'S NEXT?

Tasmania is such a prolific producer that the rest of the country could excuse it from trying any harder. But there seem to be no bounds to the innovation and energy displayed by this state. Consider these industries poised to take off, virtually as we speak. All are ideally suited to Tasmania's gentle climate and rich soil, but their attraction to overseas markets, particularly, is based on the state's 'clean green' image and record of efficiency in the produce it has established to date.

Buckwheat crops supply the Japanese soba noodle market.

Cape Barren geese are being open-range reared on Flinders Island.

Emu production is developing fast with the largest farm situated on the Tasman Peninsula.

Fallow deer have been introduced to King Island on a trial basis.

Kabocha (Japanese squash) is in great demand during winter in the northern hemisphere.

Olives are being trialled for planting.

Wallaby and bush possums are also being harvested for the gourmet meat market.

Walnuts have been planted on the east coast.

What else? Trial crops of amaranth, quinoa, wasabi, ginseng, medicinal herbs and green tea have been established. One of the largest Japanese quail breeding operations in the southern hemisphere is located in Tasmania. If you are using lavender, blackcurrant, boronia, fennel, dill, or parsley oils, they may have come from northern Tasmania as it is the main producer of essential oils in Australia.

vegetables. Makes jams, chutneys, pesto, dips and bread for sale.

■ **Open:** 9am–6pm daily, Firebird pizzas from 6pm Wed–Sun

DUNALLEY

🍎 PRODUCERS

DUNALLEY FISH MARKET
Bruce Chambers
Denison Canal, 11 Fulham Rd, Dunalley 7177
Phone: 03 6253 5428
Stop in and see a working fish processing plant or purchase fish straight out of the ocean. 'Fresh, live if possible' is the slogan of this market. Crayfish are cooked daily, and there are lobsters live or freshly cooked, oysters opened to order plus abalone. If necessary, the market will pack your purchases for travel by car or plane.

■ **Open:** 9am–5.30pm daily
● **Location:** Off Hwy C334, on the way to Port Arthur Historic Site and Tasman Peninsula

🍴 DINING

DUNALLEY WATERFRONT CAFE
Natasha Smyth and Barbara McEvoy
4 Imlay St, Dunalley 7177
Phone: 03 6253 5122
Email: nsmyth@iprimus.com.au
Licensed
Cuisine: Modern Australian
Mains average: $
Specialty: Sweet potato, pumpkin, and cashew nut burger served with sour cream and sweet chilli sauce
Local flavours: As much as possible, including Tasmanian wine and cheese
Extras: All day breakfasts, large serves, value for money and a great view over-

TAS—THE SOUTH

looking a working fishing jetty. Makes cakes, cookies, salad dressings, and tomato relish for sale. Takeaway available.
- **Open:** 10am–4pm daily (sometimes longer)
- **Location:** Follow the signs and once you enter Dunalley, turn off just before the school

FLOWERPOT
🍎 PRODUCERS
DAN AND JUDE'S HERITAGE ORCHARD
Daniel Puller and Judith Marshall
10 Beadles Rd, Flowerpot 7163
Phone: 03 6292 1611
Specialising in growing heritage apple varieties for the past nine years, this is the place to come for Spartan, Sturmer, Geeveston, Fanny and more. This couple sell the fruits of their labours through local markets or from their roadside stall.
- **Open:** Phone to arrange a farm tour
- **Location:** 5km south of Woodbridge

GARDNERS BAY
🍷 BREWERY/WINERY
HARTZVIEW VINEYARD
Robert and Anthea Patterson
70 Dillons Rd, Gardners Bay 7112
Phone: 03 6295 1623
Email: hartzviewwine@trump.net.au
www.hartzview.com.au
This vineyard, established in 1988, specialises in grape varieties such as pinot noir and chardonnay. However, a large focus is on the unique range of fruit ports and liqueurs produced here—blackberry, blackcurrant, cherry and blueberry—from fruit grown in the Huon Valley. These are also available in a number of speciality outlets around Tasmania. The property has spectacular views and outdoor seating—the ideal place to enjoy a tasting of Pig & Whistle gooseberry port and a cheese platter. Why gooseberry? 'We had too many, and had to do something with them,' say the Pattersons. You'll be glad about this.
- **Open:** 9am–5pm daily
- **Location:** Turn left at the Woodbridge police station

GLAZIERS BAY
🍎 PRODUCERS
TAS-SAFF
Terry and Nicki Noonan
156 Dillons Hill Rd, Glaziers Bay 7109
Phone: 03 6295 1921
You need around 140 000 red tipped stigmas from *Crocus sativus* to make up a kilogram of saffron, and although it should last a while, it will cost around $25,000 and Tas-Saff's annual crop is so compact that it would fit into a car glove box. The Noonans began growing crocuses in 1994 but it was around three years before they could begin harvesting their precious crop. Patience-plus is needed to hand-pick stigmas from this fragrant lilac flower, which is worth more than its weight in gold. The Noonans painstakingly hand-pick, dry, package and then market this most expensive of extras. Today they—together with a network of around thirty saffron growers in Tasmania, Victoria and southern NSW—supply the Australian domestic market and their saffron is available around the country.
- **Open:** By appointment

THE SOUTH—TAS

GLEN HUON
🍎 PRODUCERS

HUON VALLEY MUSHROOMS
Michael Brown
850 Main Rd, Glen Huon 7109
Phone: 03 6266 6333
Email: hvm@ozemail.com.au
If you are a fungi fan, this is your territory. Look for button, honey brown, black velvet wood ear, pink coral oyster, golden oyster, king oyster, shiitake, maitake and shimeji here. Although this is a serious operation producing a variety of types of cultivated mushrooms for markets all over Australia, you may buy direct and, unlike many other mushroom farms, they also offer tours by appointment.

- **Open:** 8am–5.30pm weekdays, tours by appointment on weekends and public holidays

KNOW YOUR APPLES

Golden Delicious: This mid-season standard, originally from the United States, has a greenish-yellow skin that becomes golden with an occasional pink blush, and crisp cream flesh with a sweet flavour.

Granny Smith: A very late-season apple, with green to greenish-yellow skin, crisp greenish flesh and juicy refreshing flavour.

Red Delicious: From Iowa in the United States, this mid- to late-season apple has a greenish-yellow skin with red stripes, although it may be completely crimson to dark red. The flesh is firm, crisp and creamy-white and it is sweet, juicy and highly aromatic.

Jonathan: Also from the United States, this early season fruit has a pale green skin with scattered red to blushed stripes over most of the apple. It has a crisp, white flesh and is fairly juicy and sweet.

Sundowner: A late maturing, medium-sized apple with red to green skin and sweet flavour.

Fuji: In season April to September, this is the most popular apple in Japan, China and Korea. Excellent eating quality with firm crisp flesh and blushed dull pink or red skin.

Pink Lady: A cross between Golden Delicious and Lady Williams, this is a late season apple with green-yellow skin with pink or red blush. It has a crisp dense flesh and a flavour like Golden Delicious.

Gala: A New Zealand cross between Golden Delicious and Kidds Orange Red, appears early in the season—around February. The skin may have a pale or much darker blush, but the flesh is cream, sweet and crisp.

Minor Varieties

Tassie Snow: Thin skin with whitish background, sweet and subtle flavour.

Democrat: Greenish-yellow skin with a dull flush on one side, firm, juicy white flesh.

Bonza: Green or cream skin with bright red stripes, firm, sweet white flesh.

Sturmer: Bright green skin flushed a purplish brown, juicy, firm, greenish flesh.

Legana: Yellow skin with deep crimson blush, crisp, yellow-green flesh.

Crofton: Pale yellow skin with crimson stripes, firm, white flesh.

Mutsu: Deep green to golden skin, depending on maturity, firm crisp flesh.

Cox's Orange Pippin: Golden-yellow skin flushed with red, tender, crisp yellow flesh.

GROVE

🍎 PRODUCERS

DORAN'S FINE FOODS
Mike Swinburne and Bill Button
Pages Rd, Grove 7109
Phone: 03 6266 4377
Email: dorans10@bigpond.com
www.doransjams.com
Doran's, Australia's oldest jam company, was established in 1834 and is still making jams and juices that are deservedly popular. Made from fresh raspberries, blackcurrants, cherries and other fruit, they are an ideal gift or souvenir. When you've finished browsing, you can relax in the lovely visitors centre that houses the museum and JJ Cafe, which serves snacks using Doran's products, of course.
- **Open:** 10am–4pm daily, groups and tours welcome
- **Location:** Off the A6

HUONVILLE

🍎 PRODUCERS

MINNUCCI'S GREEN GOLD
Attilio and Vera Minnucci
Cygnet Rd, Huonville 7005
Phone: 03 6223 5002
www.melvillehouse.com.au
Attilio Minnucci produced Tasmania's first olive oil in 1982, then cold pressed oil in 1993, to carry on his family tradition in Italy that dates back 500 years. Using only Tasmanian olives, the oil is available from selected Hobart stores.
- **Open:** By appointment

🍴 DINING

HUON MANOR BISTRO
Ray and Trudi Griffiths
1 Short St, Huonville 7109
Phone: 03 6264 1311
Email: huonmanor@bigpond.com
Licensed
Cuisine: Modern Australian
Mains average: $$
Specialty: Huon Manor seafood chowder
Local flavours: Salmon, berries, venison, mushrooms
Extras: The outside deck area overlooks a delightful garden setting with views to the Huon River.
- **Open:** Lunch Sun–Fri, dinner Sat
- **Location:** On the right-side approach to the bridge when coming from Hobart

---

THE HUON VALLEY APPLE AND HERITAGE MUSEUM
2064 Main Rd, Grove 7109
Phone: 03 6266 4345
Chock-a-block with donated items from descendants of the early orchardists and settlers, this is a must-see. There are old presses, graters, and antique apple peeling and coring machines, plus apple cookery demonstrations. Each March there is a display of 500 different varieties of apples, grown by the Department of Primary Industries and Fisheries experimental station, which is at the rear of the property. If you think that's a lot of different sorts of apples, England has 6000 registered varieties.
- **Open:** 9am–5pm daily (Sept–May), 10am to 4pm daily (June and Aug), closed July
- **Location:** On Grove Strait (A6), twenty-five minutes from Hobart

KETTERING
🍴 DINING

OYSTER COVE INN
Greg James
1 Ferry Rd, Kettering 7155
Phone: 03 6267 4446
Email: oyster.cove@tassie.net.au
Licensed/BYO
Cuisine: Modern Australian seafood
Mains average: $$
Specialty: Baked Atlantic salmon
Local flavours: Salmon, oysters, berries, cheeses
Extras: The inn was built by a British plantation owner, resident in Sri Lanka, as a cool escape to the Antipodes. There is an outdoor dining area overlooking the bay and marina, and a collection of wooden sculptures. Takeaway available.
- Open: Lunch and dinner daily
- Location: Twenty-five minutes south of Hobart on the D'Entrecasteaux Channel

LYMINGTON
🍎 PRODUCERS

TASSIE BLUE
Will Brubacher
Lymington Coast Rd, Lymington 7109
Phone: 03 6295 1005
Email: joereeves979@hotmail.com
You will find blue, blue blueberries here—the fruit of several high bush, cool-climate varieties, which are for sale in punnets by the kilo in season. Out of season there are frozen ones. The fruit is sold at the farm gate as well as to Hobart shops, bakers and ice-cream manufacturers.
- Open: 7am–6pm daily (Jan–Feb)
- Location: 9km south of Cygnet

POLICE POINT
🍴 DINING

EMMA'S CHOICE JAM FACTORY AND TEAROOMS
John Tibuliac
Esperance Coast Rd, Police Point 7116
Phone: 03 6297 6309
Email: emmaschoice@optusnet.com.au
BYO
Mains average: $
Specialty: Devonshire teas
Local flavours: Jams, jellies
Extras: Lovely orchard location. Michael Palin visited once and wrote in the visitors book 'the best situated tearoom in the world'. Mail orders available.
- Open: 10am–5pm Sun–Fri
- Location: South of Hobart, 55km before the 'end of Australia', good signage

PORT ARTHUR
🍴 DINING

FELONS
Port Arthur Historic Site Visitor Centre, Arthur Hwy, Port Arthur 7182
Phone: 03 6251 2314
Email: bookings@portarthur.org.au
www.portarthur.org.au
Licensed
Cuisine: Asian, Mediterranean, seafood
Mains average: $$
Specialty: Tasman Peninsula produce
Local flavours: Rannoch quail, crayfish from Dunalley, venison from Doo Town, Eaglehawk Neck fresh fish, Barilla Bay oysters, Yaxley Estate wines from Copping
Extras: Located in the beautiful grounds of Port Arthur Historic Site.
- Open: Dinner daily

RANELAGH

🍴 DINING

HOME HILL WINERY RESTAURANT
Terry and Rosemary Bennett (owners)
Craig Scarfe (head chef)
38 Nairn St, Ranelagh 7109
Phone: 03 6264 1200
Email: homehill@bigpond.com
www.homehillwines.com.au
Licensed
Cuisine: Modern Australian
Mains average: $$
Specialty: Confit of duck leg and grilled breast
Local flavours: Atlantic salmon, Huon Valley mushrooms, Snowy Range trout, saffron, beef, lamb, Bruny Island oysters, Oyster Cove buffalo, berries, cherries, apples, some fruit and vegetable suppliers are just a few paddocks away
Extras: Outdoor terrace overlooking the vineyard, and magnificent views of Mount Wellington and Sleeping Beauty. Winery/restaurant is made from rammed earth and glass. Winner of Best Tourism Restaurant, 2003. Sells Home Hill pinot noir chocolates, apple jelly, marmalade, Berries of the Valley jam, apricot jam, herb and chilli oil and tomato sauce.
- **Open:** Lunch daily, dinner Fri–Sat
- **Location:** Turn off on Lollara Rd when you see the Home Hill sign

RICHMOND

🏠 STORES

THE TASTING HOUSE
4/50 Bridge St, Richmond 7025
Phone: 03 6260 2050
Email: gavshaw@bigpond.com

Eighth-generation locals would have to be the best people to guide visitors to local produce. Think of it as a one-stop Tasmanian produce trail, although it's really a retail outlet selling and promoting only Tasmanian-made food, beverages and natural personal care products. This helps everyone as many of the producers do not have farm gate or cellar door access let alone mainland Australia or overseas market exposure. Look for local grape and fruit wines, ciders, meads, ports, liqueurs, spirits and whisky, jams, sauces, relishes, mustards, olives, mushrooms, octopus, cheese, olive oil, vinegar, dressings, honey, pesto, biscuits, lavender products, hops, fudge, quince paste and native bush spices.
- **Open:** 10.30am–4.30pm daily (closed Wed in winter)
- **Location:** In the courtyard behind Sweets and Treats (lollies) and the Richmond bakery

🍴 DINING

RICHMOND ARMS HOTEL
Clay Ackroyd
42 Bridge St, Richmond 7025
Phone: 03 6260 2109
Email: richmondarms@bigpond.com
www.richmondarms.com
Licensed
Cuisine: Pub food
Mains average: $
Specialty: Daily specials
Local flavours: Oysters, seafood, vegetables, meat, wine
Extras: Outdoor dining and a beer garden. Takeaway available.
- **Open:** Lunch and dinner daily

RICHMOND FOOD AND WINE CENTRE
Mark, Mary and Willie Black
27 Bridge St, Richmond 7025
Phone: 03 6260 2619
Licensed
Mains average: $$
Specialty: Tasmanian seafood and game
Extras: Alfresco dining and beautiful function room. Accommodation available. Willie's love of Tasmanian fine foods has been the driving force behind the establishment of this centre, aided by Mark's equal enthusiasm for the island's wines.
■ **Open:** 8am–5pm daily, dinner Wed–Sat

SNUG

🛒 STORES

SNUG BUTCHERY
Darren Martin and Robert Wallace
2203 Channel Hwy, Snug 7054
Phone: 03 6267 9127
Email: martel@iprimus.com.au
Smoked products, especially Atlantic salmon (from Aquatas) and chicken are the specialty of this business about 20km from Hobart. The salmon is filleted and smoked on the premises and the products are only available here and at R&D Meats in Cygnet.
■ **Open:** 6am–6pm weekdays

SORELL

🍎 PRODUCERS

SORELL FRUIT FARM
Elaine and Bob Hardy
180 Pawleena Rd, Sorell 7172
Phone: 03 6265 2744
Email: info@sorellfruitfarm.com
www.sorellfruitfarm.com

Think of this as 'horticultural tourism'— you can pick your own fruit from a choice of around fifteen varieties, or take a guided tour of the orchard, taste Sorell Fruit Farm's wines, liqueurs and other fruit products, then enjoy a light lunch or snacks in the cafe. Operating for fifteen years, this farm has temperate climate conditions, which are ideal for growing completely spray-free strawberries, tayberries, raspberries, boysenberries, loganberries, blackcurrants, cherries, apricots, peaches, nashi pears and apples.
■ **Open:** 8.30am–5pm daily (late Oct–June), groups by appointment at other times
● **Location:** 2km from Sorell, off the Port Arthur Rd

🍴 DINING

BLUE BELL INN
Marlene Gooding
26 Somerville St, Sorell 7172
Phone: 03 6265 2804
Email: bluebell@trump.net.au
www.rcat.asn.au/bluebell
Licensed/BYO
Cuisine: Australian with a Polish influence
Mains average: $$
Specialty: Scallop chowder, venison, lamb, Polish baked cheesecake, chocolate sauce pudding
Local flavours: Barilla oysters, Spring Bay scallops, Doo Town venison, Tasman blue mussels, Flinders Island lamb, Sorell berries, live crays on request
Extras: Located in an historic coaching inn, which has a cosy ambience with open fires during winter. Also a local artist's gallery.
■ **Open:** Breakfast daily, lunch by appointment, dinner Thurs–Tues

SWANSEA
🍎 PRODUCERS

KATE'S BERRY FARM
Catherine Bradley
Addison St, Tasman Hwy, Swansea 7190
Phone: 03 6257 8428
Email: berry@vision.net.au
'Just follow the other ten million cars—I think I've created a monster!' says Kate when asked how to locate her berry farm. The 'monster' has been twelve years a-growing and has developed a clientele who come for Devonshire teas, cakes, coffee, berries, jams, sauces, ice-cream and wine. Kate's main event is producing organically grown strawberries, raspberries, Himalayan blackberries and youngberries, and making jams and sauces from them at her farm overlooking Great Oyster Bay. Seascape strawberry wine is popular and you can find it here or by mail order.
- **Open:** 8.30am–6pm daily
- **Location:** 2km south of Swansea on the Tasman Hwy, well signposted

TARANNA
🍴 DINING

THE MUSSEL BOYS CAFE GALLERY
Susan Richards
5927 Arthur Hwy, Taranna 7184
Phone: 03 6250 3088
Licensed (Tasmanian wine and beer)
Takeaway
Cuisine: Seafood
Mains average: $$
Specialty: Bargeman's seafood platter, oysters (water to plate in fifteen minutes sometimes)

Local flavours: Oysters, mussels, fish, octopus, crayfish, berries, fruit
Grows: Oysters, blackberries, apples, apricots, fresh herbs, tomatoes
Extras: Decking with views of the bay, so diners can see the oysters being harvested.
- **Open:** Lunch daily, dinner Thurs–Sat (and Sun Jan–Feb)

TRIABUNNA
🍎 PRODUCERS

SPRING BAY SEAFOODS PTY LTD
Phillip Lamb
488 Freestone Point Rd, Triabunna 7190
Phone: 03 6257 3614
Email: plamb@springbayseafoods.com.au
Some of the purest and coldest water in the world is one of the secrets of some of the best scallops and mussels in the world. Spring Bay half shell scallops are sold around Australia in top restaurants and Spring Bay blue mussels are premium-quality graded large mussels. Operating for the past three years, this company is Australia's first and only commercial-scale scallop aquaculture operation.
- **Open:** By appointment
- **Location:** Look for a sandwich board sign at the turn-off to Spring Bay

WESTERWAY
🍴 DINING

THE POSSUM SHED CAFE AND CRAFT GALLERY
Alison Dugan and Tony Coleman
1654 Gordon River Rd, Westerway 7140
Phone: 03 6288 1477
BYO
Cuisine: Fresh, tasty, simple

Mains average: $
Specialty: Devonshire teas, sweet potato pie, lemon and lime tart
Local flavours: Mixed greens, tomatoes, raspberries, Tasmanian cheese, smoked salmon
Extras: Riverside setting with resident platypus, relaxed ambience. Takeaway available.
■ **Open:** Lunch, morning and afternoon tea Wed-Sun
● **Location:** Follow signs to Mt Field National Park

WOODBRIDGE

🍎 PRODUCERS

MIELLERIE—THE HOUSE OF HONEY
Julie Hoskinson and Yves Ginat
159 Woodbridge Hill Rd, Woodbridge 7162
Phone: 03 6267 4669
Thirty-two years producing honey adds up to a fair amount of experience, and now it's the second generation—daughter Julie, born into the honey industry—carrying on the tradition with Yves, a French beekeeper. Varieties available include leatherwood, wildflower, blue gum, and clover honeys, plus beeswax. Golden Pearl and Greenridge honey is available throughout Tasmania and at Relish Tasmania in Sydney.
■ **Open:** Door sales most days, phone for details
● **Location:** Turn right at Woodbridge general store then continue for 1.6km up Woodbridge Hill Rd and look for the 'Honey for Sale' sign

---

MARKETS AND EVENTS

✦ **SPRING BAY SEAFEST,** TRIABUNNA
03 6257 3328
Easter long weekend

✦ **TASTE OF THE HUON,** HUON VALLEY
03 6234 7844
Easter long weekend

🍴 DINING

PEPPERMINT BAY
Simon Currant (owner),
Steven Cumper (head chef)
3435 Channel Highway, Woodbridge 7162
Phone: 03 6267 4088
Email: sales@peppermintbay.com.au
www.peppermintbay.com.au
Licensed
Cuisine: Contemporary regional
Mains average: $-$$
Specialty: Strong regional focus, using local ingredients wherever possible
Local flavours: Organic beef, greens and eggs, stone fruit, shellfish, seafood, farm cheeses, lamb
Makes: Peppermint Bay Providore has a wide range of house-made products including raspberry jam, red tomato relish, nectarine chutney, crabapple jelly, pepperberry mustard, terrines, relishes, aioli, desserts, sourdough bread
Extras: Views over Peppermint Bay and the Channel to Bruny Island, outdoor deck with huge umbrellas. Built on the site of the old Woodbridge Hotel. Tourists can take a trip on a new boat called *Peppermint Bay*, have lunch and return via boat in the afternoon.
■ **Open:** Dining Room—lunch daily, dinner Sat, The Local—lunch daily, dinner Tues-Sat

THE NORTH-WEST

Wild and beautiful, this left side of Tasmania has something to offer everyone. There are wilderness and whisky, majestic mountains, rainforest and wrecks, yet some parts are richly agricultural. This is where, after all, Lactos has its base, capitalising on the fine dairying lands nearby.

Other parts though are mysterious, home to wild berries and fungi; places where you could believe a Tasmanian tiger might live. All the while the sea forms a boundary, and it is to this that Tasmania always returns. Whether it is octopus or oysters, salmon or sardines, these wild oceans deliver up produce that finds its way onto some of the state's best tables.

BURNIE

PRODUCERS

LACTOS CHEESE TASTING AND SALES CENTRE
145 Old Surrey Rd, Burnie 7320
Phone: 03 6433 9255
Come early to this home of the magical Heidi gruyere if you want to taste it, as cheese tastings finish thirty minutes before closing time. Also here you'll find Heritage Mersey Valley and Aussie Gold cheeses. Lactos' products have won many awards at cheese and agricultural shows around Australia, and it's one of Australia's older cheese producers, established in 1955. Lactos cheeses are sold Australiawide.
■ **Open:** 9am–5pm weekdays, 10am–4pm weekends and public holidays
● **Location:** Off Hwy 1 at Burnie, about 3km along Old Surrey Rd

CRADLE MOUNTAIN

DINING

THE HIGHLAND RESTAURANT
Benito Lanzarote
Cradle Mountain Lodge, Cradle Mountain Rd, Cradle Mountain 7306
Phone: 03 6492 1303
Email: foodbev@cradle.poresorts.com
www.poresorts.com
Licensed
Cuisine: Rustic modern Australian
Two courses: $$$
Specialty: Seared Bass Strait cod, ocean trout fillets
Local flavours: Fish and seafood, pepper berries, herbs, Spreyton beef, cheese (including King Island blues), almonds
Extras: Awesome mountain and lake setting. Large open fireplace in centre of restaurant, walk-in wine cellar underneath the restaurant. Cradle Mountain Lodge is a unique wilderness retreat on the edge of the World Heritage Listed Cradle Mountain–Lake St Clair National Park. Takeaway available.
■ **Open:** Breakfast and dinner daily (bookings essential)
● **Location:** Approximately two hours by road from Launceston, Burnie or Devonport. Transfers from Devonport airport available.

DEVONPORT

🍴 DINING

ESSENCE FOOD AND WINE
Steve and Jo Harding
28 Forbes St, Devonport 7310
Phone: 03 6424 6431
Licensed
Cuisine: Modern Australian
Mains average: $$
Specialty: Fresh, seasonal local produce
Local flavours: As much as possible— meat, berries, vegetables, fruit, seafood
Extras: Outdoor courtyard for summer, fires inside in winter.
■ **Open:** Lunch Tues–Fri, dinner Tues–Sat

ELIZABETH TOWN

🍎 PRODUCERS

ASHGROVE CHEESE
Bennett family
6173 Bass Hwy, Elizabeth Town 7304
Phone: 03 6368 1105
Email: info@ashgrovecheese.com.au
www.ashgrovecheese.com.au
Ashgrove Cheese is a boutique operation specialising in the production of handcrafted hard and semi-hard cheeses. It is the only cheese manufacturer in Australia specialising in traditional English styles of cheese such as Lancashire, Wensleydale, Cheshire and Double Gloucester. The Ashgrove Cheese shop provides an educational and informative experience for cheese lovers. Visitors can observe the cheesemaking and maturing areas through viewing windows. Jane Bennett originally studied cheesemaking with two of Britain's best farmhouse cheesemakers. She has developed a cheese unique to Ashgrove called Bush Pepper, featuring the spicy flavour of Tasmanian native pepperberries and has recently created Wild Wasabi cheese, using both the stem and the leaf from wasabi grown by farmer Ian Farquhar in north-east Tasmania.
■ **Open:** 7.30am–6pm daily
● **Location:** 4km north of Elizabeth Town

🍴 DINING

CHRISTMAS HILLS RASPBERRY FARM
Dornauf family
9 Christmas Hills Rd, Elizabeth Town 7304
Phone: 03 6362 2186
Email: lindidornauf@vision.net.au
www.raspberryfarmcafe.com
Licensed
Cuisine: Country
Mains average: $
Specialty: All things raspberry, particularly homemade raspberry ice-cream
Local flavours: Raspberries, local cheeses (Ashgrove, Heidi, Lactos) local venison, pork, Tasmanian seafood, fruit, vegetables, wine and beer
Extras: Sells homemade raspberry vinegar, syrup, vinaigrette, sauce, jam, relish, infused brandy, chocolate coated raspberries, double cream, raspberry ice-cream, raspberry and mint tea and raspberry-chilli sauce. There are picnic tables and decking overlooking a lake filled with waterlilies, platypus and birdlife and a backdrop of trees and the raspberry fields. The berry season is from December to May.
■ **Open:** 7am–5pm daily, evening functions by appointment for groups over twenty
● **Location:** Just off the highway halfway between Launceston and Devonport

FORTH

🏠 STORES

COATES QUALITY SMALLGOODS
Raymond and Lea Coates
666 Forth Rd, Forth 7310
Phone: 03 6428 2377
Email: pippaaus@southcom.com.au
A seachange drew the Coateses here in 1996. Now incorporating Forth Butchery it's the ideal place to stock up on all sorts of smallgoods and fresh meats. There are gourmet gluten-free sausages, as well as old-fashioned style hams and bacons, Cornish pasties and pork pies. While the pork legs come from the mainland, all the beef, lamb, chicken and pork for other uses comes from Tasmania. The smokehouse creates a wide range of options, and there are also pan-ready dishes. Also sold at IGA in Devonport and in Launceston at Davies Grand Central Station and Balfour Street Deli.
■ **Open:** 8am–6pm weekdays, 9am–12pm Sat

GOWRIE PARK

🍴 DINING

WEINDORFERS
Charlotte King
Wellington St, Gowrie Park 7306
Phone: 03 6491 1385
Licensed
Cuisine: Swedish-influenced Tasmanian
Mains average: $$
Specialty: Ice-cream, cakes, smoked trout, soup, beef and garlic with red wine
Local flavours: As much as possible, including lamb, beef, trout, eggs, organic vegetables, local berries, apples

Extras: Surrounded by mountain scenery (Mt Rowland range on three sides), there are two big tables in the gardens, and four self-contained family cabins (Gowrie Park Wilderness Cabins).
■ **Open:** Lunch, morning and afternoon tea, dinner daily (1 Oct–31 May, and weekends during June–July)
● **Location:** Fifteen minutes south of Sheffield

LATROBE

🏠 STORES

THE HOUSE OF ANVERS
Igor and Jocelyn Van Gerwen
9025 Bass Hwy, Latrobe 7307
Phone: 03 6426 2958; Freecall 1800 243 063
Email: anvers@bigpond.com.au
www.anvers-chocolate.com.au
Ten minutes from the place where the Spirit of Tasmania docks and you can be indulging in some of Tasmania's best chocolates, truffles and fudge, produced here since 1989. Set in a 1930s-style cottage, theirs is a retail shop, a museum and a cafe—and do ask for the Cradle Coast Pie using regional ingredients. Visitors can see the fine chocolates being made before indulging themselves at the tasting centre. The cafe specialises in a variety of cocoa drinks, freshly-ground coffees, French-style breakfasts and light lunches featuring Tasmanian food. A display of antiques related to chocolate-making showcases the history of chocolate. Products also sold in David Jones and Myer stores throughout Australia, as well as specialty shops.
■ **Open:** 7am–6pm daily

THE NORTH-WEST—TAS

🍴 DINING

GLO GLO'S RESTAURANT
Patrick Johnston (owner),
Xavier Mouche (chef)
78 Gilbert St, Latrobe 7307
Phone: 03 6426 2120
Email: gloglos@bigpond.com
Licensed
Cuisine: International
Mains average: $$$
Specialty: Lobster soup
Local flavours: Ocean trout, lobster, quail, beef, mushrooms, chicken, Atlantic salmon, mussels, fruit and vegetables
Extras: A lovely veranda and garden for dining on warm summer nights. Grows most herbs, chamomile for tea, salad greens, green beans, broad beans, tomatoes, zucchini, cucumbers, chillies, berries, passionfruit, lemons, limes, kaffir lime leaves—even princess lilies and roses for the tables.
■ **Open:** Dinner Sat

MOLE CREEK

🍎 PRODUCERS

R STEPHENS TASMANIAN HONEY
Shirley and Ian Stephens
25 Pioneer Dr, Mole Creek 7304
Phone: 03 6363 1170
Email: k.l.honeybee@hotkey.net.au
Each spring Ian Stephens takes himself off into pristine World Heritage areas on the west coast of Tasmania to place his hives and collect leatherwood honey. He has been doing this for fifty years, making his one of the older honey companies in Tasmania, operating as a family business since 1919. Golden Bee flora honey comes from ground flora in the Stephens's own area, and Golden Nectar is the unique leatherwood honey (surely one of the rarest honeys in the world) with a truly organic claim. The honeys are exported to other states and overseas (the United States, Hong Kong, Japan and Indonesia).
■ **Open:** 9am–4pm weekdays
● **Location:** Via Deloraine or through Sheffield Cradle Mountain

PROMISED LAND

🍴 DINING

TASMAZIA AND THE VILLAGE OF LOWER CRACKPOT
Brian and Laura Inder
500 Staverton Rd, Promised Land 7306
Phone: 03 6491 1934
www.tasmazia.com.au
BYO
Mains average: $
Specialty: Wide range of pancakes
Local flavours: Berries, cream
Extras: The honey boutique includes dream flavours like wild bramble, honey of paradise, fennel, leatherwood, lavender and an original invention—liqueur honeys—including Drambuie, Wild Turkey bourbon and Galliano. The associated lavender farm supplies honey, jams and mustards and there is also a complex of seven mazes. Takeaway available.
■ **Open:** 9am–5pm daily
● **Location:** 14km from Sheffield on the scenic route between Sheffield and Cradle Mountain

SISTERS CREEK

🍎 PRODUCERS

NATURALLY NICHOLS
Carolyn Nichols
Redbanks Farm,
152 Broomhalls Rd, Sisters Creek 7325
Phone: 03 6445 1438
Email: naturallynichols@tassie.net.au
www.naturally-nichols.com.au
Carolyn Nichols's home-style farm cooking uses only natural ingredients, and everything here—moist rich fruitcakes, shortbread, cookies, macaroons, stuffing mixes, and gourmet butters—is produced at this farm. What's more, when you buy here, part of the proceeds goes towards the conservation of one hundred hectares of the adjoining Rocky Cape National Park, a sanctuary to local wildlife. And while the Naturally Nichols brand is found throughout Australia, you'll only find Forth Valley Pork Pies in Tasmania.
- Open: 9am–5pm weekdays
- Location: 29km north-west of Burnie, 1.5km off the Bass Hwy, 18km west of Wynyard

SPREYTON

🏠 STORES

THE BIG APPLE—SPREYTON FRUIT AND MEAT MARKET
RW Squibb and Sons
77 Mersey Main Rd, Spreyton 7310
Phone: 03 6427 2263 (store);
03 6427 2065 (office)
Email: rwsquibb@southcom.com.au
This is a one-stop place to stock up on as many local and regional goodies as possible. There's an on-site butchery, bakery and greengrocer with local olives, wines, cheese and lots more. There are two generations involved here and the cherries, plums, apples and some other fruits are grown by the family, who also export them and sell them nationally. There are orchard tours by appointment for groups of thirty or more.
- Open: 7am–6pm weekdays, 7.30am–5.30pm weekends
- Location: 5km south of Devonport

STANLEY

🍎 PRODUCERS

STANLEY SEAQUARIUM
Diane Charles
Fisherman's Dock, Stanley 7331
Phone: 03 6458 2052
Email: di_charles@craigmostyn.com.au
This display of Tasmanian sea life is also a seafood sales outlet. Here you'll find, in season, lobster and giant crabs, cooked or live, and scallops.
- Open: 9am–4.30pm daily, 11am–3pm daily during winter
- Location: At the wharf (off Wharf Rd)

🍴 DINING

JULIE AND PATRICK'S RESTAURANT AND HURSEY SEAFOODS
Valerie Hursey
2 Alexander Tce, Stanley 7331
Phone: 03 6458 1103
Licensed
Cuisine: Seafood
Mains average: $–$$
Specialty: Seafood platter, beef and reef
Local flavours: Bass Strait seafood, beef
Extras: Enclosed balcony with views over the park and water. There is a restaurant

upstairs and a cafe downstairs. Hursey Seafoods is also downstairs and sells fresh seafood from the Hurseys' fishing boats. There are also giant tanks for live fish and they will cook crays to order.
- **Open:** Cafe/takeaway—9am–6pm daily (later in summer), restaurant—dinner daily

STRAHAN
🍴 DINING

FRANKLIN MANOR
Meyjitte Boughenout (co-owner/chef), Debbie Boughenout, Tony Wurf
The Esplanade, Strahan 7468
Phone: 03 6471 7311
Email: franklinmanor@bigpond.com
www.franklinmanor.com.au
Licensed
Cuisine: Innovative modern Australian–French
Four-course dinner: $$$
Specialty: The seven-course tasting menu (degustation) features sugar-cured ocean trout, a signature dish
Local flavours: As much as possible, including farmed ocean trout, organic herbs from Strahan, wild mushrooms in season, crayfish, bush pepper
Extras: Dining is available on the veranda for dinner or afternoon tea. In spring the garden is brilliant with rhododendrons. Grows own wild sorrel, miniature pansies for garnish, mushrooms, wild blackberries, apples, pears, free-range chicken and duck eggs. Makes for sale mustard dressings, raspberry vinegar, chocolates and jams.
- **Open:** Breakfast, afternoon tea and dinner daily
- **Location:** Approximately 1km from the town centre, on the left

RISBY COVE RESTAURANT
Chris Short and Rick Gumley
The Esplanade, Strahan 7468
Phone: 03 6471 7572
Email: risbycove@bigpond.com
www.risby.com.au
Licensed
Cuisine: Modern Australian Fusion
Mains average: $$
Specialty: Locally farmed salmon, seared with roasted almonds and lemon beurre blanc
Local flavours: Salmon, crayfish, beef, pork, oysters, scallops, cheese, trevalla, squid
Extras: Extensive wine list, outdoor absolute-waterfront dining around the deck of a private marina, views and sunsets over Macquarie Harbour. On the original site of the Risby family's Huon Pine Sawmill. There is also a gallery is located in the former sawmill.
- **Open:** Lunch and dinner daily
- **Location:** 700m past main town centre on the water's edge

---

CRUISING THE GORDON RIVER

Gordon River Cruises (phone: 03 6471 7187) and World Heritage Cruises (phone: 03 6471 7174). These companies operate cruises on the Gordon River and offer a smorgasbord featuring local foods and produce, such as wines, cheeses, seafood, fruit and salads.

TAS—THE NORTH-WEST

🚌 TOURS

WESTHAVEN YACHT CHARTERS
Trevor and Megs Norton
Strahan Wharf, Strahan 7468
Phone: 03 6471 7422
Email: wcyc@tassie.net.au
www.tasadventures.com/wcyc
The Macquarie Harbour sailing cruise of most interest to food lovers is the Crayfish Dinner Cruise. A two-and-a-half hour return sail with a delicious local crayfish meal, plus salad, dessert, complimentary wine or juice, tea and coffee is about as good as it gets.
■ **Open:** 6pm–8.30pm, phone for bookings

WYNYARD

🏪 STORES

WYNYARD WHARF SEAFOODS
Veronica and Kevin Pellas
3 Goldie St, Wynyard Wharf, Wynyard 7325
Phone: 03 6442 3428
Email: mpellas@bigpond.net.au
You can't buy fish much fresher than this. It comes straight from the boats, which are often unloading while you watch. The Pellases have been running this place now for five years, serving fish and chips plus fresh fish, crayfish, seafood, oysters, scallops and a range of smoked fish. Trevally is the most common, but you'll find many other fish too.
■ **Open:** 9am–6.30pm, daily

🍴 DINING

BUCKANEERS FOR SEAFOOD
St John Smith
4 Inglis St, Wynyard 7325
Phone: 03 6442 4104
BYO
Cuisine: Seafood
Mains average: $$
Specialty: Seafood, steak, chicken
Local flavours: Seafood
Extras: There's a definite nautical theme to suit the pirate name—even a boat in the dining room. Also sells a wide range of fresh fish, much of it from local fishermen.
■ **Open:** Lunch and dinner daily, takeaway 9am–7.30pm Sun–Thurs, 11.30am–8pm Fri–Sat

---

MARKETS AND EVENTS

✦ **BURNIE FARMERS' MARKET,**
BURNIE
03 6431 5882
9am–1pm first and third Sat

✦ **WYNYARD FARMERS' MARKET,**
WYNYARD
03 6438 1165
9am–12pm second and fourth Sat

✦ **SPREYTON PRIMARY APPLE FESTIVAL,** SPREYTON
03 6427 2075
Easter

✦ **TASTINGS AT THE TOP,**
CRADLE MOUNTAIN LODGE
03 6492 1303
Mid June

VICTORIA

Despite accounting for only three per cent of Australia's land mass, Victoria produces around a quarter of the country's agricultural commodities and almost a third of its food products.

The state's unique position, stretching from semi-arid areas, through mountains and lush agricultural land, and bordered by a long coastline, allows a unique diversity of produce. It's so dense and varied that, to the traveller, the state seems much larger. Maybe this is because the landscape often changes dramatically, as wheat lands give way to orchards and the roadsides are suddenly lined with fruit stalls and signs.

Due largely to the state's temperate climate and lush pastures watered by natural rainfall or irrigation, Victoria dominates Australia's dairy industry. This small state produces around two thirds of the country's milk, and seventy-five per cent of manufactured dairy products.

Victoria is also a major producer of high-quality beef, as well as sheep meat (mutton and lamb), chicken and pork. Paddocks of wheat, barley, oats and pulses, as well as the brilliant golden-flowered canola, used for oil, cover much of the rest of the state.

Predominantly cool-climate fruits such as apples and pears, stone fruits and berries, as well as vegetables, round out the balanced agricultural 'diet' of this state, and the very fine wine industry adds the final luxury.

Melbourne is not just the state capital but a mecca for gourmands and style-junkies. The ongoing contest with Sydney, as to which has the better dining, will never be adequately settled—unless you are a Victorian, in which case you know the answer is Melbourne.

> **QUARANTINE NOTE**
>
> Movement of fruit across the boundary of the Fruit Fly Exclusion Zone (FFEZ) and into South Australia is illegal. Motorists may be subjected to random road-blocks and if you are found in breach of this law, an on-the-spot $200 fine can be levied. Fresh fruit may not be taken into the FFEZ, which covers most of the Murray Outback region. If in doubt contact the tourism office.

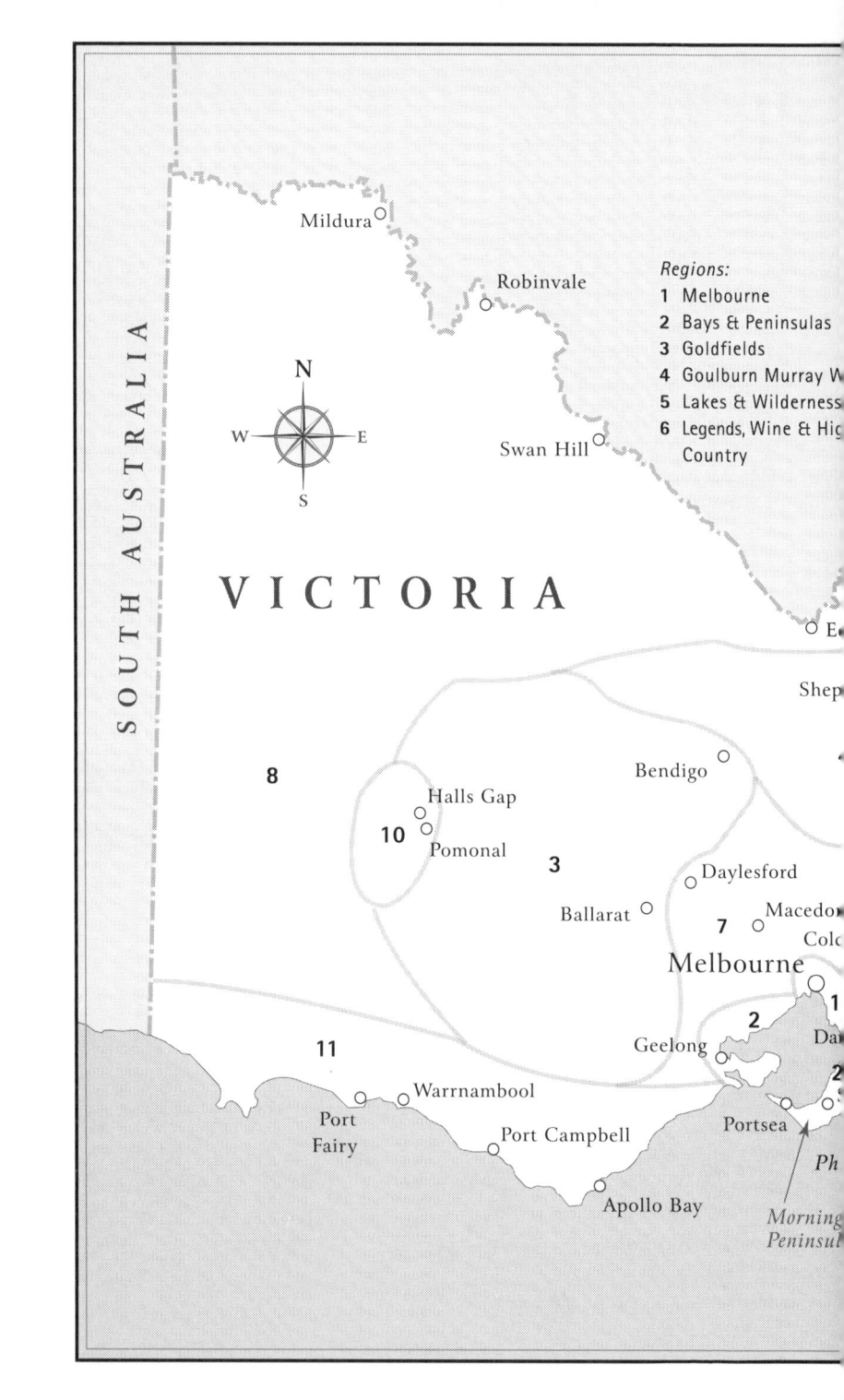

Mildura

Robinvale

Regions:
1 Melbourne
2 Bays & Peninsulas
3 Goldfields
4 Goulburn Murray W
5 Lakes & Wilderness
6 Legends, Wine & Hig Country

N W E S

Swan Hill

SOUTH AUSTRALIA

VICTORIA

E

Shep

8

Bendigo

Halls Gap
10
Pomonal
3
Daylesford
Ballarat
7   Macedo
        Col
Melbourne
2
1
Da
11
Geelong
2
Warrnambool
Port Fairy
Port Campbell
Portsea
Ph
Apollo Bay

Morning Peninsul

NEW SOUTH WALES

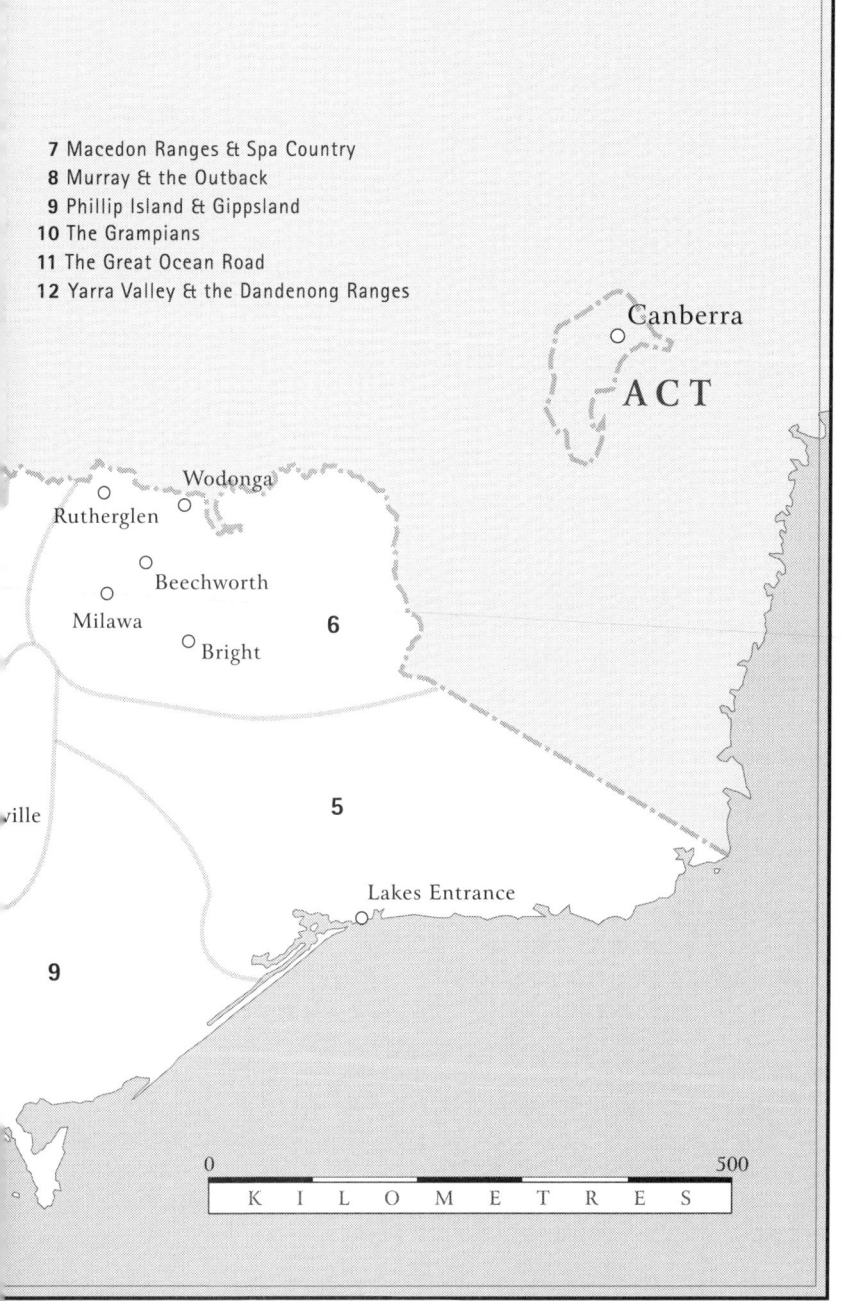

7 Macedon Ranges & Spa Country
8 Murray & the Outback
9 Phillip Island & Gippsland
10 The Grampians
11 The Great Ocean Road
12 Yarra Valley & the Dandenong Ranges

MELBOURNE

Perhaps more than in other capital cities, Melbourne's restaurants reap the benefits of their rich and productive state.

Melbourne has such a fine reputation for elegant and eclectic dining, and most of the more prominent and popular restaurants use Victorian produce almost exclusively because of its proximity and high standard. For this reason, and because of space limitations in this book, it would be impossible to list the very many restaurants that would qualify as superbly showcasing the state's produce. Consult the websites, newspapers or books mentioned for more complete and objective reviews and information. Better still, enjoy the various eating precincts of the city.

With over 3000 restaurants embracing more than seventy international cuisines, Melburnians are extremely spoilt by the wide selection of exotic cuisines. Eating out is a part of daily life, whether it's an Italian gelato, a sizzling souvlaki or a gourmet dinner in a five-star restaurant. No matter where you go in this city, at any hour of the day or night, you will stumble across fabulous cafes, bars and bistros that will tempt you with their offerings.

Melbourne has entire streets devoted to establishments of particular ethnicity: Lygon Street in Carlton is a little piece of Italy, offering delicious Italian pasta, pizza and gelati. Chinatown, in Little Bourke Street, offers sticky buns, yum cha and Peking duck. Swan Street, Richmond and Lonsdale Street in the city offer a great variety of Greek tavernas serving pita, Greek dips, souvlaki and char-grilled seafood. Victoria Street, in Abbotsford, is Little Saigon, with a staggering array of Vietnamese restaurants serving hundreds of dishes at bargain prices. For the flavours of Spain, try Johnston Street, in Fitzroy, which has side-by-side tapas bars and restaurants, and smell the aromas of Turkey along Sydney Road, Brunswick. Victoria Market Place at Queen Victoria Markets has a wide range of ethnic specialties too.

Then there are the areas ideal for general dining. Enjoy Southbank's variety of excellent restaurants, cafes and bars, with views of the city, or wander through St Kilda. Fitzroy Street is the centre of al fresco dining, but people-watching is also a favourite pastime at chic sidewalk cafes in Chapel Street, Prahran and Toorak Road in South Yarra. Otherwise Brunswick Street in Fitzroy and Acland Street in St Kilda, or NewQuay offer multiple grazing stops.

🍎 PRODUCERS

DONNYBROOK FARMHOUSE CHEESE
Con Monteleone
915 Donnybrook Rd, Donnybrook 3064
Phone: 03 9745 2315
Email: cmonteleone@optusnet.com
Con Monteleone runs his own small dairy herd and uses the milk to produce a range of cheeses including pecorino, parmesan, romano, bleu vein, brie and ricotta. Donnybrook also makes chevre and fetta using local goats milk.
■ **Open:** 8am–5pm Tues–Sun

MELBOURNE–VIC

🏠 STORES

DISCOVER VICTORIA
Paul Ryan
Shop RM09 and Shop 11B Qantas Domestic Terminal, Tullamarine 3043
Phone: 03 9334 5201 OR 03 9334 5401
These two stores carry a wide range of Victorian produce including olive oil, jams, herbs, lavender products, chutneys, sauces, jams, biscuits, and cheeses. The ideal gift to pick up as you jet off to another state.
■ Open: 7am–9pm Sun–Fri, 7am–6pm Sat

MILAWA CHEESE SHOP
David and Anne Brown
665 Nicholson St, Carlton North 3054
Phone: 03 9381 1777
Anne and David Brown of Milawa Cheese Company opened this business in 1998 to provide a city outlet for their products. Besides Milawa cheeses, they sell cheeses that you wouldn't usually find in the supermarket and non-cheese products, such as preserves, Milawa Mustards, olive oils, coffee, and light lunches.
■ Open: 10am–6pm weekdays, 9am–2pm Sat

QUEEN VICTORIA MARKETS
Cnr Elizabeth and Victoria Streets, Melbourne 3000
Phone: 03 9320 5835
Email: qvmtours@melbourne.vic.gov.au
www.qvm.com.au
Australia's biggest fresh food market and the largest in the Southern Hemisphere has over 1000 stalls selling everything from bargain clothing and garden ornaments to produce and multicultural food.
■ Open: 6am–2pm Tues and Thurs, 6am–6pm Fri, 6am–3pm Sat, 9am–4pm Sun
Office: 513 Elizabeth St, Melbourne, 3000.

MARKETS AND EVENTS

✦ **BORONIA FARMERS' MARKET,** BORONIA
03 5664 0096
8am–1pm third Sat

✦ **BOROONDARA FARMERS' MARKET,** EAST HAWTHORN
03 5664 0096
8am–12.30pm third Sat

✦ **BUNDOORA PARK FARMERS' MARKET,** BUNDOORA
03 5664 0096
8am–1pm first Sat

✦ **COLLINGWOOD CHILDREN'S FARM FARMERS' MARKET,** ABBOTSFORD
03 5657 2337
8am–1pm second Sat

✦ **QUEEN VICTORIA MARKETS,** MELBOURNE
03 9320 5835
6am–2pm Tues and Thurs, 6am–6pm Fri, 6am–3pm Sat, 9am–4pm Sun

✦ **VEG OUT FARMERS' MARKET,** ST KILDA
Fourth Sat

✦ **MELBOURNE FOOD AND WINE FESTIVAL,** VARIOUS VENUES AROUND MELBOURNE AND VICTORIA
03 9823 6100
Mid Mar–early Apr

VIC–MELBOURNE

🚌 TOURS

ABORIGINAL HERITAGE WALK
Visitor Centre, Observatory Gate,
Royal Botanic Gardens, Birdwood Ave,
South Yarra 3141
Phone: 03 9252 2300
Email: visitor.centre@rbg.vic.gov.au
www.rbg.vic.gov.au
Because the Royal Botanic Gardens Melbourne are located on a traditional camping and meeting place for the local custodians of the area—the Boonwurrung and Woiworung people—the Aboriginal Heritage Walk is a particularly stirring cultural experience. Now you may share in the wealth of local plant lore and see this land through the eyes of an Aboriginal guide. Bookings essential.
- **Open:** 11am–12:30pm Thurs–Fri, 10:30am–12pm alternate Sun

FOODIES' DREAM TOUR
Natasa Palamiotis
Tour Centre, Queen Victoria Markets,
cnr Elizabeth and Victoria Streets,
Melbourne 3000
Phone: 03 9320 5835
Email: qvmtours@melbourne.vic.gov.au
www.qvm.com.au
Guides take you on a two-hour tour of all that is tasty and wonderful at this most interesting of city markets. Tour includes food samplings, a coffee and a calico bag (to carry your purchases home in of course!).
- **Open:** 10am–12pm Tues and Thurs–Sat

BAYS AND PENINSULAS

Wrapping Port Phillip in easily accessible gastronomic excellence, the bays and peninsulas offer another dimension to Melbourne and its environs. Whether you head east down the Mornington Peninsula or west towards Geelong and its reborn wine region accompanied by zappy restaurants and bars, the sheer abundance is awesome.

As the Mornington Peninsula has developed as one of Australia's newer wine regions, so has the food scene matured, offering restaurants and food that are a fabulous foil for the wines.

It seems that the sunny seaside ambience of the area even permeates menus, as on both sides of the bay, fish and seafood is in abundance and served with enthusiasm at all sorts of local eateries, from the humblest fish-and-chippery to the most elegant silver service restaurant.

ANAKIE
DINING

STAUGHTON VALE VINEYARD
Lyn and Paul Chambers
20 Staughton Vale Rd, Anakie 3221
Phone: 03 5284 1477
Email: staughtonvalewines@bigpond.com.au
Licensed
Cuisine: Modern Australian
Mains average: $$
Specialty: Stylish food to accompany the wines

Extras: Dining room has views over the vineyard, outdoor dining amongst the vines, two motel-style units and a self-contained long-term unit.
- **Open:** Lunch Fri-Mon; dinner last Fri of the month (other times by appointment)
- **Location:** Cnr Ballan and Staughton Vale Roads, halfway between Geelong and Bacchus Marsh

BELLARINE
DINING

BELLARINE ESTATE
Peter and Liz Kenny
2230 Portarlington Rd, Bellarine 3222
Phone: 03 5259 3310
Email: info@bellarineestate.com.au
www.bellarineestate.com.au
Licensed
Cuisine: Modern Australian
Mains average: $$
Specialty: Roast pumpkin tortellini and carpet-bag steak
Local flavours: Mussels, seafood
Extras: Nestled in the vineyard with lovely views of the vineyard and over Corio Bay, there is also a large, grassed area for car club events, outdoor functions and wedding ceremonies.
- **Open:** Lunch Fri-Mon, dinner Sat
- **Location:** Halfway between Portarlington and Drysdale

BELLBRAE
🍴 DINING

BELLBRAE HARVEST
Sarah Locke
45 Portreath Rd, Bellbrae 3228
Phone: 03 5266 2100
Email: bellbraeharvest@bigpond.com.au
www.bellbraeharvest.com.au
Licensed
Cuisine: Modern Australian
Mains average: $$
Local flavours: Meredith dairy products, seafood, meat, organic herbs, local wines
Extras: Makes a range of relishes, chutneys and jams for sale. Tuscan-style mud brick cafe, large paved dining area with water feature and views across the farm. Three accommodation units available.
- **Open:** Breakfast and lunch Thurs–Sun, dinner Fri–Sat
- **Location:** Signposted directly off the Great Ocean Rd

BITTERN
🍴 DINING

BITTERN COTTAGE
Jenny and Noel Burrows
2385 Frankston–Flinders Rd, Bittern 3918
Phone: 03 5983 9506
Email: info@bitterncottage.com.au
www.bitterncottage.com.au
BYO
Cuisine: French/Italian Provincial
Set menu: $$$
Specialty: Slow-cooked food combined with simple, quality vegetables (often steamed)
Local flavours: Used as much as possible
Extras: The set menu is for a half season (six or seven weeks) and then it changes. Dine in a cottage-garden setting.
- **Open:** Lunch Sun, dinner Sat and second Mon of the month (set menu)

DROMANA
🍴 DINING

HICKINBOTHAM OF DROMANA
Terryn and Andrew Hickinbotham
194 Nepean Hwy, Dromana 3936
Phone: 03 5981 0355
Email: info@hickinbotham.biz
www.hickinbotham.biz
Licensed (for own wines)
Cuisine: Country style
Mains average: $$
Specialty: Winemakers platter—cold meat, cheese, seasonal fruit, chutney, quandong paste, served with wine-bread (made from the yeast 'lees' of the winery's wines)
Local flavours: Eggs, seasonal fruit and vegetables, homemade bread made with estate wine, cheeses
Extras: Outdoor area overlooks Port Philip Bay. Surrounded by gum trees, open grassy woodland area with two very friendly Clydesdale horses, Bill and Ernie, who like visitors with carrots, apples or bread. Open fire in winter, live music on weekends.
- **Open:** Lunch Wed–Sun (Nov–Mar), weekends only the rest of the year

DRYSDALE
🍎 PRODUCERS

TUCKERBERRY HILL
Christine and David Lean
35 Becks Rd, Drysdale 3222
Phone: 03 5251 3468
Email: tuckerberry@iprimus.com.au

The blueberry season is short and sweet. Fortunately, it coincides with some summer spare time, so you might have a moment to pick your own berries (said to have the highest anti-oxidant activity of all fresh fruit) and enjoy the views from the Murradoc hill slopes at the same time. You'll also find pre-picked berries, blueberry recipe books, jams, chutney, toppings, juice, muffins and frozen blueberries—and blueberry plants so you can plant your very own.
- **Open:** 9am–5pm daily (26 Dec–31 Jan), 9am–5pm weekends (Feb)

FLINDERS
DINING

SAFI BULUU
Mark Goodrem and Vikki Aitken
41 Cook St, Flinders 3929
Phone: 03 5989 1165
Email: safibuluu@hotmail.com
Licensed
Cuisine: Modern Australian/Asian
Mains average: $$
Local flavours: Mussels, Flinders Farm tomatoes, Flinders Bakery bread
Extras: Large deck with beanbags and couches, also picnic rugs on the lawn.
- **Open:** Lunch Thurs–Sun, dinner Fri–Sat

STEAM PACKET PLACE
Geelong's newest waterfront precinct, Steam Packet Place, has a new lease of life with cafes, restaurants, sea baths and the marina. A great place for local seafood.

GEELONG
STORES

WHOLEFOODS COOPERATIVE LTD
2 Baylie Place, Geelong 3220
Phone: 03 5221 5421
Come to this shop that has been operating for twenty-five years, for organic bulk products, local fruit and vegetables, nuts, grains and groceries.
- **Open:** 10am–5pm weekdays, 10am–1pm Sat

DINING

FISHERMEN'S PIER RESTAURANT
Rick Munday (owner) Simon Parrott (executive chef)
Bay end of Yarra St, Eastern Beach, Geelong 3220
Phone: 03 5222 4100
Email: eatfresh@fishermenspier.com.au
www.fishermenspier.com.au
Licensed
Cuisine: Seafood
Mains average: $$$
Specialty: Hot and cold seafood feast for two
Local flavours: Fish, seafood, olive oil,
Extras: Right on the water's edge overlooking Corio Bay.
- **Open:** Lunch and dinner daily

MAIN RIDGE
PRODUCERS

SUNNY RIDGE STRAWBERRY FARM
Mick and Anne Gallace
244 Shands Rd, Main Ridge 3928
Phone: 03 5989 6273
Email: info@sunnyridge.com
www.sunnyridge.com

VIC–BAYS AND PENINSULAS

Pick your own strawberries then relax in the Strawberry Cafe with homemade ice-cream, smoothies, Devonshire tea (with strawberry scones) and a range of strawberry wines and port. How decadent! When you are finished doing that, watch a video that explains strawberry growing, then browse through a range of other Mornington Peninsula specialities such as Main Ridge Connections, Eureka Farm herb vinegars, and Flinders Fruit Sauces. You can pick your own strawberries from November to April at this, Australia's largest strawberry producer, or purchase seasonal fresh fruit and vegetables grown on the property.
- **Open:** 9am–5pm daily (Nov–Apr), 11am–4pm weekends (May–Oct)
- **Location:** Cnr Mornington-Flinders and Shands Roads

TAMARILLO FRUIT FARM
Jean Corbett
100 Barkers Rd, Main Ridge 3938
Phone: 03 9592 8075
Despite the name, there are apples (some of them heritage varieties such as Ribston pippins), limes and olives here too, as well as those lovely tree tomatoes, aka tamarillos. Begun as a retirement hobby because of an interest in horticulture, this has evolved into a busy place. You can pick your own or there are ready-picked apples, with the remainder of the crop turning up at wineries and restaurants in the area.
- **Open:** Weekends Mar–June (during harvest)
- **Location:** Look for signs on the roadside

DINING

LA BARACCA TRATTORIA
Nicole Koppelman (chef)
T'Gallant Winery, 1385 Mornington-Flinders Rd, Main Ridge 3928
Phone: 03 5989 6565
Email: tgallant@satlink.com.au
Licensed
Cuisine: Rustica, homestyle cooking
Mains average: $$
Specialty: Wild Italian herb and ricotta gnocchi with duck ragu, crisp sage and mustard fruit
Local flavours: Most dishes use local produce
Extras: There is an outside pizza and wine bar with live music. The restaurant has views of the vineyard.
- **Open:** Lunch daily, dinner Sat
- **Location:** Near the corner of Shands Rd

MERRICKS NORTH
DINING

SALIX RESTAURANT AT WILLOW CREEK WINERY
Michael and Rebecca Cook (owners), Sean Duggan (chef)
166 Balnarring Rd, Merricks North 3926
Phone: 03 5989 7640
Email: salixrestaurant@bigpond.com.au
www.willow-creek.com.au
Licensed
Cuisine: Modern Australian
Mains average: $$$
Specialty: Twice-cooked duck served on fresh rice noodles with Asian vegetables
Local flavours: Cherries, berries, lettuce, meats, cheese
Extras: Outdoor decking overlooks the vineyard. Takeaway available.
- **Open:** Lunch daily; dinner Fri-Sat

MOOROODUC

🍴 DINING

JILLS AT MOOROODUC ESTATE
Jill (chef) and Richard McIntyre
501 Derril Rd, Moorooduc 3933
Phone: 03 5971 8506/7
Email: moorooduc@ozemail.com.au
www.moorooducestate.com.au
Licensed
Cuisine: Seasonal organic, Mediterranean influenced
Mains average: $$
Specialty: Small, seasonal à la carte menu, changing six-weekly
Local flavours: As much as possible
Extras: Grows much of its own seasonal produce as well as eggs, verjuice, and wine. Architect-designed building and accommodation with great views.
■ **Open:** Lunch weekends, dinner Fri–Sat

MORNINGTON

🏠 STORES

HOUGHTON'S FINE FOODS
Mark and Sue Houghton
7/59 Barkly St, Mornington 3931
Phone: 03 5975 2144
Email: houghtons@bigpond.com
Specialising in quality, take-home foods, Houghton's (established in 1989) relies greatly on the superb local produce available in this region. Houghtons' Range includes casseroles, curries, pies, quiches and flans, as well as cakes and slices. Ideal complements are their own pickles and jams, and some quality local olive oils. There are local and imported speciality cheeses—with knowledgeable staff to assist you in your choice—and locally produced breads (French-style and traditional sourdough). There are cakes and flans, homemade curries and casseroles and innovative, freshly made salads. Five Senses coffee is roasted daily in Mornington to order, and there is now an extensive range of Mornington Peninsula wines.
■ **Open:** 9am–5.30pm weekdays, 9am–5pm Sat, 9am–2pm Sun

MOUNT ELIZA

🍴 DINING

MORNING STAR ESTATE
Judy Barrett
1 Sunnyside Rd, Mount Eliza 3931
Phone: 03 9787 7760
Email: cellardoor@morningstarestate.com.au
www.morningstarestate.com.au
Licensed
Cuisine: Modern Australian
Mains average: $$$
Local flavours: Vegetables, seafood, wines
Extras: One of the best views on the peninsula, extensive garden with over 30 000 roses. Hotel accommodation available. Range of seasonal preserves available at the cellar door.
■ **Open:** Lunch daily, dinner Fri–Sat

PORTARLINGTON

🍴 DINING

KATIALO RESTAURANT
Steven (chef) and Alexandra Souflas
98 Newcombe St, Portarlington 3223
Phone: 03 5259 3580
Licensed/BYO

Cuisine: Modern European
Mains average: $$
Specialty: Goat, seafood dishes
Local flavours: Calamari, seafood, vegetables, olive oil
Extras: Outdoor seating, views through the park to the bay beyond.
- **Open:** Breakfast and lunch Wed–Sun, dinner Wed–Sat

PORTSEA
DINING

PIERS RESTAURANT
The Grange at Portsea Village, 3765 Point Nepean Rd, Portsea 3944
Phone: 03 5984 8484
Email: portseavillage@grangecc.com.au
www.grangecc.com.au
Licensed
Cuisine: Modern Australian
Mains average: $$$
Local flavours: Mornington Peninsula beef, seafood, mussels, strawberries, cheeses, wines
Extras: Apartment-style accommodation and resort facilities. Terrace extension of the dining room for outdoor dining.
- **Open:** Breakfast, lunch and dinner daily (in-house guests), hours vary seasonally for others

RED HILL
PRODUCERS

ELLISFIELD FARM
Barry and Elizabeth Pontifex
109 McIlroys Rd, Red Hill 3937
Phone: 03 5989 2008
Email: ellisfieldfarm@bigpond.com
Not many cherry growers will let you pick your own cherries, but here pick-your-own sweet cherries are available from late November to early January. Then there are sour morello cherries from January to mid February, quinces from late March to April, as well as jams, preserved quince and even a morello cherry fortified dessert wine. All this plus a rose covered cottage ideal for two, and farm-fresh breakfast provisions. Look for the products in local restaurants and regional produce stores on the Mornington Peninsula.
- **Open:** 8.30am–5pm daily (late Nov–mid Feb and late Mar–early May, phone to check availability)

RED HILL CHEESE
Jan and Trevor Brandon
81 William Rd, Red Hill 3937
Phone: 03 5989 2035
Email: rhr@alphalink.com.au
www.redhillcheese.com.au
A little off the beaten track, but worth locating to find handmade cheeses sourced from 'the best organic cows and goats milk available'. The cheese makers use a vegetarian rennet, and the cows milk comes from a single herd farm in Gippsland. Four years ago, with the burgeoning wine industry surrounding the family's farm, the Brandons sensed a need for a local cheese to complement the wine and set up the only cheesery on the Mornington Peninsula. Trevor has a food microbiology degree and had always made the family's yoghurt and cheese, so it was a comparatively easy step from there. One-day home cheese-making classes are available regularly, but book early as they are popular. Accommodation at Red Hill Retreat B&B is available all year.
- **Open:** 12pm–5pm daily

- **Location:** Look for the large Main Ridge Estate winery sign on the left and the finger sign to Red Hill Cheese, which is 800 metres down William Rd on the left

RED HILL SOUTH

🏠 STORES

RED HILL COOL STORES AND ART GALLERY
Gillian Haig
165 Shoreham Rd, Red Hill South 3937
Phone: 03 5931 0133
Email: rhcs@pen.hotkey.net.au
Full of a wide range of local produce, jams and relishes, and also functioning as a cellar door for local wineries, there is also local pottery, art and craft, and jewellery here. 'I have a great love of the Peninsula', says Gillian Haig, the owner, explaining the eclectic mix. 'Why not show it all off at once?' She stocks olive oils—VGO, The Stream, Montalto—as well as Flinders Sauce, jams and chutneys, Red Hill Mud, Merricks marmalade, Ellisfield Farm conserve and local honey.
- **Open:** 10am-5pm Sun-Fri, 9am-6pm Sat

🍴 DINING

MONTALTO
John Mitchell
Montalto Vineyards and Olive Grove,
33 Shoreham Rd, Red Hill South 3937
Phone: 03 5989 8412
Email: restaurant@montalto.com.au
www.montalto.com.au
Licensed
Cuisine: Regional Australian
Mains average: $$$
Local flavours: Fish from Westernport Bay, berries, wild mushrooms, Red Hill cheese

Extras: Cooking classes throughout the year, sculpture exhibition each year, casual dining space in piazza courtyard. Grows much of its own produce including herbs, salad greens, tomatoes, zucchini, fruit and nuts as well as olive oil and verjuice. The olive grove provides oil, and there are cellar door sales of the estate's wine. Gourmet picnic hampers available.
- **Open:** Lunch daily, dinner Mon-Sat (summer) Fri-Sat (other months)

MAX'S AT RED HILL ESTATE
Max Paganoni (owner-chef)
Red Hill Estate, 53 Shoreham Rd,
Red Hill South 3937
Phone: 03 5931 0177
Email: info@maxsatredhillestate.com.au
www.maxsatredhillestate.com.au
Licensed
Cuisine: Modern International with Mediterranean and Asian influences
Mains average: $$$
Specialty: Local Flinders mussels served in a Red Hill Estate chardonnay cream sauce
Local flavours: Flinders mussels, vegetables, Red Hill strawberries, cheese, muesli, herbs, mushrooms, Max's olive oil
Extras: Outdoor terrace area featuring the 'best view on the Mornington Peninsula' overlooking Westernport Bay to Phillip Island and beyond. Award-winning wines, and winner of the 2002 American Express Best Restaurant in Victoria Award.
- **Open:** Lunch daily, dinner Fri-Sat

LINDEN TREE RESTAURANT
Tony Ryan
Lindenderry Country House Hotel,
142 Arthurs Seat Rd, Red Hill South 3937
Phone: 03 5989 2933
Email: restaurant@lindenderry.com.au
www.lindenderry.com.au

Licensed
Cuisine: Modern Australian
Mains average: $$$
Local flavours: Bread, fruit, berries, wines, preserves, duck, seafood
Extras: Alfresco dining courtyard, vineyard views, cellar door.
- **Open:** Breakfast, lunch and dinner daily

ROSEBUD
🍴 DINING

PENINSULA OCCASIONS
Shane Clarke
1489 Pt Nepean Rd, Rosebud 3939
Phone: 03 5982 0900
Email: clarkey46@optusnet.com.au
Mains average: $
Specialty: Red Hill roast coffee, focaccias
Local flavours: Local relishes and condiments
Extras: Outdoor dining, Peninsula produce (including homemade caramelised onions, pesto, sauces and relish), gourmet hampers, homewares, gifts, and deli. Takeaway available.
- **Open:** 9am–5pm daily (until 6pm Fri)
- **Location:** Between Boneo Rd and Rosebud Hospital

SAFETY BEACH
🍎 PRODUCERS

DROMANA BAY MUSSELS
Michael Hunder
211 Dromana Pde, Safety Beach 3936
Phone: 03 5987 3808
Email: mussels@alphalink.com.au
Michael Hunder specialises in farmed blue mussels, which take eighteen months to mature, and require a lot of work. The season for fresh sea mussels is from August to June, so don't bother coming in July. Bad weather can see this small place shut up shop because it creates conditions inhospitable to healthy mussels.
- **Open:** 10.30am–5pm daily in season

SORRENTO
🏠 STORES

STRINGERS STORES OF SORRENTO
Helen and Warwick Fairlie
2–8 Ocean Beach Rd, Sorrento 3943
Phone: 03 5984 2010
Email: stringer@elite.net.au
The specialty of this licensed grocery and cafe is its take-home prepared foods. Just about any need is covered. Stringers have a range of wines from over forty local wineries and lot of specialty foods including many brands from local producers. The cafe has coffee, cakes, and biscuits. The Fairlies say, 'We always shopped here and loved it, so we bought the store!'
- **Open:** 8am–5.30pm daily (later in summer)

🍴 DINING

SMOKEHOUSE SORRENTO
Elizabeth Blan and David Stringer
182 Ocean Beach Rd, Sorrento 3943
Phone: 03 5984 1246
Email: smokehousesorrento@ozemail.com.au
Licensed
Cuisine: Mediterranean
Mains average: $$
Specialty: Wood-fired pizza, fish
Local flavours: Seafood, Mt Emu Creek sheeps milk fetta, Bay mussels
Extras: Close to the beach.
- **Open:** Dinner daily

ST LEONARDS
🏠 STORES

WHITE FISHERIES PTY LTD
Dennis and Greg White
Shop 3, Murradoc Rd, St Leonards 3223
Phone: 03 5257 1611
Email: whitefisheries@hotmail.com
These wholesale distributors of local King George whiting and snapper source the fish directly from local fishermen for absolute freshness and sell it live, cooked or frozen (on a daily basis).
■ **Open:** 9am–12pm weekdays, 9am–11am Sat

WALLINGTON
🍎 PRODUCERS

WALLINGTON BERRIES
William Wieland
440 Wallington Rd, Wallington 3221
Phone: 03 5250 1541
This farm has been growing strawberries since 1972. There are masses of them—three or four acres—says William, and you can pick your own. But if you have a bad back, or not enough time, you can buy your strawberries loose, already picked, for an extra charge.
■ **Open:** 9am–5pm daily

WAURN PONDS
🍴 DINING

PETTAVEL WINERY AND RESTAURANT
Mike and Sandi Fitzpatrick
65 Pettavel Rd, Waurn Ponds 3216
Phone: 03 5266 1120
Email: pettavel@pettavel.com
www.pettavel.com
Licensed
Cuisine: Modern Australian
Mains average: $$$
Local flavours: Varies seasonally
Extras: The restaurant's floor-to-wall glass windows overlook thirty-five acres of shiraz vines. There is a fully operational winery with a viewing window so you can see the workings of the winery.
■ **Open:** Lunch daily and dinner Fri (Nov–Feb)
● **Location:** Cnr of Princes Hwy

MARKETS AND EVENTS

✦ **GEELONG FARMERS' MARKET,**
GEELONG
03 5272 2777
7.30am–12.30pm second Sat

✦ **OCEAN GROVE PRIMARY SCHOOL**
Apple Fair, Ocean Grove
03 5255 1340
Sun of Labour Day long weekend in Mar

✦ **OTWAY HARVEST FESTIVAL,**
GEELONG
1800 620 888
Apr

✦ **TOAST TO THE COAST,** GEELONG
03 5227 0270
Nov

✦ **WALLINGTON STRAWBERRY FAIR,**
WALLINGTON
03 5250 1841
Nov

GOLDFIELDS

Built solidly, built well. The Goldfields region has a comfortable air of abundance about it. Banks built to withstand bushrangers, hotels large enough to celebrate the discovery of a new reef of gold, land so fertile it can feed all comers. Ballarat is the centre of this prosperous region, an affluent yet up-to-date city with a buzzing cafe culture. More established restaurants guard the cuisine of the region, with menus that rely on today's natural wealth—local lamb, lavender, wheat, fruits and vegetables, yabbies and trout. The nearby Pyrenees region balances the dining experience with its fine wines, which combine with the food to give diners a priceless experience.

ARARAT
DINING

VINES CAFE AND BAR
Sandra Alsop (chef) and Ron O'Malley
74 Barkly St, Ararat 3377
Phone: 03 5352 1744
Email: stnicholas@ozisp.com.au
Licensed
Cuisine: Modern regional
Mains average: $$
Specialty: Local, fresh produce
Local flavours: Organic greens, eggs, meats, jams, honey, bread
Extras: Courtyard dining, and a blackboard menu. Takeaway available.
■ Open: Breakfast and lunch daily, dinner Fri–Sat

AVOCA
DINING

BLUE PYRENEES ESTATE
Julie Driscoll (chef)
Vinoca Rd, Avoca 3467
Phone: 03 5465 3202
Email: info@bluepyrenees.com.au
www.bluepyrenees.com.au
Licensed
Cuisine: Seasonal, regional
Mains average: $$
Specialty: Eclectic mix of dishes
Local flavours: Lamb, beef, duck
Extras: Outdoor dining area.
■ Open: Lunch weekends and public holidays
● Location: 7km west of Avoca

BENALLA
DINING

GEORGINA'S RESTAURANT
Leigh Hall and Elly Havers
100 Bridge St, Benalla 3672
Phone: 03 5762 1334
Licensed
Cuisine: Modern Australian
Mains average: $$
Specialty: Regular specials
Local flavours: As many as possible, varies seasonally
Makes for sale: Tomato chilli jam
Extras: Wine and food matching groups (of over eighteen people) can be organised, featuring guest winery and winemaker, while chef Leigh Hall explains matching wine with food.
■ Open: Lunch Thurs–Fri, dinner Tues–Sat

BENDIGO

🍎 PRODUCERS

VICTORIAN OLIVE GROVES
Peter Caird (owned by the Caird, de Pieri, Hart, and Zito families)
Area 15, Mayfair Park, McDowells Rd, Bendigo 3550
Phone: 03 5441 5388
Email: vog@iinet.net.au
www.victorianolivegroves.com
The purchase of an established olive grove in 1996 and the desire to make a quality product was the beginning for these four families. Sold under the Victorian Olive Groves or VOG label, the multiple award-winning range includes infused oils (garlic, lemon, mandarin), three varietals and two blends, which are sold in all Australian capital cities plus Germany, Canada, and the United Kingdom. Stefano de Pieri uses these oils exclusively in his restaurants.
■ **Open:** By appointment

🍴 DINING

BAZZANI
Brendan Tuddenham (chef), Kathie Bolitho
2-4 Howard Place, Bendigo 3550
Phone: 03 5441 3777
Email: office@bazzani-bendigo.com
www.bazzani-bendigo.com
Licensed
Cuisine: Modern Australian
Mains average: $$
Local flavours: Figs, game, local cheeses, various herbs and vegetables
Extras: Many awards have been won by this restaurant. Stunning outdoor area at the front, located off the street and overlooking a park.
■ **Open:** Lunch daily, dinner Mon–Sat
● **Location:** Overlooking Rosalind Park in the heart of the city—off Bridge St

QUILLS RESTAURANT
Brendan and Judy Peterson (owners), Aaron Jose (chef)
286 Napier St, Bendigo 3550
Phone: 03 54455344
Email: lvmi@bigpond.net.au
www.lakeviewbendigo.com.au
Licensed
Cuisine: Modern
Mains average: $$
Specialty: Bread, pasta and ice-creams
Local flavours: Source as much produce locally as possible
Extras: Lake views.
■ **Open:** Dinner Mon–Sat, or by appointment
● **Location:** Opposite Lake Weeroona

WHIRRAKEE RESTAURANT AND WINE BAR
Nikki Halleday (chef)
17 View Point, Bendigo 3550
Phone: 03 5441 5557
Email: whirrakee@whirrakee.com.au
www.whirrakee.com.au
Licensed
Cuisine: Modern Australian
Mains average: $$
Specialty: Kangaroo dishes
Local flavours: Eggs and fruits
Extras: A huge window and an outdoor dining area overlook the Alexandra Fountain and Rosalind Park. There is a range of ice-creams made for the restaurant and for sale.
■ **Open:** Lunch Wed–Fri, dinner Tues–Sat
● **Location:** Opposite the historic Alexandra Fountain in the centre of Bendigo

CASTLEMAINE

🍴 DINING

THE GLOBE RESTAURANT
Gary Tugball
81 Forest St, Castlemaine 3450
Phone: 03 5470 5055
Email: globerestaurant@telstra.com
www.castlemaineweb.com/globe
Licensed
Cuisine: Modern Australian
Mains average: $$
Specialty: Duckling—double-cooked duck Maryland with a port cherry sauce
Local flavours: Freshwater trout and local game, fruit and vegetables
Extras: The Globe is set in an historic building dating from the 1850s. The restaurant overlooks a garden courtyard, ideal for evening dining during warmer months.
- **Open:** Dinner daily (Mon and Tues takeaway pizza/pasta only)

HARCOURT

🍎 PRODUCERS

DEUMER'S HARCOURT VALLEY ORCHARD
Alan and Pam Deumer
3389 Calder Hwy, Harcourt 3453
Phone: 03 5474 2181
Email: deumer@bigpond.com.au
Unwaxed apples are the drawcard here, although fresh pears and fresh, unpasteurised apple juice and pear juice are also popular at this roadside stall.
- **Open:** 10am–5pm daily (Mar–Nov)
- **Location:** 1.5km south of Harcourt

🍷 BREWERY/WINERY

HENRY OF HARCOURT
Drew, Irene and Michael Henry
219 Reservoir Rd, Harcourt 3453
Phone: 03 5474 2177
Harcourt apples are widely regarded as some of the best in Victoria, and the Henrys are capitalising on the unique flavour by turning apples from their orchard into cider. They have Pink Lady cider made from apples, and perry, which is made from pears. Look for them at Castlemaine Cellars or at the farm.
- **Open:** 10am–5pm daily

HEATHCOTE

🍴 DINING

EMEU INN RESTAURANT, B&B AND WINE CENTRE
Fred (chef) and Leslye Thies
187 High St, Heathcote 3523
Phone: 03 5433 2668
Email: info@emeuinn.com.au
www.emeuinn.com.au
Licensed
Cuisine: International
Mains average: $$
Specialty: International cuisine, using native Australian bush products
Local flavours: Yabbies, pigeons, turkey
Extras: Grows vegetables, herbs and fruit (including oranges, plums, peaches and crab apples). Makes preserves, orange slices dipped in chocolate, oils, sauces and other products, which are for sale in the wine shop. Sells wines from around fifty local producers, and there are select tastings of products each week.
- **Open:** Lunch, afternoon tea and dinner Thurs–Mon

REDBANK
🍴 DINING

THE FLYING PIG DELI AT REDBANK WINERY
Neill and Sally Robb
'Redhill', 1 Sally's Lane, Redbank 3478
Phone: 03 5467 7255
Email: info@sallyspaddock.com.au
www.sallyspaddock.com.au
Licensed
Cuisine: Light lunches, platters, antipasto
Mains average: $
Specialty: Neill's smoked lamb, Sally's soups and delicious cakes
Local flavours: Estate-grown items, fresh bread from the Avoca bakery
Extras: Outdoor seating available with views of the Pyrenees ranges and the Sally's Paddock vineyards. Sells homemade smoked lamb, prosciutto, estate-grown olives, quince paste, jams and chutneys.
- **Open:** Lunch daily
- **Location:** At the 200km post on Sunraysia Hwy (B220), 18km north of Avoca

SAILORS FALLS
🍴 DINING

SAULT RESTAURANT AND FUNCTIONS
Troy McDonagh and Elissa Elliott, Benjamin O'Brien (chef)
2349 Ballan-Daylesford Rd, Sailors Falls 3461
Phone: 03 5348 6555
Email: elle@sault.com.au
www.sault.com.au
Licensed
Cuisine: Middle Eastern-influenced Modern Australian
Mains average: $$
Specialty: Use of local and home-grown produce
Local flavours: Farm fresh eggs, Istrian smallgoods, regional potatoes, mushrooms
Extras: Grows own herbs, spices and seasonal vegetables as well as trout from own lake and seasonal freshwater yabbies. Sells homemade elderberry jam, blackberry jam and a range of Sault salts. Takeaway available. Deck with 270-degree views over the property.
- **Open:** Breakfast, lunch and afternoon tea Fri-Sun, dinner daily
- **Location:** 5km from Daylesford

SMEATON
🍴 DINING

TUKI TROUT FARM
Robert D Jones
Stoney Rises, Smeaton 3364
Phone: 03 5345 6233
Email: info@tuki.com.au
www.tuki.com.au
Licensed
Mains average: $$
Specialty: Farm raised trout and lamb
Extras: Guests may fish if they wish. Homemade smoked trout pâté and lamb sausages are available for sale. Self-contained bluestone cottages, available with fly-drive or other packages. The complex has won a number of tourism awards. Products are also available at Daylesford Naturally Fine Foods, and Cliffy's, Daylesford.
- **Open:** Lunch, dinner (by arrangement) daily
- **Location:** 8km north of Smeaton on the Newstead-Castlemaine Rd

SUTTON GRANGE
🍎 PRODUCERS

COLLIERS CHOCOLATES
Jill and Ray Collier
1221 Sutton Grange-Elphinstone Rd,
Sutton Grange 3448
Phone: 03 5474 8232
Email: chocolate@collierschocolates.com
www.collierschocolate.com
These chocolates, handmade using the finest Swiss couverture, have won the hearts of all who have tried them. There are layered cocktail chocolates, pralines, and more.
■ **Open:** 11am-5pm, weekends (groups by appointment at other times)
● **Location:** At Elphinstone Flyover, turn right onto Sutton Grange Rd, then continue for 12km

MARKETS AND EVENTS

✦ **BENDIGO FARMERS' MARKET,** BENDIGO
03 5444 4646
Second Sat

✦ **CENTRAL VICTORIA FARMERS' MARKET,** HARCOURT
03 5470 6340
10am–3pm first Sun

✦ **LANCEFIELD AND DISTRICT FARMERS' MARKET,** LANCEFIELD
03 5429 2115
9am–1pm fourth Sat (except third Sat in Dec)

✦ **AUSTRALIAN COUNTRY LIFE FESTIVAL,** GREAT WESTERN
03 5361 2239
Sun of long weekend in Jan

✦ **BENDIGO EASTER WINE FESTIVAL,** CASTLEMAINE
1300 656 650
11am–5pm Easter Sun

✦ **HARCOURT APPLEFEST,** HARCOURT
03 5474 2104
Long weekend in Mar

✦ **HEATHCOTE WINE AND FOOD FESTIVAL,** HEATHCOTE
03 5433 3121
July

✦ **HEATHCOTE'S DEEP WINTER WINE DINNER,** HEATHCOTE
03 5433 2668
Third Sat in Aug

✦ **PYRENEES PINK LAMB AND PURPLE SHIRAZ COUNTRY RACE MEETING,** AVOCA
03 5465 3231
Sat of long weekend in Mar

✦ **PYRENEES VIGNERONS GOURMET WINE AND FOOD RACE MEETING,** AVOCA
03 5465 3231
Anzac Day

✦ **TALBOT YABBIE FESTIVAL,** TALBOT
1800 356 511 or 03 5463 2555
10am–4pm Easter Sat

✦ **TALTARNI AVOCA CUP,** AVOCA
03 5465 3231
Oct

GOULBURN MURRAY WATERS

If one river symbolises Australia, then the mighty Murray would have to be it. Here the river is broad and majestic and its waters create a productive green belt along the banks. Riverside, there are wineries and fine restaurants, and of course paddleboats by the score that take visitors out on day trips or holiday cruises. Many, vying for ways to impress their passengers, offer meals, and often these include the local produce. Irrigation has been the basis for the establishment of major fruit growing in this area and some of the country's largest fruit canneries are located here. Dairying is also an important industry. Indeed you could be well fed eating just what comes from within a few kilometres of this famous waterway.

ALEXANDRA
DINING

STONELEA COUNTRY ESTATE
George and Cathie Watkins
Connelly's Creek Rd, Alexandra 3714
Phone: 03 5772 2222
Email: stay@stonelea.com.au
www.stonelea.com.au
Licensed
Cuisine: Victorian
Mains average: $$$
Specialty: Baked Buxton trout
Local flavours: Trout, salmon, vegetables, fruits

Extras: Recently named one of the top six restaurants in regional Victoria. Serves 'Victorian regional cuisine', a fusion of European and Pacific styles relying upon fresh produce and local ingredients. Set on 1500 acres with extensive gardens, ornamental lakes and recreational facilities including an eighteen-hole golf course and expansive views.
■ **Open:** Breakfast, lunch and dinner daily
● **Location:** Corner Maroondah Hwy and Connelly's Creek Rd, Acheron near Alexandra

AVENEL
DINING

HARVEST HOME COUNTRY HOUSE HOTEL
Suzi McKay
1 Bank St, Avenel 3664
Phone: 03 5796 2339
Email: info@harvesthome.com.au
www.harvesthome.com.au
Licensed
Cuisine: Regional Australian
Fixed price multi-course meal: $$$
Specialty: Fine regional food
Local flavours: Everything—the hotel has a three-acre organic garden
Extras: Grows all its own produce. The complex spans eleven properties (almost the whole of the main street of Avenel) and also has about 2.5 acres of flower gardens, ponds, and hideaways. Multi-award-winning chef. Self-catering cottages and rooms in the hotel.

■ Open: Breakfast, lunch and dinner daily (for in-house guests), lunch and dinner Thurs–Sun, or by prior arrangement

PLUNKETT WINES
Plunkett family, Heather Hammond (chef)
Lambing Gully Rd, Avenel 3664
Phone: 03 5796 2150
Email: max@plunkett.com.au
www.plunkett.com.au
Licensed
Cuisine: Country
Mains average: $$
Specialty: Platters
Local flavours: Strathbogie Ranges beef
Extras: Mediterranean-style courtyard; great views of the Strathbogie ranges. Wine tasting and winery tours.
■ Open: Lunch Thurs–Mon
● Location: Corner Hume Hwy and Lambing Gully Rd

KIALLA WEST
DINING

BELSTACK STRAWBERRY FARM
Peter and Margaret Tacey
80 Bennetts Rd, Kialla West 3631
Phone: 03 5823 1324
Email: info@belstack.com.au
www.belstack.com.au
Licensed
Cuisine: Light lunches and teas
Mains average: $
Specialty: Scones, strawberry sundaes
Local flavours: Farm fruit/vegetables
Extras: This strawberry farm also grows boysenberries and blackberries—pick your own available or buy from the orchard shop in season. Visitors can also enjoy bushwalks on the property. Sells a range of homemade jams, pickles and sugar-free foods, as well as frozen berries.
■ Open: Lunch and teas daily
● Location: 10km south of Shepparton, off the Goulburn Valley Hwy

LEMNOS
STORES

CAMPBELL'S SOUPS AUSTRALIA
55 Lemnos North Rd, Lemnos 3631
Phone: 03 5833 3444
Email: heather_homes@campbellsoup.com
www.campbellsoup.com
You can't miss this place under the massive Campbell's soup can, poised a dozen or so metres above the buildings. Inside the small factory shop there are masses of discounted Campbell's products—stocks, soups, sauces, Fray Bentos pies and corned beef, and Kettle Chips with discounts at up to fifty per cent off retail prices.
■ Open: 8am–4.30pm weekdays, 9am–12.30 Sat, product tastings 9am–4pm weekdays

MOOROOPNA
PRODUCERS

SPC ARDMONA FACTORY SALES
Jim Andreadis, manager
91–95 McLennan St, Mooroopna 3629
Phone: 03 5825 2444
Email: info@ardmonafactorysales.com.au
www.ardmonafactorysales.com.au
The Goulburn Valley's first orchards were planted on riverbanks near Shepparton in 1884. Since then, especially with the spread of irrigation, the area has become

widely known for the quality of its fruit. SPC was Victoria's first country fruit cannery, established in 1918. Today it employs 1500 staff and turns out millions of cans of fruit, tomatoes and other foods annually using local produce. The stock sold at the factory shop may be dented or unlabelled, but of course the quality of the fruit inside conforms to the same high standard of all the products sold under the SPC, Ardmona, Goulburn Valley, Campbell's Soups, IXL, Golden Circle, Cadbury, Mars, Heinz/ Greenseas, Berri and Cerebos brand names in supermarkets all over Australia, and it is heavily discounted—twenty to seventy per cent less.
- **Open:** 9am–5pm daily
- **Location:** 4km west of Shepparton on Midland Hwy

NAGAMBIE
DINING

MITCHELTON WINERY RESTAURANT
Melissa and Brett Dobson,
Matt and Adele Aitken
Mitchelton Winery, Mitchellstown-Tahbilk Rd, Nagambie 3608
Phone: 03 5794 2388
Email: restaurant@mitchelton.com.au
www.mitchelton.com.au
Licensed
Cuisine: Modern regional
Mains average: $$
Specialty: Eye fillet of beef
Local flavours: Murray cod
Extras: Freshly made bread is an added bonus at this restaurant that specialises in matching food and wine, and has a dining area that overlooks the Goulburn River. Antipasto picnic platters available.
- **Open:** Lunch daily
- **Location:** Entrance off Goulburn Valley Hwy, 4km south of Nagambie

RUFFY
STORES

RUFFY PRODUCE STORE
Helen McDougall and Doug Maclean
26 Nolans Rd, Ruffy 3666
Phone: 03 5790 4387
Email: ruffystore@mcmedia.com.au
What a great find. This produce store in the Strathbogie Ranges may be off the beaten track but it's well placed to pick up picnic supplies. You'll find light lunches using local produce that will keep you going as you travel in the area, and the shop also sells local produce and Strathbogie wines.
- **Open:** 8am–6pm weekends and public holidays
- **Location:** 21km off the Hume Hwy, north of Avenel

SHEPPARTON
PRODUCERS

BERRY SWEET AUSTRALIA AND TUMBARUMBA BLUEBERRIES
Nicky Postma
7 Grant Court, Shepparton 3630
Phone: 03 5822 0007
Email: berrysweet@berrysweet.com.au
www.berrysweet.com.au
Berries, fruit sorbets, and blueberry 'shrivels' can be found here, as well as

blueberry juice, raspberry juice, and a mix of the two, plus strawberry puree, and frozen berry punnets. It's all grown at this Tumbarumba property established about ten years ago. In the last two years Postma has been producing the only 100 per cent blueberry juice in the world. The farm is now certified organic.
- Open: 9am-3pm, weekdays

DUCAT'S FOOD SERVICE
Ray Ducat
116-124 Corio St, Shepparton 3630
Phone: 03 5823 4800
Email: michelle@ducatsfood.com.au

For around ninety years this company has been producing, marketing and distributing fresh milk throughout the Goulburn Valley. In the early days, Percy Ducat and his two sons carted milk to Shepparton from the family farms at Grahamville and Ardmona. Over the years the company grew, from a milk bar in Shepparton, to a milk depot, and finally settled at the current factory site in 1945. Today's managing director is the founder's grandson, making this a true family business. Milk and fruit juices, pastries, desserts and other dairy products are all available from the showroom or at outlets throughout the Goulburn and Murray Valley region.
- Open: 8.30am-5.30pm weekdays, 9.30am-12pm Sat

SOUTH SHEPPARTON

TOURS

EMERALD BANK HERITAGE FARM
Dianne Scott
Goulburn Valley Hwy,
South Shepparton 3632
Phone: 03 5823 2500
Email: scotty2@austarnet.com.au

Set on the banks of the Seven Creeks and smack-bang in the centre of Victoria's food bowl, this farm is ideally located to present such an innovative concept. On this ten-hectare working farm, presented as if in the 1930s, you can experience farming as it once was here. Poultry raising, vegetable-growing in season, and butter-making demonstrations bring to life the history of this rich agricultural region. Ideal for school groups with bookings for weekends and school holidays. A kiosk serves light snacks and drinks.
- Open: 10am-4.30pm weekdays, 10am-4.30pm daily (during Victorian school holidays and public holidays)
- Location: About five minutes south of Shepparton

TONGALA

🍎 PRODUCERS

GOLDEN COW DAIRY EDUCATION CENTRE
Sharon Goodwin-Wicks
Cnr Henderson and Finlay Roads,
Tongala 3621
Phone: 03 5859 1100
Email: goldencow@mcmedia.com.au
This is the only working dairy attached to an interpretative centre, so it is able to offer dairy industry displays as well as daily milking demonstrations. It is appropriately located at Tongala, the centre of Australia's largest milk-producing region. There are videos and displays, tours, a working irrigation model, dairy museum and a cafe/milk bar serving, as you would imagine, excellent milk shakes in thirty-nine flavours. Milking is at 11am daily.
- **Open:** 10.30am–4pm Wed–Mon
- **Location:** 6km off the Murray Valley Hwy, twenty minutes from Echuca, at the south end of Tongala

MARKETS AND EVENTS

✦ **AVENEL PRODUCE MARKET,** AVENEL
03 5796 2339
12.30pm–5pm, second Sun

✦ **MOOROOPNA FRUIT SALAD DAY,** SHEPPARTON
03 5831 4400/1800 808 839
Feb

✦ **TASTE OF TATURA WINE AND FOOD FESTIVAL,** TATURA
1800 808 839
11am–5pm first Sun in Mar

LAKES AND WILDERNESS

This is the visitor's first exposure to the countryside when entering from coastal New South Wales—a beautiful yet sometimes daunting landscape, rising swiftly to mountains inland.

Those who have been before and have, quite literally, tasted what is on offer here, have no difficulty in returning. The plenitude of places to the west may be lacking, but the quality more than makes up for quantity. For here you will find a vast array of seafood and all things fishy, with restaurants more than equipped to deal with them. There is a winery or two (including fruit wines), orchards, ice-cream and of course the magic lakes and mysterious wildernesses to explore.

CASSILIS
BREWERY/WINERY

MT MARKEY WINERY
Christine and Howard Reddish
1344 Cassilis Rd, Cassilis 3896
Phone: 03 5159 4328
Email: mtmarkey@tpg.com.au
www.omeoregion.com.au/winery
If the idea of a sip of elderberry port before bedtime makes your eyes glaze over, you'd better get acquainted with Christine and Howard Reddish. They make this and a host of other old-style tipples, plum and other fruit wines, and a unique mead—Olde Horny wood-aged fruit and honey mead, mixed with elderberry, blackberry and grape. All the fruit is from their own orchard, and even the honey—although collected by an apiarist friend—is from hives placed under their own trees. Howard had been a hobby winemaker for years before finally taking the plunge commercially. At their cellar door, they also sell other local products such as fruit and herb vinegars, honey, natural spring water, pottery, and occasionally homemade jams and pickles.
- **Open:** 9am–5pm daily
- **Location:** Now Omeo–Swifts Creek Rd (formerly known as Cassilis Rd)

GIPSY POINT
DINING

GIPSY POINT LODGE AND RESTAURANT
Ian and Libby Mitchell (owners), Matthew Allen (chef)
Macdonalds St, Gipsy Point 3891
Phone: 03 5158 8205
Email: info@gipsypoint.com
www.gipsypoint.com
Licensed
Cuisine: Modern Australian
Three courses: $$
Specialty: Set menu changes every night
Local flavours: Fish, seasonal vegetables, berries
Extras: This wilderness lodge is located on Mallacoota Inlet, boat hire available.
- **Open:** Dinner daily, bookings essential
- **Location:** Near Mallacoota

JOHNSONVILLE

◆ PRODUCERS

THE FRUITFARM JOHNSONVILLE
Graeme and Elaine Jenkins
Bumberrah Rd, Johnsonville 3902
Phone: 03 5156 4549
Email: elaine@fruitfarm.com.au
www.fruitfarm.com.au
'Plan to stay a while,' say the Jenkinses, 'it's such a lovely area.' These fourth-generation orchardists are doing their part to make it even lovelier by having available quality farm-fresh and flavourful peaches, nectarines, apples, pears and cherries. Many are old varieties from Graeme's grandfather's garden, all picked, graded, sorted and packed by hand. Graeme also operates a weekly growers' stand at the Footscray wholesale markets. Their fruit is tree ripened for the best flavour and while you are buying fruit in the sales shed, less than seventy-two hours after it was picked, take time to admire Elaine's magnificent art works, as the area doubles as her gallery.
■ **Open:** 9am–6pm daily Nov–Easter, 10am–5pm daily after Easter
● **Location:** 500 metres north of Princes Hwy, halfway between Lakes Entrance and Bairnsdale

LAKES ENTRANCE

🏠 STORES

LAKES ENTRANCE FISHERMENS' CO-OP
Tom Davies, Manager
Bullock Island, Lakes Entrance 3909
Phone: 03 5155 1688
Email: lefcol@lefcol.com.au
The fish and seafood here is straight from local fishermen who catch 'a bit of everything'—ocean fish, shellfish and trawl fish. No cooked seafood or fish.
■ **Open:** 8.30am–5pm daily

RIVIERA ICE-CREAM PARLOUR
Peter and Beth Evans, Liz and David Evans
583 Esplanade, Lakes Entrance 3909
Phone: 03 5155 2972
This locally-made, award-winning gourmet ice-cream has won both state and national awards. It is a high quality product with dozens of unique flavours, made from rich, creamy organic milk.
■ **Open:** 9.30am–6.30pm daily (until 10pm during holidays and summer)
● **Location:** Opposite the footbridge, Lakes Entrance

🍴 DINING

THE BOATHOUSE
Paul and Melissa Coggan
Quality Inn Bellevue, 201 Esplanade, Lakes Entrance 3909
Phone: 03 5155 3055
Email: info@bellevuelakes.com
www.bellevuelakes.com
Licensed
Cuisine: Seafood
Mains average: $$
Specialty: Seafood platter for two, seafood buffet Friday nights
Local flavours: Seafood
Extras: Both cafe and restaurant overlook the harbour which is home to the Lakes Entrance Fishing Fleet—local traders that fish offshore or within the lakes. Takeaway available.
■ **Open:** Breakfast, lunch and dinner daily (restaurant open for dinner only)

LAKES AND WILDERNESS–VIC

HENRY'S WINERY CAFE, WYANGA PARK WINERY
Geoffrey and Lyndel Mahlook
Wyanga Park Winery, 222 Baades Rd, Lakes Entrance 3909
Phone: 03 5155 1508
Email: wyangapark@i-o.net.au
Licensed
Cuisine: Fresh local produce
Mains average: $$
Specialty: Homemade bread, desserts and scones
Local flavours: Fish, poultry, steaks, jams, sauces, seafood, cheeses, and condiments
Extras: Come by car or take a river cruise to this riverfront property with lake views from walking track, and resident artist.
■ **Open:** Lunch and morning and afternoon tea daily, dinner Fri–Sat

L'OCEAN FISH AND CHIPS
Constance and Guy Plateau
19 Myer St, Lakes Entrance 3909
Phone: 03 5155 2253
Email: sales@crispybatter.com.au
www.crispybatter.com.au
Cuisine: Seafood
Mains average: $
Specialty: Wheat- and gluten-free food, also free of dairy, yeast, colourings, and additives
Local flavours: Fish and seafood
Extras: Health conscious takeaways with the frying done in Heart Foundation approved brown rice oil. The local fish and seafood is prepared in a gluten- and wheat-free batter. Takeaway available.
■ **Open:** 11am–8pm Tues–Thurs, 11am–8.30pm Fri–Sun

NAUTILUS FLOATING DOCKSIDE RESTAURANT
McKenzie family
Cunninghame Arm, Western Boat Harbour, Lakes Entrance 3909
Phone: 03 5155 1400
Licensed
Cuisine: Regional seafood
Mains average: $$$
Specialty: Seafood platter for two
Local flavours: All fish, scallops, prawns, bugs, lobster (only mussels and oysters sourced from nearby Eden and Pambula)
Extras: The floating location ensures spectacular views of the water, the birds, fish and local fishermen.
■ **Open:** Dinner Tues–Sat (varies in peak periods)
● **Location:** Melbourne end of town

NOWA NOWA
🍴 DINING

MINGLING WATERS
Jenny and Mike Ryan
42 Princes Hwy, Nowa Nowa 3887
Phone: 03 5155 7247
Email: minglingwaters@bigpond.com
www.lakes-entrance.com/minglingwaters
BYO
Mains average: $
Specialty: Vegetarian dishes, curries, homemade bread and pita
Local flavours: Fresh, seasonal organic produce
Extras: A glass gazebo overlooks the lake, and the Ramsdell Gallery showcases a collection of unique, tree-root sculptures. Mike Ryan runs eco-tours in the area. Takeaway available.
■ **Open:** Breakfast and lunch daily, dinner Fri–Sat

STRATFORD

🍴 DINING

WA-DE-LOCK CELLAR DOOR
Graeme and Astrid Little
76 Tyers St (Princes Hwy), Stratford 3862
Phone: 03 5145 7050
Email: wadelock@netspace.net.au
Licensed
Mains average: $
Specialty: Grazing platters, smoked salmon and Tarago chevre, smoked trout platter
Local flavours: Cheese, smoked meats, chutneys, preserves, wines, beers
Extras: Gippsland's largest range of Gippsland wine under one roof—around fifty wineries represented. Factory outlet for Maffra Farmhouse Cheeses. Specialising in wine and food from all over Gippsland. Sells a large range of cheeses, smoked meats, chutneys, preserves, wine and beer.
■ **Open:** Breakfast and lunch Thurs–Tues

MARKETS AND EVENTS

✦ **BAIRNSDALE FARMERS MARKET,**
BAIRNSDALE
03 5657 2337
8am–1pm first Sat

✦ **LAKES ENTRANCE FISH TASTING,**
LAKES ENTRANCE
03 5155 1966 or 1800 637 060
Good Friday

LEGENDS, WINE AND HIGH COUNTRY

This region has a rich history, literally, with a solid foundation of gold mining. The towns with their prosperously strong buildings reflect this too.

Today's wealth, though, comes from other sources, much of it liquid in the form of wine, as this is one of Victoria's oldest vine-growing and winemaking regions. As so often happens, wine brings food, and visitors who require board and lodging. This lovely rural area now has some excellent B&Bs, many of them in lovingly restored homes from the mid 1800s.

Local cheeses, jams, pickles, trout and honey appear on the best tables everywhere. This is truly a land where the legends of this high country live on, transmuted today into fine wines and sublime food. Tomorrow's legends.

ALLANS FLAT
BREWERY/WINERY

SCHMIDT'S STRAWBERRY WINERY
Martin Schmidt
Yackandandah-Wodonga Rd,
Allans Flat 3691
Phone: 02 6027 1454

These fine strawberry wines are available at the cellar door for tastings and sales. Picked just metres away, the fresh flavour is captured perfectly, so it is no surprise to learn that Schmidt's wines have won international awards. Martin Schmidt is following a family history in winemaking. The wines are available dry, sweet, or semi-sweet, and there is a new strawberry liqueur. From October to Christmas, you can also buy fresh strawberries from the farm.
■ **Open:** 9am–5pm Mon–Sat, 10am–4pm Sun
● **Location:** Allans Flat turn-off, 10km north of Yackandandah

BEECHWORTH
STORES

BEECHWORTH BAKERY
Tom and Christine O'Toole
27 Camp St, Beechworth 3747
Phone: 03 5728 1132
Email: enquiries@beechworthbakery.com
www.beechworthbakery.com

Drop in here at the weekends and you will be amazed at the popularity of this award-winning bakery in such a tiny town. There is seated dining upstairs and downstairs for 250, and even then there are times when it is full house. Famous for 'bee stings', a type of cake, there are more than 250 different varieties of breads, cakes and pastries available, baked fresh daily on the premises. This is the second time Tom O'Toole has owned Beechworth Bakery, where pastry cooks have worked since 1857.
■ **Open:** 6am–7pm daily

BEECHWORTH PROVENDER
Ray Pond
18 Camp St, Beechworth 3747
Phone: 03 5728 2650
Email: raypond@yahoo.com

Just across the street from the Beechworth Bakery, the Beechworth

Provender also attracts attention. Ray Pond sells wines from eighty north-east Victorian wineries, olive oils, local cheeses, jams, pickles and chutneys, and smoked meats from this region. Proudly showcasing the rich bounty of this area, Ray is happy to advise you on various products, and a local wine is always available for tasting. This is the ideal spot to stock up your picnic basket for lunch in this most beautiful of regions.
- **Open:** 10am–5.30pm daily

GOLDFIELDS GREENGROCERS
Fay and Peter Mim
61–63 Ford St, Beechworth 3747
Phone: 03 5728 2303
Local hydroponic tomatoes, corn, a large variety of local potatoes, apples and apple juice, honey, pears, chestnuts, hazelnuts, walnuts, almonds, and home-made preserves cram the shelves at this greengrocer's shop which has been open for seven years and expanded significantly in that time to now include Milawa bread, local cordials, organic products and gluten-free foods. The Mims grow their own vegetables when possible.
- **Open:** 8am–6.30pm daily

DINING

GIGI'S OF BEECHWORTH
Luigi Cipolato
69 Ford St, Beechworth 3747
Phone: 03 5728 2575
Licensed/BYO
Cuisine: Italian regional
Mains average: $$$
Specialty: Tiramisu, chocolate brioche, great coffee
Local flavours: Chestnuts, tomatoes, figs, quinces, berries, handmade charcuterie

Extras: Takeaway available. Sells home-made muesli, pasta and porcini risotto.
- **Open:** Breakfast and lunch Thurs–Tues, dinner Thurs–Sat and Mon–Tues

THE SPIRITED CHEF FOODSTORE AND COMESTIBLES
Jayne Thatcher
9 Bridge Rd, Beechworth 3747
Phone: 03 5728 3227
Email: beechworthpreserves@dragnet.com.au
www.beechworthpreserves.com.au
BYO
Cuisine: Seasonal regional
Mains average: $
Local flavours: Fruits and vegetables, herbs, cheeses, smoked trout, smallgoods, free-range chicken
Extras: Outdoor courtyard dining and view of Red Hill. Manufacturers, retailers and wholesalers of premium quality preserves and regional ingredients. Tasting rooms and historic rustic produce shop. Takeaway available.
- **Open:** 10am–5pm weekdays

SPRING CREEK CAFE AND COURTYARD
Maree Rennie
78 Ford St, Beechworth 3747
Phone: 03 5728 2882
Email: rats48@optusnet.com.au
BYO
Mains average: $
Specialty: Speciality salads, gluten-free items including cakes and biscuits
Local flavours: Organic focaccias, gluten-free products
Extras: Breakfast is served all day. Takeaway available. Shaded garden courtyard.
- **Open:** 9am–5.30pm daily

LEGENDS, WINE AND HIGH COUNTRY—VIC

THE BANK RESTAURANT AND MEWS
Wayne and Denise McLaughlin
86 Ford St, Beechworth 3747
Phone: 03 5728 2223
Email: info@thebankrestaurant.com
www.thebankrestaurant.com
Licensed
Cuisine: Modern regional Australian
Mains average: $$
Specialty: Seasonally changing menus
Local flavours: Varies seasonally—asparagus, berries, orchard fruits, high country beef, lamb, cheeses, wines
Extras: B&B accommodation, courtyard dining for Sunday lunches, pre-dinner drinks or after-dinner fortifieds.
■ **Open:** Breakfast daily (for guests), lunch Sun, dinner Wed–Sun

THE GREEN SHED BISTRO
James Loveridge and Therese Shanley
37 Camp St, Beechworth 3747
Phone: 03 5728 2360
Licensed
Cuisine: Modern European
Mains average: $$
Specialty: Seafood, duck confit with apple, pear, and walnut salad
Local flavours: Apples, pears, local wines, cheeses, bread
Extras: Located in an historic printing office, built in the 1860s, with green trim—hence the name. Own cleanskin red wine available for sale.
■ **Open:** Lunch Fri–Sun, dinner Wed–Sun

BOORHAMAN
BREWERY/WINERY

BUFFALO BREWERY
Leanne and Greg Fanning, Steve Hill
Boorhaman Hotel, Boorhaman 3678
Phone: 03 5726 9215
Email: buffalobrewery@netc.net.au
www.buffalobrewery.com.au
Greg Fanning has taken over this microbrewery, conveniently situated at the Boorhaman Hotel, and brews a Buffalo lager, ginger stout and dark ale, plus a wheat beer and stout, using good Ballarat grain. The beer has been brewed here for seven years and began as a scheme to lure tourists to the region. Not only has this happened, but the brewery won two gold medals at the International Beer Awards in Melbourne, against over 400 competitors, even though it holds the distinction of being the smallest brewery in Australia. So far the beers are only sold at this hotel and at the Rutherglen Tourism Centre.
■ **Open:** Dinner Fri–Sat, 10am–8pm Mon–Thurs, 11am–11pm Fri–Sat, 11am–9pm Sun
● **Location:** Between Wangaratta and Rutherglen, off the Murray Valley Hwy

BRIGHT
DINING

KELLY'S HUMBLE SPUD
Yvonne Kelly
13 Ireland St, Bright 3741
Phone: 03 5755 2077
Cuisine: Country
Mains average: $
Specialty: Baked potatoes, coffee, ice-cream
Local flavours: Potatoes
Extras: Grows own potatoes. Outdoor dining area, photographs of the area, jigsaw puzzles. Takeaway available.
■ **Open:** 8.30am–5.30pm daily (later in holiday season)

SASHA'S OF BRIGHT AND SASHA'S WINE BAR
Sasha Cinatl
2D Anderson St, Bright 3741
Phone: 03 5750 1711
Email: mrsasha@netc.net.au
Licensed
Cuisine: Classic European and Australian
Mains average: $$
Specialty: Duck, chateaubriand, goulash
Local flavours: Poultry, squab, beef, duck, venison, goat, fresh berries and vegetables, and great range of local wines
Extras: Outdoor dining, wine list of eighty wines mainly regional, good music.
- **Open:** Dinner daily (breakfast and lunch by arrangement)

SIMONE'S OF BRIGHT
George and Patrizia Simone (chef)
98 Gavin St, Bright 3741
Phone: 03 5755 2266
Email: gsimone@brightvic.com
Licensed/BYO wine
Cuisine: Modern Italian
Mains average: $$$
Specialty: Homemade pastas
Local flavours: Rabbit, goat, pigeon, trout, nuts, chestnuts
Extras: Outdoor dining area, private function room. Sells homemade pickled cherries and marinated baby pears.
- **Open:** Dinner Tues–Sun (Mon during holidays)

SMOKO'S BIG SHED CAFE AND FRUIT AND VEG MARKET
Faith and Frank Russo
1438 Great Alpine Rd, Smoko, Bright 3741
Phone: 03 5759 2672
Email: ffrusso@netc.net.au
Licensed
Cuisine: Country food
Mains average: $
Specialty: Nanna's croquettes
Local flavours: Varies seasonally—tomatoes, cucumbers, zucchini, most vegetables, stone fruit
Extras: Spectacular views of snow-capped Mt Feathertop. Outdoor dining and takeaway available.
- **Open:** 10am–5pm Wed–Sun

VILLA GUSTO—LA DOLCE VITA
Maureen Hogan
630 Buckland Valley Rd, Bright 3741
Phone: 03 5756 2000
Email: villagusto@bigpond.com
www.villagusto.com.au
Licensed, regional wines only
Cuisine: Italian, l'Osteria-style
Four-course set menu: $$
Specialty: Daily changing menu
Local flavours: Smoked trout, free-range squab, veal, Milanese salads with local ingredients, Milawa cheeses, local wines only from Alpine and King valleys
Extras: The menu changes nightly and depends entirely on what is available to the chef. This highly luxurious villa is modelled on opulent Tuscan villas and there is an abundance of marble, antiques, art, gracious gardens, and fountains. The owner and chef are passionate advocates of the Slow Food philosophy and everything on the table each night is local and freshly sourced that day. The wines are exclusively regional. Chef will frequently travel 100km in the morning just to procure the perfect product for that day. It is not surprising that it has been voted in the top ten accommodation places in Australia.
- **Open:** Dinner Thurs–Sun

CHESHUNT
🍎 PRODUCERS

KING VALLEY OLIVES
Julie and James Findlay
Meringa, Ryans Lane, Cheshunt 3678
Phone: 03 5729 8413
Email: jandjfindlay@netc.net.au
Begun in 2003, this olive grove is already producing organically grown olives. There is oil and marinated olives, which are sold mainly at local markets.
▪ **Open:** By appointment

WARRAWEE ORCHARD
Hans and Anne Arnoldussen
Rose River Rd, Cheshunt 3678
Phone: 03 5729 8331
Email: arnoldussen@iprimus.com.au
The century-old Warrawee orchard has two remaining eighty-year-old King Cole apple trees that are still bearing fruit. The Arnoldussens grow 3000 apple trees, using very low amounts of chemicals, and producing around ten varieties—both old and new—including the most popular apple, Pink Lady, and also the flavoursome Cox's Orange Pippin.
■ **Open:** 10am–5pm daily, shed gate available twenty-four hours on an honesty system
● **Location:** On the Whitfield–Myrtleford back road, 2km from Cheshunt on the Myrtleford side, at Paradise Falls turn-off

CHILTERN
🍎 PRODUCERS

CHILTERN HONEY FARM
Simon Knowles and Kay Lavender
76 Conness St, Chiltern 3683
Phone: 03 5726 1286
Email: chfarm@netc.net.au
A general fascination with bees and the wide variety of honey flavours led Simon Knowles and Kay Lavender into this business. Their local honeys are sold only at the farm.
■ **Open:** Sales and tastings by appointment, honesty box system available daily
● **Location:** 100m from the centre of Chiltern

HOTSON'S CHERRIES
Bill and Lois Hotson
143 Old Cemetery Rd, Chiltern 3683
Phone: 03 5726 1358
Email: hotson@netc.net.au
Cherry season only lasts for about six weeks, so you need to move fast to catch these luscious fruits at their peak. At Hotson's you'll discover delicious freshly picked cherries, direct from the orchard, available in boxes for your convenience. The Hotsons have about fifteen acres of cherries, which equates to around 2500 trees and they've been growing them here for twenty years, but it's only in the last twelve years they have increased production. You can buy cherries from the shed or look for them at Wodonga farmers' markets, and a few other places locally.
■ **Open:** 9am–5pm weekends (Nov–Dec), weekdays by appointment
● **Location:** 4km south of Chiltern

EDI UPPER
🍎 PRODUCERS

PASSCHENDAELE VERJUICE
Lynda and Peter Jackson
1309 Cheshunt–Edi Rd, Edi Upper 3678
Phone: 03 5729 8362

Email: jackson@netc.net.au
This business was named after a grandfather who died in the war at Passchendaele, France. The Jacksons have been making this wonderfully tart verjuice for more than four years. They grow the grapes, which are picked in January before they are ripe and then processed in a similar fashion to wine, but without any resulting alcohol, preservatives or additives. The acidic juice can be used anywhere you would use vinegar or lemon—ideal to marinate steak, they say. Also made here is a tofu and a plain mayonnaise, as well as a tarragon and an orange salad dressing. Sold in delis and providores in north-east Victoria, Melbourne, and at shops in Sydney.
- **Open** 10am–5pm, weekends and public holidays, or by appointment
- **Location:** About 50km from Wangaratta

EUROBIN

🍎 PRODUCERS

BRIGHT BERRY FARM AND MT BUFFALO VINEYARD
Col and Lorraine Leita
RMB 5650, Great Alpine Rd, Eurobin 3739
Phone: 03 5756 2523
Email: lleita@bigpond.com
The patriotic colours of blueberries and bright red raspberries make this pick-your-own farm really live up to its name. There are fresh blueberries, blackberries and raspberries about ten varieties in all, and jams, syrups, and Scarecrow berry wines (raspberry, blackberry, blackcurrant, apple and blueberry), plus frozen berries. Cherries are sold roadside, and Mt Buffalo Vineyards has a range of red wines.

- **Open:** 8am–5pm daily (Dec–Apr), 10am–5pm Fri–Mon (at other times)
- **Location:** 19km from Myrtleford towards Bright on the Great Alpine Rd

WESTON'S WALNUTS
Michael Weston
Great Alpine Hwy, Eurobin 3737
Phone: 03 5756 2320
Michael Weston has been growing walnuts for thirty-seven years in this scenic area. You can buy them in the shell or as kernels, and Weston also sells almonds, hazelnuts, macadamias, and pecans from other growers, as well as his own home-grown fresh chestnuts in season. Sold in half-kilogram and kilogram packs, including mixed nuts, all are available from the shop on the premises, along with slices, nut and date loaf, pan forte, pickled walnuts, ligo (an eastern-Mediterranean walnut in syrup), jams and preserves.
- **Open:** 9am–5pm daily
- **Location:** About halfway between Bright and Myrtleford

GAPSTED

🍎 PRODUCERS

VALLEY NUT GROVES
Gillian Gasser
RMB 3360, Schlapps Rd, Gapsted 3737
Phone: 03 5752 1251
Although this farm sells principally in-shell walnuts, there are also pickled walnuts, made to Mrs Schlapp's (Gillian Gasser's mother's) own recipe, and walnut timber crafts. Until you have tasted really fresh nuts, you can never realise how addictive they are. Gillian says, 'Our nuts are always fresh, and people always want more once they are hooked!' Most of the

walnut crop is sold direct to the wholesale market, so count yourself lucky if you can buy some from the farm.
- **Open:** 9am–4pm Apr–June, coach groups welcome by appointment
- **Location:** 2.5km off the Ovens Hwy, 10km from Myrtleford

GLENROWAN
🍎 PRODUCERS

WHITE COTTAGE HERBS
Neil and Dawn Aird
30 Hill St, Glenrowan 3675
Phone: 03 5766 2285
Email: aird@whitecottageherbs.com.au
www.whitecottageherbs.com.au
White Cottage Herbs has been an organic herb grower for fifteen years. The Airds have some rare herbs too, such as astragalis, a Chinese medicinal herb, and around thirty mainstream culinary ones too, all organically grown and sold in small or medium pots. There are very few organic sellers of potted herbs in Australia, and White Cottage is organically certified with the Organic Herb Growers of Australia. There is also a display garden and the Airds take groups on tours and run courses in herb growing.
- **Open:** 9am–5pm Thurs and Fri

GOORAMADDA
🍎 PRODUCERS

GOORAMADDA OLIVES
Kathy and Jos Weemaes
RMB 1715, River Rd, Gooramadda, via Rutherglen 3685
Phone: 02 6026 5658
Email: olives@albury.net.au
www.olivesandoil.info
Several varieties of olive trees make up this grove of around 1000 that have been growing here since 1998. The fruit is not machine harvested; instead it is lovingly hand picked. There is a processing plant for table olives, and a traditional cold press for the oil. There is an oil-tasting room and a tour is available. The green and black table olives are prepared using the Greek method of natural fermentation, and both the oil and table olives have won several awards in the northeast Victoria Olive Growers Association competition (2001 and 2003). Gooramadda also makes pure olive oil soap and skin cream.
- **Open:** 11am–5pm, Fri–Mon and public holidays
- **Location:** 13km from Rutherglen

HARRIETVILLE
🍎 🍴 PRODUCERS/DINING

LAVENDER HUE
Bill and Verona Sullivan
20 Great Alpine Rd, Harrietville 3741
Phone: 03 5759 2588
Lavender products produced from these century-old gardens perfume every corner of this mud-brick tearoom and B&B, which was opened in 1997. Here you can indulge in scones, biscuits and ice-cream or purchase specialty table vinegars. Originally the lavender was planted simply to attract bees so they would pollinate the organically grown vegetables, but it became a business in its own right. The Sullivans distil lavender oil on the property, and it is sold only here (mail orders available). They have also devel-

oped the site to take advantage of the serene and lovely views.
- **Open:** 10am–5pm Thurs–Mon
- **Location:** 28km south of Bright, second property on the right after Howards Bridge

KERGUNYAH
🍴 DINING

WADDINGTONS AT KERGUNYAH
Rod, Jan and Rhys Waddington (chef)
RMB 1059, Kiewa Valley Hwy,
Kergunyah 3691
Phone: 02 6027 5393
Email: waddington@iprimus.com.au
Licensed
Cuisine: Modern Australian regional
Mains average: $$
Local flavours: Berries, olive oil, vegetables, rabbit, venison, farm eggs, Gundowring ice-cream, buffalo, Milawa cheese, wines
Extras: Dine in a large, hand-built timber barn with a deck area and 180-degree views over the Kiewa River Valley and the three-acre garden. Jan Waddington is a specialist garden designer and the family has lived on the property since 1868. There is a 500-year-old Red Box tree in the garden, with an Aboriginal scar mark.
- **Open:** Lunch Wed–Sun, dinner Fri–Sat (bookings only, functions)
- **Location:** Twenty minutes from Wodonga on the Mt Beauty–Falls Creek Hwy

MANSFIELD
🏠 STORES

NOLAN'S BUTCHERY
Des and Paul Nolan
52 High St, Mansfield 3722
Phone: 03 5775 2029
Email: meat@nolansbutchery.au.com
www.nolansbutchery.au.com
Des and Paul are brothers, and the family celebrated the centenary of running this shop in 2002. They carefully select, slaughter, and hang the meat that they sell, and have won three National Sausage King awards (all sausages are wheat and gluten-free). In addition, they cure all their own hams and bacons, and make marinades from scratch. Nothing is bought in—it's all produced on the premises. Now there is a range of pies and other products such as home-cooked tripe and locally grown high country beef and lamb.
- **Open:** 7am–5.30pm weekdays, 7am–12.30pm Sat

MARKWOOD
🍎 PRODUCERS

EV OLIVES
Maureen Titcumb and Eberhard Kunze
RMB 2248, Everton Rd, Markwood 3678
Phone: 03 5727 0209
Email: ekunze@netc.net.au
www.evolives.com
Olives have been grown here for the past five years, and EV Olives has oil, table olives, and tasting rooms. Recently a German gourmet magazine selected oil from this property as among the top 200 oils in the world. Australians seem to like it too, as it also won silver at the Canberra Show, and was the 2002 People's Choice in the north-east Victoria Golden Olives award. Also sold elsewhere in the region.
- **Open:** 10am–4pm daily
- **Location:** Via Wangaratta

MILAWA

🍎 PRODUCERS

MILAWA MUSTARD PTY LIMITED
Anna and David Bienvenu
The Old Emu Inn, The Cross Roads,
Milawa 3678
Phone: 03 5727 3202
Email: milmustd@iinet.net.au

These handmade seeded mustards are made from home-grown ingredients. In fact, you can see the flowering mustard just nearby. The Bienvenus see their product through from beginning to end, even growing most of the other ingredients they use as well. In 1983 there were very few Australian mustards. Anna believed there was a place for such a unique product, especially if it could be made without preservatives and additives. Together, she and David set out to make their dream come true. The cool and spacious Old Emu Inn is an ideal location to present the mustards, the bar perfect for tastings. The Bienvenus now make many mustards, plus fruit and herbal salad vinegars, a range of jellies, fruit pastes, pickles, chutneys and salad dressings, and stock other local products as well. Milawa Mustards are widely available throughout Australia.
■ **Open:** 10am-5pm daily

WALKABOUT APIARIES
Jennifer and Rod Whitehead
Snow Rd, Milawa 3678
Phone: 03 5727 3468

Walkabout Honey is a good name for this honey of a business, as Rod Whitehead takes his hives all around southern New South Wales, and northern Victoria chasing blossoming red gum, stringy bark, grey box and ironbark trees. Less common, but yielding delightful honey is the apple box tree. The family shop at Milawa offers tastes of all of these plus their own blends, depending on what is available at the time. Try their meads, and stock up on beeswax candles and honeycomb while you're there.
■ **Open:** 9am-5pm Wed-Mon
● **Location:** Next to the Milawa church

🏠 STORES

THE OLIVE SHOP
Robyn and Graham Barrow
Snow Rd, Milawa 3678
Phone: 03 5727 3887
Email: oliveshop@ozemail.com.au

Barrow olive oil comes from the Barrows' own groves, and this oil-oriented shop now has local olive oils, table olives, tapenades, dukkahs, balsamic vinegars, books and olive oil soap. There are accessories too such as pourers and plates.
■ **Open:** 10am-5pm Thurs-Sun, 10am-4pm Mon

🍴 DINING

MERLOT RESTAURANT
Matthew Lello
Lindenwarrah Country House,
Bobinawarrah Rd, Milawa 3678
Phone: 03 5720 5777
Email: info@lindenwarrah.com.au
www.lindenwarrah.com.au
Licensed
Cuisine: International
Mains average: $$$
Specialty: Caramelised olive oil and raisin rubbed breast of chicken

VIC–LEGENDS, WINE AND HIGH COUNTRY

Local flavours: Figs, olives, olive oil, breads, cheeses, herbs, mushrooms, beef, mustards
Extras: Courtyard with fountain and outdoor dining, and views of the vineyard and Mt Buffalo.
■ **Open:** Breakfast daily, lunch Fri–Sun, dinner Thurs–Sun
● **Location:** Just across the road from Brown Bros

MILAWA CHEESE FACTORY BAKERY AND RESTAURANT
Robyn McDonald
Factory Rd, Milawa 3678
Phone: 03 5727 3589
Email: milawacheese@netc.net.au
www.milawacheese.com.au
Licensed/BYO wine
Cuisine: Modern Australian
Mains average: $$
Specialty: Cheese plates
Local flavours: Milawa free-range chicken, Milawa cheeses, Samaria free-range pork, trout, Kiewa olive oil
Extras: Located in the Old Butter Factory, the bakery under the same roof also sells Milawa Cheeses. All cheeses are handmade using traditional methods and no preservatives. There are almost a dozen cows-milk cheeses, including the famed Milawa Gold, and King River Gold as well as goats milk cheeses, affiné (Best New Product at the Sydney Regent Cheese Show) and four ewes-milk cheeses. The shop offers free tastings of Milawa cheeses as well as the affiliated Hunter Valley Cheese Company cheeses. You can dine on the terrace, and gourmet platters are available for picnics.
■ **Open:** Lunch daily, dinner Thurs–Sat

THE EPICUREAN CENTRE
Chris Lee
Brown Bros Winery, Bobinawarrah Rd, Milawa 3678
Phone: 03 5720 5540
Email: browns@brown-brothers.com.au
www.brown-brothers.com.au
Licensed
Cuisine: Regional
Wine and food matched dishes (wine included): $$
Local flavours: Varies seasonally—asparagus, Milawa cheese, mustards, free-range chicken, trout
Extras: All dishes on the menu are matched to wines. The Lounge also serves light snacks, antipasto, coffee and cake. Outdoor dining with views across gardens.
■ **Open:** Lunch daily, dinner by arrangement (for groups/functions)

MT BEAUTY
🏠 STORES

KIEWA VALLEY MEAT SUPPLY
Gerry and Sue Frawley
22 Hollands St, Mt Beauty 3899
Phone: 03 5754 4026
Some sorts of smoking are okay, and Gerry Frawley has hit on one that is applauded by his customers. Although this butcher shop has only been open since 2000, his traditional smoked hams and kabana, as well as smoked chicken, turkey, trout, scallops, venison, quail and other meats under the High Country Smokehouse brand, have attracted plenty of fans. Look for them at restaurants and providores in the area, or at the shop.
■ **Open:** 6am–6pm weekdays, 6am–1pm weekends

🍴 DINING

MT BEAUTY BAKERY–PASTICCERIA–CAFE
Jason Keogh and Kiona Best
Cnr Kiewa and Hollands Streets,
Mt Beauty 3699
Phone: 03 5754 4870
Email: bakery@mtbeauty.albury.net.au
Licensed
Mains average: $
Specialty: Ham and cheese press with homemade mayo; chocolate mud muffins; beef, cheese, bacon and tomato pies
Local flavours: Beef, chicken, sun-dried tomatoes, artichoke hearts, mustards, chutneys, wine list only features northeast Victorian wines
Extras: Architecturally designed building with lots of rock, glass, and granite, large, citrus lined courtyard with views of parks and stunning Mt Bogong. All food produced fresh on the premises. Gravity espresso coffee and leaf teas.
■ **Open:** 6.30am–6.30pm daily

MT BRUNO
🍎 PRODUCERS

CHERRYBROOK CHERRY FARM
Tony and Marion Rak
RMB 4400, Jones Rd, Mt Bruno, via Glenrowan 3675
Phone: 03 5765 2331
Email: ammerak@netc.net.au
During cherry season each year, over twenty growers in the area may band together as a cooperative. The Raks have been growing cherries for fifteen years, and Tony is a third-generation grower. The fruit is sold already picked, from the packing shed.
■ **Open:** 8am–6pm daily (Nov–Dec)
● **Location:** 22km north of Glenrowan on the western side of the Warby Ranges

OVENS
🍎 PRODUCERS

SELZER PASTORAL
John, Meta, Mark and Janene Selzer
'Myrtlebrae', 176 Selzers Lane, Ovens 3738
Phone: 03 5751 1340
Remember how tomatoes used to taste? If you don't, these tomatoes grown hydroponically in a greenhouse—allowing the Selzers to sell produce almost year round—may remind you. Sold here and in the local district.
■ **Open:** 10am–5pm daily

OXLEY
🍎 PRODUCERS

BLUE OX BERRIES
Wayne Pegler and Helen Taylor
Smith St, Oxley 3678
Phone: 03 5727 3397
Email: blueox@netc.net.au
As the name suggests, blueberries are the mainstay here and you are welcome to pick your own or buy from the kiosk beside the orchard. Mind you, there are plenty of other berries to choose from, including brambleberries, raspberries, lawtonberries, boysenberries, and youngberries. Picking season is from mid December to late January. There is also a range of jams, chutneys, relishes, preserves and sauces available as well as

frozen fruit all year round. Mail order is also an option.
- **Open:** Daily, or according to signs
- **Location:** Next to Oxley Estate Wines

🍴 DINING

KING RIVER CAFE
Ben and Judy Bonwick
Snow Rd, Oxley 3678
Phone: 03 5727 3461
Email: bon@netc.net.au
www.kingrivercafe.com.au
Licensed/BYO wine
Cuisine: Mediterranean
Mains average: $$
Specialty: Lemon tart
Local flavours: Free-range chicken, cheeses, pork, squab, Ramelton duck
Extras: Grows own tomatoes, herbs, vegetables and melons in the kitchen garden. Outdoor paved courtyard, lawn for functions, views of the river. Wine cellar and cellar door selling local wines next door. Takeaway available.
- **Open:** Breakfast and dinner Wed–Sun, lunch Wed–Mon

POREPUNKAH

🍴 DINING

BOYNTON'S WINERY AND RESTAURANT
Kel Boynton (winery), Reo Tautari (restaurant)
Great Alpine Rd, Porepunkah 3741
Phone: 03 5756 2356
Email: boyntons@bright.albury.net.au
Licensed
Cuisine: Modern Australian
Mains average: $$
Specialty: Antipasto with selections from the vineyard garden

Local flavours: Breads, vegetables, herbs
Extras: One of the most spectacular cellar-door views in Australia with the breathtaking backdrop of Mt Buffalo, Boynton's has a broad range of premium quality wines and Boynton's pale ale and pilsener are available too. Outdoor area in summer, open fire for winter.
- **Open:** Lunch daily
- **Location:** Just before Porepunkah when travelling from Myrtleford

RUTHERGLEN

🍴 DINING

BEAUMONT'S CAFE
Peter and Birgit Weir
84 Main St, Rutherglen 3685
Phone: 02 6032 7428
Email: beaumontscafe@bigpond.com
www.beaumontscafe.com.au
Licensed/BYO
Cuisine: Mediterranean
Mains average: $$
Specialty: 'It's all special', they say
Local flavours: Milawa chicken, Somers lettuce, local wines
Extras: Dine in a Tuscan-style courtyard.
- **Open:** 3pm–11pm Tues–Thurs, 10am–11pm Fri–Sat, 10am–5pm Sun

🏠 STORES

FOOD WINE FRIENDS
Fiona Reddaway
Shop 2, 6 Ireland St, Bright 3741
Phone: 03 5750 1312
Email: indulge@foodwinefriends.com.au
www.foodwinefriends.com.au
'The valleys of Victoria's north-east are bursting with fabulous food and wine,' says Fiona Reddaway. That's why she

decided a year ago to open Food Wine Friends and bring together a range of fine foods and wines from the area. She complements the local delicacies with selected food and wine accessories, and essential foods from beyond the region. You'll find Milawa Cheese, Milawa Mustards, King Valley Fine Foods, Beechworth Preserves and Traditional Plum Pudding Company on the shelves, and local wines to match.
- **Open:** 10am–5pm daily
- **Location:** Opposite the post office

PARKER'S PIES AND PASTRIES
Fred Parker
86-88 Main St, Rutherglen 3685
Phone: 02 6032 8497
Email: pieman@dragnet.com.au
Fred Parker is a well-known Rutherglen identity who makes fabulous pies. If you find the Aussie pie hard to resist, then you must try his special version with homemade chutney, fresh garden salad and the locally brewed Bintara beer. He has some twenty-four varieties of pie with up to a dozen available on any one day. Favourites are lamb, mint and rosemary; buffalo, bacon, onion and garlic; or chicken, cheese, ham and mustard. Interestingly, the buffalo is sourced from a local award-winning herd. Parker's Pies is fully licensed, so you can enjoy Rutherglen wines with your fare. But the menu doesn't only include pies, there are coffee and cakes as well.
- **Open:** 8.30am–4.30pm, weekdays; 8am–5pm, weekends

TUILERIES
Tony Lamb
35 Drummond St, Rutherglen 3685
Phone: 02 6032 9033
Email: info@tuileriesrutherglen.com.au
www.tuileriesrutherglen.com.au
Licensed
Cuisine: Modern Australian
Mains average: $$$
Local flavours: As much as possible including Milawa Chickens and Hume Weir trout
Extras: A modern, glass-walled restaurant integrated into a historic 1886 winery overlooking five acres of vines. Nine luxury accommodation units for those who wish to extend the experience.
- **Open:** Cafe 8am–2.30pm daily, dinner daily

STANLEY
🍎 PRODUCERS

SNOWLINE FRUITS
Henry and Rita Hilton
507 Stanley-Myrtleford Rd, Stanley 3747
Phone: 03 5728 6584
Email: snowlinefruits@bigpond.com
A family owned and operated business with world-class orchards. The orchards are constantly changing with many different varieties of tree-ripened apples including Pink Lady, Sundowner, Red Fuji and Royal Gala. Sold at farmers' markets, commercial markets and from the farm. A fun sideline is 'printed apples'—ideal for corporate functions, special occasions and conferences.
- **Open:** 9am–5.30pm daily
- **Location:** On the left 5km from Stanley

VIC—LEGENDS, WINE AND HIGH COUNTRY

TAGGERTY
🍴 DINING

CLEARSTREAM OLIVE FARM AND WOOLSHED CAFE
Ken Silvers
Webbs Lane, Taggerty 3714
Phone: 03 5774 7523
Email: clearstream.olives@bigpond.com
www.clearstreamolives.com
Licensed/BYO
Cuisine: Local, fresh and healthy
Mains average: $
Specialty: Wendy's gourmet platter
Local flavours: Smoked trout, meats
Extras: Dining in the old sheep pens is a unique experience at this converted shearing shed, now a rustic cafe. The olive farm cellar door is in the cafe, which is managed by Wendy Watt, formerly from the Taggerty Herb Farm. Wendy's Workshop creates colourful gourmet products such as berry coulis, peach chilli sauces, and green tomato relishes from local produce as well as Gingko Gold Elixir.
■ Open: 10am–4pm Thurs–Sun
● Location: Off Maroondah Hwy, take Eildon Rd from Taggerty, 100m east of Taggerty store

TAMINICK
🍴 DINING

AULDSTONE CELLARS
Nancy and Michael Reid
RMB 4270, Booths Rd, Taminick, via Glenrowan 3675
Phone: 03 5766 2237
Email: winery@auldstone.com.au
www.auldstone.com.au

Licensed (for Auldstone wines)
Cuisine: Luncheon
Mains average: $
Specialty: Wine-taster's platters, Nancy's homemade pâté and pickles
Local flavours: Mustards, cheeses and chutneys
Extras: Set in 110-year-old granite stone winery with open fireplace in winter and shady landscaped gardens for summer.
■ Open: Lunch Thurs–Sun (daily during school holidays)

TAWONGA SOUTH
🍴 DINING

ANNAPURNA WINES AND CAFE
Lynda Fox
Simmonds Creek Rd, Tawonga South 3698
Phone: 03 5754 1356
Email: enquiries@annapurnawines.com.au
www.annapurnawines.com.au
Licensed (for Annapurna Wines)
Mains average: $$
Specialty: Gourmet antipasto
Local flavours: Organically grown vegetables and fruits, smoked meats, cheeses
Extras: Gazebo dining with stunning views of Mt Bogong. Takeaway available.
■ Open: 10am–5pm Wed–Sun
● Location: 3km from Mt Beauty

WAHGUNYA
🍴 DINING

THE PICKLED SISTERS CAFE
Ali McKillop
Cofield Wines, Distillery Rd, Wahgunyah 3687

Phone: 02 6033 2377
Email: pickled_sisters@hotkey.net.au
Licensed
Cuisine: Modern regional
Mains average: $$
Specialty: Pickled Sisters vineyard platter
Local flavours: Milawa free-range chicken, local yabbies, north-east venison, Hume Weir trout, Milawa cheeses, Blue Ox berries, Lyric olive oil, Milawa mustards, Passchendaele verjuice, Murray cod, Marjie's eggplant
Extras: This cafe has regional wines, as well as Cofield Wines, and you can dine in the cellar or outdoors on the deck. Winner Best Family Establishment, Restaurant and Caterers Association, 2003. Home catering menu available.
■ **Open:** Breakfast and lunch daily, dinner Fri-Sat

THE TERRACE RESTAURANT
All Saints Estate, All Saints Rd, Wahgunyah 3687
Phone: 02 6035 2209
Email: terrace@allsaintswine.com.au
www.allsaintswine.com.au
Licensed
Cuisine: Modern Australian
Mains average: $$
Local flavours: As much as possible
Extras: Vineyard views and award-winning gardens. The Cafe has snacks.
■ **Open:** Lunch daily, dinner Sat

WANGARATTA
🍴 DINING

QUALITY HOTEL WANGARATTA GATEWAY
Wendy and Peter Lester
29-37 Ryley St, Wangaratta 3677
Phone: 03 5721 8399
Email: info@wangarattagateway.com.au
www.wangarattagateway.com.au
Licensed
Cuisine: Steak and seafood
Mains average: $$
Local flavours: Milawa cheese and mustard
Extras: Courtyard and outdoor seating. A wide range of north-east Victorian wines to complement every dish.
■ **Open:** Breakfast and dinner daily, lunch large groups only

WHITFIELD
🍎 PRODUCERS

NEWTON'S ORGANIC PRICKLEBERRY FARM
Jim and Barbara Newton
118 Gentle Annie Lane, Whitfield 3733
Phone: 03 5729 8272
What a great name for this Certified Level A organic berry-growing farm, which evolved after hop growing stopped being a viable industry for the region. Eighteen years later, the general public is happy with the switch. After all, berries taste better than hops, and you can also pick your own, or have your order supplied. The farm grows most berries and also has Barb's berry jams available.
■ **Open:** Daily, when the Open sign is out (Dec-Feb) or phone first
● **Location:** Look for the farm sign at the cattle grid along Gentle Annie Lane, 3km from Whitfield.

🍴 DINING

MOUNTAINVIEW HOTEL– KING VALLEY
Alfred and Katrina Pizzini
Cnr Mansfield and Cheshunt Roads, Whitfield 3733
Phone: 03 5729 8270
Email: mountainviewhotel@bigpond.com
Licensed
Cuisine: Country-style hotel fare
Mains average: $$
Specialty: Pan-fried peppered sirloin with a creamy mustard sauce
Local flavours: Black Range trout, Passchendaele verjuice, Black Range honey and pestos, chutneys and other preserves from King Valley Fine Foods, organic olives and breads from the Milawa Cheese Factory bakery
Extras: Grows own tomatoes, herbs and lemons in the family's gardens. Surrounded by beautiful gardens, with some trees more than one hundred years old, beside a creek lined with tree ferns. The grounds are used for many events and festivals including a petanque festival organized by Pizzini wines and the Hotel's petanque social club. The hotel has a piste and lends boules to anyone interested in learning how to play.
■ **Open:** Lunch Tues–Fri, dinner Tues–Sat

WODONGA

🍎 PRODUCERS

HUME WEIR TROUT FARM
Matthew Benfield
RMB 5029, Riverina Hwy, Wodonga 3690
Phone: 02 6026 4334
Email: trout@humeweirtrout.com.au
www.humeweirtrout.com.au

This well-established trout farm has been providing high quality smoked trout pâté, smoked trout, seasoned smoked fillets, and fresh trout to many places throughout the country for the past thirty-plus years. Fresh trout is processed here daily.
■ **Open:** 7am–6.30pm daily
● **Location:** Below the Hume Weir wall, fifteen minutes from Albury, check the website for directions

KIEWA ESTATE OLIVE GROVE
Sveti Ignjatovic
RMB 1109, Redbank Rd, Coralbank, Wodonga 3699
Phone: 02 4267 3722
Email: kiewaestate@optusnet.com.au
www.kiewaestate.com.au

These quality olives are grown on the banks of the Kiewa River and their oil is extracted using the cold pressed method then naturally decanted and bottled on the estate. The oils, soaps and massage oils are available here, from quality retailers, and via the website.
■ **Open:** 9am–5pm daily by appointment
● **Location:** Turn off Kiewa Valley Hwy at Mongans Bridge

🏠 STORES

DARE'S FRESH FRUIT AND VEGETABLES
Dare family
13 Elgin Boulevard, Wodonga 3690
Phone: 02 6024 3997

Lots of fresh produce, 'local if available', say the Dares, who have been weighing it out here for fourteen years. They also stock Milawa Mustards, Beechworth Preserves, Spirited Chef products, Murray Breweries cordials, King Valley pestos, Yackandandah Jam Co jams, Green Grove

LEGENDS, WINE AND HIGH COUNTRY—VIC

Organic licorice, Table Top fig products, and local almonds and walnuts. Plus fresh vegetables, rhubarb and fresh herbs from their own garden.
- Open: 8am–6pm weekdays, 8am–1pm Sat

KENNEDY'S MEATS
Geoff McDonald
10 Osburn St, Wodonga 3690
Phone: 02 6024 4955
Email: kennedy@albury.net/au
This specialist butcher's shop has all the usual mainstream meats but also offers crocodile, emu, kangaroo, bison, venison, pheasant or buffalo, and all types of seafood, as well as sausages made on the premises. The bison is local, from Corryong, but the other meats are sourced from elsewhere and sold at wholesale prices.
- Open: 7am–6pm weekdays, 8am–4pm weekends

DINING

HARVEYS FISH AND FUN PARK
Garry and Darren Harvey
Lincoln Causeway (Hume Hwy),
Wodonga 3690
Phone: 02 6021 2070
This unusual place has smoked fish, eels and oysters for sale, and meals are served in the licensed beer garden or the large function room. Promoted as Australia's only indoor native fish farm, visitors can see rainbow trout and Atlantic salmon through the viewing windows, then move on to outdoor ponds and catch fish for themselves using free bait. Keep the kids happy with a train ride, mini-golf or a visit to see the farm animals.
- Open: 9am–5pm Wed–Mon (daily in NSW and Victorian school holidays)

- **Location:** Between Albury and Wodonga

ZILCH—FOOD STORE AND CAFE
Liz Escott (owner) and Jodie Jones (chef)
8/1 Stanley St, Wodonga 3690
Phone: 02 6056 2400
Licensed
Mains average: $
Specialty: Twenty-two types of sandwiches on Valentine's breads
Local flavours: Butt's smoked trout, Ross & Em's Pickled Onions, Blue Ox Preserves, Beechworth Preserves, cheeses, Parker's Pies from Rutherglen, and Herb Barn teas
Extras: Takeaway available and there is also a catering outlet, deli and two outdoor courtyards.
- Open: 8.30am–5pm Mon–Wed, 8.30am–6pm Thurs and Fri

YACKANDANDAH

STORES

YACKANDANDAH GENERAL STORE
Glenn Clark
15 High St, Yackandandah 3749
Phone: 02 6027 1230
Email: gaclark@tpg.com
Here's the place to find local Victorian cheeses and pâtés, smoked trout, olives and dips, fruit and vegetables, special locally-grown hydroponic lettuces, and strawberries picked minutes away. Then there's the Yackandandah Jam Factory range of jams and simmer sauces, as well as local free-range eggs, peaches, apples and local apple juice from Beechworth. The store is a magnificent building, 145 years old.
- Open: 8.30am–5.30pm weekdays, 8.30am–1pm Sat

MARKETS AND EVENTS

✦ **BEECHWORTH COUNTRY CRAFT MARKET,** BEECHWORTH
03 5728 3350
9am–3pm four Sat per year

✦ **COROWA FARMERS' MARKET,** COROWA
02 6024 5662
8am–12pm first Sun

✦ **HUME MURRAY FOOD BOWL FARMERS' MARKET,** WODONGA
02 6058 2996
8am–12pm (summer),
9am–1pm (winter) alternate Sat from 31 Aug

✦ **LOCAL PRODUCERS' MARKET,** MILAWA
03 5727 3589
9am–1pm Sat fortnightly from first weekend in Nov to end of summer

✦ **MYRTLEFORD PRODUCE MARKET,** MYRTLEFORD
03 5751 2996
8am–12pm Sat Jan–mid Apr

✦ **TATONG FARMERS' MARKET,** TATONG
03 5767 2210
8am–1pm first Sat

✦ **A TASTE OF FALLS CREEK,** FALLS CREEK
03 5758 3224
Second weekend in Jan

✦ **ALBURY–WODONGA WINE AND FOOD FESTIVAL,** ALBURY–WODONGA
02 6058 2996
Oct long weekend

✦ **ALPINE VALLEYS FOOD AND WINE FESTIVAL,** MYRTLEFORD
1800 500 117/03 5751 1575
Mar

✦ **MYRTLEFORD ALPINE VALLEYS GOURMET WEEKEND,** BRIGHT
03 5755 2275
Mid Jan

✦ **BEECHWORTH HARVEST FESTIVAL,** BEECHWORTH
03 5728 3233
Third Sun in May

✦ **BROWN BROTHERS WINE AND FOOD WEEKEND,** MILAWA
03 5721 5711
Second weekend in Nov

✦ **TASTES OF RUTHERGLEN,** RUTHERGLEN
02 6033 6302
Mar

✦ **WANGARATTA JAZZ FESTIVAL,** WANGARATTA
03 5722 1666
First weekend in Nov

MACEDON RANGES AND SPA COUNTRY

Flowing on from the Goldfields, the spa country offers a change of pace, a healthy reminder that too much of a good thing is, well, in this case anyway, just what the doctor ordered.

While the health-giving properties of the waters from the natural springs in this region have been highly regarded for many years, today's visitors use them as an adjunct not a cure. Signs everywhere direct visitors to springs where crystal-clear water sparkles up from the earth, and you are advised to come armed with bottles and flasks to fill.

Come too with a good appetite as the region also produces some great food, and the superb local restaurants know just what to do with it. Take the waters, but sample the food too.

BACCHUS MARSH
🏠 STORES

BIG APPLE TOURIST ORCHARD
Paul Reivers
432 The Avenue of Honour,
Bacchus Marsh 3340
Phone: 03 5367 4752
The Big Apple orchard grows and sells apples, peaches, cherries, apricots, nashis, brown pears and nectarines. More recently it has introduced the peacharine, a cross between a peach and a nectarine, which ripens in late summer and early autumn. The shop also sells other fresh market produce, local honey, and a very good apple juice that they make themselves, under the Big Apple brand. It is 100 per cent pasteurised, unfiltered juice, and contains no chemicals or sugar. Everything sold here is local.
■ **Open:** 9am–7pm daily

DAYLESFORD
🍎 PRODUCERS

COUNTRY CUISINE (AUSTRALIA) PTY LTD
Phillippa Wooller and Geoff West
6 Midland Hwy, Daylesford 3460
Phone: 03 5348 4141
Email: country@netconnect.com.au
www.countrycuisine.com.au
Ornate bottles of herbed olive oils, sweet sauces, marinades and mustards—these are the stock-in-trade of Country Cuisine. Phillippa Wooller never seems to run out of nifty ideas, constantly adding to her wide range of vinaigrettes, all types of preserved fruits in alcohol, Christmas fare, and fresh berry jams that

> **BACCHUS MARSH BARGAINS**
> There are many growers and orchardists along Main Road at Bacchus Marsh. Watch out for Gibbons Nursery & Orchard and other signs all the way along. Some only show the signs when fruit is available. Best buys are stone fruit throughout summer, then apples and pears later in the year.

VIC—MACEDON RANGES AND SPA COUNTRY

include mouth-watering choices like wild blackberry, blueberry, fruits of the summer forest, raspberry, strawberry and champagne sold in many places around the country. Don't forget the vinegars, salad dressings, chutneys and sauces either, all made from fresh local ingredients. It's abundantly apparent that local berry farms contribute heavily to Phillippa's efforts. All this began in Trentham, but more recently Country Cuisine has transferred to these premises which, appropriately, were once used as the Old Hepburn Spa Bottling Plant.
- **Open:** 8am-4pm weekdays
- **Location:** On the outskirts of Daylesford, towards Ballarat, on the left-hand side with a large sign

STORES

CLIFFY'S EMPORIUM
Mary Ellis and Geoffrey Gray
30 Raglan St, Daylesford 3460
Phone: 03 5348 3279
Email: liberty@netconnect.com.au
This one stop shop for all things local produce, craft and wines, is more than just that. It has become an institution for shoppers and tourists, a labour of love for its owners, and an amazing and fitting showcase for the best of what the local region has to offer. There is too much to describe—just get yourself there and see it first hand.
- **Open:** 9.30am-5pm Sun-Thurs, 9.30am-10pm Fri-Sat

SPA CENTRE MEATS
Ron Layfield
37 Vincent St, Daylesford 3460
Phone: 03 5348 2094

If you're a fan of bullboar sausages—those robust snags made from a mixture of pork and beef, herbs, spices and red wine and a legacy of the early Swiss-Italian settlers of this district—then this is the place to go. Ron has been butchering for thirty years and he sells to many local hotels and restaurants.
- **Open:** 7am-5pm weekdays, 7am-12pm Sat

DINING

BAD HABITS CAFE
Tina Banitzka
Convent Gallery, Daly St, Daylesford 3460
Phone: 03 5348 3211
Email: marija@conventgallery.com.au
Licensed
Cuisine: Mediterranean
Mains average: $$
Specialty: Homemade pies
Local flavours: Fruit and vegetables, berries, fish, lamb, eggs, cheeses
Extras: This delightfully restored convent and gallery has a conservatory-style cafe. Function rooms available for weddings and other functions. Art exhibitions every six weeks.
- **Open:** Breakfast, lunch and morning and afternoon teas daily, dinner Fri-Sat

FRANGOS AND FRANGOS
Dianne and James Frangos
82 Vincent St, Daylesford 3460
Phone: 03 5348 2363
Licensed
Cuisine: Mediterranean
Mains average: $$
Specialty: Regional cuisine
Local flavours: Tuki trout, lamb, goat, vegetables, Kyneton olive oil, eels

MACEDON RANGES AND SPA COUNTRY—VIC

Extras: Outdoor area and walled courtyard, two restaurants—one with a wood-fired oven that serves pizza.
■ **Open:** Breakfast, lunch and dinner daily

LAKE HOUSE
Alla Wolf-Tasker (executive chef)
King St, Daylesford 3460
Phone: 03 5348 3329
Email: marketing@lakehouse.com.au
www.lakehouse.com.au
Licensed
Cuisine: Contemporary Australian
Mains average: $$$
Specialty: The 'Lake House changing plate' featuring charcuterie, condiments and other delicacies all made in the Lake House kitchen
Local flavours: Heathcote yabbies, Lake House charcuterie, Tuki trout, Kyneton ducks, organic greens.
Extras: A regional and seasonally based menu drawing on the fine local produce has merited numerous dining awards. This luxury accommodation is situated on the shores of Lake Daylesford surrounded by six acres of manicured gardens, waterfront views and terrace dining in warmer months. Gourmet picnic hampers also available. Makes and sells duck liver parfait, beetroot relish, tomato kasundi, elderberry glaze, chewy almond nougat, biscotti, sienna cake and seasonal jams.
■ **Open:** Breakfast, lunch and dinner daily
● **Location:** Turn left off King St, the main street

GISBOURNE
🍎 PRODUCERS
MACEDON GROVE OLIVES
Chris and Julie Green
Macedon Grove, 129 Peters Rd,
Gisbourne 3437
Phone: 03 5428 3350
Email: mac.grove@hotkey.net.au
www.macedongrove.com.au
Macedon Grove's black tapenade won first prize at the Royal Canberra Show and the champion olive exhibit in 2003. This grower and processor of certified organic table olives, which began eight years ago because of a belief in organic principles, also produces a range of tapenade and olive pastes. These olives are produced using only spring water, natural rock salt

---

DAYLESFORD MACEDON PRODUCE GROUP
Hepburn Shire Tourism Offices,
98 Vincent Street, Daylesford 3460
Phone: 03 5321 6111 or 03 5348 2306
Email: www.dmproduce.com

DAYLESFORD TRAILS
The Macedon Ranges and Spa Country region offers the well travelled foodie a myriad of interesting, quality food and wine experiences. Seasonal, and frequently organic, farm gate produce is advertised on roadside signs. Potatoes, chestnuts, olives, berries, wild mushrooms, organic herbs, fruit and vegetables, yabbies, trout, smoked eels and excellent honey are just some of the regional produce offered on local menus by the talented cooks of the region.

and organic vinegars. They are sold at seventy outlets throughout Victoria and New South Wales, through the 'cellar door' and at shows and markets.
- **Open:** By appointment
- **Location:** Turn right off Kilmore Rd into Campbells Rd then first right into Peters Rd

HEATHCOTE
PRODUCERS

CENTRAL VICTORIAN YABBY FARM
Greg and Helen Williams
RSD 1920, Northern Hwy, Heathcote 3523
Phone: 03 5433 2332
Email: yabby@netcon.net.au
This yabby farm, the longest running one in Victoria, has a large selection of live, different sized, good eating yabbies available direct to the public. It also provides stock for dams, bait for fishing and, for the kids, pet yabbies.
- **Open:** 9am–6pm Thurs–Mon (shorter hours possibly in winter)
- **Location:** 5km from Heathcote on the Melbourne side

KYNETON
STORES

KYNETON COUNTRY FRESH
Sue and Jeremy Glassel
59 Mollison St, Kyneton 3444
Phone: 03 5422 2906
Email: sue.glassel@bigpond.com
Another case of liking the business so much, they bought it. The Glassels had been customers for some time, and took over this shop in mid 2003, continuing and expanding the emphasis on market-fresh, local fruit and vegetables. There

are Harcourt apples, fresh flowers, nuts, and jams made by Patricia, whose marmalade has a lot of fans too. Now there's also organic produce, as well as pasta and sauces and award-winning Kyneton olive oil.
- **Open:** 8.30am–6pm weekdays, 8am–1pm Sat

KYNETON PROVENDER
Fran Wigley
30 Piper St, Kyneton 3444
Phone: 03 5422 3745
Email: provender@bigpond.com
'Provisions for a finer life' is how Fran Wigley describes the goodies in this place, which is so well-suited to browsing. Along with second-hand and new books and gifts, there are specialty biscuits and preserves and a wealth of other foodstuffs from regional Victoria. It's all here: Kyneton Olive Products, Beattie's Biscuits, Warrens Jams, Kez's cookies, Marion's Kitchen sauces and chutneys, Country Cuisine, Red Kelpie Relish, Milawa Mustards, Yackandandah Jams and Sauces.
- **Open:** 10am–5pm daily
- **Location:** Northern end of shopping precinct

MT EGERTON
PRODUCERS

YUULONG LAVENDER ESTATE
Edythe Anderson and Rosemary Holmes
58 Sharrocks Rd, Mt Egerton 3352
Phone: 03 5368 9703
Email: yuulong@tpgi.com.au
www.ballarat.com/yuulong
The sweet scent from the Australian national collection of 128 different

MACEDON RANGES AND SPA COUNTRY—VIC

varieties of lavender wafts out to greet you on arrival. In summer you can wander amongst the flowering rows and watch sickle harvesting, compare varieties, even taste the lavenders in different foods. With seventeen lavender gourmet food products to choose from (it's important, they say, only to use *blue* lavender—lavandula angustifolia—in cooking) there are light refreshments in the tearooms, including lavender cheese, cheesecake, biscuits, fruitcake, and muffins.

- **Open:** 10am–4.30pm Wed–Sun (Oct–Apr), 10am–4.30pm daily (Jan)
- **Location:** Take the Gordon exit from the Western Freeway (Melbourne to Ballarat) or Yendon No 2 from Midland Hwy

MUSK

PRODUCERS

MUSK BERRY FARM
Kerry and Greg Henderson
19 Cantillons Rd, Musk 3461
Phone: 03 5348 5593
Email: gkah@ozemail.com.au

The wealth of berries grown here include raspberries, blueberries, boysenberries, silvanberries (which are like a wild blackberry), blackberries, youngberries, marionberries (resembling a sweet blackberry), morello cherries, elderberries, black, red and white currants, and gooseberries, plus nashi pears. There is a shop on site, which is the only place to buy this property's seasonal fresh berries, as well as the farm's frozen berries, jams, and vinegars. Make sure you try the delicious Kerry's Very Berry ice-cream, too.

- **Open:** 11am–5pm weekends and public holidays

- **Location:** 8km from Daylesford towards Trentham. A sign is displayed when the farm is open

ROMSEY

DINING

THE CONISTON DINING ROOM
Judy Cope-Williams
Cope-Williams Winery, Glenfern Rd, Romsey 3434
Phone: 03 5429 5428
Email: enquiries@cope-williams.com.au
www.cope-williams.com.au
Licensed
Cuisine: Modern Australian
Mains average: $$
Specialty: Weekend platter lunches
Local flavours: All fruits and vegetables
Extras: A village cricket green, a Real tennis (that's the ancient form) court and various accommodation options alongside the winery.

- **Open:** Lunch weekends and dinner Sat (or by prior arrangement for functions at other times)

HESKET HOUSE
Peter Watsford and Sean Cadzow (chef)
1201 Romsey-Woodend Rd, Romsey 3434
Phone: 03 5427 0608
Email: heskethouse@ssc.net.au
www.heskethouse.com.au
Limited license/BYO
Cuisine: Modern Australian
Mains average: $$
Specialty: Braised dishes in the wood-fired Aga
Local flavours: Cheeses, green produce
Extras: Not far from the legendary Hanging Rock, this earthen homestead is set in 150 acres of untouched natural

forest, home to koalas, kangaroos, wallabies, wombats, echidnas and native birds. Huge communal living room, open fire and vaulted thatched ceiling. Indoor rainforest atrium, with waterfall, in-house professional massage service, seven bedrooms and bunk room accommodation available for six. Outdoor dining with lake view.
- **Open:** Breakfast (continental for guests), lunch, dinner daily
- **Location:** Twenty-five minutes from Melbourne airport, see website for directions

SHEPHERDS FLAT
DINING

LA TRATTORIA
Carol White
Lavandula Swiss Italian Farm,
350 Hepburn-Newstead Rd,
Shepherds Flat 3461
Phone: 03 5476 4393 (farm);
03 5476 4347 (cafe)
Email: carolwhite@lavandula.com.au
www.lavandula.com.au
Licensed/BYO
Cuisine: Swiss-Italian/French provincial
Mains average: $
Specialty: Lavender ice-cream, lavender scones and lavender 'champagne'
Local flavours: Cheese, meats, wines, olives, herbs, fruit and vegetables
Extras: Lavandula is a working farm with a strong Swiss-Italian heritage, and painstakingly restored gardens. Several festivals take place throughout the year: Lavender Harvest (January); Autumn Swiss-Italian Festa (early May); In The Deep Midwinter (July); La Primavera (October). There is also a range of lavender body products and aromatherapy blends that is only sold here.
- **Open:** 10.30am-5.30pm daily (Sept-May), 10.30am-5.30pm weekends and school holidays (June-July), closed Aug
- **Location:** Via Daylesford, 5km from Hepburn Springs

SKIPTON
PRODUCERS

OSS EELS
Ken, Pauline and Ben Osbourne
3 Anderson St, Skipton 3361
Phone: 03 5340 2147
Silver Lake Smoked Eel is the other name you may associate with these slippery customers—that's the eels of course, not the Osbournes. They're hardworking people who saw this as an ideal alternative industry. The eels are grown from juvenile stock—elvers if you are interested in the vocab—in the wild, then smoked using traditional wood-fired methods to achieve a high quality product that is now sold all over Australia. All eels are sourced from the Osbourne's own lakes and from other local fishermen.
- **Open:** By appointment

WOODEND
DINING

CAMPASPE HOUSE
Milton and Liz Collins
Goldies Lane Woodend, (P.O. Box 472),
Woodend 3442
Phone: 03 5427 2273

Email: sales@campaspehouse.com.au
www.campaspehouse.com.au
Licensed
Cuisine: Modern Australian
Mains average: $$$
Specialty: A menu that changes with what is fresh and available
Local flavours: Musk smallgoods, prosciutto and salami, olive oils, yabbies, beef, lamb, game
Extras: The gracious 1920s house is set in the beautiful Edna Walling Garden. There are a tennis court, golf course, and swimming pool, twenty rooms, a lovely courtyard for relaxation, and two conference rooms. Milton Collins describes this grand B&B as a 'food venue'. An active member of the Daylesford Macedon Produce Group, Campaspe sources the best food and wine from this fertile corner of Victoria.

■ **Open:** Breakfast and lunch Sun, dinner Fri–Sat
● **Location:** Look for the blue sign at the only set of traffic lights. Follow Ashbourne Rd till you reach Goldies Lane

MARKETS AND EVENTS

✦ **DAYLESFORD MACEDON MARKET DAY,** DAYLESFORD
03 5348 3329
10am–4pm Sun, three times annually

✦ **FOOD, WINE AND JAZZ FESTIVAL,** SMEATON
03 5321 6123
First weekend of Apr

✦ **THE AGE HARVEST PICNIC AT HANGING ROCK,** WOODEND
03 5427 2033
Late Feb

MINERAL WATERS

Spring water, as distinct from artesian or bore water, comes originally from the atmosphere and falls to the earth as snow or rain. In many cases it then seeps far underground, often travelling many kilometres as it soaks down. When it later bubbles out again, it is both purified and impregnated with minerals from the soil and rocks through which it passes.

What is the nutritional value of the different minerals?

Calcium is essential for blood coagulation, muscle contraction and relaxation and the transmission of nerve impulses. Lack of calcium can cause rickets and osteoporosis among other conditions.

Magnesium is involved with the synthesis of protein, fats, complex carbohydrates and nucleic acid. It is vital for the correct transmission of nerve impulses and muscle health. While there is conflicting evidence, many believe that it is also particularly important to the health of the heart muscle and in the treatment of heart arrhythmias and blood pressure.

Potassium is related to sodium in its work in the body, and they are both vital for health: they assist in maintaining an acid/base balance, regulate fluid balance, blood volume and pressure, and control muscle contraction and relaxation.

A good ratio of both minerals is found in most naturally occurring foods such as fresh fruits and vegetables, cereals, nuts and seeds.

Unless the mineral water has been sweetened, there are no kilojoules, making it an ideal drink for those watching their weight. Perhaps the greatest factor to consider with mineral water, especially for those on a restricted-sodium diet, is its sodium level, which may be up to double that of household tap water.

MURRAY AND THE OUTBACK

As the mighty Murray River slows and widens, so the surrounding countryside flattens and becomes more open. Outside of the irrigated areas, you could call it barren, or only suitable for large acreages to support sheep and wheat.

Yet, put water on this land and it turns lush and fertile, producing citrus fruits and grapes, which in turn become juices and wine that are equal to any in the country.

Mildura is the economic centre of the region, and because of its early background of European immigration, has a distinctly cosmopolitan feel about its eateries. A beautifully planned city, it is bordered by kilometres of orchards and vineyards and, of course, the massive river that makes it all possible.

BEARII
PRODUCERS

AINTREE ALMONDS AND APIARY
Trinity and Mariea Richards
RMB 2231 Ferris St, Bearii 3641
Phone: 03 5868 2203
Email: aintree@hotmail.com
www.bellarte.com.au/aintree
These enterprising people have been growing, producing and distributing their distinctive almond and honey products since 2000. There is dukkah, bread, nougat, flavoured and spiced almonds and almond honey for sale, plus B&B accommodation (for one couple only) available. Almonds are sold under the Howzyanutz label. Products available at local wineries, tourist outlets, and the farm shop.
- **Open:** By appointment
- **Location:** Near Strathmerton

BOORT
PRODUCERS

SIMPLY TOMATOES
Marilyn Lanyon
479 Parkers Rd, Boort 3537
Mobile: 0428 554 235
Email: marilyn@simplytomatoes.com.au
www.simplytomatoes.com.au
Marilyn Lanyon has been making this unique green tomato product for her friends for many years. More recently she has been marketing it as an antipasto called Simply Green Tomatoes, preparing it in the commercial kitchen on her tomato-growing property. One-hour tours for small groups and coaches by appointment
- **Open:** 9am–5pm daily by appointment
- **Location:** Off the Boort–Kerang Rd, approximately fifteen minutes from Boort

COBRAM
PRODUCERS

RIVERVIEW JUICES
Charlie Gattuso, Bev and Noel Fisher
Schubert St, Cobram 3644
Phone: 03 5872 2535
Email: njplumbing@bigpond.com
If you have never tried blood orange juice, there could hardly be a better place to

begin than right here. Most people know Valencia orange juice, which is bottled here year round, and naturally sweet. Blood orange juice (called Riverview Red) has a complex flavour—a full bodied citrus with a hint of raspberries. There is also tangelo, grapefruit and lemon juice available, all sourced from local orchards. They are available at local farmers' markets, or you can pick them up direct from the factory or in local cafes.
- **Open:** 9am–5pm Mon–Fri, 9am–12pm Sat, tours by appointment only
- **Location:** In new industrial estate in Cobram

DINING

FLORIST CAFE
Rosalind Forrest
Cobram Florist, 77 Punt Rd, Cobram 3644
Phone: 03 5872 1923
Mains average: $
Specialty: Creative pitas, cakes, punch
Local flavours: Richglen olive oil, cheeses, wines, chutneys
Extras: A small cafe located at the front of a florist with good coffee and outdoor dining. Grows its own herbs, organic vegetables and fruit and sells a range of produce.
- **Open:** 8am–4pm weekdays

ECHUCA
PRODUCERS

BEECHWORTH BAKERY
Tom and Christine O'Toole
513 High St, Echuca 3564
Phone: 03 5480 6999
Email: echuca@beechworthbakery.com
www.beechworthbakery.com

This sibling to the hugely popular and long-established Beechworth Bakery in Beechworth, opened in 2001 with the same range and high quality of goods made on the original premises.
- **Open:** 6am–7pm daily

WILLMULLIE FARM
Doug and Julia Mulley
83 Rowe Rd, Echuca 3564
Phone: 03 5480 6343
Email: willmull@mcmedia.com.au
As Doug puts it 'after dairy farming for forty years and using 405 million litres of water a year, it appealed to me to produce good food for 1.5 million litres, and still make a living.' He adds that he likes tomatoes, which is lucky as this hydroponic farm, five minutes south of Echuca, produces a lot of them year round. Expect to see these tomatoes in local cafes and at farmers' markets but they are also sold at the farm gate. The rest go to Sydney and Melbourne markets.
- **Open:** 9am–6pm daily
- **Location:** 6km on Kyabram Rd from Echuca turn left into Rowe Rd, then continue for 800m

DINING

OSCAR W'S
Dean Oberin
Murray Esplanade, Echuca Wharf, Echuca 3564
Phone: 03 5482 5133
Email: oscarws@bigpond.com
www.oscarws.com.au
Licensed
Cuisine: Contemporary Australian
Mains average: $$
Specialty: Crispy skinned duck, seasonal and regional flavours

Local flavours: Yabbies, Murray cod, aged local beef
Extras: Outdoor dining and views of the mighty Murray river. Sells a range of homemade tapenades, jams, chutneys and spice mixes.
- **Open:** Lunch, dinner daily

IRYMPLE

🍎 PRODUCERS

ANGAS PARK FRUIT COMPANY
Andrée Wilksch
1554 Koorlong Ave, Irymple 3498
Phone: 03 5024 7077
Email: kevans@chiquita.com.au
www.angaspark.com.au
This centre is one of three Angas Park promotions centres in Australia, selling a wide range of quality dried fruits under the Angas Park and Anchor labels. There is also fruit confectionery, nuts, and chocolate for sale, and unique regional cuisine. Other centres are at Angaston and Berri in South Australia.
- **Open:** 9am–5pm weekdays, 11am–4pm Sat and public holidays

KOONOOMOO

🍎 PRODUCERS

SCENIC DRIVE STRAWBERRIES
Michael, Lorraine and Darren Hayes
1541B Torgannah Rd, Koonoomoo 3644
Phone: 03 5871 1263
Email: scenicdrive@bigpond.com
www.strawberryfarm.com.au
Just the word 'strawberries' can make a mouth water. And when you add in a scenic drive, the chance to sample fresh strawberries, home-made jams and condiments, or maybe some fruit ice-creams, it begins to sound even better. There is a full range of fruit wines and liqueurs, plus fresh or frozen strawberries in various sizes and grades. Pick your own strawberries in season, farm tours, strawberries and ice-cream, tea and coffee—it's all here.
- **Open:** 8am–6pm daily
- **Location:** 2km off Tocumwal-Benalla Rd

MILDURA

🍎 PRODUCERS

MURRAY RIVER SALT
Alan Hutcheon
4 Bothroyd Crt, Mildura 3500
Phone: 03 5021 5355
Email: murrayriversalt@sunsalt.com.au
www.sunsalt.com.au
These gourmet salt products, which capitalise on and draw attention to inland salt and salinity, have been manufactured since 2002 and have won a devoted fan club. The salt flakes are locally produced at Mourquong and Hattah from safe, saline groundwater, and are ideal to use as a garnish for dishes. Sold in many gourmet food stores, including 27 Deakin in Mildura.
- **Open:** 8am–5pm, weekdays
- **Location:** Between Eleventh and Fourteenth Streets, off Benetook Ave

🏪 STORES

THE CITRUS SHOP
41A Deakin Ave, Mildura 3500
Phone: 03 5021 1678
If it has citrus fruit in it—that means foodstuffs such as marmalades, choc-dipped dried orange slices, relishes, pick-

les, and jams—then you'll find it here in the citrus capital of Victoria.
- **Open:** 9.30am–5pm weekdays, 9.30am–12pm Sat

HUDAK'S BAKERY 15TH STREET STORE
Michael and Melissa Hudak
Fifteenth St, Mildura 3500
Phone: 03 5023 5906
Email: michael@hudaksbakery.com.au
www.hudaksbakery.com.au
This family bakery, established in 1947, and winner of many awards, is the birthplace of the Sunraysia orange loaf. Owner Gary Hudak was instrumental in urging and convincing local retailers and restaurateurs to support the local citrus industry and only stock Australian orange juice. Each orange loaf contains the juice of a whole orange and Hudak has been sharing the recipe with other bakers—all to promote the local produce he cares so passionately about. A second store is the 8th Street Store in the city heart.
- **Open:** (15th St Store) 7am–5.30pm daily (8th St Store) 7am–6pm weekdays, 7am–5pm Sat, 7am–4pm Sun and public holidays
- **Location:** Opposite Mildura Centre Plaza, 8th St

ORGANICS ON ELEVENTH
Gary Schirr (Manager)
200 Eleventh St, Mildura 3500
Phone: 03 5021 2230
Email: info@au.fcoop.org
http://au.fcoop.org
This organic and biodynamic shop, which opened in mid 2002, sells a range of local produce and other organic foodstuffs. It's a volunteer-run, not-for-profit business encouraging community participation and providing organic and biodynamic produce (including fresh fruit and vegetables, grains, seeds, nuts, legumes and dairy) and is sourced from local farmers and Australiawide. It is the only shop in Sunraysia specialising in organic and biodynamic produce.
- **Open:** 10am–6pm Tues–Wed, 4pm–7pm Fri, 10am–2pm Sat
- **Location:** Between Deakin and San Mateo Avenues

DINING

AVOCA
Stefano de Pieri
Mildura 3500
Phone: 03 5022 1444
Email: 27deakin@ncable.com.au
www.stefanosonthemurray.com.au
Avoca is an 1877 paddleboat that now has been transformed into a relaxing, informal riverfront cafe probably set for another hundred years of service, and moored on one of the most beautiful stretches of the Murray River within walking distance of the Art Gallery, Mildura Weir, Lock Island and the city centre. The licensed Gondola Cafe is downstairs, and the Squid Bar's alfresco area is located on the top deck.
- **Open:** 11am–late Tues–Sat, 11am–4pm Sun
- **Location:** Near the Mildura tennis club

BLUE SEAFOOD AND CHAR-GRILL
Tony Tyson
Rockford Mildura, 373 Deakin Ave, Mildura 3500
Phone: 03 5023 3823
Email: mildura@rockfordhotels.com.au
www.rockfordhotels.com.au
Licensed/BYO

VIC–MURRAY AND THE OUTBACK

Cuisine: Contemporary Australian
Mains average: $$
Specialty: Beef fillet stuffed with pancetta
Local flavours: From local suppliers such as Malley Foods, Angas Park and Sunbeam Foods
Extras: Oasis-style landscaped gardens.
■ **Open:** Breakfast and dinner daily

STEFANO'S

Stefano de Pieri and Donata Carrazza
Grand Hotel Resort, Seventh St,
(cnr Langtree Ave),
Mildura 3500
Phone: 03 5023 0511
Email: info@milduragrandhotel.com.au
www.stefanosonthemurray.com.au
Licensed
Cuisine: Italian
Set menu including coffee: $$$$
Local flavours: Murray cod, asparagus, citrus, wine, olive oil, Mildura salt, own preserves
Extras: Two other restaurants located in this hotel are The New Spanish Bar and Grill (lunch, weekdays) serving aged Mallee beef and lamb cooked on a real charcoal grill, and The Dining Room (dinner, Mon–Sat) offering a range of Italian dishes. De Pieri continues to vigorously champion local produce and slow food principles in his restaurants, books and on TV.
■ **Open:** Dinner Mon–Sat

27 DEAKIN

Stefano de Pieri and Donata Carrazza
27 Deakin Ave, Mildura 3500
Phone: 03 5021 3627
Email: 27deakin@ncable.com.au
www.stefanosonthemurray.com.au
Licensed
Cuisine: Mediterranean
Mains average: $
Local flavours: Murray cod, asparagus, citrus, wine, olive oil, Mildura salt, own preserves
Extras: Jams, bread, and other products are made and sold here, including the unique pink lake salt from nearby Hattah Lakes, plus Stefano de Pieri's own olive oil made in Bendigo. Hampers can also be made up. Takeaway available.
■ **Open:** Breakfast and lunch daily

OUYEN

🏠 STORES

MUNRO'S PRIME MALLEE MEATS

Scott Munro
56 Oke St, Ouyen 3490
Phone: 03 5092 1063
Email: lennys@iinet.net.au
Lamb raised here in the Mallee is much like the 'pre-sel' lamb that is famous in Brittany because the sheep eat salt bush, which infuse the meat with a salty flavour. Scott Munro breeds lamb and beef and his lambs are fed salt from the nearby salt lake. Munro's also makes all its own sausages and smallgoods.
■ **Open:** 8.30am–5.30pm weekdays, 8.30am–12pm Sat

> **GREAT AUSTRALIAN VANILLA SLICE TRIUMPH**
>
> If you love this sweet, sticky, addictive slice, make sure you are in Ouyen when bakers from around the country contest to see who can make the best.

ROBINVALE
🍎 PRODUCERS

ROBINVALE ESTATE OLIVE OIL
Ralph and Glenda Gallace and family
RMB 2270, Tol Tol Rd, Robinvale 3549
Phone: 03 5026 3814
Email: oil@robinvaleestate.com.au
www.robinvaleestate.com.au

This olive grove on the banks of the Murray River, established in the 1940s, produces olives as well as a good affordable olive oil. In 2003 the estate was awarded Gold in its class at the Australian Olive Association Show for its Murray Gold Blend. Now here's a sea change—or is it simply an oil change? This family used to grow strawberries on the Mornington Peninsula and operate fruit shops. When an opportunity arose to purchase 600 acres of irrigated olives and citrus in Robinvale, they jumped at it and since then the property has added carrots, wheat, table grapes and wine grapes. Also available are kalamata, manzanillo (black) and sevillano (green) olives pickled in brine, plus handmade Robinvale Estate olive oil soap. Also sold on site are Murray River Salt and Australian pistachios, both produced locally.

- **Open:** 9am–5pm, weekdays (closed 12pm–1pm) or by appointment
- **Location:** 10km on the Swan Hill side of Robinvale, left into Tol Tol Rd, then 3km further

ROBINVALE WINES
Steve and Bill Caracatsanoudis
Block 43B, Sealake Rd, Robinvale 3549
Phone: 03 5026 3955
Email: robwine@iinet.net.au
www.organicwines.com.au

This vineyard is noted for its organic and biodynamic wines, and you may also buy dried fruits and table grapes from the cellar door.

- **Open:** 9am–5.30pm daily

SWAN HILL
🍎 PRODUCERS

OLSON PHEASANT FARM
Kevin and Ros O'Bryan
2167 Chillingollah Rd, Swan Hill 3585
Phone: 03 5030 2648
Email: gamebirds@bigpond.com
www.gamebirds.com.au

Kevin and Ros specialise in supplying pheasant, guinea fowl, partridge and pheasant sausages to restaurants under the Olson Game Birds name. The business has been established for around forty years, but for the O'Bryans it was a welcome change of lifestyle ten years ago, after operating a retail business for seventeen years. All their birds are free-range and fed a chemical-free diet with no growth hormones. The sausages are unique—100 per cent pheasant meat and preservative- and gluten-free. Even the fresh herbs that season them are home grown. Most products are available from the farm, or the O'Bryans may be contacted for information on retail and wholesale outlets. The farm itself is an attraction, featuring displays of ornamental pheasants, and a walk-through garden aviary. All visitors can take a guided tour of the operation.

- **Open:** 10am–4pm daily
- **Location:** Turn left off the Murray Valley Hwy, onto Chillingollah Rd 10km north of Swan Hill, then continue for 23km more, farm is on the left

VIC—MURRAY AND THE OUTBACK

🍴 DINING

SPOONS DELI
Sandra and Chris Jewson
387 Campbell St, Swan Hill 3585
Phone: 03 5032 2601
Email: spoons@bigpond.net.au
Licensed
Cuisine: Regional
Mains average: $
Specialty: Terrines—changing daily, some days pheasant some days chicken, Jamie's poppy seed dressing
Local flavours: Olive oils, olives, tapenade, pistachios, almonds, stone fruit, grapes, tomatoes, carrots, onions, pheasant, local dressings, condiments, wine, Murray River salt flakes, dried fruit, organic fruit juices and citrus
Extras: Makes a wide range of marmalades, chutneys and jams for sale as well as pickled pears, cakes, biscuits and almond bread. Seating outside and inside. Picnic boxes, hampers and takeaway available. Cookery demonstrations.

TRENTHAM CLIFFS

🍴 DINING

TRENTHAM ESTATE
Anthony, Nola and Pat Murphy (owners)
Nathan Smith (chef)
Sturt Hwy, Trentham Cliffs 2738
Phone: 03 5024 8888
Email: rebeccaw@trenthamestate.com.au
www.trenthamestate.com.au
Licensed
Cuisine: Modern Australian
Mains average: $$
Specialty: Use of native herbs and spices
Local flavours: All fruit and vegetables

MARKETS AND EVENTS

✦ **MURRABIT COUNTRY MARKET,** MURRABIT
03 5457 2205
8.30am–1pm first Sat, also Easter Sat and third Sat in Dec

✦ **AUSTRALIAN INLAND WINE SHOW,** SWAN HILL
03 5032 3033
Oct

✦ **ITALIAN FESTA,** SWAN HILL
03 5032 3048
July

✦ **LINGA LONGA LUNCH FESTIVAL,** YARRAWONGA
03 5744 1989
June

✦ **MIGHTY MURRAY TART TRAWL,** ECHUCA
02 6058 2996
Mar

✦ **RIVERBOATS, JAZZ, FOOD AND WINE FESTIVAL,** ECHUCA, MOAMA
03 5480 7555 or 1800 804 446
Feb

✦ **SUNRAYSIA JAZZ, FOOD AND WINE FESTIVAL,** MILDURA
03 5021 4424
Weekend before Melbourne Cup

Extras: Winery with wine tastings and sales, situated on the banks of the Murray River with spectacular views.
■ **Open:** Lunch Tues–Sun (available for evening functions)
● **Location:** Fifteen minutes from Mildura

PHILLIP ISLAND AND GIPPSLAND

West Gippsland adopted the 'gourmet deli' tag some time back for very good reasons. When you see the wealth of smallgoods produced here, you have to envy Melburnians who can be here after a ninety-minute drive.

Better yet, the food has not sat for days in some refrigerated cabinet, or for weeks on a shelf. Here you can often pick your own fruit or berries, or at the very least buy some that were harvested in the cool of the same morning.

Bring a picnic hamper to this region, but make sure it is empty as, rounding out the deli department of this region you will find organic vegetables, farm cheese, wood-fired breads, venison, emu, kangaroo and trout—enough to satisfy any lunch requirements.

All that remains then is to find a tranquil park or riverbank, beach or waterfall. And, you guessed it, these are in abundance too.

BASS
DINING

WILDLIFE WONDERLAND
Robert Jones and Chris Cohen
Bass Hwy, Bass 3991
Phone: 03 5678 2222
Email: info@wildlifewonderland.com
www.wildlifewonderland.com
Licensed/BYO
Mains average: $$
Local flavours: Venison, King Island cheese

Extras: This large tourist park draws visitors from around the world, and appropriately, is a tourist information centre as well. Raises own venison. Takeaway available.
- **Open:** Breakfast, lunch and dinner daily
- **Location:** Turn onto Woolmer Rd from the main Bass Hwy then look for the signs

BEACONSFIELD
STORES

OFFICER ORGANICS
Kristina Morgan
27–33 Woods St, Beaconsfield 3807
Phone: 03 9796 2414
Email: officer.organics@bigpond.com
A good place to come for Jindivick Smokehouse meats, Piano Hill Farm cheeses and a selection of certified organic and locally produced products. Kristina likes to describe her products as 'real food'. 'My passion', she says, 'is food that tastes like it should'.
- **Open:** 9am–5pm weekdays, 9am–12pm Sat
- **Location:** Within gym complex

CALDERMEADE
DINING

CALDERMEADE FARM CAFE
Phil Malcolm (owner), Amanda Grundy (chef)
4385 Sth Gippsland Hwy, Caldermeade 3984
Phone: 03 5997 5000
Licensed
Cuisine: Modern Italian
Mains average: $

Specialty: Hearty breakfasts, cakes and pastries
Local flavours: Extensive use of vegetables, eggs, cheeses, wines, honey
Extras: Located on a fully-operational dairy farm with a baby animal nursery, open fireplace, enclosed deck from which to view milking at 3.30pm. Winemakers' dinners, musical festivals and jazz afternoons.
- **Open:** Breakfast and lunch (all day) daily

CHILDERS
PRODUCERS

SUNNY CREEK FRUIT AND BERRY FARM
Philip Rowe and Cathie Taylor
Tudor Rd, Childers 3824
Phone: 03 5634 7526
Email: sunnycrk@dcsi.net.au
Philip Rowe's NASAA certified organic orchard nestled in the Strzelecki ranges specialises in berries—there are around a hundred varieties—especially raspberries, although he does grow strawberries and blueberries. He also has several apple varieties, including Jonathans and Cox's Orange Pippin, hazelnuts and lots of chestnuts available from Easter onwards. The berries mainly appear from December to January although some go on until Easter. Collecting varieties again, Rowe has a full set of raspberries—around forty kinds—including red, yellow, purple (a black and red cross) and black raspberries, as well as a shatoot white mulberry. You can pick your own berries, although the packing shed sells them already picked, many of them earmarked for wholesale or export. Rowe also processes jams and toppings, under the Sunny Creek brand, on site.

- **Open:** School hours by appointment, 8am–6pm weekends
- **Location:** Off Sunny Creek Rd, 7km south of Princes Hwy, 1km west of Trafalgar, follow the signs

COWES
DINING

THE CASTLE–VILLA BY THE SEA
Harley and Jennifer Boyle
7–9 Steele St, Cowes, Phillip Island 3922
Phone: 03 5952 1228
Email: info@thecastle.com.au
www.thecastle.com.au
Licensed
Cuisine: International
Mains average: $$$
Specialty: Hot pot (such as beef bourguignon)
Local flavours: Spatchcock, eggs, seafood
Extras: Magnificent seaside garden beside a north-facing beach, bird life, art gallery.
- **Open:** Only for in-house guests or functions by pre-booking

THE JETTY RESTAURANT AND STAR CLUB BAR
Vivian Viglietti (owner),
Wayne Foster (chef)
11–13 The Esplanade, Cowes 3922
Phone: 03 5952 2060
Email: jetty@nex.net.au
Licensed
Cuisine: Modern Australian
Mains average: $$
Specialty: Local seafood
Extras: Waterfront views of Westernport, with outside dining available.
- **Open:** Lunch weekends, dinner daily

DROUIN WEST
🍴 PRODUCERS/DINING

BERRY GOOD CAFE AND BERRY FARM
Joanne Butterworth-Gray and Colin Gray
Drouin West Fruit and Berry Farm,
315 Fisher Rd, Drouin West 3818
Phone: 03 5628 7627
Email: berry@dcsi.net.au
www.visitvictoria.com/drouinberryfarm
Licensed
Mains average: $
Specialty: Blackberry pie
Local flavours: Gippsland cheeses, meats, trout, fruit and vegetables
Extras: Calling itself 'Victoria's most diverse farm', with 180 varieties of tree-ripened, pesticide-free fruit and berries on its twenty-five hectares, the ripening season extends all year. Pick your own, farm talks and a twenty-minute train ride around the property with commentary, bellbird-serenaded meals overlooking the orchard, plus fruit wines (blueberry and raspberry). Two B&B cottages.
■ **Open:** 10am-5pm daily, shorter hours in winter
● **Location:** Follow the signs from the Princess Freeway

FOSTER
🍎 PRODUCERS

AMEYS TRACK BLUEBERRIES
Ian and Chris Sandiford
670 Ameys Track, Foster 3960
Phone: 03 5681 2273
Email: ameystrk@tpg.com.au
The five different varieties of blueberries grown at Ameys Track ripen early and, the Sandifords say, have a superior taste and size, because they are of the high bush variety. They also grow raspberries and have some 200 hazelnut trees on the property. There is no pick your own available as all fruit and nuts are prepacked.
■ **Open:** 9am-6pm daily (Jan)
● **Location:** 6.7km from South Gippsland Hwy

🏠 STORES

FOSTER SEAFOOD
Andy and Sandy Collett
35 Main St, Foster 3960
Phone: 03 5682 2815
The shop has been here several years and is a 'bit of an institution'. And why wouldn't it be when it only sells local seafood—fish, prawns, oysters, mussels, scallops and crayfish—all fresh and straight off the boats every morning.
■ **Open:** 8.30am-5pm Tues-Fri, 8.30am-12pm Sat

FRENCH ISLAND
🐄 FARMSTAY

LA TAVERNE DU NATURALISTE
Mark Cunningham
McLeod Eco Farm, McLeod Rd,
French Island 3921
Phone: 03 5678 0155
Email: ecofarm@bigpond.com
www.McleodEcoFarm.com
Licensed/BYO
Cuisine: Mediterranean
Mains average: $$
Specialty: Steaks and vegetarian meals
Local flavours: Meats, vegetables and some fruit, homemade bread

Extras: The restaurant has an organic and biodynamic vegetable garden which supplies most of the vegetables used in the restaurant. It produces its own berries and jams, and hopes to eventually raise and butcher its own cattle. Surrounded by National Park, there is also a beach and plenty of native flora and fauna, scenic cycling tracks, nine-hole golf course, and fishing. The 222-hectare organic farm uses permaculture and biodynamic techniques. Comfortable accommodation.

- **Open:** Breakfast, lunch and dinner daily (only for guests because of the island's isolation)

GARFIELD

🍎 PRODUCERS

GARFIELD BERRY FARM
Maria Doherty and Francesca Ferraro
2895 Princes Hwy, Garfield 3814
Phone: 03 5629 2520
Email: berryfarm@nex.net.au

Begun as the sort of place where weekend travellers from Melbourne could stock up on their way to the country, this place has become a respected supplier of regional produce and foods including Country Style smoked meats, Tony's Own chutneys and Summer Snow Apple Juice. The farm grows sweet varieties of strawberries, capsicum, tomatoes, zucchini, squash and chillies. Other locally grown fruit and vegetables are sold in the produce market. There's a large range of gourmet products too, including locally produced cheeses—Tarago River and Piano Hill—and gourmet smoked products produced in Garfield, as well as gourmet jams, conserves, mustards, pickles, relishes, chutneys, and dessert sauces. And of course Flavourama ice-cream, which is full of berries.

- **Open:** 8am–6.30pm daily (DST), 8.30am–5.30pm daily (non-DST)
- **Location:** 4.5km past Gumbuya Park, Tynong on Princes Hwy

GARFIELD FISH FARM
Brian and Elaine Fox
20 Railway Ave, Garfield 3814
Phone: 03 5629 1166
Email: garfieldfishfarm@dcsi.com.au

One day a week this place is open for sales of fresh-farmed barramundi raised in a recirculated system. Available year round, the main business is growing barramundi up to around 400 grams for restaurants in Gippsland and Phillip Island.

- **Open:** 1pm–6pm Thurs (sales only)

🏠 STORES

COUNTRY STYLE MEATS
John and Kim Preston
Main St, Garfield 3814
Phone: 03 5629 2593
Email: csmeats@nex.net.au

This local butcher smokes a delicious selection of meats including chicken, pork, kangaroo, venison and lamb, and also stocks Gippsland beef and lamb. Smallgoods made on the premises from meat bought locally include strasbourgs, smoked chicken breasts, smoked boneless leg nuggets, twelve varieties of kabanas, rindless middle bacon, and pastrami. Local delis and restaurants also snap these up.

- **Open:** 7am–5.30pm weekdays, 7am–12.30pm Sat

KOONWARRA

🍴 DINING

KOONWARRA FINE FOOD AND WINE STORE
Melissa Burge and Maria Stuart
South Gippsland Hwy, Koonwarra 3954
Phone: 03 5664 2285
Email: tkstore@tpg.com.au
www.koonwarrastore.com.au
Licensed/BYO wine
Mains average: $
Specialty: Vanilla slices
Local flavours: Mainly uses organic produce, free-range eggs, chicken, pork, Gippsland Natural beef
Extras: Roses and cottage gardens, heritage fruit trees, cooking classes, farm tours, courtyard dining, takeaway available and a garden room which is a wine bar and used for private functions. Gippsland Slow Food Convivium operates from this store. Also represented at two Gippsland and two Melbourne farmers' markets.
- **Open:** 8am–5.30pm daily, dinner Fri–Sat
- **Location:** 6km from Leongatha

KOO-WEE-RUP

🚌 TOURS

GIPPSLAND GREEN ASPARAGUS TOURS
Kerri Burge
210 Railway Rd, Koo-wee-rup 3981
Phone: 03 5997 2202
Mobile: 0408 565 187
Email: kerri@asparagustours.com.au
www.asparagustours.com.au
This farm-based tourism business, operating since 2000, runs tours on the property throughout the spring harvest.

Morning or afternoon tours must be booked ahead, and include a look at the computerised packing shed, homestead, cooking demonstrations and paddock walk, including introduction to the farm's Clydesdale horses.
- **Open:** Twice daily Sept–mid Dec

KORUMBURRA

🏠 STORES

KELLY'S BAKERY
Gilbert, Lyn, Jason and Tarli Kelly
63-67 Commercial St, Korumburra 3950
Phone: 03 5655 2061
Knowing that they have been consistent medal winners in the National Great Aussie Pie Competition over the past few years makes the idea of buying a pie from this bakery, which uses only local meat, a tempting one. Top it off with a thick shake using local milk and you will start to see why people live in the country.
- **Open:** 6am–6pm weekdays (until 5.30pm in winter), 7am–3.30pm Sat, 8am–3.30pm Sun

LEONGATHA

🏠 STORES

MURRAY GOULBURN TRADING
Grant Lethborg
1 Yarragon Rd, Leongatha 3953
Phone: 03 5662 2308
Located next to the factory which is visible from most parts of the area, this is the place to come for Devondale butter, Devondale spreads, full-cream milk powder, smoothies made with full- and reduced-fat cream, and cheeses such as tasty, mild, trim tasty, vintage, Cobram

vintage tasty, shredded and slices. And here you will find them generally priced lower than elsewhere. The main selling point though is that all of these products are based on local Gippsland milk, some of the best in the country. Cheeses are not manufactured on site but come from the Murray Goulburn Cooperative in the north of the state.

- **Open:** 8.30am–5pm weekdays, 9am–12pm Sat

MIRBOO NORTH
🍎 PRODUCERS

CASALARE SPECIALITY PASTA PTY LTD
Barry and Nancy Hewitt
38-40 Burchell Lane, Mirboo North 3871
Phone: 03 5668 1746
Email: casalare@netspace.net.au
www.casalare.com.au

Some of these specialty pastas are wheat-free or gluten-free, and the other pastas are made from certified organic whole-wheat flour. Then there is the durum wheat semolina pasta that incorporates Australian native bush flavours such as lemon myrtle, roasted wattle seed, mountain pepper, and river mint, some of which the Hewitts are able to source locally. The company also manufactures ancient grains including spelt, kamut and Pharaoh pastas.

- **Open:** 9am–5pm weekdays
- **Location:** Turn right off Princes Hwy to Mirboo North, turn left at supermarket in main street and left again into Burchell Lane

NEERIM SOUTH
🍎 PRODUCERS

JINDIVIK SMOKEHOUSE
141 Main Rd, Neerim South 3831
Phone: 03 5628 1797

Changes for Jindivik. The tearooms have closed and the butchers have moved to Neerim South, but the smoking continues. Free-range heritage breeds of pigs become hams, salamis, smoked racks, bacon, and other smallgoods.

- **Open:** 8am–5.30pm weekdays, 9am–4pm weekends

SAN REMO
🏠 STORES

SAN REMO FISHERMEN'S CO-OP
Pam Ward
Marine Parade, San Remo 3925
Phone: 03 5678 5206

Renowned for superbly fresh crayfish, fish, scallops and shellfish straight off the boats, this is the place, right on the waterfront, to come for take-home or ready-cooked fish and seafood.

- **Open:** 8am–5.30pm daily for fish sales), 8am–8pm daily for takeaway food

THORPDALE
🚌 TOURS

POTATOES, PADDOCK TO PLATE
Val Murphy
RMB 1310, Thorpdale 3835
Phone: 03 5634 6267

Val offers a selection of tours of the Gippsland area, ranging from a one-day

coach tour to a six-day comprehensive tour. These are mainly for senior citizens, Probus clubs and similar groups, and include food of the district as well as garden and craft attractions. The one-day tour with a potato theme is a fun day out visiting the potato-growing areas and can include a potato-cooking demonstration, potato lunch and finish with potato ice-cream in Val's garden. 'Heaps of jokes, poems, ensuring good old belly laughs', says Val.
- **Open:** Phone for details

TRAFALGAR

STORES

THE SPUD SHED
Jim and Megan Abrechts
Princes Hwy, Trafalgar 3824
Phone: 03 5633 1410
This great barn is easy to notice as it is located right on the highway and sells all sorts of local fruit and vegetables, and of course different varieties of potatoes, from possibly the richest potato-growing area in the state. There is also a nursery with a wide selection of plants.
- **Open:** 8.30am–6.30pm daily
- **Location:** 1km Melbourne side of Trafalgar

WARRAGUL

FARMSTAY

CLEARVIEW FARM RETREAT
Jim and Carole Lewis
Van Ess Rd, Ferndale, Warragul 3820
Phone: 03 5626 4263
Email: clearviewretreat@hotkey.net.au
www.clearviewretreat.com

The restaurant is no longer open to the public except for group bookings or for guests staying on the farm. Which creates a good excuse to stay here and enjoy the free-range eggs, vegetables and fruit, most of which is grown on this organic property. Accommodation is both B&B and self contained, near Mt Worth State Park and the beautiful Strzelecki Ranges. The Lewises also stock a range of Clearview Range Preserves—homemade chutneys, jams and pickled eggs from their own produce.
- **Open:** Breakfast, lunch and dinner daily (house guests only)
- **Location:** Signposted from Warragul–Korumburra Rd

STORES

THE GRANGE CAFE AND DELI
Garry and Rowena Crow
15 Palmerston St, Warragul 3820
Phone: 03 5623 6698
This is where everyone comes for local produce including Tarago cheese (it's the factory outlet), local wines and beer, preserves, smoked meats, and other locally made, grown or raised produce. The licensed deli–cafe specialises in gourmet snacks, and gift hampers.
- **Open:** 7am–5.30pm weekdays, 8am–2pm Sat

YARRAGON

DINING

GIPPSLAND FOOD AND WINE
Alan and Angela Larsen
123 Princes Hwy, Yarragon 3823
Phone: 03 5634 2451 OR 1300 133 309
Email: yarragonvic@dcsi.net.au

VIC—PHILLIP ISLAND AND GIPPSLAND

Licensed
Cuisine: Modern Australian
Mains average: $
Specialty: Devonshire teas, berry thick shakes, muffins
Local flavours: Potatoes, berries, cheeses
Extras: Veranda dining, enclosed in winter. Takeaway available. The gourmet deli stocks local ice-cream, cheeses, and wines. Fully accredited tourist information centre.
■ **Open:** 7.30am–5.30pm weekdays, 7.30am–6pm weekends

STICCÁDO CAFE

*Fred and Inge Mitchell (owners),
Susan Hardie (chef)
Shop 6, The Village Walk, Yarragon 3823
Phone: 03 5634 2101
Email: sticcado@vic.australis.com.au
www.yarragonvillage.com*
Licensed
Cuisine: Australian
Mains average: $
Specialty: Beef and stout pies; beef, mushroom and red wine pies; beef rendang
Local flavours: Tarago, Jindi, Piano Hill cheeses, tomatoes, chutneys, olives
Extras: Outdoor seating, views to the Strzelecki Ranges to the south and Great Divide to the north. This cafe is owned by beef farmers. Takeaway available and there are also homemade, beef-based main meals, including pies, for sale.
■ **Open:** 10am–5.30pm Sat–Mon and Wed–Thurs, 10am–8pm Fri

MARKETS AND EVENTS

✦ **CARDINIA RANGES FARMERS' MARKET,** PAKENHAM
03 5629 6259
8am–12pm second Sat

✦ **DROUIN FARMERS' MARKET,** DROUIN
03 5664 0096
8am–1pm third Sat

✦ **PHILLIP ISLAND FARMERS' MARKET,** CHURCHILL ISLAND
03 5664 0096
8am–12.30pm fourth Sat

✦ **SOUTH GIPPSLAND FARMERS' MARKET,** KOONWARRA
03 5664 0096
8am–1pm first Sat

✦ **TYERS (TRARALGON) FARMERS' MARKET,** TYERS
03 5664 0096
8am–1pm fourth Sat

✦ **WELLINGTON FARMERS' MARKET,** SALE
03 5142 3380
8am–12pm third Sat

✦ **KILCUNDA LOBSTER FESTIVAL,** KILCUNDA
1300 366 422
Australia Day long weekend

✦ **TASTES OF PROM COUNTRY,** FOSTER
03 5682 2035
Early Jan

THE GRAMPIANS

Inland from the Great Ocean Road, the rich farming lands just contribute to the surfeit of goodies. There is lamb, local trout, smoked meats, and some of the best cheese and dairy products you could hope to find.

DUNKELD

DINING

ROYAL MAIL HOTEL
Martina Harris (manager), Jo Frost (chef)
Parker St (Glenelg Hwy), Dunkeld 3294
Phone: 03 5577 2241
Email: info@royalmail.com.au
www.royalmail.com.au
Licensed
Cuisine: Modern Australian
Mains average: $$$
Specialty: Hopkins River beef, Royal Mail fish and chips
Local flavours: Varies seasonally—sheeps milk fetta, yoghurt, smoked trout, Shaw River mozzarella, beef
Extras: Courtyard dining with mountain views, private dining room, an award-winning wine list (cellar with 120 000 bottle capacity), demonstration kitchen.
■ **Open:** Breakfast, lunch and dinner daily

GLENTHOMPSON

PRODUCERS

GRAMPIANS PURE SHEEP MILK PRODUCTS
Bruce and Elisabeth Cuming
'Stirling', Glenthompson 3293
Phone: 03 5577 4223
The Cumings produce specialty farmhouse yoghurts—but with a twist. They use milk from their own flocks of sheep and have been doing this since 1993 when wool prices dropped. These deliciously different yoghurts are sold nationwide, or you could pop into the farm, and see the Cumings hard at work making them. Milking is at 8am and 5.30pm.
■ **Open:** 8am–5.30pm daily
● **Location:** Glenelg Hwy, 6km from Glenthompson, forty-five minutes from Hamilton, look out for the signs

HALLS GAP

DINING

BRAMBUK LIVING ABORIGINAL CULTURAL CENTRE
Kaye Harris
Grampians Rd, Halls Gap 3381
Phone: 03 5356 4452
Email: brambuk@netconnect.com.au
www.brambuk.com.au
This is the ideal way to learn about Koori culture, with a visit to this excellent Cultural Centre that features five local Aboriginal communities. A snack in the Bush Tucker Cafe will introduce you to

kangaroo, emu, crocodile and various bush plants. Try the kangaroo steak sandwiches topped with a variety of native sauces or crocodile or emu kebabs served on a native herb rice.
- **Open:** Cafe—9am–4pm daily, centre—9am–5pm daily
- **Location:** 3km from Halls Gap

THE KOOKABURRA RESTAURANT
Rick and Vonne Heinrich
Grampians Rd, Halls Gap 3381
Phone: 03 5356 4222
Licensed
Cuisine: International
Mains average: $$
Specialty: Thai style prawns, French baked duckling with peppercorn brandy cream sauce
Local flavours: Mostly Victorian produce, ducklings from Nhill
Extras: Located in a lovely valley with box windows framing the view. Meals complemented by Grampian wines.
- **Open:** Lunch and dinner daily

HAMILTON
DINING

HAMILTON STRAND RESTAURANT
Sandra and Paul Donnelly (chef)
100 Thompson St, Hamilton 3300
Phone: 03 5571 9144
Licensed
Cuisine: Modern Australian
Mains average: $$
Specialty: Seafood dishes, lamb
Local flavours: All local, except seafood
Extras: Enclosed and heated patio area, gardens. Grows own vegetables and herbs. Sells homemade bread.
- **Open:** Breakfast, lunch and dinner daily

LAHARUM
DINING

MT ZERO OLIVES CAFE
Jane and Neil Seymour
Cnr Winfield and Mt Zero roads,
Laharum 3401
Phone: 03 5383 8280
Email: neil@mountzeroolives.com
www.mountzeroolives.com
Licensed
Cuisine: Mediterranean Australian
Mains average: $
Specialty: Focuses on regional products
Local flavours: Olives, olive oil, chick peas, hummus, lentils, lamb, goat, specialty cheeses
Extras: This building, which also houses a farm shop, was built from transported and restored school buildings. It is now used as a cafe and as a store in which to sell products from the Demeter biodynamic grove of 7000 trees adjacent to the Grampians National Park. There is an outdoor area for sitting, meals or functions. There are usually some examples of outstanding local artists' work on show and for sale. The original granite olive crushing stones used in this region forty years ago are also on display. In 2002 the Australia Fine Food Awards selected Mt Zero EVOO as the Champion extra virgin olive oil of Australia.
- **Open:** Cafe—lunch and teas Thurs–Mon, shop—10am–4.30pm daily

LANDSBOROUGH

🍎 PRODUCERS

BLUE VIEW OLIVES
Peter and Peg Bowles
31 Rifle Butts Rd, Landsborough 3384
Phone: 03 5356 9295
Email: pepebow@netconnect.com.au
Wonderful small producers of olives and olive oil are popping up everywhere lately. Here, the Bowleses are making small amounts of pickled olives as well as EVOO. Begun as a hobby on this small acreage four years ago, like many hobbies, though, it took off, and now there are three varieties of olive oils sold at this olive grove at the west end of the Pyrenees Ranges, or at local outlets and cellar doors, B&Bs, and farmers' markets and festivals.
- **Open:** By appointment

POMONAL

🍎 PRODUCERS

HARRISON'S ORCHARD
Lyn and Richard Harrison
1970 Lake Fyans Rd, Pomonal 3381
Phone: 03 5356 6245
Orchard growing must be in the Harrison blood. Richard has been growing fruit for over fifty years, carrying on his father's work. This area of Victoria is ideally suited climatically for growing stone and pome fruits (fruit with pips such as apples, pears and quinces) and the Harrisons have capitalised on both this and their lengthy experience. They cultivate over ninety varieties of these fruits in Pomonal's sandy loam soil that they say imparts an 'incredible flavour' to the fruit. 'Our customers think our fruit is the best', they say, and have even developed their own variety of apple, a Goldsmith, that usually sells out in three weeks.
- **Open:** 10am–5pm daily (when fruit available)
- **Location:** 1.3km from the Pomonal intersection

POMONAL BERRY FARM
Lorraine and Ray Rowe
Halls Gap Rd, Pomonal 3381
Phone: 03 5356 6393
Strawberries, four sorts of raspberries, loganberries, youngberries, boysenberries and silvanberries—all spray-free—are grown here, as a way of diversifying the farm after the wool market collapsed. Lorraine also makes jams, ice-creams and vinegars with seconds of the berries, sells fresh berries in season, and frozen year round. Also diabetic jams, seedless jams, and gluten-free relishes and pickles, so no-one needs to miss out on something lovely to enjoy.
- **Open:** 10am–5pm daily (Nov–Apr)
- **Location:** 2km from Pomonal store towards Halls Gap, opposite Grampians Lavender Patch

RED ROCK OLIVE OIL
Greg Aimer and Christian Freestone
Cnr Halls Gap–Pomonal Rd and Tunnel Rd, Pomonal 3381
Mobile: 0401 700 868
Email: info@redrockolives.com.au
www.redrockolives.com.au
Greg Aimer began growing olives in 1997 and he now has a range of cold-pressed EVOO, table olives—kalamata or manzanillo depending on the time of year—dressings,

VIC—THE GRAMPIANS

tapenades, herbal infused oil and other olive products in his 'olive shop' that also carries olive wood products and soaps.
- **Open:** 10am–5pm, weekends and public holidays
- **Location:** 2km north of Pomonal

STAWELL

🍎 PRODUCERS

GRAMPIANS LAVENDER PATCH
Graham and Jocelyn Fuller
RMB 2063, Pomonal, Stawell 3380
Phone: 03 5356 6285
Email: fullers@netconnect.com.au

Originally choosing lavender as a business to complement their small herd of Scottish Highland cattle, the Fullers have seen it grow into a business all of its own, with a sweet smell of success. The farm shop is right beside the planting and you can find dried lavender, as well as lavender-flavoured ice-cream ('Very special' says Jocelyn), biscuits (also available gluten-free), jams, jellies, honey, marmalade, cakes and scones for special occasions, plus mustard and chutney. Also available from Carisbrook post office, Seppelts and Vines Cafe and Bar in Ararat.
- **Open:** 10am–5pm Tues–Sat, 1pm–5pm Sun–Mon
- **Location:** Route 31, 2km from Pomonal store towards Halls Gap, on the right

---

MARKETS AND EVENTS

✦ GRAMPIANS GRAPE ESCAPE—
THE FOOD AND WINE FESTIVAL,
HALLS GAP
03 5358 2314
Early May

THE GREAT OCEAN ROAD

Stunning views, easy driving, great places to stop and fill the picnic basket or take a break. Ocean breezes are your appetiser, local fish and seafood, honey, berries, sheep yoghurt, herbs and local cheeses are the meal itself.

If Gippsland is the gourmet deli of Victoria, perhaps this, its western counterpart, should be termed the state's gourmet smorgasbord, as the goodies are spread on this table-like cliff-top region.

The coastal eateries serving fine local fish and seafood are too numerous to mention, but linger there, talk to the owners and discover what is fresh and in season. Chances are they'll know the person who caught the fish you're eating!

Take a trip here and you might just start to understand why this is called the *Great* Ocean Road.

ALLANSFORD

STORES

CHEESE WORLD
Kim Kavanagh
Great Ocean Rd, Allansford 3277
Phone: 03 5563 2130
Email: kimk@wcbf.com.au
www.cheeseworld.com.au
Licensed
Cuisine: Country style
Mains average: $
Specialty: Cheese platters, ploughman's lunches, milk shakes and ice-creams
Local flavours: Cheeses, fruit and vegetables, meats, dairy products

Extras: A fascinating peek into the world of cheese making, and the local dairy industry. The cheese and wine cellar offers tastings of over a hundred varieties of Australian cheeses and local honeys, plus sales of local wines and other specialty products. There is also a museum with displays of early cheese and butter-making equipment, and a video about cheese making. Takeaway available.
- **Open:** 8.30am–5pm weekdays, 8.30am–4pm Sat, 10am–4pm Sun
- **Location:** Across the road from the cheese factory

APOLLO BAY

PRODUCERS

OTWAY HERBS
Ken and Judi Forrester
155 Biddles Rd, Apollo Bay 3233
Phone: 03 5237 6318
Email: Forrester1@bigpond.com.au
The Forresters' biodynamically grown quality plants, with fresh herbs cut on request, are in great demand at Apollo Bay community markets and for use in local restaurants. They also make homemade marmalades, herb mustards and oils, and sell dried herbs. Visitors are generously supplied with information on planning and selection of herbs, based largely on Ken and Judi's twenty years of experience. Just a stroll around the large terraced garden with its magnificent views is worth the trip. Mail order available.
- **Open:** 9am–5pm daily

VIC—THE GREAT OCEAN ROAD

- **Location:** Turn off Ocean Rd at Wild Dog Rd and follow the signs for 8km. 12km from Apollo Bay

🍴 DINING

CHRIS'S BEACON POINT RESTAURANT AND VILLAS
Christos Talihmanidis
280 Skenes Creek Rd, Apollo Bay 3233
Phone: 03 5237 6411
Email: chrisbeaconpoint@bigpond.com.au
www.visitvictoria.com/christos
Licensed/BYO
Cuisine: Greek, southern Mediterranean
Mains average: $$$
Specialty: lobster
Local flavours: Fish, aged beef, roast duck, lamb and pork, Otway berries
Extras: Overlooking Bass Strait, set high in the Otways above the Great Ocean Rd, with on-site villas.
- **Open:** Lunch and dinner daily

BIRREGURRA

🍎 PRODUCERS

PENNYROYAL RASPBERRY FARM
Katrine and Mike Juleff
115 Division Rd, Murroon, Birregurra 3242
Phone: 03 5236 3238
Email: mkjuleff@bigpond.com
Fifteen years after discovering the farm as visitors themselves, the Juleffs took the place over, and now in this hidden valley they grow chemical-free raspberries, boysenberries, marionberries, youngberries, jostaberries and strawberries. Visitors can taste real country food at the new teahouse and feast on berries and preserves made from home-grown produce. There is also a pair of two-bedroom, self-contained B&B cottages on the property and you can guess what features on the breakfast menus—fresh home-grown fruit and vegetables, of course.
- **Open:** 10am–5pm daily (Dec–Jan), B&B year round
- **Location:** 12km south-east of Birregurra

DEANS MARSH

🍎 PRODUCERS

GENTLE ANNIE BERRY GARDENS
Russell and Janie Carrington
520 Pennyroyal Valley Rd,
Deans Marsh 3235
Phone: 03 5236 3391
Email: gentleannie@bigpond.com
www.gentleannie.com.au
The picturesque Pennyroyal Valley runs back into the Otway Mountains from Deans Marsh just twenty minutes drive from the coast. The Carringtons have retained an old tradition of berry growing in an area justly famous for its fruit. They now raise raspberries, strawberries, gooseberries, red and black currants, silvanberries, thornless loganberries, boysenberries, thornless blackberries, apples, nashi pears, plums, apricots and blueberries. Their country wines 'confirm that grapes do not have a monopoly on quality wine', and they serve Devonshire teas, lunches and Gentle Annie homemade preserves, chutneys, and sorbets in the tearooms. And as a diversion for the enthusiastic angler, there is trout fishing.
- **Open:** 10am–5pm daily (1 Nov–30 Apr)

KOROIT

STORES

KOROIT TOWER HILL GOURMET LARDER
Angela and Peter Carey
162 Commercial Rd, Koroit 3282
Phone: 03 5565 9117
Mobile: 0412 056 722
Email: grady.carey@bigpond.com
Sourcing regional foods and cheeses directly from the producer, this larder is stocked with wonderful things, and there is no comparable store in the area. Come here for pestos, mustards, organic oils, preserves, curry pastes, chutney and salsas from regional Victoria (Yarra Valley, King Valley, Beechworth, Milawa) and specialty goat, sheep and water buffalo cheeses and yoghurts. Angela makes marinated olives with preserved lemon, seasonal jams, organic sourdough, and wholemeal and white breads, baked Friday to Sunday to order.
- **Open:** 10am–6pm Wed–Sat, 10am–5pm Sun
- **Location:** About 5km from Warrnambool

PORT CAMPBELL

STORES

TIMBOON FINE ICE-CREAM
Tim Marwood
41 Lord St, Port Campbell 3269
Phone: 03 5595 0390
Email: timboon@ansonic.com.au
www.timboonicecream.com.au
Fresh local milk and cream is what turns this ice-cream into a super premium product with around fourteen to fifteen per cent fat. Owner Tim Marwood has a background in the dairying industry and began this business in 1999. He admits that he loves to play around with unusual flavours—hence Turkish delight, organic rhubarb sorbet, blackberry, and white chocolate ice-creams—around twenty in all. Buy cones or packs here, or at Leo's Fine Food and Wine Stores in Kew and Heidelberg.
- **Open:** 12pm–late daily (summer), 12pm–4pm daily (school holidays)

DINING

WAVES PORT CAMPBELL
William and Elizabeth Kordupel
29 Lord St, Port Campbell 3269
Phone: 03 5598 6111
Email: waves@standard.net.au
www.wavesportcampbell.com.au
Licensed
Cuisine: Modern Australian
Mains average: $$
Specialty: Fish dishes that vary daily depending on availability
Local flavours: Timboon strawberries, cheeses, Lower Gellibrand organic rhubarb, farmed fish when in season
Extras: Two outside eating areas.
- **Open:** Breakfast, lunch and dinner daily

PORT FAIRY

DINING

WISHARTS AT THE WHARF
Farmhill family (owners),
Brett Holmes (chef)
29 Gipps St, Port Fairy 3284
Phone: 03 5568 1884
Email: finewine@netconnect.com.au
www.cathcartwines.com.au
Licensed

VIC—THE GREAT OCEAN ROAD

Cuisine: Seafood
Mains average: $$
Specialty: Seafood platters
Local flavours: Fish, cheese, vegetables
Extras: Fresh fish sales and fish and chips to take away. Outdoor dining and spectacular views of the Moyne River and fishing fleet. Sells own wines and takeaway is available.

- Open: Breakfast, lunch and dinner daily
- Location: At the wharf

MERRIJIG INN
CJ and KM Robertson
1 Campbell St, Port Fairy 3284
Phone: 03 5568 2324; Freecall 1800 682 324
Email: info@merrijiginn.com
www.merrijiginn.com
Licensed/BYO
Cuisine: Modern Australian
Mains average: $$$
Specialty: Seafood risotto
Local flavours: Fish, crayfish, cheese, wines
Extras: Surrounded by beautiful cottage gardens, directly opposite the working wharf.

- Open: Breakfast daily (house guests only), dinner daily

TIMBOON
🍎 PRODUCERS

BERRY WORLD
Alan and Joy Kerr
24 Egan St, Timboon 3268
Phone: 03 5598 3240
www.berryworld.com.au
'We have been told that we grow the best strawberries in Australia', say the Kerrs, 'and we agree!' Thirty-four years ago their small acreage was hard-hit by drought, but just when they needed it, they received information on strawberry growing, and have not looked back. The Kerrs' berries and berry products are sold to small shops and restaurants in the local area. Pick-your-own addicts love the wide grassed picking paths between the rows, which allow pusher and wheelchair access, and make for cleaner berries. In addition to strawberries, there are hybrid blackberries, boysenberries, loganberries and youngberries. When you are finished picking, homemade jams, ices and ice-creams are waiting to refresh you.

- Open: 10am–4pm Tues–Sun (Nov–Mar)
- Location: Follow the Berry World fingerboard signs in Timboon

LARA SMALLGOODS
Over the years migrants to Australia have made an immeasurable contribution to this country's plates and palates. It is often the small operator like Angel Cardoso who, through passion and persistence, has provided this excellence. Dubbed the 'King of Ham', for twenty years he searched for that elusive recipe that is so easily taken for granted. He proudly tells you that his hams are his contribution to the future of Australia. This book pays tribute to him, and all like him in Australia, whose passion for fine food has contributed to this being the lucky country. His products are available at city markets.

🏠 STORES

THE MOUSETRAP
Deborah Calvert
Timboon Farmhouse Cheese,
Cnr Ford and Fells Roads (RMB 4130),
Timboon 3268
Phone: 03 5598 3387
Email: kmckenzie@kidairy.com.au

The Mousetrap is the place to come for sales and tastings of Timboon's extensive range of cheeses. In this little licensed cafe there is soup in winter, and cheesecake, cheese platters, Devonshire teas, gardens with seating outdoors and a gazebo. Using fine, biodynamic milk produced in their dairy, Timboon Farmhouse Cheese makes boutique cheeses without stabilisers or animal rennet. Sold throughout Australia in supermarkets, delis, and health-food stores, the range includes camembert, brie, gourmet fetta, brie spread, fresh cheese with herbs, fresh cheese with pepper, Timboonzola blue, and St Joseph's blue.

- **Open:** 10.30am–4pm daily (closed Mon and Tues early May–late Sept)
- **Location:** Signpost on Timboon–Peterborough Rd

WARRNAMBOOL

🍎 PRODUCERS

WARRNAMBOOL TROUT FARM
Jenny Paton and Peter Kavanagh
221 Wollaston Rd, Warrnambool 3280
Mobile: 0409 943 396 (business hours)
Email: pkavanagh@datafast.net.au

This is what all timid anglers need—a guarantee that they will catch a fish. That, and all the equipment and bait supplied. And having the fish cleaned by someone else. If you still choose not to fish, even with all this encouragement, you can enjoy the smoked fish and trout pâtés, created by Jenny Paton and Peter Kavanagh, or try hand feeding the fish—they will come within centimetres of you. If you are travelling the Kavanaghs will season and wrap the fish so you can take it home for dinner or a picnic barbecue. And now there is also wholesale Murray Cod available. Catch Cook and Eat is a group event for a minimum of twenty people. All fish that are caught are barbecued and served in the viewing room, along with a trophy for the biggest fish, certificates, and yabby races for the kids.

- **Open:** 10.30am–5pm daily
- **Location:** Two minutes from Warrnambool

🏠 STORES

ALLFRESH SEAFOODS AND POULTRY
John Hearn
47 McMeekin Rd, Warrnambool 3280
Phone: 03 5562 6633
Email: allfresh@bigpond.com

John Hearn was a professional prawn fisherman for twenty years, so he knows the industry. This wholesale operation sells, direct to the public, live and cooked local crays, fresh local fish, live Tasmanian oysters (opened on the premises), and a full variety of scallops, prawns and other seafood. Seafood is also processed and distributed in conjunction with the local fish trawling industry and live crays are exported to Taiwan and China

- **Open:** 9am–5.30pm weekdays, 9am–12pm Sat
- **Location:** In the Warrnambool Industrial Estate

🍴 DINING

PIPPIES BY THE BAY
Peter and Jane McLauchlan (owners),
Chris Considine (chef)
Flagstaff Hill, 91 Merri St,
Warrnambool 3280
Phone: 03 55612188
Email: pippiesbythebay@bigpond.com
Licensed
Cuisine: Italian-influenced modern Australian
Mains average: $$$
Specialty: Crayfish with three-cheese sauce
Local flavours: Crayfish, crabs, wines, buffalo milk cheese, seasonal fruit and vegetables, trout
Extras: An award-winning restaurant with a balcony dining area and exceptional views over Lady Bay, a themed 1870 maritime village, parkland and the fishing fleet.
■ Open: Breakfast, lunch and dinner daily

MARKETS AND EVENTS

✦ **PORT FAIRY FARMERS' MARKET,**
PORT FAIRY
03 5568 2421
7am–1pm, third Sat

✦ **WARRNAMBOOL GROWERS' MARKET,** WARRNAMBOOL
03 5562 7030
8am–12pm first Sat

YARRA VALLEY AND THE DANDENONG RANGES

The word gourmet is bandied around these days, its meaning almost lost. Yet just an hour or so from Melbourne is a region that perhaps revives the term in a practical way. Here you will find fine wines of course, but the food grown on these fertile slopes and valleys offers a perfect foil.

There is cheese, from goats and sheep as well as cows, orchards growing every type of stone fruit and cool climate fruit, lavender, herbs and honey. Berries abound in these cool temperate conditions and many places even allow you the tactile enjoyment of getting down and dirty as you pick your own.

To match the wealth of produce, of course there are restaurants and accommodation—everything from economy to elegant, suiting all needs. For this really is the key to the region. An urban getaway for some, a back-to-nature dip for others, but always a gastronomic experience.

BUXTON
PRODUCERS

BUXTON TROUT AND SALMON FARM
Mitch MacRae
2118 Maroondah Hwy, Buxton 3711
Phone: 03 5774 7370
Email: buxtontrout@austarnet.com.au
Victoria's oldest commercial trout hatchery has sales of fresh trout or mountain ash smoked trout. It also grows Atlantic salmon, which may seem strange, as most of us think of this as a saltwater fish, but salmon are adaptable to fresh water. In fact they are spawned and spend their first couple of seasons in freshwater before migrating to the sea, says Mitch MacRae. This farm offers public fish-outs and a barbecue area to deal with the catch. The trout and salmon also find their way to many restaurant tables in the state.
■ **Open:** 9am–5pm daily

CHIRMSIDE PARK
DINING

BELLA RESTAURANT
Marc Brown (chef)
The Sebel Lodge Yarra Valley, Heritage Golf and Country Club, Heritage Ave, Chirmside Park 3116
Phone: 03 9760 3355
Email: restaurants@hgcc.com.au
www.hgcc.com.au
Licensed
Cuisine: Modern regional
Mains average: $$$
Specialty: Regional produce including Buxton trout, Yarra Valley free-range pig and chicken
Local flavours: Yarra Valley Pasta, Yarra Valley free-range pork, Lillydale free-range chicken, Buxton trout, Yarra Valley Dairy cheese, fruit and vegetables form Harvest Fruit and Vegetables, Cunliffe and Waters jams, Fruition breads

VIC—YARRA VALLEY AND THE DANDENONG RANGES

Extras: Located in luxurious hotel beside a golf course.
- **Open:** Breakfast, lunch and dinner daily

COLDSTREAM

🍎 PRODUCERS

MAROONDAH ORCHARDS
Jennifer and Alan Upton
713-719 Maroondah Hwy, Coldstream 3770
Phone: 03 9739 1041
Email: marorch@hotlinks.net.au
The motive behind the Uptons' involvement with this orchard is, they say, 'To allow people to appreciate fully ripened fruit, and also to sell a range of fruit not sold to supermarkets because of small production lines.' To this end, they grow seven types of apples and seven varieties of pears, and make fresh, pure apple juice according to variety. Sold in Melbourne markets and local fruit shops and restaurants.
- **Open:** 10am-5.30pm weekends (Sept-June)

🍴 DINING

ROCHFORD'S EYTON
Helmut and Yvonne Konecsny
Cnr Maroondah Hwy and Hill Rd, Coldstream 3770
Phone: 03 5962 2119/1800 622 726
Email: info@rochfordwines.com.au
www.rochfordwines.com.au
Licensed
Cuisine: Asian-influenced Modern Australian
Mains average: $$
Local flavours: Wines, oil, beef, salmon, venison, cheeses

Extras: Outside courtyard and observation tower with views of Yarra Valley.
- **Open:** 10am-5pm, lunch daily

TOKAR ESTATE
Rita and Leon Tokar
6 Maddens Lane, Coldstream 3770
Phone: 03 5964 9585
Mobile: 0418 353 611
Email: tokar@bigpond.com.au
www.tokarestate.com.au
Licensed
Cuisine: Mediterranean
Mains average: $$
Specialty: Richard's stuffed chicken breast
Local flavours: As much as possible Yarra Valley produce
Extras: Large terrace overlooking vineyard with 360-degree views of surrounding hills.
- **Open:** Lunch daily (à la carte lunches Wed-Sun, light lunches Mon-Tues)

DIXONS CREEK

🍴 DINING

SHANTELL RESTAURANT
Turid and Shan Shanmugam (owners)
Jane O'Connor (chef)
Shantell Vineyard, 1974 Melba Hwy, Dixons Creek 3775
Phone: 03 5965 2155
Email: shantell@shantellvineyard.com.au
www.shantellvineyard.com.au
Licensed
Cuisine: Modern Australian
Mains average: $$
Local flavours: Yarra Valley salmon, berries
Extras: Outdoor dining on verandas overlooking the vineyard with great views.
- **Open:** Lunch Thurs-Mon

HEALESVILLE

🍎 PRODUCERS

FORGOTTEN FRUITS
Fiona Dickson
PO Box 1663, Healesville 3777
Phone: 03 5962 5264
Fiona Dickson's skills at coaxing delicate flavours out of grapes and rose petals and transferring them to shimmering jellies is exceptional. These unusual fine fruit jellies utilise several grape varieties including pinot and cabernet sauvignon, as well as rose petals. Even though you cannot buy them directly from Fiona Dickson, they are sold at the Yering Farmers' market, held on the third Sunday of the month in Yarra Glen.

🍴 DINING

ADA AT STRATHVEA
Dianne Clarke
55 Myers Creek Rd, Healesville 3777
Phone: 03 5962 4109
Email: stay@strathvea.com.au
www.strathvea.com.au
Licensed/BYO
Cuisine: Modern country cooking
Three-course menu: $$
Specialty: Local Yarra Valley produce
Local flavours: As much as possible
Extras: Ada at Strathvea is a historic guest house built in the late 1930s. Breakfast with local rosellas and cockatoos on the windowsill while by evening possums and sugar gliders watch from the trees.
■ **Open:** Breakfast daily (house guests only), lunch daily, dinner Fri–Sat (group bookings)

HEALESVILLE HOTEL
Michael Kennedy and Kylie Balharrie
256 Maroondah Hwy, Healesville 3777
Phone: 03 5962 4002
Email: info@healesvillehotel.com.au
www.healesvillehotel.com.au
Licensed
Cuisine: Modern country/French provincial
Mains average: $$$
Specialty: Classic country pub
Local flavours: Salmon, pork, cheese, fruits and vegetables
Extras: A produce store, Healesville Harvest, is located in the old bottleshop and sells local produce, fresh as well as packaged. Winners of the Tucker Seabrook Best Pub Wine list of the Year in Australia 2003. Takeaway available.
■ **Open:** Breakfast, lunch and dinner daily

3777 RESTAURANT
Sean Lee and Ross Parker
Mt Rael Mountaintop Retreat,
140 Healesville–Yarra Glen Rd,
Healesville 3777
Phone: 03 5962 1977
Email: mtrael@zarliving.com.au
www.zarliving.com.au
Licensed/BYO
Cuisine: Moroccan influenced
Mains average: $$
Local flavours: Yarra Valley fig ice-cream, pasta, spatchcock, salmon
Extras: Balcony dining overlooking the Yarra Valley, luxury accommodation.
■ **Open:** Brunch and lunch Fri–Sun, dinner Thurs–Sun

HODDLES CREEK

🍎 PRODUCERS

THE BIG BERRY
Nikki and Paul Casey
925 Gembrook Rd, Hoddles Creek 3139
Phone: 03 5967 4413
Email: margalit@bigpond.com
www.thebigberry.com
Fresh and frozen fruit—blueberries, blackberries and raspberries—available one way or another, all year. Fresh berries and pick your own from December to April, and frozen berries in the other months. There are pure berry juices, purees and jams as well. Fresh berries go to supermarkets nationally, while the manufactured products are sold in Victoria.
- **Open:** 10am–5pm daily

KALORAMA

🍎 PRODUCERS

WHISPERS FROM PROVENCE
Ann Creber
41–45 Barbers Rd, Kalorama 3766
Phone: 03 9728 4475
Email: provence@bluedandenongs.com.au
Ann began making these lovely products because she had been so impressed by the markets of France where local farmers arrive at the market each week with a couple dozen jars of different products, seasonally determined. Now her organic gourmet products include marinated fetta, marinated olives, preserves, chutneys, relishes (made to her grandmother's recipes) and jams, all using organic ingredients from her own organic garden, where possible. There are French soaps and also possible are garden tours, cookery classes, and afternoon teas (bookings necessary). A guest speaker on herbs is available. Products also sold at selected country markets and food outlets.
- **Open:** By appointment
- **Location:** Off Mt Dandy Tourist Rd

KANGAROO GROUND

🍴 DINING

EVELYN COUNTY ESTATE
Robyn and Roger Male
55 Eltham-Yarra Glen Rd,
Kangaroo Ground 3097
Phone: 03 9437 2155
Email: wine@evelyncountyestate.com.au
www.evelyncountyestate.com.au
Licensed
Cuisine: Contemporary Australian
Mains average: $$
Specialty: House smoked chicken and fish using chardonnay vine clippings
Local flavours: As members of the Yarra Valley Regional Food Group, local produce is promoted on the menu as often as possible
Grows: Kalamata olives from own grove, lemon and lime trees
Extras: Winner of Tourism Victoria's Tourism Restaurant and Catering Services, 2004 award. Promotes regional excellence in wine, food and art in an RAIA award-winning building overlooking the vineyard. In the process of developing a herb and vegetable garden which will be accessible to the public. Picnic sites are planned, with picnic hampers available.
- **Open:** Breakfast Sun, lunch Wed–Sun, dinner Thurs–Sun

LILYDALE

🍴 DINING

RESTAURANT AT THE LILYDALE HERB FARM
John Corcoran
61 Mangans Rd, Lilydale 3140
Phone: 03 9739 6899
Email: JGCorcoran@bigpond.com
www.theherbfarm.com.au
Licensed/BYO wine
Cuisine: Modern Australian
Mains average: $$

Specialty: Antipasto using ingredients from the Yarra Valley
Local flavours: Mostly Yarra Valley produce
Extras: Alfresco dining amongst the wild herb garden established for twenty-seven years, and growing around 500 different culinary, medicinal, insect repellent and fragrant herbs. Also a nursery and gift shop. Catering available on request.
- **Open:** Lunch Tues-Sun, dinner Wed-Sat

ROSEHILL LODGE B&B
George and Catherine Hill
Kalorama Terrance, Kalorama 3766
Phone: 03 9761 8889
Email: geh@ozemail.com.au
www.rosehill-lodge.com.au
You can't describe this B&B benchmark without using superlatives to describe the creative genius behind it. George Hill is a giant when it comes to cuisine. A chef of forty-eight years standing, educator, author and grand wizard with food, his knowledge is only exceeded by his passion for the finest things edible. Over the years he estimates that he has trained around 28 000 chefs in the culinary arts. In spite of his awards and achievements he chose, with his wife Catherine, to establish a B&B with a difference. Whatever suits the guests, is the mantra here. The guest is the centre of attention and food is the focus. Dinner is a grand event with the meal crafted around the guest's tastes. Desserts are the pièce de résistance and always an amazing surprise. And wines? Only the finest. The Hills are active in promoting the produce of the Dandenong region and concentrate on providing local flavours.

MACCLESFIELD

🍎 PRODUCERS

YARRA VALLEY FREE RANGE PORK
Christine Ross
25 Coopers Rd, Macclesfield 3782
Phone: 03 5968 8664
Mobile: 0429 309 525
Email: christine@largeblackpigs.com.au
www.largeblackpigs.com.au
Also operating as Eastwind Rare Breeds Farm, this farm is dedicated to the preservation of rare breeds of domestic farm livestock and, in particular, Large Blacks, which are raised free-range without the use of hormones, antibiotics or pesticides and graze happily in the paddock. This variety of pig provides pork the way it used to taste. Ten years ago, Christine Ross became concerned at the rapid loss of diversity in farm livestock. In the past century, one third of the world's native cattle, sheep and pig breeds have either become extinct or are nearly so. There are only sixty breeding sows of Large Black pigs left in Australia. Yarra Valley Free Range Pork produces pork and smallgoods, such as ham, kabana, and Polish sausage, and makes prosciutto 'to die for!' says Christine. The products are used by restaurants throughout the Yarra Valley

and sold by butchers in Emerald and Healesville. Orders can also be placed through the website.
- Open: By appointment

MARYSVILLE
🍴🍎 DINING/PRODUCERS

GILBERTS RESTAURANT AT FRUIT SALAD FARM
Patrick, Patrica and Jeffrey Jennings
30-32 Albury Cuzins Dr, Marysville 3779
Phone: 03 5963 3232
Email: info@fruitsaladfarm.com.au
www.fruitsaladfarm.com.au
Licensed
Mains average: $$
Local flavours: As much as possible, including local trout, and heirloom varieties of apples, plums and strawberries that grow in the old orchard
Extras: Self-contained cottages in addition to the farm and restaurant.
- Open: Lunch Sat, dinner Fri-Sun

MOUNT DANDENONG
🍎 PRODUCERS

ENCHANTED COTTAGE PRESERVES
Ruth D'Alessandro
149-151 Ridge Rd, cnr Ridge and Yarrabee Roads, Mount Dandenong 3767
Phone: 03 9751 1892
Email: enchantedcottagepreserves@dodo.com.au
When you've been making preserves for eighteen years and fudge for fifteen, you must love cooking. Ruth admits she does, and her all-natural, handmade goodies are best found at farmers' markets, although she will sell from her workplace. Generally she specialises in filling small jars (she makes around 200 varieties of jams, marmalades, jellies, butters, honey, mustards, chutneys) and her bulk raspberry jam is in demand all over the Dandenongs for decadent Devonshire teas.
- Open: By appointment

OLINDA
🏠 STORES

HERBICIOUS DELICIOUS
Trish Jonescu
Shop 3, Parsons Walk, 10 Parsons Lane, Olinda 3788
Phone: 03 9751 0026
This shop has lots of light and a very open feel, with everything from utensils to a nice line of preserves, toppings and flavoured vinegars. Great for browsing, and Trish, wherever possible, makes tastings available of the foods she sells. A 'sea change in reverse' she calls her move here 'to the prettiest mountains in the world'. As a member of the Yarra Valley Food Group, she primarily stocks local produce direct from farms and properties in the area.
- Open: 10.30am-5.30pm daily (opening slightly later in winter)

PHEASANT CREEK
🍎 PRODUCERS

KINGLAKE RASPBERRIES
Gavin and Tracey Molloy
Tooheys Rd, Pheasant Creek 3757
Phone: 03 5786 5360
Email: rasp@kinglake-raspberries.com.au

YARRA VALLEY AND THE DANDENONG RANGES—VIC

www.kinglake-raspberries.com.au
With frozen berries and berry products available all year, Kinglake Raspberries opens for pick your own during raspberry season. The Molloys have over 36 000 metres of raspberry rows, making this farm one of the largest pick your own operations in Victoria.

- **Open:** 8am–5pm daily (Dec–Jan), 9am–4pm daily (late Mar–May), 9am–5pm weekdays and weekends by appointment (between seasons)

SASSAFRAS
🍎 PRODUCERS

RIPE—AUSTRALIAN PRODUCE
Drew Colpman and Gary Cooper
376–788 Mount Dandenong Tourist Rd, Sassafras 3787
Phone: 03 9755 2100
Email: gcd@netspace.net.au
www.ripesite.com
Crammed with local produce, and a real boon for those who care about good food, this place has the talents of two chefs. Drew Colpman trained under Guy Grossi from Florentinos, while Gary Cooper is formerly of Chateau Yering, which had all three hats in The Age Good Food Guide for the five years that Gary was there. In addition to its own food and fine coffee, this produce store carries the largest collection of Yarra Valley and Mt Dandenong small producers including: Kennedy and Wilson, Berry King, Woolaway Farm, Yarra Valley Icecream, Yarra Valley Pasta, Rare Breed Pigs, Silvan Estate Raspberries, Forgotten Fruit and dozens more.

- **Open:** 8.30am–6.30pm daily

SILVAN
🍎 PRODUCERS

VICTORIAN STRAWBERRY FIELDS
Nina and Sam Corrone
Lot 1, 160–162 Lilydale-Monbulk Rd, Silvan 3795
Phone: 03 9737 9549
This part of Victoria is lush with green and growing things—so much so, even the locality is called 'silvan'. For the past ten years the Corrones have been selling their strawberries already picked, as well as cherries, raspberries, blackberries, blueberries, loganberries, silvanberries and youngberries fresh in season, with frozen berries out of season. Just for good measure they have also planted some vegetables—zucchini, squash, and tomatoes—so you can almost get your meal here. Even the prices are special.

- **Open:** 9am–5pm daily

RL CHAPMAN AND SONS
Mark and Steve Chapman
21 Parker Rd, Silvan 3795
Phone: 03 9737 9534
Email: admin@chappieschoice.com.au
www.upick.com.au
The Chapmans' father brought the U-Pick concept to Australia from America in 1970, and so it seems right that you can still pick your own cherries, apples, lemons and persimmons according to the season here on this farm. There is only first-class fruit available for you to pick or buy from the on-site farm outlet. Chapman and Sons are a large-scale, commercial grower supplying all markets in Australia. There are also frozen berries (blackberries and raspberries) year round which are ideal for catering and jam

making, or you can simply buy up some of the jams made on the property.
- **Open:** 8.30am-4.30pm daily 25 Nov-31 Dec (except Christmas Day)

SILVAN ESTATE RASPBERRIES
Jeremy Tisdall and Pam Vroland
70 Hollis Rd, Silvan 3795
Phone: 03 9737 9415
Email: silvan@raspberries.com.au
www.raspberries.com.au

Silvan Estate grows not just raspberries, but also blueberries, red currants and 'brambles'—a term which covers a whole range of the darker berries—all chosen for flavour and quality, and which are unsprayed and picked at full ripeness. Most go to market, but there are shed-door sales in season, and in winter the over-ripe fruit which has been frozen since summer can be purchased, but phone first. There is also a small range of products such as jams and vinegars.
- **Open:** 8am-6pm daily (Dec-Apr)

WANDIN NORTH
🍴 DINING

RUSTIC CHARM RESTAURANT
Jeynelle and Scott (chef) Forrest
Warburton Hwy, Wandin North 3139
Phone: 03 5964 3694
Email: rustic@xtreme.net.au
Licensed
Cuisine: Regional Australian bush foods
Mains average: $$
Specialty: Kangaroo
Local flavours: Venison, quail, zucchini flowers, yabbies, trout, buffalo, cheeses, apples, peaches, strawberries, cherries, wines and preserves

Extras: The restaurant won awards in 2002 for Best Menu and Best Restaurant in the Yarra Valley at the Salon Culinaire, the judges commenting on the excellent use and promotion of local produce. Outdoor veranda and courtyard, cosy log fire.
- **Open:** Lunch Wed-Sun, dinner Wed-Mon

WANDIN YALLOCK
🍎 PRODUCERS

WARRATINA LAVENDER FARM
Peter and Annemarie Manders
Quayle Rd, Wandin Yallock 3139
Phone: 03 5964 4650
Email: enquiries@warratinalavender.com.au
www.warratinalavender.com.au

More than ten years ago a long-term interest in horticulture and a desire to operate a cottage industry led the Manderses to set up this lavender farm. Their dried, home-grown lavender, plants, culinary, cosmetic and gift items (that run the gamut from mustard to massage oil, polish to lavender pillows) are great favourites at specialty markets, and shops and galleries around Australia. There is a nursery section for plants, a shop and licensed tearooms where light lunches, cheese platters, lavender scones, and produce from the Yarra Valley are served.
- **Open:** 10am-5pm Wed-Mon

WARRANWOOD

🍴 DINING

VINES RESTAURANT OF THE YARRA VALLEY
Stuart (chef) and Kellie Harvey
1 Delaneys Rd, Warranwood 3134
Phone: 03 9876 4044
Email: vinesrestaurant@bigpond.com.au
www.vinesrestaurant.com.au
Licensed
Cuisine: Modern international
Mains average: $$
Specialty: Yarra Valley produce
Local flavours: Yarra Valley pork, Wandin quail, Yarra Valley venison, Yarra Valley Pasta Shop spaghetti, Yarra Valley freshwater salmon, Lilydale free-range chicken, Cunliffe and Waters beetroot and orange relish, Yarra Valley raspberry icecream, Wilson & Kennedy chocolates, local berries and cheeses
Extras: The restaurant has a relaxed atmosphere in an elegant setting, an open fire for winter, and French windows opening to overlook the vines and sweeping lawns in summer.
■ **Open:** Lunch Fri–Sun, dinner Thurs–Sun
● **Location:** Delaneys Rd is on a sharp bend in Croydon Rd

YARRA GLEN

🏠 STORES

REGIONAL FARE
Wendy Eastman and Ron Harper
26 Bell St, Yarra Glen 3775
Phone: 03 9730 1007
Open over two years, and part of Yarra Valley Regional Food Trail, this is the ideal one-stop place to pick up a selection of the local goodies—about 800 items—plus wines from twenty-four wineries and, if you're counting, goods from around seventy local producers. Regional Fare provides picnic and gift hampers, and there is now a cafe here as well.
■ **Open:** 9am–5pm daily

🍴 DINING

CHEESEFREAKS AND YARRA VALLEY FUDGE
Ian Balmain
21/36 Bell St, Yarra Glen 3775
Phone: 03 9730 1122
Email: cheesefreak@bigpond.com

YARRA GLEN
Yarra Valley Regional Food Group and Food Trail, PO Box 260, Yarra Glen 3775
Phone: 03 9513 0677
www.yarravalleyfood.com.au
The Yarra Valley Regional Food Group was established in early 1998 by food-based businesses in the Yarra Valley to identify small and large specialist growers and producers. Their stated aims are to involve all levels of food, wine, and hospitality businesses in the region, and to encourage purity and freshness in the local food products as well as providing a centralised and effective method of promoting and publicising the primary production of the region.
    People with well-known names for wine lovers, Leanne De Bortoli and Suzanne Halliday, along with Carolyn Burgi, have been vital to the creation and progress of what has become a promotions model for food groups in other areas of the country.
    Look for the distinctive blue signs on farm gates and shops.

Licensed
Cuisine: Cafe-style
Mains average: $
Specialty: Lemon tart (Ian says they have 'probably the best lemon tart in the universe'), Devonshire teas
Local flavours: As many as possible— berries, fruit (seasonally), cheeses, trout, yabbies, venison, crème fraîche, sour cream, clotted cream
Extras: Since 1998, Ian Balmain (ex-Yarra Valley Dairy), a self-confessed 'cheese freak', has been running this memorably named cafe. The store sells cheese from Yarra Valley growers, homemade fudge, jams, relishes, clotted cream, ice-cream and cakes, plus around forty Yarra Valley wines.
- **Open:** 9am–5pm daily

ROUNDSTONE WINERY AND CAFE
John and Lynne Derwin
54 Willow Bend Dr, Yarra Glen 3775
Phone: 03 9730 1181
Email: roundstonewine@ozemail.com
www.visitvictoria.com/roundstonewine
Licensed
Cuisine: Mediterranean
Mains average: $$
Specialty: Yarra Valley salmon hot smoked in a wood-fired oven on the premises
Local flavours: Yarra Valley cheeses, fruits and vegetables, salmon, figs
Extras: Cellar door for the winery and restaurant nestled amongst the vines. Veranda dining overlooking the lake. Petanque court.
- **Open:** Lunch Wed–Sun (functions by appointment at other times)

YARRAMBAT
DINING
RIVERS CAFE
Ian and Merilyn Moad
28 Kurrak Rd, Yarrambat 3091
Phone: 03 9436 1466
www.riversgardenandcafe.com.au
Licensed
Cuisine: Regional
Mains: $
Specialty: Rivers berry crepes, homemade cakes and desserts
Local flavours: Cheese, fresh herbs, rhubarb, apples, and lemons
Extras: Located in a garden centre with gift shop, conference facilities, natural wildlife and domestic ducks. Makes a range of sauces and chutneys for sale.
- **Open:** 9am–3pm daily

YERING
DINING
ELEONORE'S AT CHATEAU YERING
42 Melba Hwy, Yering 3770
Phone: 03 9237 3333
Email: info@chateau-yering.com.au
www.chateau-yering.com.au
Licensed
Cuisine: Modern European
Mains average: $$$
Specialty: Soft-boiled, free-range egg with smoked salmon and black truffle infused-yolk
Local flavours: Dairy products, fruit and vegetables

YARRA VALLEY AND THE DANDENONG RANGES—VIC

Extras: Luxury accommodation. This property is closely connected with the very active Yarra Valley Regional Food Group. Chateau Yering also operates Sweetwater Cafe serving meals during the day.

■ **Open:** Lunch weekends, dinner daily

YARRA VALLEY DAIRY
Ben Mooney
McMeikans Rd, Yering 3770
Phone: 03 9739 0023
Email: mail@yvd.com.au
www.yarravalleydairy.com.au
Licensed
Cuisine: Modern Australian
Mains average: $$
Specialty: Antipasto platter featuring cheeses
Local flavours: All local Yarra Valley produce
Extras: Located in a working dairy farm with a viewing window and fabulous views over the Yarra Valley. The excellent, handmade farm cheeses are available for sale here too.

■ **Open:** 10.30am–5pm daily

VIC WEBSITES
www.visitvictoria.com
www.yarravalleyfood.com.au
www.winemakers.com.au
www.northeastvalleys.info
www.milawagourmet.com
www.thehighalpinecountry.info
www.beechworth.com
www.geelongfarmersmarket.com
www.kvv.com.au
www.wangaratta.vic.gov.au

MARKETS AND EVENTS

✦ **TEMPLESTOWE FARMERS' MARKET,**
TEMPLESTOWE
03 5664 0096
8am–1pm second Sat

✦ **YARRA VALLEY FARMER'S MARKET,**
YERING
03 5964 9494
10am–3pm third Sun (except Feb)

✦ **GLADYSDALE APPLE AND WINE FESTIVAL,**
GLADYSDALE
03 5966 6202
First Sun in May

✦ **KALORAMA CHESTNUT FESTIVAL,**
KALORAMA
03 9728 1480
First weekend in May

✦ **KELLYBROOK CIDER FESTIVAL,**
WONGA PARK
03 9722 1304
May

✦ **KINGLAKE RASPBERRY FAIR,**
PHEASANT CREEK
03 5786 5360
11am–5pm second Sun in Jan

✦ **LAVENDER FEST,**
WANDIN YALLOCK
03 5964 4650
Mid Nov

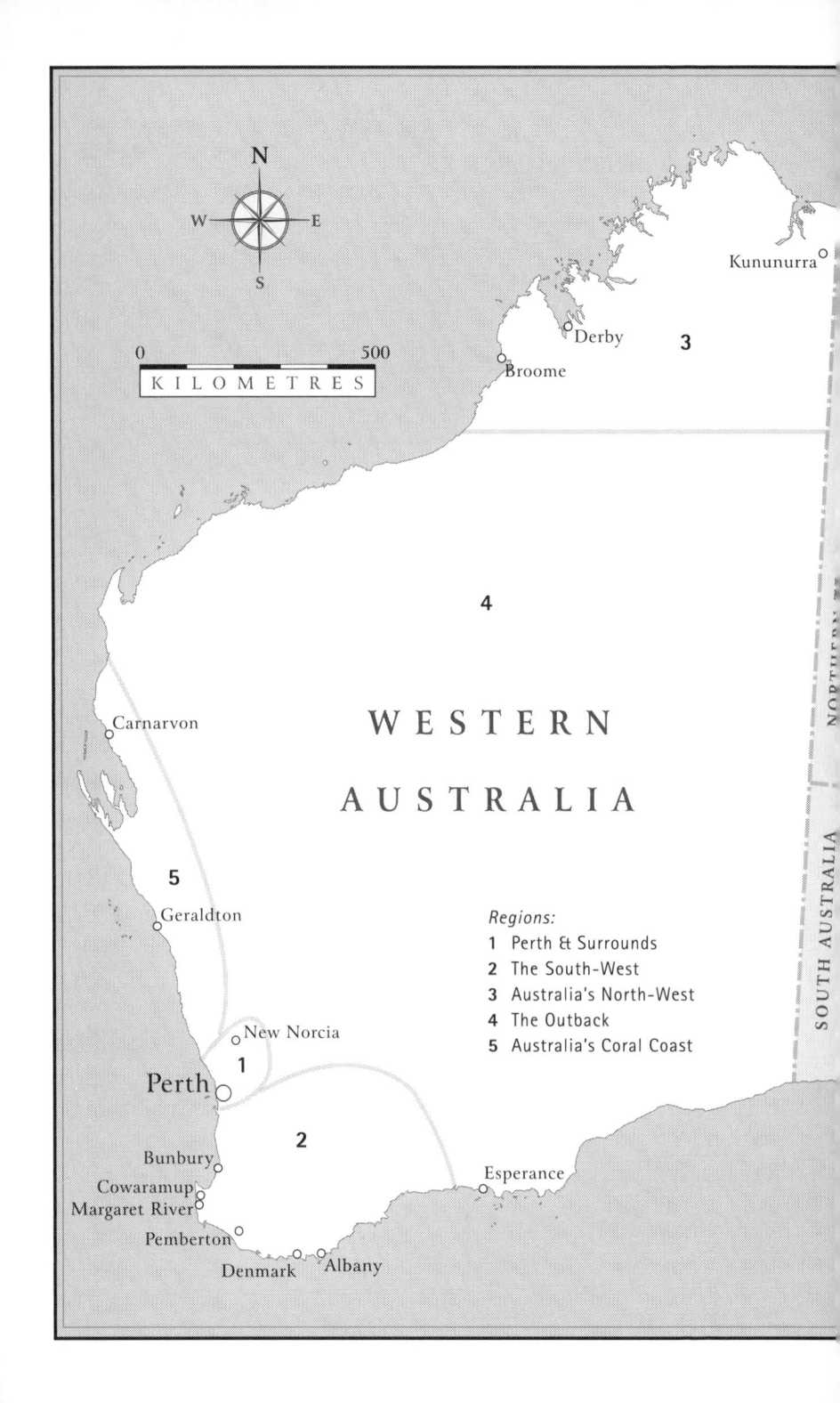

Regions:
1 Perth & Surrounds
2 The South-West
3 Australia's North-West
4 The Outback
5 Australia's Coral Coast

WESTERN AUSTRALIA

Covering approximately one-third of the country, Western Australia measures 2400 kilometres north to south, and 1600 kilometres across. Produce is limited to an extent by the harsh and often arid conditions found throughout much of the state.

The Ord River Irrigation Project attempted to address the problem and has, in its own area, been successful, although only affecting a comparatively small area. It is the highest-yielding sugar cane area in Australia, though.

The south-west of the state is the most temperate, climatically, and supports a wide range of crops and animals. There is everything from dairying, berries and orchards to beef and sheep-growing, and of course a burgeoning wine industry. Further north, in drier conditions, wheat, barley, oats and canola thrive. Western Australia produces around one-third of Australia's wheat, and two-thirds of the country's pulse crops.

Western Australia's 12 500-kilometre coastline is rich with seafood and fish, including abalone, and the $300 million western rock lobster fishery represents the most valuable single-species fishery in Australia, exporting primarily to the United States and Asia.

Perth's many seafood restaurants showcase the fruits of the coastline most effectively. The city also has a multicultural history, and Italian and other ethnic dining spots reflect this, as well as the prolific market garden industry that brings morning-fresh produce to city tables. Perth's climate is hot and hints of the Mediterranean, and this no doubt influenced early settlers to plant wine grapes. The Swan Valley still is known for its wines, but Margaret River and parts of the Great Southern region are most important too.

> **QUARANTINE NOTE**
> Fruit and vegetables, honey, used fruit and produce containers, plants and flowers, aquatic plants, soils, weeds and seeds, hay and fodder, livestock, birds and animals must not be brought into Western Australia. If in doubt, contact the Western Australian Quarantine Service (Phone: 08 9353 2757).

PERTH AND SURROUNDS

Lovely Perth is small as capital cities go, with a population of just over a million, yet its size makes it more accessible. Fremantle is the port for riverside Perth, and is different again. Where there is a cheery elegance and cosmopolitan feel to Perth with its sun-loving cafes and northern city trattorias, Fremantle bustles with exuberance. Refurbished some years ago when it was the site of the battle for the America's Cup, the place has never looked back. Pubs, cafes, nightclubs and restaurants abound, many of them serving the fabulous local seafood. You may not get a parking spot easily, but when you do, a fine meal is assured.

The hills around Perth hide dozens of orchards, many of which sell fruit from their packing sheds. Drive through the hill suburbs in summer and autumn and watch for the signs that pop out like mushrooms after rain.

The Swan Valley is worth visiting too. One of the oldest wine-producing areas in Australia, its riverside ambience makes it an ideal day-trip that should include a picnic or meal at one of the winery restaurants.

While Italian food dominates the dining scene there is still something for everyone. Wander off to some of the upper-income leafy Perth suburbs and you will find elegant small eateries in every shopping centre; and the city's five-star hotels all provide excellent international-standard dining. Best of all, look for waterside dining with two advantages—the fresh local seafood and usually a stunning view of city lights across water.

PERTH

🍎 PRODUCERS

SWAN VALLEY EGGS
BJ and T Cocking
60 Cheltenham St, West Swan 6055
Phone: 08 9274 2502
Free-range eggs are available every day here, and also through some local outlets.
■ Open: 8.30am–1.30pm daily

🏠 STORES

FRESH PROVISION STORES
Cnr Leura Ave and Stirling Hwy, Claremont 6010
Phone: 08 9383 3308
Email: fresh@provisions.com.au
www.provisions.com.au
Fresh fruit and vegetables, most of them local, regional produce, gourmet groceries, many WA-made fancy chocolates, cakes, bread, and a gourmet deli—a stunning 20 000 lines in total. Catering for the top end of the market, it seems they also cater to the time-poor, as they are open twenty-four hours a day.
■ Open: Twenty-four hours daily

HERDSMAN FRESH ESSENTIALS
Allan and Dennis Cerinich
9 Flynn St, Churchlands 6018
Phone: 08 9383 7733
Email: hfe@herdsman-fresh.com.au
Herdsman Fresh Essentials is a family-owned specialty fresh food market, which has been operating since 1979. The market is made up of the finest

PERTH AND SURROUNDS—WA

fresh Western Australian produce, plus gourmet grocery and deli items, cheeses and convenience foods. There's even a quality butcher, seafood outlet, bakery, and exotic flower shop on site.

- **Open:** 8am–8pm daily

KINGS CHOICE
Darren Salkilld
16 Emplacement Cres, Hamilton Hill 6163
Phone: 08 9433 2000
www.kingschoice.com.au

The standard of beef from the Harvey region is very high, and the abattoirs in the town add to its value by vacuum-packing many cuts. The greatest advantages of this process is that meat has a prolonged shelf life while being convenient to store, and allows for meat ageing, which dramatically improves both tenderness and flavour by up to fifty per cent. The tours through the plant at Harvey are no longer available but you can visit this shop and see how good Harvey beef is. The only catch is that there is a fifteen-kilogram minimum order for any one product.

- **Open:** Phone ahead for orders

MONDO DI CARNE PTY LTD
Vince and Anne Garreffa
824 Beaufort St (cnr Sixth Ave), Inglewood 6052
Phone: 08 9371 6350
Email: mail@mondo.net.au
www.mondo.net.au

Vince Garreffa has been butchering since 1979, but more recently his name has become synonymous with White Rocks Veal, and Eagle Hill Organic. His shop, Mondo di Carne, offers a world of meat, as its name suggests. Veal is especially raised for the shop near Brunswick Junction, and other meats come from trusted and selected suppliers. Mondo carries (or can access) all game meats, such as kangaroo, ostrich, emu, buffalo, and venison. 'If they want it, I'll find it', says Vince Garreffa. He also makes fresh sausages, and has manufacturers that produce smallgoods to his specifications.

- **Open:** 8am–6pm Tues–Fri, 7am–2pm Sat

NEW NORCIA WOODFIRED BAKERY
The Cloisters, 225 Bagot Rd, Subiaco 6008
Phone: 08 9381 4811
Email: service@newnorciabaker.com.au
www.newnorciabaker.com.au

This bakery bakes and sells New Norcia breads and cakes, biscotti and pan chocolatti. Also available at the New Norcia Woodfired Bakery Cafe in Mt Hawthorn (phone: 08 9443 4114).

- **Open:** 7.30am–6pm daily

TOURS

EPICUREAN TOURS
Barb Humphrey
14 Macarthur Ave, Padbury 6025
Mobile: 0427 766 717
Email: info@epicurean.com.au
www.epicurean.com.au

This company can arrange cooking classes with top chefs in wineries in Margaret River and the Swan Valley. Tours specialise in premium wine and food experiences with a five-star touch. So you could find yourself cycling around wineries then being treated to a gourmet feast featuring local produce before being chauffeured on to the next destination or a luxury hotel for the night. It's like a food trail with someone else doing all the organising and planning.

- **Tours:** On demand

DINING AND ENTERTAINMENT

Perth's nightlife has something to suit most tastes and budgets. For information about live theatres, bands, movies, orchestral, opera and ballet performances check the West Australian newspaper daily. For a rock music and nightclub gig guide see the local free music paper, *X-Press*.

Fremantle: The city is well known for its big variety of Italian and fish restaurants, thanks mostly to its cultural heritage as a fishing port. Pasta, pizza and fish 'n' chips are the local fare, but you can also find quality seafood eateries at Fishing Boat Harbour and along South Terrace. Cafes and pubs are everywhere. Try some locally brewed beer, or grab a latte and spend time people-watching. The Fremantle markets bustle on Fridays and weekends, with fresh produce, crafts and buskers who provide street-side entertainment.

Leederville: Oxford Street is always busy in the evenings. Full of after-work professionals, families and students, the crowd is as diverse as the entertainment and dining options. Many outlets here are café style, serving Mediterranean food, but you can also find Asian and fish restaurants, pizza parlours and kebab takeaways. The Leederville Hotel is a happening place, or check at the newly refurbished Oxford Hotel, at the northern end of the street.

Mt Lawley: This stretch of Beaufort Street is fast gaining a reputation for fine dining with some of Perth's best new restaurants establishing themselves here. There is also a good mix of noodle and pizza bars, as well as busy pubs serving bar food and live music.

Northbridge: This is Perth's nightlife central. The area bordered by William, James, Aberdeen and Parker Streets is where you'll find restaurants offering every cuisine imaginable, as well as pubs, bars and clubs. On weekends the streets fill with people enjoying the multicultural and party atmosphere.

Perth: The west end of the central city area, along Hay, Murray and King Street, offers the biggest range of options. The busiest time is Friday night which pulls the after-work crowd and late-night shoppers. There are several pubs offering casual dining, as well as theatres and bars. Just a few minutes' drive from the city, Burswood International Resort Casino boasts a choice of nine restaurants, six bars and a two-storey nightclub, the Ruby Room.

Subiaco: This area buzzes on weeknights and weekends alike. Rokeby Road and Hay Street is where you'll find the most action. While Subiaco is perhaps best known for its quality restaurants there is a wide choice—from fine dining to takeaway, Italian to Asian. The Subiaco Hotel is the central drinking zone, but there are other smaller bars and clubs that get busy once the hotel closes.

LABEL AND TABLE
Russell Jordan
PO Box 332, Greenwood 6924
Phone: 08 9240 9850
Email: russell@labelntable.com
www.labelntable.com
Russell Jordan describes his company this way: 'The key ingredient is that my guests are able to truly go behind the scenes and evaluate Western Australia's great wine and food lifestyle, as a different experience with a bit of fun!' These tours to the Swan Valley, Perth CBD, Margaret River and Great Southern, and most food and wine regions of Western Australia are available during food and wine festivals, and for conference groups.
Tours: On demand

OUT AND ABOUT WINE TOURS
Claude and Linda Rossetto
PO Box 95, Bassendean 6054
Phone: 08 9377 3376
Mobile: 0419 954 402
Email: oaat@multiline.com.au
www.outandabouttours.com.au
This winery experience tour includes five wineries, plus a boutique brewery, the Swan Valley Cheese company and the Margaret River Chocolate Factory.
Tours: 9.30am–4.30pm Tues–Sun

SWAN VALLEY WINERY TOURS
Kim and Gregg Boalch
PO Box 3, Darlington 6070
Phone: 08 9299 6249
Email: kim@svtours.com.au
www.svtours.com.au
Food and wine tours in Perth's Swan Valley—'the Valley of Taste'. Some gourmet tours include a river cruise.
■ **Open:** 10am–5pm daily

BASKERVILLE
BREWERY/WINERY
FERAL BREWING COMPANY
Neil and Gillian Lamont
152 Haddrill Rd, Baskerville 6056
Phone: 08 9296 4657
Email: gillian@feralbrewing.com
www.feralbrewing.com
This beer oasis in the heart of Perth's wine country, which began in 2002, even has a feral look to it with rustic decor and wide verandas overlooking the vineyard. In the first year of production Feral Brewery was the most highly awarded microbrewery in the Draught section at the 2003 Australian International Beer Awards. Dine inside and you'll be able to able to observe the workings of the modern microbrewery. The restaurant caters for indoor and outdoor dining and offers a wide selection of premium quality West Australian meat, fish and local produce. Beers include Feral's Belgium White, Organic Pilsener, Pale Ale, German Red, Monty's Mild, Farmhouse Ale and Strawberry Porter. The beers are also sold in selected local bars and hotels around Perth.
■ **Open:** 11am–6pm Mon and Thurs–Sun, dinner Thurs–Sun, Bar only—11am–6pm Tues and Wed
● **Location:** About 1.6km along Haddrill Rd, on right

🍴 DINING

UPPER REACH WINERY AND CAFE
Laura and Derek Pearse
Memorial Ave, Baskerville 6056
Phone: 08 9296 0078
Email: info@upperreach.com.au
www.upperreach.com.au
Licensed
Cuisine: Modern Australian/Mediterranean
Mains average: $$
Specialty: Tapas platters
Local flavours: All fruit and vegetables, seafood, free-range eggs
Extras: Indoor and outdoor dining, views of the river, cellar door adjacent.
- **Open:** Breakfast Sun, lunch and afternoon tea Thurs–Mon and public holidays

BELHUS
🍴 DINING

EDGECOMBE BROTHERS
Alf Edgecombe
12130 West Swan Rd, Belhus 6069
Phone: 08 9296 4307
Email: edgecombe_brothers@telstra.com
Licensed
Cuisine: Homestyle
Mains average: $
Specialty: Asparagus dishes
Local flavours: Asparagus, muscat grapes
Extras: Outdoor dining overlooking Lake Yakine. Weekend horse and cart tours of the Swan Valley. This family has been selling their fruit from the farm for over thirty years.
- **Open:** 9am–6pm, lunch daily (longer in peak grape season)
- **Location:** Cnr Gnangara Rd

BINDOON
🍎 PRODUCERS

APRICOT ACRES FARM
Bruce and Jenny Wharton
124 Toy Rd, Bindoon 6502
Phone: 08 9576 1030
Email: apricot_acres@bigpond.com
Fairly unique in this area, this family-owned orchard began twelve years ago at the suggestion of a nearby farmstay operator. There are tours and sales of its stone fruits, citrus and grapes, which are available dried or fresh at the packing shed in season.
- **Open:** 8am–5pm daily (phone first)
- **Location:** Turn left on Bindoon-Moora Rd and proceed for 1km then turn right onto Toy Rd and the farm is just over 1km along on your right

CHITTERING
🍴 DINING

STRINGYBARK WINERY AND RESTAURANT
Mary (chef) and Bruce Cussen
2060 Chittering Rd, Chittering 6084
Phone: 08 9561 6547
Mobile: 0418 856 290
Email: brucec@philmac.com.au
Licensed/BYO
Cuisine: Country style
Mains average: $$
Specialty: Duck, seafood
Local flavours: Duck, seafood, wines
Extras: Situated in fourteen acres of vines and bushland, with wildflowers in spring, outdoor dining on a deck and a log fire in winter. Sells medal-winning wines.

DWELLINGUP

🍴 DINING

MILLHOUSE CAFE AND CHOCOLATE CO
Ami Davis
Main St, Dwellingup 6213
Phone: 08 9538 1122
Licensed/BYO wine
Cuisine: Modern Australian
Mains average: $$
Specialty: WA barramundi with local fruit salsa on a hot potato cake, Harvey beef sirloin
Local flavours: Harvey beef, Harvey cheese, fruit, cherries in season, barramundi
Extras: Outdoor dining on the veranda, bushland views, campers welcome, no dress code—just come as you are. Chocolate shop inside the cafe. Takeaway available.
■ **Open:** 10am–4pm Thurs, 9am–5pm Fri, 9am–late Sat, 8am–5pm Sun

FALCON

🏠 STORES

MOOKA FISHING CO
Irene Pinny and Alan Butler
25 Pleasant Grove Circle, Falcon 6210
Phone: 08 9582 2101
Email: blueking@westnet.com.au
If you are looking for some very fresh seafood, this is your place. There is freshly trawled fish and squid, Mandurah prawns (Nov–June) and Mandurah crabs. Mooka has been in the business of wholesaling seafood since 1988, but since 2000 it has become a retail business as well. The small trawler operates each night from November to June and the fresh catch is on sale the next day.
■ **Open:** 8am–5pm daily (closed Aug)
● **Location:** Off Old Coast Rd, South Mandurah between Miami shopping centre and Dawesville Channel

GIDGEGANNUP

🍎 PRODUCERS

KERVELLA CHEESE
Gabrielle Kervella and Alan Cockman
9 Clenton Rd, Gidgegannup 6083
Phone: 08 9574 7160
Email: kervella@gidgenet.com.au
Kervella Cheese set the bench mark in Australia twenty years ago for handmade, soft curd goat cheeses made in the traditional French way. It is now a certified biodynamic farm and still produces those same high quality goat cheeses, but they are unashamedly organic. After studying traditional goat cheese making in France, Kervella adapted it to Australian conditions. She grows most of the food for the 200 goats, and farms in the traditional biodynamic way with the addition of special minerals for depleted Australian soils. A new dairy has just been completed (with special viewing areas for the public) and they will shortly be opening facilities for coffee.
■ **Open:** By appointment
● **Location:** 15km along O'Brian Rd, from junction of Toodyay Rd

■ **Open:** Lunch and dinner Thurs–Sat (and Sun on long weekends)
● **Location:** Fifty minutes from Perth

KYTREN FINE QUALITY GOATS CHEESE

Margaret and Ken Vinicombe
1243 McKnoe Dr, Gidgegannup 6083
Phone: 08 9574 7147
Email: kytrengoatdairy@bigpond.com

The Vinicombes have been operating their own business for seven and a half years with a goat herd and cheese room and do not need to buy in any other milk. All of their cheeses are handmade with non-animal rennet, and no preservatives are used, other then a small amount of salt. The cheeses are made to order to give maximum shelf life. Kytren also produces Gidgie fresh pasteurised goats milk, which is sold to major supermarkets in Western Australia, and the cheeses are sold in gourmet shops throughout Australia.

- **Open:** By appointment

HENLEY BROOK
🍎 PRODUCERS/DINING

HANSON'S SWAN VALLEY
Jon and Selina Hanson
60 Forest Rd, Henley Brook 6055
Phone: 08 9296 3366
Email: stay@hansons.com.au
www.hansons.com.au
Licensed
Cuisine: French-influenced Modern Australian
Mains average: $$
Specialty: Oxtail
Local flavours: Asparagus, wines, melons, fruit, grapes, vegetables, Kervella goats cheese
Extras: Set on twenty-five acres with Swan River frontage, views and outdoor dining. Takeaway hampers available.

RESTAURANTS AND ACCOMMODATION

The choice of places to eat in Perth and its suburbs is so varied that you would need many weeks to sample everything the city has to offer. Because of the early Italian migration to this state, there are dozens of well-established Italian restaurants and bars. Many began in the Northbridge area, which has now evolved into a buzzing cosmopolitan centre for every type of dining. Here there are food stores, footpath dining, cafés, bars, grills, brasseries and restaurants.

A distinct cafe culture has arisen in relaxed suburbs such as Cottesloe and Subiaco, and Fremantle has it all—markets, cafes, pavement dining, pubs and serious restaurants—plus water views from many spots.

- **Open:** Lunch and dinner daily
- **Location:** Off West Swan Rd

MERRICH ESTATE OLIVE FARM AND MEDITERRANEAN KITCHEN
Shirley Richardson (owner) and Christian Montagne (chef)
11111 West Swan Rd, Henley Brook 6055
Phone: 08 9296 0750 or 08 9296 0752
Email: merrich@bigpond.com
www.merrichestateolivefarm.com.au
Licensed (only Swan Valley wines and Swan Valley-brewed Feral Beer on tap)
Cuisine: Mediterranean
Mains average: $$
Specialty: Prawn gambas, lamb tagine and mezze platters
Local flavours: Asparagus, marron, fruit in season, goats cheese, olives and olive oil from Merrich Estate
Extras: Gold Plate Award winner 2004, open-plan kitchen, wood-fired oven.

Upstairs balcony overlooking the Swan Valley, outdoor lawn area, downstairs beer and wine cellar—cool in summer, romantic at night. There are champagne breakfasts on Sunday and takeaway pizzas and deli hampers are available. Grows a wide range of produce, including olives, green vegetables, tomatoes, oranges (for orange cake), eggplant, rocket, figs, grapes and stone fruit. The olive shop opened four years ago selling products made entirely from olives grown on the estate. There are regular cooking demonstrations, combined with olive oil appreciation and olive grove tours in season. Richardson started the shop to educate people about the uses and health benefits of extra virgin olive oil. You'll find freshly made tapenades, dukkah, herbs and spices, fresh pasta and sauces and a range of infused oils (vanilla, chilli, lemon and lime). There is also guest accommodation in a restored heritage early settlers cottage (Thyme and Again Cottage), which includes a hot Mediterranean breakfast.

- **Open:** Breakfast Sun, lunch Wed–Sun and public holidays, dinner Thurs–Sat, olive farm—11am–4pm Wed–Sun and public holidays

---

HARVEY VISITOR CENTRE
South West Hwy, Harvey 6220
Phone: 08 9729 1122
Email: info@harveytourism.com
www.harveytourism.com
The Harvey Visitor Centre celebrates the rich dairying industry of the region with an innovative 'Moo Shop'. Here you will find all sorts of local produce for sale, along with free district information.
Open: 9am to 5pm weekdays, 9.30am–4.30pm weekends
Location: Near the Harvey River bridge

HERNE HILL

PRODUCERS

CAPRINO FARM
John Martin
139 Lennard St, Herne Hill 6056
Phone: 08 9296 2008
John Martin has been making this chemical-free goats milk' for eighteen years, and produces it using organic principles. The resulting cheese is without the strong flavour often associated with goats milk products and is ideal for both children and adults with allergies and special dietary needs. The Caprino cheese is only available from the farm.
- **Open:** Mon–Sat, by appointment
- **Location:** Off Great Northern Hwy

STORES

SWAN VALLEY CENTRAL
Sally Zannino
Cnr Great Northern Hwy and Lennard St, Herne Hill 6056
Phone: 08 9296 4219
Email: zannino@iinet.net.au
There are over one hundred different varieties of dried fruit, nuts and confectionery here at this store that specialises in supplying all the ingredients for fruitcakes and baking. It was originally set up by Swan Settlers Co-op over thirty years ago, and still features local sultanas, currants, honey, jams and fruitcakes.
- **Open:** 7.30am–5pm daily

JARRAHDALE
🍴 DINING

MILLBROOK WINERY
Peter and Lee Fogarty (owners),
Sal Davis (chef)
Old Chestnut Lane, Jarrahdale 6124
Phone: 08 9525 5796
Email: tastingroom@millbrookwinery.com.au
www.millbrookwinery.com.au
Licensed
Cuisine: Modern Australian
Mains average: $$
Specialty: Home-grown marron, preserves made from home-grown fruit and vegetables
Local flavours: Fruit, vegetables, trout, marron
Grows: Marron, fruit and vegetables
Extras: The winery overlooks a large lake and the grounds are stunning with beautiful fifty-year-old fruit trees—the perfect place for enjoying a Millbrook gourmet picnic basket. Bush walks, forest and wildflowers in the vicinity. Sells a range of homemade produce including apricot, plum and tomato jams, dukkah, preserved lemons, pickles, chutneys, salted chilli almonds, tapenade, candied cumquats and quince paste.
■ **Open:** Cellar door—10am–5pm daily, lunch—Wed–Sun
● **Location:** Forty-five minutes from Perth

MANDURAH
🍴 DINING

MANDURAH QUAY RESTAURANT
Scott Spicer (chef)
1 Marina Quay Dr, Mandurah 6210
Phone: 08 9582 8300
Email: resort@aceonline.com.au
www.mandurahquay.com.au
Licensed
Cuisine: Regional Australian
Mains average: $$
Specialty: Pan-fried chicken livers on New Norcia sourdough toast
Local flavours: Mandurah whiting
Extras: Wonderful views over Peel Inlet and the marina. Takeaway available.
■ **Open:** Breakfast and lunch daily, dinner Tues–Sat

MIDDLE SWAN
🍎 PRODUCERS

FAZZOLARI OLIVE OILS
Antonio Fazzolari
33 Campersic Rd, Middle Swan 6056
Phone: 08 9274 6204
Email: tonyfazz@hotmail.com
Tony Fazzolari purchased this property thirty minutes from Perth in 1981. His father had pressed olive oil in Italy and so they decided to make it here. Over twenty-three years Fazzolari has planted many trees and now has 150. He produces cold pressed olive oil—mission, kalamata, correreggiola, a blended Hardy's mammoth and Spanish queen—and pickled olives. He also has macadamias and pecans.
■ **Open:** 9am–5pm daily

MILLENDON
🍴 DINING

LAMONT'S
Kate, Fiona and Corin Lamont
85 Bisdee Rd, Millendon 6056

Phone: 08 9296 4485
Email: winery@lamonts.com.au
www.lamonts.com.au
Licensed (Lamont Wines only)
Cuisine: Modern Australian
Mains average: $$$
Specialty: Marron
Local flavours: Swan Valley marron, Swan Valley goats cheese
Extras: Restaurant balcony overlooks the vineyards. Sells homemade tomato relish, apricot sauce, strawberry jam, olives, prunes in port and handmade chocolates.
- **Open:** Brunch Sun (summer), lunch Wed–Sun
- **Location:** North of Midland, off the Great Northern Hwy

MUNDARING
DINING

THE LOOSE BOX RESTAURANT
Alain and Elizabeth Fabregues
6825 Great Eastern Hwy, Mundaring 6073
Phone: 08 9295 1787
Email: loosebox@ozemail.com.au
www.loosebox.com
Licensed
Cuisine: French
Set-price four courses: $$$$
Specialty: Pigs trotters, beef cheeks, duck dishes, famous for desserts
Local flavours: All produce grown in own garden
Extras: A multi-award-winning restaurant and all tables have a view of the garden. There are six luxury cottages in the grounds, as wells as monthly cooking classes and twice monthly gourmet evenings.
- **Open:** Lunch Sun, dinner Wed–Sat

NEW NORCIA
PRODUCERS

BENEDICTINE COMMUNITY OF NEW NORCIA
Great Northern Hwy, New Norcia 6509
Phone: 08 9654 8056
Email: museum@newnorcia.wa.edu.au
www.newnorciabakery.com.au
New Norcia is Australia's only monastery town, established in 1846 as a mission to the Aboriginal community. The monks have been making olive oil here for over one hundred years using traditional European methods. This cold pressed olive oil from the community's 130-year-old olive grove is only sold at the New Norcia Museum and Art Gallery shop, and this is the only place to get the New Norcia Abbey wines and muscat. In 1993 the 150-year tradition of bread-making at the monastery was rekindled, using the wood-fired oven, and a second wood-fired oven was opened in Mount Hawthorn in 1996 to supply New Norcia Bread to the Perth market. The famous New Norcia nut cake, the Dom Salvado pan chocolatti and the almond biscotti are all made at the monastery bakehouse, but biscuits, slices and delicious wood-fired-oven roasted Mediterranean vegetables are made in Perth.
- **Open:** 9.30am–5pm daily (Aug–Oct), 10am–4.30pm daily (Nov–July)

ROLEYSTONE

🍎 PRODUCERS

RAEBURN ORCHARDS
Peter and Paul Casotti
95 Raeburn Rd, Roleystone 6111
Phone: 08 9397 5325

In 1948, Peter and Paul Casotti's father established an orchard which still produces peaches, plums, nectarines, cherries, persimmons, apples and pears. Although visitors stream in to buy the fruit, there is still a very local element. The honey comes from a local man, and a sister-in-law makes jams using fruit from the orchard. There are always dried fruits and nuts and chocolates too. Not all their fruit goes off in shopping bags, though. The sixteen-hectare orchard sells much of the crop to market, and also produces export-quality stone fruit for Asia. Apart from buying fruit, you can wander through the orchards, (although you may not pick your own), and there are tastings, and tractor rides for groups.
- **Open:** 9am–5pm daily

WEST TOODYAY

🍴 DINING

PECAN HILL TEA ROOM-MUSEUM
Craig and Suzanne Lomax
Lot 59, Beaufort St, West Toodyay 6566
Phone: 08 9574 2636
Email: info@pecanhill.com.au
www.pecanhill.com.au
BYO
Mains average: $
Specialty: Pecan pie
Local flavours: Pecans

Extras: Situated on top of the hill with great views, the balcony dining overlooks the Avon Valley and the pecan grove. There are four B&B rooms, all with ensuite, and a small memorabilia museum is open to groups.
- **Open:** Lunch, morning and afternoon teas, Fri–Tues
- **Location:** 4km west of Toodyay along the Julimar (Chittering) Rd, signposted

WOKALUP

🍎 PRODUCERS

HARVEY CHEESE
Robert and Penny St Duke
Lot 36, South West Hwy, Wokalup 6221
Phone: 08 9729 3949
Email: jupero@bigpond.com.au
www.harveycheese.com

Harvey Cheese began in January 2004, and is proud to be the first biodynamic cheese and milk manufacturer in Western Australia. Why Harvey? 'Simple', says Penny, 'good cows make good milk, which makes good cheese!' And this is the heart of fine dairy country. The company manufactures seventeen varieties of handmade, Mediterranean-style cheeses such as fetta (both plain and flavoured), ricotta, romano, and romano-style savoury hard cheese (plain and flavoured), which are sold under the Ha~Ve brand by independent retailers and health and whole food outlets. The milk comes from a Bridgetown dairy farm and the organic herbs from Margaret River. There are free tours, a viewing area, and cheese tasting available hourly. There is also fresh local produce for sale, light lunches daily, BYO bistro on Friday and Saturday nights and Sunday lunch,

and a children's activity area. There are many plans for this former deer park, which is set to become one of the state's premier cheese making and tourist facilities.
- **Open:** 9.30am–5pm daily and public holidays
- **Location:** 3km south of Harvey

MARKETS AND EVENTS

✦ MIDLAND FARMERS MARKET,
MIDLAND
08 9576 1234
8am–3pm Sun

✦ PERTH ROYAL SHOW,
CLAREMONT
08 9384 1933
Late Sept–early Oct

✦ SPRING IN THE VALLEY,
SWAN VALLEY
08 9379 9400
Second weekend in Oct

✦ TASTE OF THE VALLEY,
SWAN VALLEY
08 9379 9400
Apr

✦ WESTERN AUSTRALIAN WINE AND FOOD EXHIBITION,
PERTH
08 9279 2559
June

YORK
DINING

SETTLERS HOUSE
Des Mullins (owner), Trevor Randell (chef)
125 Avon Tce, York 6302
Phone: 08 9641 1096
Email: sales@settlershouse.com.au
www.settlershouse.com.au
Licensed
Mains average: $$
Specialty: Large selection of seafood dishes
Local flavours: York olives and olive oil, emu
Extras: Built in 1845, the exterior remains unchanged, memorabilia and old photos inside, two courtyards for dining.
- **Open:** Breakfast Sun, lunch daily, dinner Wed–Sat

THE SOUTH-WEST

The tiny Peel district nestles just south of Fremantle on the coast. Predictably, the local seafood, especially the blue manna crabs, is worth stopping to enjoy—or catch, as many visitors can't resist the area, staying on for days.

Yet the lure of the south-west ultimately prevails as the great Margaret River wine region beckons. Although only established for around thirty years, the standard of wines produced here is exceptional. Of course the local food matches the standard of the wine with superb cheese, marron, berries, fruit wines, and organic lamb showcased at restaurants that would rate well anywhere in the world.

The rich agricultural heartlands of the state, raising wheat, sheep and cattle, are the backbone of Western Australia's primary industries, established and solid, although more recently some farms have diversified, and are now running emus and goats, raising trout or planting nuts.

As you head further south, the terrain becomes just a little greener, the hills slightly higher, but by the time you arrive at the coast you will have passed through lush orchard and wine-growing areas. At Albany, the Southern Ocean yields rich fishy harvests that feature on local menus or turn up wrapped in paper with chips at take-away shops.

This is the place to find berries and venison, trout, and other cool-climate produce. The scenic environs are a bonus, as you eat outdoors overlooking crystal-clear bays, or a crescent of sandy beach.

ALBANY

🍎 PRODUCERS

ACACIA OSTRICH BREEDERS
Ian and Kath Fryer
Lot 131, Robinson Rd, Albany 6330
Phone: 08 9841 4045
Mobile: 0419 949 037
Email: acaciaostrich@bigpond.com
Ostrich meat is not as well-known in this country as it could be, so the Fryers are trying to change that. A red meat with a texture similar to beef, they say that ostrich meat is extremely heart-friendly being low in fat, cholesterol and kilojoules, as well as high in iron. After processing is done at an abattoir in the south-west of the state, the meat is sold in a variety of cuts and smallgoods. While there are some export sales, the products are mainly available from the property, at the Albany farmers' market and to local restaurants.
■ **Open:** By appointment

DEER-O-DOME (AUST)
Tony and Vivienne Davis
Link Rd, Albany 6331
Phone: 08 9841 7436
Email: tonydaviswa@hotmail.com
www.jrc.net.au/~deerdome/
There is so much happening here: velvet down, deer products such as venison and antlers, and now there is unique deer-antler craft and jewellery as well. There are three types of gourmet sausages as well as sausage rolls made with venison from the farm's deer. From the antlers,

THE SOUTH-WEST—WA

there is a range of unique lotions and creams. The products are also available at the Albany farmers' market every Saturday morning (8am–12pm). As if that's not enough, Guinea Pig World is also at Deer-O-Dome, featuring over 150 guinea pigs, many that run free around the park—and in answer to your next obvious question, no, they don't eat them!

- **Open:** 10am–4pm Sun–Fri, 2pm–4pm Sat, 10am–4pm daily (in school holidays), phone to confirm times
- **Location:** On Link Rd, 7km north of Albany, just off Albany Hwy, before the Albany airport

EDEN GATE BLUEBERRY FARM
Greg and Susan Luke
Eden Rd, Youngs Siding, Albany 6331
Phone: 08 9845 2003
Email: blueberry@edengate.com.au
www.edengate.com.au

'A great experience for all the family', is how the Lukes describe this farm. They have been growing spray-free blueberries for twenty years and have fresh berries (including pick your own) plus jams and a very interesting blueberry wine, produced under the Eden Gate label. The fresh blueberries are available from selected regional outlets but the other lines—wine, muffins, ice-cream, jams—are only available from the farm.

- **Open:** 10.30am–4pm Thurs–Mon (Dec–Apr)
- **Location:** Between Albany and Denmark

HL AND BJ SHAPLAND
Bev and Howard Shapland
RMB 9054A, South Coast Hwy, Albany 6330
Phone: 08 9845 1064
Email: shap@omninet.net.au

The Shaplands have been growing vegetables here for twenty-three years, and selling them from the roadside, at farmers' markets and of course for city markets. Look for hydroponic lettuce, herbs, English spinach, eggplants, peas, capsicum—basically anything that's fresh and good for you.

- **Open:** 7am–7pm daily
- **Location:** 25km west of Albany

🍴 DINING

RISTORANTE LEONARDOS
Dean Blanchard
166 Stirling Tce, Albany 6330
Phone: 08 9841 1732
BYO
Cuisine: Northern Italian
Mains average: $$$
Local flavours: Marron, veal, lamb, abalone, goat, fish, fruit and vegetables
Extras: Three separate intimate dining rooms.

- **Open:** Dinner Mon–Sat

BALDIVIS

🍎 PRODUCERS

MACNUTS WA
Nick and Sindhu Dobree
213 Doghill Rd, Baldivis 6171
Phone: 08 9524 2223
Email: info@macnutswa.com.au
www.macnutswa.com.au

The only commercial macadamia processing plant in Western Australia, MacNuts produces and packages macadamias in eight different flavours. They also make a wide variety of macadamia products—spread, oil, cookies, nougat, nut crackers, chocolate coated nuts, and a range of

macadamia cosmetics—all mostly sold from the shop, by mail order, or from the website. The Dobrees planted their first trees seventeen years ago and started the processing plant nine years ago, opening the shop in 2000.
- **Open:** 10am–5pm Wed–Sun and public holidays, daily for large group tours (up to fifty people, bookings necessary)
- **Location:** Exit Kwinana Freeway at Mundijong Rd

BOYANUP
BREWERY/WINERY
JOSHUA CREEK FRUIT WINES
Jan and Joan Brouwer
42 South West Hwy, Boyanup 6237
Phone: 08 9731 5937 (shop)
08 9731 5184 (winery)
Email: joshuacreek@bigpond.com
www.joshuacreek.com.au
Joshua Creek Fruit Wines' range includes apple, pear, satsuma plum, Narrabeen plum, ruby red plum, apricot, and persimmon wines, as well as a fortified ruby red called Ruby Royale, a warming almost twenty per cent alcohol, and Golden Peach brandy (twenty-nine per cent alcohol). Jan Brouwer had always wanted to make his own wine and when he entered an amateur wine competition years ago and won, the winemaking turned serious. Today, Joshua Creek wines are fermented solely from fruit picked from the orchard, which is situated next to a large lake that attracts abundant birdlife and is home to the Brouwers' flock of domestic geese. The shop also sells the orchard produce plus local arts and pottery.
- **Open:** 10am–4pm weekends or by appointment at other times

BRIDGETOWN
DINING
NELSON'S RESTAURANT
Ray Murphy, chef
38 Hampton St, Bridgetown 6255
Phone: 08 9761 1641
Email: nelsons@westnet.com.au
www.nelsonsofbridgetown.com.au
Licensed
Cuisine: Local and regional
Mains average: $$
Local flavours: Fruit and vegetables, herbs
Extras: Outdoor dining on the veranda.
- **Open:** Breakfast, lunch and dinner daily
- **Location:** On the South West Hwy

BRUNSWICK COW STATUE
Built by the Brunswick Lions, this cow is not as imposing as some big statues around the country, but it does pay tribute to the local, and vital, dairy industry. You will find the cow on the corner of South West Highway and Beela Road at Brunswick Junction.

BRUNSWICK
PRODUCERS
WHITE ROCKS FARM
WS Partridge and Sons
South West Hwy, Brunswick 6224
Phone: 08 9726 1085
Email: whiterocks@wn.com.au
This dairy farm milks 500 cows, and specialises in milk-fed veal. You'll be pleased to know that the veal industry in

Australia is under very strict controls. These youngsters are raised for four to five months, still retaining a light coloured flesh, and David Partridge assures us that they are fed well and spoiled rotten during that time. Mondo di Carne meats in Perth has the sole licence to market this product which is sold nationally and also exported to Singapore and Mauritius. You can't buy meat from the farm, but you can watch the milking, take a tour of the farm or visit the dairy museum.
- **Open:** 2pm–4pm by appointment only
- **Location:** 4.5km north of Brunswick

OLD BUTTER FACTORY MUSEUM
Peel Tce, Busselton 6280
Phone: 08 9752 2892
This museum has a collection of butter and cheese-making equipment that includes other historical items.
- **Open:** 2pm–5pm Wed–Mon

COWARAMUP
🍎 PRODUCERS

CANDY COW
*Ed and Frances Mangano (owners),
Jane Bennett (manager)
Shop 3, cnr Bussell Hwy and Bottrill St,
Cowaramup 6284
Phone: 08 9755 9155
Email: sweets@candycow.com.au
www.candycow.com.au*

The Candy Cow is the most 'wicked sweet shop in the south-west'. It's a busy place with a wonderful aroma and a range of old fashioned lollies, fine fudge, nougat and chocolate. Great quantities of fresh, golden, crunchy Honeycrunch are made with Western Australian and local honeys from the Margaret River region. In 1990, despite having no experience in confectionery making, the Manganos made the decision to launch 'Lollipop Lane', at the Perth Royal Agricultural Show. The entire family (including five children) became involved, creating thousands of show bags and making lollipops, chocolates, honeycomb and fairy floss, to the delight of thousands of children and families visiting the show. Now located in the country, there are tastings, free demonstrations and presentations and the products are also sold in Perth.
- **Open:** 10am–5pm daily
- **Location:** Ten minutes north of Margaret River and twenty minutes south of Busselton on the Bussell Hwy

MARGARET RIVER GOAT CHEESE
*Shannon and Stephanie Dempster
Cowaramup 6284
Phone: 08 9755 5353*

This fledgling goat dairy and cheese producer, which began in late 2003, is not open to the public just yet, but phone them if you are really interested in seeing what they are doing. Currently they are making a plain and an ashed soft fresh cheese, and we're told 'it's yummy' and there has been good feedback. Shannon is a farmer and his wife Stephanie is a chef, so that is a great start. Look for the cheese at Margaret Riviera providore and at other local places selling fine foods.
- **Open:** By appointment only

OLIO BELLO ORGANIC EXTRA VIRGIN OLIVE OIL
*Lot 1 Armstrong Rd, Cowaramup 6284
Phone: 08 9755 9771*

Email: oliobello@westnet.com.au
www.oliobello.com
Estate-pressed and grown, organic extra virgin olive oils are shown off to advantage in the stylish tasting room, visitors centre and deli. There, the range of olive oils—romanza, kurunba, kalamata, as well as lime-pressed, chilli-garlic and parmesan infused styles—are available. The olives are hand-harvested and pressed the same day to ensure the finest quality oils, displaying rich fruitiness, subtle pungency and lingering flavours. The olives are also available through selected delis and fine food stores nationally, and local winery restaurants.
- **Open:** 10am–4.30pm daily
- **Location:** Fifteen minutes North of Margaret River, off Cowaramup Bay Rd

STELLAR RIDGE ESTATE
Colin and Helene Helliar
2950 Clews Rd, Cowaramup 6284
Phone: 08 9755 5635
Email: stellar@netserv.net.au
In the business of making wines since 1994, and olive oil since 1998, this began as a lifestyle change that took the Helliars from Sydney to the other side of the country and on a stellar rise to become highly respected olive oil producers in just a few years. They produce seven varieties of wines, including a port, and four separate varieties of extra virgin olive oil, as well as olive tapenade and grape jellies from estate-grown fruit . The oil is also available at David Jones, Australiawide.
- **Open:** 11am–4.30pm daily, coach groups by appointment only
- **Location:** Off Miamup Rd

STORES

MARGARET RIVIERA
Kim Murray and Marcia King
3 Bottrill St, Cowaramup 6284
Phone: 08 9755 9333
Email: kim@mronline.com.au
www.mronline.com.au/riviera
This provedore specialises in local produce and showcases the wares of many small producers. Come here for one-stop shopping for local olive oil, olives, pickles, preserves, cheeses, condiments and much more from this rich corner of the state.
- **Open:** 10am–5pm daily

DINING

CULLEN WINES
Gordon W Parkes (chef)
Caves Rd, Cowaramup 6284
Phone: 08 9755 5656 (restaurant);
08 9755 5277 (winery)
Email: enquiries@cullenwines.com.au
www.cullenwines.com.au
Licensed
Mains average: $$
Specialty: Vineyard platter featuring fresh local produce
Local flavours: Mount Barker chicken, Busselton beef, Dunsborough dhufish, Wirring Road goose and goats cheese, Albany oysters, venison, baked goods, fruit and vegetables.
Extras: The winery and restaurant are built using stone with the restaurant featuring an undercover and tree-shaded outdoor area overlooking the vineyard.
- **Open:** Lunch, morning and afternoon teas daily
- **Location:** 15km from Margaret River off Caves Rd

THE SOUTH-WEST—WA

VASSE FELIX
Aaron Carr (chef)
Cnr Caves Rd and Harmans Rd South, Cowaramup 6281
Phone: 08 9756 5000
Email: info@vassefelix.com.au
www.vassefelix.com.au
Licensed (for Vasse Felix wines)
Cuisine: Modern Australian
Mains average: $$
Specialty: Fresh, regional cuisine
Local flavours: As much as possible, menu is designed around local produce
Extras: The upstairs balcony restaurant overlooks the original vines that were planted in 1967 and Willyabrup Brook. Vasse Felix is justly proud that this vineyard was the first commercial vineyard and winery in the Margaret River region. The gallery has regular Holmes à Court art collection exhibitions.
- **Open:** 10am–5pm daily for lunch, morning and afternoon tea
- **Location:** Entrance off Caves Rd

DENMARK

BREWERY/WINERY

BARTHOLOMEW'S MEADERY
Bart Lebbing and Barbara Thayne
RMB 1067, South Coast Hwy, Denmark 6333
Phone: 08 9840 9349
Email: info@bartholomewsmeadery.com.au
www.bartholomewsmeadery.com.au
A honey ice-cream cone on a hot day is worth the drive. But you probably won't leave with just that. Bartholomew's (named for the patron saint of the honey crop and mead making) also produces honey and other bee products, such as bee pollen, honeycomb, wax, honey wines and liqueur, and there is also art and craft in this rustic, mud-brick building. Bart Lebbing has been an apiarist since 1975 and released his first mead in 1989. Apitherapy—healing from the hive—is promoted here too. When available, Bart sells cappings honey, which is honey that is drained from the wax layer covering the comb, and is said to be great for healing many injuries and ailments.
- **Open:** 9am–4.30pm daily
- **Location:** 16km west of Denmark, and 1km west of William Bay turn-off

PRODUCERS

BRIDGART'S ORCHARD AND GOURMET PRESERVES
Margaret and Keith Bridgart
Barry Rd, Denmark 6333
Phone: 08 9840 9205
Email: hi@bridgarts.com
www.bridgarts.com
Margaret and Keith Bridgart's orchard bustles with activity. The jams, chutneys, relishes, and dessert toppings all come from their own crops, which include all the popular summer fruits, as well as sub-tropical ones like tamarillos and feijoas—try these made into ice-cream! The Bridgarts have been orchardists for over twenty years, and the products based on the thirteen varieties of tree-ripened apples are very popular, but top sellers are chilli asparagus, chilli Peppas in sweet marinade, and zingy tamarillo chutney. Sold at Discover WA (Perth domestic airport), Bay Merchants (Albany), by mail order and at the farm or Albany farmers' market.
- **Open:** 11am–4pm Thurs–Sun, 8am–12pm Sat (Albany farmers' market)
- **Location:** 9km from Denmark on the Scottsdale Tourist Drive, signposted

🍴 DINING

AQUA BLUE MARRON FARM AND RESTAURANT
Vic Grazulis and Anthony Turich
RMB, Tindale Rd, Denmark 6333
Phone: 08 9840 8008
Email: bluelobster@westnet.com.au
BYO
Cuisine: Freshwater crayfish
Mains average: $$
Specialty: Gourmet local marron dishes
Extras: Relaxed outdoor dining, tours of the facilities for 'blue' marron, raised for export on the property. You can also buy cooked or live marron to take away.
- **Open:** Lunch Tues–Sat
- **Location:** Follow the signs, 5km down Tindale Rd

MARY ROSE RESTAURANT
Karen and Paul Sydney-Smith
11 North St, Denmark 6333
Phone: 08 9848 1260
Email: sydney-smith@wn.com.au
Licensed
Mains average: $
Specialty: Traditional pies (steak and kidney, beef burgundy), vegetarian cuisine, home-baked cakes
Local flavours: Mushrooms, trout, berries, vegetables in season, Mt Barker free range chickens, Albany seafood, asparagus
Extras: Situated next to the Old Butter Factory Galleries, there is an alfresco terrace under shade sails and there are tables under the palm tree. No takeaway, but deliveries of prepared food can be made to guests of local accommodation providers.
- **Open:** Breakfast (all-day), lunch, morning and afternoon tea, daily

TINGLEWOOD WINES AND PUZZLES
Judy Wood
Glenrowan Rd, Denmark 6333
Phone: 08 9840 9218
Email: tinglewoodpuzzles@bigpond.com
www.tinglewoodwines.com.au
Licensed
Mains average: $
Specialty: Vineyard platters and a children's menu
Extras: All tables painted with games boards—chess, backgammon, snakes and ladders, draughts, tic-tac-toe—there are also puzzles to try while dining. Children's playground, view over farm.
- **Open:** 10am–4pm Thurs–Mon (school holidays only)

DONNYBROOK

🍴 DINING

OLD GOLDFIELDS ORCHARD AND CIDER FACTORY
Steve and Russell Trigwell
Goldfields Rd, Donnybrook 6239
Phone: 08 9731 0311
Email: trigs@oldgoldfields.com.au
www.oldgoldfields.com.au
Licensed/BYO
Mains average: $
Specialty: Old Goldfields caesar salad
Local flavours: Gourmet lettuce, avocados (in season), fruit, farm-fresh eggs, cakes, Heritage Country cheeses, olives
Extras: There is a restaurant, cider factory, museum of local history, the old minehead from the 1898 gold rush and marron ponds on this 125-year-old farm. All the prize-winning ciders, as well as the 100 per cent fresh Old Goldfields apple juice are made on site entirely from the orchard's own apples (no concentrates).

THE SOUTH-WEST—WA

There is also outdoor dining overlooking the orchard.
- **Open:** 9.30am–4.30pm Wed–Sun, daily during school and public holidays
- **Location:** 6km south of Donnybrook, then follow signs off Marmion St at South West Hwy turn-off

DUNSBOROUGH
🍎 PRODUCERS

THE OLIVE PIT
Maree Goulden
Cnr Commonage Rd and Brushbrook Dr, Dunsborough 6281
Phone: 08 9755 2438
Email: theolivepit@westnet.com.au
This is a regional centre for olive products and grew out of a desire to educate the public about EVOOs (extra virgin olive oils). You can taste the oils before buying, and there are other olive products too, such as pickled olives, tapenades, soaps and lotions, as well as other regional products. Basically it's a gallery in the heart of an olive grove.
- **Open:** 10.30am–4.30pm daily

🍴 DINING

WISE VINEYARD RESTAURANT
Heath Townsend
Eagle Bay Rd, Dunsborough, 6280
Phone: 08 9755 3331
Email: wisefood@compwest.net.au
www.wisewine.com.au
Licensed
Cuisine: Modern Australian
Mains average: $$$
Specialty: Soy glazed half duck, daily changing specials
Local flavours: Duck, venison, marron, Albany rock oysters, olive oil, Vasse asparagus, dhufish
Extras: Vineyard setting overlooking bushland to Geographe Bay.
- **Open:** Breakfast weekends, lunch daily, dinner Fri–Sat

FOREST GROVE
🍎 PRODUCERS

FOREST GROVE OLIVE FARM
Jill and Tony James
RMB 317, Harrison Rd, Forest Grove 6286
Phone: 08 9757 6428
Email: forestgrove@westnet.com.au
Estate-grown Forest Grove extra virgin olive oil is produced from the fruits of this six-year old grove. You'll find varietal oils here, as well as dukkah and lemons preserved in EV oil. There are also tastings and cooking classes related to using olive oil. Also sold in Margaret River, at the Fremantle markets, stores in Perth and at some local wineries and cafes.
- **Open:** 11am–4pm Fri–Sun or by appointment
- **Location:** 10km south of Margaret River, turn right at west Calgardup and follow the signs

KALGAN
🍎 PRODUCERS/DINING

ALBANY MARRON FARM AND NIPPERS CAFE
Karl and Sharon Rost
Two People's Bay Rd, Kalgan 6330
Phone: 08 9846 4239
Email: albanymarronfarm@bigpond.com
Marron is a delicious freshwater crayfish that is farmed mainly in the south-west

of Western Australia. It is a protected species and is not always available in restaurants, so where else but a marron farm would you find a cafe called Nippers Cafe where you can sample marron, yabbies and freshwater rainbow trout (all of which are farmed here) plus fresh salads and local fruits? There is also local Naked Bean coffee and homemade cakes and scones. The cafe is proud to be very child friendly and caters for all types of diets. Smoked trout pâté and pickled yabbies should be on the menu soon.

- **Open:** 11am–4pm Tues–Sun, 11am–4pm daily during school and public holidays, closed July–Aug
- **Location:** 25km from Albany. Turn left into Two Peoples Bay Rd; the farm is about 3km further on the right.

KARRIDALE

PRODUCERS

CLOVERDENE DAIRY
Trevor and Deb Dennis
RMB 200A, Karridale 6288
Phone: 08 9758 5579
Email: Cloverdene@bigpond.com
www.cloverdene.com.au

Twelve years ago, the Dennises began this dairy as a diversification from prime lamb production, and now it is producing a fine range of sheep milk products. Sheeps milk is easier to digest, they say, so people with dairy allergies can tolerate it. Better still, these products are organic, so look for Old Karridale, Karridale Maritime cheese, fetta and yoghurt at health food shops and local restaurants.

- **Open:** 2.30pm–5.30pm Wed–Fri, some weekends (phone for appointment)

KOJONUP

PRODUCERS

PERFECT POULTRY
Greg and Di Hill
RMB 587, Albany Hwy, Kojonup 6395
Phone: 08 9832 8018
Email: frisostud@westnet.com.au

Perfect Poultry sells its free-range chickens fresh or frozen, but these are no ordinary chickens. They are raised without antibiotics or hormones, and the huge ones are processed at thirteen weeks, even though they are genetically bred to be processed at six weeks. There are also turkeys at Christmas and there are gourmet chicken products such as boned rolls or marinates. Sold at butchers, specialised grocery stores and Albany farmers' market.

- **Open:** 9am–5pm daily, phone for large orders
- **Location:** 4km south of the Beaufort River Tavern, 27km north of Kojonup

DINING

KOJONUP COUNTRY KITCHEN
Peta Zadow
Unit 1/88 Albany Hwy, Kojonup 6395
Phone: 08 9831 1338
Email: peta_zadow@westnet.com.au
BYO
Cuisine: Country style
Mains average: $
Specialty: Soups, quiche, specials—all made from scratch
Local flavours: Albany Naked Bean coffee, Perfect Poultry chicken
Extras: Everything is made on the premises. Takeaway available.

- **Open:** 8am–5pm weekdays

MANJIMUP

🍎 PRODUCERS

THE WINE AND TRUFFLE COMPANY
Hazel Hill Ltd
Seven Day Rd, Manjimup 6210
Phone: 08 9228 0328
Email: markq@wineandtruffle.com.au
www.wineandtruffle.com.au
This company's first truffle 'crop' in 2003 weighed 160 grams. However, acting on faith, a 'cellar door' sales facility opened in mid 2004. Truffles or not, there is all sorts of educational and interesting truffle paraphernalia here, as well as samples of wine and other products, cellar door sales and there will be trufferie visits in due course.
- **Open:** Phone for details
- **Location:** Just south of the Manjimup township

MARGARET RIVER

🍎 PRODUCERS

MARGARET RIVER VENISON
Morrison family
Caves Rd, Margaret River 6285
Phone: 08 9755 5512, Shop: 08 9755 5028
Email: mrv@deermark.com
www.deermark.com
Step into the new shop on Caves Rd and have a chat with a member of the Morrison family and you'll soon see why this venison producer is so well respected. They sell all cuts of venison, as well as mince and sausages, and smallgoods that you'll only find here. There is also leather, jam and handicrafts, and there are tastings too. This company supplies approximately ninety-five per cent of the state's venison, which is used in many major restaurants, including Margaret River wineries, and the meat is distributed through a wholesaler to Perth butchers.
- **Open:** 9am–5pm daily (later in summer)
- **Location:** Between Ellensbrook and Cowaramup roads

MARGARET RIVER DAIRY COMPANY
Bussell Hwy, Margaret River 6280
Phone: 08 9755 7588
Email: mriver@manassen.com.au
A need for premium Western Australian cheeses and yoghurts was the reason this company began fifteen years ago and quickly expanded into the national market. And aren't we glad they did! The milk is sourced from local Margaret River dairy farmers and turned into premium, reduced-fat, fat-free, flavoured and natural yoghurts, quality mould-ripened cheeses (brie and camembert, including double cream and peppered), and a range of fresh and flavoured farmhouse cheeses (including premium vintage cheddars, club cheddars in several flavours and baked ricotta). These are available through supermarket and retail outlets in Western Australia and the rest of the country.
- **Open:** 9am–5pm daily
- **Location:** 30km south of Busselton and 15km north of Margaret River

THE BERRY FARM
Eion and Andrea Lindsay
Bessell Rd (RMB 222),
Margaret River 6285
Phone: 08 9757 5054
Email: info@berryfarm.com.au
www.berryfarm.com.au
Over seventeen years this business has grown from a two-person operation to a

busy business with many staff and it is now a destination in its own right. The farm's produce is turned into fruit and berry wines and ports, fermented fruit wine vinegar, jams, preserves and chutneys under the labels of The Berry Farm, Thornhill Wines and Cahoots, and sold nationally. There is also, by day, a busy restaurant that showcases the produce from the farm, and is set in beautiful cottage gardens with abundant bird life to enjoy.
- **Open:** 10am–4.30pm daily
- **Location:** Follow signs off the Bussell Hwy just south of Margaret River

DINING

LEEUWIN ESTATE RESTAURANT
Donna Seed
Leeuwin Estate, Stevens Rd,
Margaret River 6285
Phone: 08 9759 0000
Email: info@leeuwinestate.com.au
www.leeuwinestate.com.au
Licensed
Cuisine: Modern Australian
Mains average: $$
Local flavours: Margaret River marron, Pemberton trout, venison, cheeses, local vegetables
Extras: Outdoor dining with views of vineyard grounds. Annual concert in a large amphitheatre on the estate grounds.
- **Open:** Lunch daily, dinner Sat

MARGARET RIVER HOTEL
Keith Rose (chef)
139 Bussell Hwy, Margaret River 6285
Phone: 08 9757 2655
Email: mrhotel@netserv.net.au
www.margaretriverhotel.com.au
Licensed/BYO
Cuisine: Bistro style
Mains average: $$
Specialty: Gourmet pizzas
Extras: Alfresco dining.
- **Open:** Lunch and dinner daily

VAT 107
Gary Cream and Jenny Spencer (owners),
Dany Angove (chef)
107 Bussell Hwy, Margaret River 6285
Phone: 08 9758 8877
Email: vat107@highway1.com.au
www.vat107.com.au
Licensed/BYO wine
Cuisine: Modern Australian with Asian influences
Mains average: $$$
Specialty: Szechuan duck
Local flavours: Marron, oysters, cheese, venison, lamb
Extras: Ideally located for wine tasting in the area.
- **Open:** Breakfast, lunch and dinner daily

VOYAGER ESTATE WINERY
Nigel Harvey (chef), Michael Wright (owner)
Lot 1, Stevens Rd, Margaret River 6285
Phone: 08 9757 6354
Email: cellardoor@voyagerestate.com.au
www.voyagerestate.com.au
Licensed
Cuisine: Modern Australian
Mains average: $$
Specialty: Marron, Margaret River venison
Local flavours: Harvey beef, Mt Barker chicken, marron, Vasse asparagus, Margaret River venison, fish
Extras: Renowned for its Cape Dutch architecture, manicured grounds and amazing rose gardens. Sells Voyager

Estate jams, pickles and other produce lines.
- **Open:** Lunch, morning and afternoon tea, daily

XANADU MARGARET RIVER RESTAURANT AND CELLAR DOOR
Wayne Booth (chef)
Boodjibup Rd, Margaret River 6285
Phone: 08 9757 3066
Email: restaurant@xanadunormans.com
www.xanadunormans.com
Licensed
Cuisine: Modern Australian/Asian
Mains average: $$$
Specialty: Taste of the Dragon—a tapas plate of seared scallops, house smoked salmon, venison pie, grilled chorizo and olives
Local flavours: Seafood, chicken
Extras: Finalist in the 2003 Gold Plate Awards. Large covered playground and dedicated kids menu. Grows herbs and supplies winery barrel shavings, used to smoke salmon. Fine art gallery focusing on Western Australian artisans, with regular exhibitions.
- **Open:** Lunch daily, dinner Sat
- **Location:** Turn off the Bussell Hwy

METRICUP

PRODUCERS

THE MARGARET RIVER DAIRY CO
Denis Hann
Bussell Hwy, Metricup 6282
Phone: 08 9755 7588
Email: dhann@manassen.com.au
Fonti Dairy Products began the business of producing quality gourmet cheeses and yoghurts here and built up a fine reputation and a loyal following. The business changed hands in 2000, but the factory still makes the same quality dairy products—camembert, brie, handmade farmhouse cheeses and pot-set yoghurts and baked or fresh ricotta—and these are now sold in supermarkets in most states. There are free tastings at both shops, which are located close to each other, cooler bags for your cheeses, souvenirs, and a viewing window into the factory so you can see just how the cheese is made.
- **Open:** 9.30am–5pm daily

MT BARKER

DINING

TASTE MOUNT BARKER WINE CAFE
Joan Bath
26 Langton Rd, Mt Barker 6324
Phone: 08 9851 2500
Email: tastemb@westnet.com.au
Licensed (Great Southern/WA wines only)
Cuisine: Artisan (as defined by Australian Country Style magazine)
Mains average: $
Specialty: Gourmet platters
Local flavours: Venison, Mt Barker free-range chicken, local preserves and antipasto
Extras: Shady outdoor courtyard, fires in winter, classical background music. Visitors can taste and purchase wines from more than ten wineries in the region. Sells homemade pickled walnuts, mustard, pickled ginger, chutney and Plantagenet prunes.
- **Open:** 10.30am–5pm Wed–Mon

NANNUP
🍴 DINING

BLACKWOOD CAFE
Ken Darnell and Jo Clews
24 Warren Rd, Nannup 6275
Phone: 08 9756 1120
BYO
Mains average: $$
Specialty: Burgers
Local flavours: Beef
Extras: Outdoor garden setting. Takeaway available.
■ Open: 8am–8pm daily

NORTH TORBAY
🍎 PRODUCERS

SUSSEX BARN
Andrew and Liz Marshall
RMB 9055, South Coast Hwy,
North Torbay (via Albany) 6330
Phone: 08 9845 1081
Email: marshall@omninet.net.au
This well-established (that's thirty-two years, if you're counting) fruit and vegetable farm grows asparagus, sweet corn, passionfruit and kiwifruit, with raspberries ripening, just when you need them, over Christmas. The fruit is grown as organically as possible in the cleanest and best of environments, close to the southern ocean, and you can taste the difference at the farm gate, at local outlets in Albany and Denmark, or at Herdsmans Fresh Essentials in Perth.
■ Open: 7am–7pm daily (Sept–June)
● Location: About 22km west of Albany

NORTH WALPOLE
🍴 DINING

THURLBY HERB FARM
Penny Jewell (owner),
Stephen Roberts (chef)
Lot 3, Gardiner Rd, North Walpole 6398
Phone: 08 9840 1249
Email: thurlbyherb@wn.com.au
www.thurlbyherb.com.au
BYO
Cuisine: Country
Mains average: $
Specialty: Homemade brioche and other breads, rosemary scones and homemade jams, lavender ice-cream, homemade lime and lemon verbena cordial
Local flavours: Smoked trout and locally grown mushrooms
Extras: Herb display gardens in a peaceful, picturesque, rural setting. Sells hams and pickles from home-grown produce, including pear and cashew chutney, quince pickle, brandied lemon and ginger marmalade and port cinnamon plum jam.
■ Open: Lunch, morning and afternoon tea daily
● Location: From Walpole travel 11km up North Walpole Rd, turn left onto Gardiner Rd and continue for 3km

PEMBERTON
🍎 PRODUCERS

FOREST FRESH MARRON
Sue Bamess
Pemberton Aquaculture Producers,
Pump Hill Rd, Pemberton 6260
Phone: 08 9776 0099
Email: wamarron@wn.com.au
www.wn.com.au/wamarron

This live marron wholesaler and retailer started up six years ago to coordinate market supply and demand. The marron is served in some very fine restaurants locally such as Lamont's, Voyager Estate, Gloucester Ridge, and Flutes as well as in Perth. They have just started running tours and you can buy marron here (order ahead so there are a few days to purge them) and there are even marron souvenirs and marron jewellery.
- **Open:** 9am–3pm or by appointment; tours hourly 10am–2pm weekdays, or by appointment
- **Location:** Turn off Club Rd then look for the signs on Pump Hill Rd

🍴 DINING

GLOUCESTER RIDGE WINERY AND RESTAURANT
Don Hancock (owner),
Gregg Burdon (chef)
Burma Rd, Pemberton 6260
Phone: 08 9776 1035
Email: gridge@karriweb.com.au
www.gloucester-ridge.com.au
Licensed
Cuisine: Modern Australian
Mains average: $$
Specialty: Marron and trout dishes
Local flavours: Marron, trout, beef and vegetables
Extras: Outside deck and glass conservatory, fantastic views, relaxed atmosphere, kid friendly. Grows own marron.
- **Open:** Lunch daily, dinner Sat

KING TROUT CAFE
Egon (chef) and Shelley König
Cnr Old Vasse and Northcliffe Roads,
Pemberton 6260
Phone: 08 9776 1352

Email: kingtrout@westnet.com.au
www.kingtrout.com.au
BYO
Cuisine: Modern Australian
Mains average: $$
Specialty: Farm-grown rainbow trout, marron, smoked trout, homemade pâté
Local flavours: Farm-grown rainbow trout, marron
Extras: The restaurant overlooks a lake surrounded by towering Karri trees. Dine outdoors or, in winter, you can dine inside by a wood fire. Catch your own fresh fish from the lake and have it prepared for you. Takeaway is available and there are counter sales of trout (whole and fillets), smoked trout and trout pâté.
- **Open:** Lunch, morning and afternoon tea, daily

LAKESIDE RESTAURANT
Karri Valley Resort, Vasse Hwy,
Pemberton 6260
Phone: 08 9776 2020
www.karrivalleyresort.com.au
Licensed
Cuisine: Modern Australian
Mains average: $$
Specialty: Marron and trout
Local flavours: Marron, trout
Extras: Resort accommodation beside trout-filled Lake Beedelup, floodlit at night.
- **Open:** Breakfast and dinner daily, lunch weekends and holidays

PEMBERTON MILL HOUSE CAFE
Morrie and Cas Mills
14 Brockman St, Pemberton 6260
Phone: 08 9776 1122
Email: marronblue@westnet.com.au
BYO
Mains average: $$

Specialty: Marron, including whole marron on the plate
Local flavours: Marron, fruit, vegetables
Extras: Timber museum featuring the history of local timbers. You can't get marron any fresher than this—what you eat here comes from the pond at the owner's marron farm out of town. Takeaway marron platters are also available.
- **Open:** 9.30am–5pm daily

SADIES
Ian Leaning
Gloucester Motel, Ellis St, Pemberton 6260
Phone: 08 9776 1266
Email: glousmot@karriweb.com.au
www.gloucestermotel.com.au
Licensed/BYO
Mains average: $$
Specialty: Dishes from Europe and Bermuda, award-winning fish chowder, pastries
Local flavours: Trout, marron, potatoes, avocados, strawberries, cherries, apples, beef, lemons, herbs
Extras: Nestled in the Karri forest with alfresco dining. Takeaway available.
- **Open:** Breakfast, lunch and dinner daily

PORONGURUP
🍴 DINING

MALEEYA'S THAI
Maleeya Form
Porongurup Rd, Porongurup 6324
Phone: 08 9853 1123
BYO
Cuisine: Thai
Mains average: $$
Specialty: Panang curries
Local flavours: Pork, organic beef and eggs, Mt Barker chicken, fruit, chilli mixes

Extras: Outdoor tables on a veranda and under a tree. Grows ten kinds of chillies.
- **Open:** Lunch and dinner Fri–Sun

REDMOND
🍎 PRODUCERS

ALBANY FARM FRESH EGGS
Keith Amess
Albany Hwy, cnr Redmond Rd,
Redmond 6330
Phone: 08 9845 3106
Email: albanyeggs@yahoo.com.au
About two-thirds of the eggs produced here are barn eggs, which means the hens are free to range in an enclosure. Already raising around 6000 hens, the numbers are due to increase as the poultry farm expands. All eggs are collected daily and can be bought at the farm shop, along with local jams and honeys.
- **Open:** 8am–5pm daily
- **Location:** 20km north of Albany

STIRLING RANGE
🍴 DINING

THE LILY STIRLING RANGE DUTCH WINDMILL
Hennie (chef) and Pleun Hitzert
Chester Pass Rd, Stirling Range National Park 6338
Phone: 08 9827 9205; Freecall 1800 980 002
Email: thelily@ozemail.com.au
www.thelily.com.au
Licensed
Cuisine: Dutch, French, Australian
Three-course set menu: $$
Specialty: Soups, Dutch apple cake

Local flavours: Mount Barker free-range chickens, Butterfield beef
Extras: There is an authentic, five-storey, sixteenth-century replica Dutch windmill and a fully operational windmill producing organically grown, stone ground wheat and spelt flour. Nearby, there are traditional Dutch houses and the restaurant is in the original, relocated and reconstructed, 1924 Federation-style former Gnowangerup railway station. Guided windmill tours, accommodation, private airstrip, panoramic views of the Stirling Ranges.
- **Open:** Lunch, morning and afternoon tea and dinner Tues–Sun and public holidays

WANNUNUP
DINING

THE JOLLY FROG RESTAURANT
Jean-Daniel Ichallalene
8 Rod Court, Wannunup 6210
Phone: 08 9534 4144
Email: jollyfrog@westnet.com.au
www.jollyfrog.com.au
Licensed
Cuisine: French
Mains average: $$$
Specialty: Duck à l'orange
Local flavours: Fish, crustaceans, lobster, crab and prawns
Extras: The restaurant has a great view of the estuary and Indian Ocean, and is a perfect place to look out for dolphins. Takeaway available.
- **Open:** Breakfast, lunch and dinner daily

WILLYABRUP
DINING

CAPE LAVENDER
Bev and Jeff Clarke
Carter Rd, Willyabrup 6280
Phone: 08 9755 7552
Email: info@capelavender.com.au
www.capelavender.com.au
Licensed
Mains average: $$
Specialty: Pie of the day
Local flavours: Organic lamb, venison, fish and anything associated with lavender
Extras: Views and a great place to stop while wine-tasting. Sells a range of homemade lavender preserves, jams, chutneys, kasundi and mustards.
- **Open:** 10am–5pm, lunch daily
- **Location:** Off Metricup Rd

FLUTES RESTAURANT
François Morvan
Brookland Valley Vineyard, Caves Rd, Willyabrup 6280
Phone: 08 9755 6250
Email: flutes@flutes.com.au
www.flutes.com.au
Licensed
Cuisine: Modern Australian
Mains average: $$$
Specialty: Seafood trio
Local flavours: Venison, lamb
Extras: Outdoor dining on a deck, with views overlooking the brook.
- **Open:** Lunch daily, dinner Fri–Sat in winter and Wed–Sat in summer

MARKETS AND EVENTS

✦ ALBANY FARMERS' MARKET,
ALBANY
08 9841 4312
8am–12pm Sat

✦ MANJIMUP FARMERS' MARKET,
MANJIMUP
08 9772 3555
Third Sat 8am–12pm, First Sat Feb–Mar and Easter Sat

✦ MARGARET RIVER FARMERS' MARKET, MARGARET RIVER
08 9755 8788
8am–12pm first Sun

✦ WANNEROO LOCAL HARVEST FARMERS' MARKET, WANNEROO
08 9575 1107.
First Sat 8am–12pm

✦ BOYANUP FARMERS' MARKET,
BOYANUP
0417 097 538
Fourth Sun 8am–12pm

✦ QUINNINUP MARKET DAY,
QUINNINUP
08 9773 1292
Oct

✦ BULL AND BARREL FESTIVAL,
DARDANUP
08 9728 1121
Oct

✦ CAMBINATA EXTRAVAGANZA— SILVER SERVICE DINNER IN THE SHEARING SHED, KUKERIN
08 9864 6054
Oct

✦ FOOD LOVERS CLUB,
MOORE RIVER
Several times each month

✦ FRANKLAND RIVER FINE WINE, FOOD AND MUSIC FESTIVAL,
FRANKLAND
08 9855 2240
Nov

✦ MANJIMUP CHERRY HARMONY FESTIVAL, MANJIMUP
08 9771 2316
Oct

✦ MARGARET RIVER WINE REGION FESTIVAL,
MARGARET RIVER
08 9757 9990
Nov

✦ PEEL ESTATE JAZZ, BALDIVIS
08 9524 1221
11am–5pm first Sun in Dec

✦ PEMBERTON MARRON AND WINE FESTIVAL, PEMBERTON
08 9776 1202
Nov

YALLINGUP
🍴 DINING

LAMONT'S MARGARET RIVER
Kate, Fiona and Corin Lamont
Gunyulgup Valley Dr, Yallingup 6282
Phone: 08 9755 2434
Email: margaretriver@lamonts.com.au
www.lamonts.com.au
Licensed
Cuisine: Regional modern Australian
Mains average: $$$
Specialty: Marron dishes
Local flavours: Fish, beef, marron, Vasse asparagus, whitebait from Geographe Bay, King George whiting, local oils, Pemberton trout
Extras: Great views over the lake, outdoor seating and wine tastings. Takeaway available.
- **Open:** Breakfast and lunch daily, dinner Fri–Sat
- **Location:** Marrinup Dr, off Caves Rd, northern end of Margaret River

AUSTRALIA'S NORTH-WEST

One of the world's last great wilderness areas, the Kimberley region covers an area nearly three times the size of England, yet has a population of just 25 000.

Once mainly the home of Indigenous people and beef station owners and staff, the Ord River Irrigation Scheme changed this, and created an oasis in an otherwise hostile land. Today there is dairying and sugarcane, chick peas, melons and many other crops are grown in the area.

Further south at Broome, the pearling industry has an interesting by-product—the immensely expensive and sought-after pearl meat. This area also produces fruit (including mangoes) and vegetables and a wealth of local seafood.

BROOME
🍎 PRODUCERS

BROOME'S OWN
Errol and Jenny de Marchi
Broome 6725
Phone: 08 9192 1731
Email: homepres@westnet.com.au
Do try to get to the Broome courthouse markets, if only to see the de Marchis and their homemade preserves and breads, which number a mind-boggling 127 varieties. They are also sold in the town and by mail order. The jam operation began as a means of using up their orchard's marked fruit. Rather than dump it, they hauled out the pots and started cooking. Their product contains no added flavours or colours and is made by natural preserving methods using chemical-free ingredients, full fruit and no fruit fillers.
■ **Open:** Broome Markets, or phone for directions to the property

Ⓜ MARKETS

BROOME COURTHOUSE MARKETS
Margaret Lowe, coordinator
Broome Courthouse, Broome 6725
Phone: 08 9487 5598
Email: olivia.galipo@broome.wa.gov.au
This is the place to come for great multicultural food. There are dozens of stalls and the food ones include fruit and vegetables (including organic), honey, jams, breads, noodles, The Icecreamery van, pickles, chutneys, and bush foods.
■ **Open:** 8am–1pm Sat

🏠 STORES

THE ICECREAMERY
Vicki Robinson and Bob Meek
Shop 3 Carnarvon St, Chinatown, Broome 6726
Phone: 08 9193 5400
How about a cool treat in a hot town? The Icecreamery, right in the middle of Chinatown, makes all its own cool fare on the premises. Bob Meek, the owner and ice-cream maker, keeps it seasonal by utilising local fruits such as mango, coconut, passionfruit and rockmelon and now has two shops in town as well as a stand at the Broome courthouse markets. Much valued is the black sapote (chocolate-pudding fruit) ice-cream.
■ **Open:** 9am–5pm daily

AUSTRALIA'S NORTH-WEST—WA

🍴 DINING

AARLI BAR
Nick Wendland
Shop 2/6, Palm Court Arcade,
6 Hamersley St, Broome 6725
Phone: 08 9192 5529
Email: aarlibar@bigpond.com
BYO
Cuisine: Middle Eastern, Mediterranean
Mains average: $$
Specialty: Tapas and whole fish
Local flavours: As much as possible, including fish and seafood
Extras: Alfresco dining in tropical surrounds.
■ Open: Breakfast, lunch and dinner daily

CAFE CARLOTTA
Michael Whatley and Charlotte Brown, Jason Dean (chef)
Jones Place, Old Broome, Broome 6725
Phone: 08 9192 7606
Email: cafecarlotta@westnet.com.au
BYO
Cuisine: Authentic Italian
Mains average: $$
Specialty: Veal saltimbocca, wood-fired pizzas
Local flavours: Pearl meat, WA chilli mussels
Extras: A coffee school on Saturday mornings in the dry season, alfresco dining under the stars, tropical gardens. Takeaway available.
■ Open: Dinner Mon–Sat

CHARTERS RESTAURANT
Ken and Lola Fitzgerald
The Mangrove Resort—'Broome with a View',
47 Carnarvon St, Town Beach, Broome 6725
Phone: 08 9192 1303
(reservations 1800 094 818)
reservations@mangrovehotel.com.au
www.mangrovehotel.com.au
Licensed
Cuisine: Modern Australian, European
Mains average: $$$
Specialty: Fish of the day
Local flavours: Barramundi
Extras: Tides bar and garden restaurant overlooking Roebuck Bay. Three nights each month between March and October the resort plays host to the remarkable Stairway to the Moon, when the tide and moon combine to create the illusion of an incredible stairway to the moon.
■ Open: Breakfast, lunch and dinner daily

BROOME PEARL MEAT
Pearl meat is very rare, very expensive and very desirable. It is the muscle that forms around the pearl and when flash-fried or marinated is a unique taste-treat. John Goodman from Paspaley Pearls in Chinatown (Phone: 08 9192 2203), is the best person to contact, but be warned, he only has some to spare occasionally, and it can fetch around $70/kilogram. Otherwise, watch out for it on better menus around town.

MATSO'S BROOME BREWERY
Daniel Williams
60 Hamersley St, Broome 6725
Phone: 08 9193 5811
Email: drink@matsosbroomebrewery.com.au
www.matsosbroomebrewery.com.au
Licensed/BYO
Cuisine: Modern Australian
Mains average: $$

Specialty: Hunka Chunka—450 gram steak on the bone, honey roasted pumpkin
Local flavours: Beef, prawns, reef fish, fruits and vegetables
Extras: Outdoor dining, beer garden, views of Roebuck Bay. Gallery featuring the work of many local and regional artists. A range of boutique beers are brewed on site, the secret, they say, is the water, drawn from deep underground, plus the local wheat and barley.
- Open: 8am-late daily, brewery tours by appointment
Restaurant: Breakfast, lunch and dinner daily

TOWN BEACH CAFE
Andrew and Linda Tonner
End of Robinson St, Broome 6725
Phone: 08 9193 5585
BYO
Mains average: $$
Specialty: Fish dishes and pancakes
Local flavours: As much as possible, including herbs, fish and seafood, eggs, Asian herbs and spices
Extras: Fantastic views of Roebuck Bay. Takeaway available.
- Open: Breakfast, lunch and dinner daily

2-RICE
Nancy Gibson
26 Dampier Tce, Broome 6725
Phone: 08 9192 1395
Email: tworice@iinet.net.au
BYO
Cuisine: Asian
Mains average: $
Specialty: Balinese curry, Nonya curry and June Oscar's famous *balachan*
Local flavours: Green papaya, barramundi, threadfin salmon, fresh turmeric, galangal, Thai basil, lemongrass and Broome-chillies (for curries)
Extras: Cooking school and food shop. Takeaway available. Food shop features curry pastes and jams, plus that *balachan* and Sue Ferrari's Ferrari Fare.
- Open: 9am-5pm dry season, check first during the wet season, lunch bar—11am-2.30pm year round

WHARF RESTAURANT
Craig Douglas
Port of Pearls House, Broome 6725
Phone: 08 9192 5700
Email: wharf@westnet.com.au
Licensed
Cuisine: Modern Australian seafood
Mains average: $$
Specialty: Barramundi, chilli crab, lobster
Local flavours: The largest range of local seafood on the west coast
Extras: Views over Roebuck Bay. Takeaway available. All sauces and dressings are made on the premises.
- Open: 10am-10pm daily

🚌 TOURS

KOOLJAMAN AT CAPE LEVEQUE
Djarindjin and One Arm Point Communities, PMB 8, Cape Leveque, Broome 6725
Phone: 08 9192 4970
Email: leveque@bigpond.com
www.kooljaman.com.au
Kooljaman wilderness camp offers the opportunity to participate in bush-tucker tours to learn about local bush foods and their traditional uses. A special treat is mud-crabbing with a local guide, then cooking and eating your catch fresh off the coals. Fishing with the local people is also a great way to learn about other

AUSTRALIA'S NORTH-WEST—WA

traditional foods taken from the ocean. A full day chartered boat trip from Kooljaman includes snorkelling, swimming, fishing, reef walking and lunch on Sunday Island, along with billy tea and damper. Winner of 2003-04 tourism awards for Aboriginal and Torres Strait Islander tourism, as well as Indigenous business of the year.
- **Open:** Breakfast (baskets to rooms), lunch and dinner daily (Apr-Oct), accommodation all year
- **Location:** 220km north of Broome, follow the Cape Leveque Rd, four-wheel drive track only

PEARL LUGGERS
Vanessa Hayden
31 Dampier Terrace, Broome 6725
Phone: 08 9192 2059
Email: pearlluggers@bigpond.com
Visitors can try three different dishes of pearl shell meat—a marinated dish and the other two lightly wok fried—which are served throughout the evening while enjoying an insight into the world of pearling and the heritage of Broome at this amazing display centre. Other condiments are also served, and there is a bar.
- **Open:** 6pm-8.30pm Thurs (Apr-Nov)

CABLE BEACH
DINING

OLD ZOO CAFE
Peter and Susie Watterston
2 Challenor Dr, Cable Beach 6726
Phone: 08 9193 6200
Email: dining@zoocafe.com.au
Licensed
Cuisine: Modern cafe food with Asian influences
Mains average: $$$
Specialty: Daily changing specials
Local flavours: Crocodile, kangaroo, local seafood
Extras: Located in the former feed shed of the famous Pearl Coast Zoo, set up by Lord McAlpine. Newly renovated premises with indoor and large outdoor deck for tropical alfresco dining. Takeaway available.
- **Open:** Breakfast, lunch and dinner daily
- **Location:** Behind the Broome Crocodile Park and Cable Beach Club Resort, on the Pearl Town bus route

DERBY
FARMSTAY

MORNINGTON WILDERNESS CAMP
Abby Elberni, Simone Phillips
Off Gibb River Rd (PMB 925), Derby 6728
Phone: 08 9191 7406
Bookings 1800 631 946
Email: mornington@australianwildlife.org
www.australianwildlife.org
Licensed
Cuisine: Modern Australian
Two-course meal: $$
Specialty: Grilled barramundi
Local flavours: Barramundi, bush tucker
Extras: Run by the not-for-profit Australian Wildlife Conservancy. One million acres of spectacular country, accommodation in safari-style tents and shady campsites. Outdoor dining under the stars in bush 'restaurant' with Kimberley views.
- **Open:** Breakfast, lunch and dinner daily
- **Location:** 250km from Derby, then 90km off Gibb River Rd

MT HART WILDERNESS LODGE
Taffy and Annabelle Abbotts (owners/cooks), Lorette Evans (cook)
Gibb River Rd, Derby 6728
Phone: 08 9191 4545
Email: mthart@bigpond.com
www.mthart.com.au
Licensed
Cuisine: Modern Australian
Meals included in package $$
Specialty: Homemade bread
Local flavours: Seafood (threadfin salmon and barramundi), Kimberley fruit
Extras: A unique location on the Gibb River Rd, with an emphasis on nature-based tourism, old-fashioned bush hospitality and modern cuisine using fresh, seasonal produce. Self-guided nature walks, canoeing, or relaxing at the homestead. Eight acres of tropical gardens, exclusive camp sites. Packed lunches.
■ **Open:** Breakfast, lunch and dinner daily (Apr-Nov)
● **Location:** 240km from Derby, along Gibb River Rd, turn-off signposted. Follow directions on sign, and radio call to homestead using UHF radio located in a wooden hut at the turn-off (directions for operation on radio)

DINING

WHARF RESTAURANT
Craig Douglas
Port of Derby, Derby 6728
Phone: 08 9191 1195
Email: wharf@westnet.com.au
Licensed
Cuisine: Modern Australian seafood
Mains average: $$
Specialty: Barramundi, chilli crab, lobster
Local flavours: Seafood

Extras: Magnificent views, outdoor dining, garden setting. Takeaway available.
■ **Open:** Lunch and dinner daily

KUNUNURRA

PRODUCERS

BARRA BARRA PTY LTD
Chris and Tina Demeo
Ivanhoe Rd, Kununurra 6743
Phone: 08 9168 2098
Email: demeo@westnet.com.au
Depending on the season, there are locally grown fruit and vegetables, smoothies, homemade jams, chutneys, pickles, dried local fruit and choc-coated fruit at this farm shop that takes advantage of the wide range of produce growing in the Ord River irrigation area. 'Best smoothies in Australia', they tell us.
■ **Open:** 9am-5pm daily (Apr-Sept)
● **Location:** 9km from Kununurra

KIMBERLEY BOAB KREATIONS
Melissa Boot
1528 Poinsettia Way, Kununurra 6743
Phone: 08 9168 1816
Email: w.mboot@bigpond.com
Must be the climate that makes ideas grow almost as fast as the plants! Melissa Boot has devised a wonderful use for the fruit of the boab tree, which only grows in this region. A true 'bush tucker',

---

REGIONAL QUARANTINE NOTE
If travelling from other parts of Western Australia, do not bring fruit or palms, or if travelling to other parts of Western Australia do not take citrus plants or cuttings and livestock.

the seed has a unique flavour and texture. Melissa uses the fruit, which has an unusual citrus flavour and a vitamin C content, in dark, milk or white chocolates and turns the seed itself into tableware creations. Melissa was the first person to experiment with using this fruit commercially and has had it passed by ANZFA (Australian and NZ Food Authority). You can buy from her or look for the chocolates in gourmet delis, local B&Bs and stations in the Kimberley and Darwin.

■ **Open:** 9am-4pm weekdays, or by appointment
● **Location:** Off Weaber Plains Rd, just past Hidden Valley Caravan Park

BREWERY/WINERY

HOOCHERY DISTILLERY
Spike Dessert
Block 300, Weaber Plains Rd,
Kununurra 6743
Phone: 08 9168 2122
Distillery: 08 9168 2467
Email: hoochery@hoochery.com.au
www.hoochery.com

MARKETS AND EVENTS

✦ **COURTHOUSE AND STAIRCASE MARKETS,** BROOME
08 9192 2510
Apr-Oct three days each month at the full moon

✦ **DERBY MARKETS,** DERBY
08 9191 1611
8am-1pm Sat, Apr-Oct

✦ **MANGO FESTIVAL,** BROOME
08 9192 2222
Nov

Not many people can claim to add a word to the language, but the hoochery's owner, Spike Dessert, coined from the word from 'hooch', the Alaskan word for liquor. The distillery—the only legal still in WA—produces rum from local sugar cane, which has only been grown in the area in the past few years. And that was his own idea too. There are cellar door sales of Cane Royale Liqueur, Aguardiente (a sort of Oz ouzo), Ord River Rum and Kimberley Cane Spirit.

■ **Open:** 9am-5pm weekdays (tours 11am and 2pm), 9am-12pm Sat (tour at 11am), open Sun during dry season. Phone distillery for updates on opening hours.
● **Location:** Approximately fifteen minutes from Kununurra on Weaber Plains Rd

FARMSTAY/CLASSES

FARAWAY BAY, THE BUSH CAMP
Robyn Ellison
Kununurra 6743
Phone: 08 9169 1214
Email: farawaybay@bigpond.com
www.farawaybay.com.au
Cuisine: Modern Australian
Tariff includes meals
Specialty: Catch of the day
Local flavours: Local bush tucker, fresh Ord Valley produce, including vegetables and tropical fruits
Extras: Eagle Lodge, the dining and lounge area, has magnificent views over the bay. Four-day cooking classes held annually in May taught by visiting chef.
■ **Open:** Breakfast, lunch and dinner daily
● **Location:** Flights only from Kununurra

THE OUTBACK

Wild, rugged and exceptionally beautiful, the coastline of the south-east region flows on across the Great Australian Bight. Fishing is a major industry, and diving for abalone for export is operated from Esperance. Inland, sheep and wheat continue as the main agricultural ventures, although one farm has diversified and now raises water buffalo and deer.

Sheep, beef and wheat have been traditionally farmed inland for generations, but today, emus and bush tucker also feature as agricultural options.

ESPERANCE

PRODUCERS

TELEGRAPH FARM
John and Gwen Starr
Telegraph Rd, Esperance 6450
Phone: 08 9076 5044
Email: johnstarr@wn.com.au
www.telegraphecofarm.com
Telegraph Farm raises deer and water buffalo. So you appreciate how fine their product is, they serve their home-grown venison and buffalo crumbed steaks and venison burgers to visitors who come on a tour to see the farm's animals and ten hectares of proteas. Coach groups enjoy a buffet lunch. Nutritionally, the meat is very low in fat, and high in protein and Telegraph Game Meats are sold all over the state in half-sides and sides by mail order. Soon the Starrs (and their visitors) may have wine to go with the venison, as 600 vines were planted in 2003.

■ **Open:** 10am–5pm Thurs–Mon, 10am–5pm daily during school holidays
● **Location:** 20km west of Esperance

MOUNT MAGNET

FARMSTAY

WOGARNO STATION
Lesley-Jane Campbell
Wogarno Homestead, Mount Magnet 6638
Phone: 08 9963 5846
Mobile: 0428 635 846
Email: wogarno2@bigpond.com
www.wogarno.com.au
BYO
Cuisine: Bush tucker
Meals included in accommodation tariff
Specialty: Indigenous flavours
Local flavours: Native seeds, seafood from Geraldton, mutton
Extras: Visitors staying overnight have a real experience—silver-service dinner under the stars at Lizard Rock or Wogarno Hill. Wogarno has been operating a station-stay program for the past six years and specialises in promoting and showcasing regional foods to its guests. Lesley-Jane Campbell (LJ) is extremely interested in bush tucker. Her studies have led her to experiment with bush tomatoes, wattleseeds, wild vines, bush foods, nearby Geraldton's fish and crustaceans, Three Rivers beef, Wogarno mutton, and meat from a nearby emu farm. Accommodation is available at the homestead, in the shearers' quarters or camping, but meals are available for all.
● **Location:** 520km from Perth on Great Northern Hwy

WAGIN
🍎 PRODUCERS

CORRALYN EMUS
Arthur and Wendy Pederick
Corralyn
Bolt Rd, Wagin 6315
Phone: 08 9861 1136
Email: corralyn@wn.com.au

Like so many others, in 1991 the Pedericks diversified from sheep and grain growing, choosing to go into emu raising. Now their prime emu meat is mainly exported, but you can learn more about this versatile, high-protein, low-cholesterol red meat that is prepared in licensed abattoirs under strict controls for the domestic and export market. There are tours of the emu farm to see the incubation process from June to November, and you can fall in love with those striped chicks from August to December. Or you can visit the product shop selling emu meat, emu leathergoods, emu eggs, emu oil and other emu oil products.

■ **Open:** 9am–5pm Mon–Sat, by appointment only
● **Location:** 3km north of Wagin, take Tudor St from Wagin, signposted

---

EMU ON THE MENU

In recent years, emu meat has progressed from being illegal in many states, to prized by chefs and appearing on top menus around the country. It is a fascinating and unusual talking point for overseas visitors and, when exported, attracts great interest.

A lean and healthy meat, low in cholesterol, and with a distinctive and palatable flavour, emu meat is gaining popularity. Emu farming has become something of a trendy niche industry, and often an ideal use of land that has struggled to produce more traditional European-style crops, or raise mainstream animals.

Obviously, emus are better adapted to this country's terrain and climate and generally thrive in captivity. The biggest problem is often convincing a nation more used to consuming beef, lamb, pork or chicken that they need to try a new meat.

Because the meat is so low in fat, it is best served rare, but many suggest that the flavour is better if not masked with strong sauces or marinades. Emu meat can be used in a variety of dishes, much as any other meat can be used.

While a variety of cuts have been on the market for some time (usually frozen or cryovac-packed for long shelf life) some producers are now testing new smallgoods on the markets such as terrines, salami, smoked emu meat and prosciutto. Classic cuts include drum cuts, thigh cuts, and forequarter.

Emu meat contains between 470 and 530 kilojoules per 100 grams, well below beef (658 kilojoules) and similar to poultry, venison or rabbit.

AUSTRALIA'S CORAL COAST

So much of these parts of Western Australia appear inhospitable yet, after rain, the desert blossoms, and in spring, these areas are carpeted with pink, white and yellow everlastings.

There is bush tucker too, but you need expert help to find it, just as you would to locate the gold and other minerals that lie below the surface of this red land, just waiting to be discovered.

This is an area of huge cattle stations and long distances between towns. Carnarvon is rich in tropical fruits, especially bananas, as well as seafood, and is the main centre for the Gascoyne region to the south of the Pilbara. The locals call this area 'the outback coast'.

The mid-west is within easy reach of Perth. Its coast has an abundant share of seafoods, including rock lobsters, which have created a major industry in Geraldton and nearby coastal towns.

MARKETS AND EVENTS

✦ GASCOYNE GROWERS MARKETS, CARNARVON
08 9941 1803
8am–12pm Sat, May–Nov

CARNARVON
🍎 PRODUCERS

WESTOBY BANANA PLANTATION
Milena Ritchie, Joyce and Paul Nevill
Robinson St, Carnarvon 6701
Phone: 08 9941 8003
Email: westoby@wn.com.au
BYO
Cuisine: Australian
Mains average: $
Specialty: Banana and mango smoothies
Local flavours: Seafoods, fruit
Extras: Buy bottles of Westoby's tropical fruit wines to enjoy with meals at this relaxing tropical garden and cafe. In season you may purchase bananas, mangoes, pawpaw, avocados, star fruit, eggplant, chillies, tomatoes, watermelon, rockmelon, passionfruit, beans and citrus fruit. Carnarvon is situated at the mouth of the Gascoyne River 'where the desert meets the sea'.
■ **Open:** 10am–4pm Wed–Mon (closed Mon Oct–May), 10am–4pm daily (July school holidays), Tours–11am and 2pm, daily (June–Sept), one tour, daily (Oct–May)
● **Location:** On the left of the main road into Carnarvon

KALBARRI
🍴 DINING

GILGAI TAVERN
Shane Hart (Manager)
Russell Berry, Ross Palmer and Cheyne Fletcher (chefs)

Lot 398, Porter St, Kalbarri 6536
Phone: 08 9937 1083
Email: gilgaitavern@wn.com.au
Licensed
Cuisine: Seafood and steaks
Mains average: $$
Specialty: Guinness and seafood pies
Local flavours: Fish caught locally, delivered from the boat—north-west snapper, Kalbarri pink snapper, dhufish, red emperor, baldchin groper, coral trout, Spanish mackerel
Extras: Outside courtyards and balcony, all with ocean and river aspects, great views of the Murchison River mouth.
- **Open:** Lunch and dinner daily
- **Location:** Off Grey St, in front of the shopping centre

NANSON
DINING

CHAPMAN VALLEY WINES
Robert and Karen Pederick, Rick and Pam Pederick
Lot 14, Howatharra Rd, Nanson 6532
Phone: 08 9920 5148
Email: cvw@westnet.com.au
www.chapmanvalleywines.com.au
Licensed
Cuisine: Local seafood, chicken and yabbies
Mains average: $$
Specialty: Yabby pies
Local flavours: Abrolhos Island fish, yabbies
Extras: Most northerly winery in Western Australia with own yabby dams. Outdoor eating in gazebos with a large lawn area.
- **Open:** 10am–5pm daily, lunch, morning and afternoon tea
- **Location:** Turn onto the Howatharra-Nanson Rd and continue for 3km

---

FISHERIES IN WESTERN AUSTRALIA

Western Australia's coastline measures many thousands of kilometres, with waters rich with dozens of varieties of fin-fish and crustaceans. It is inevitable that fishing is one of the state's major industries.

An estimated fifty per cent of all commercially caught fish products are exported, with the majority being sold to Asian markets.

Aquaculture is a major and much newer branch of fisheries in Western Australia. Lake Argyle in the Kimberley region may soon be used to raise barramundi, aiming at an eventual five-to-ten thousand tonne annual harvest. Research continues into other types of aquaculture, too, including a bizarre project to produce a strain of yabby that only produces male offspring. This is because keeping one-sex yabbies in dams has been shown to result in increased growth rates.

Work is also being carried out to reduce the 'bycatch' of commercial trawlers, which may also catch untargeted fish and marine life when fishing. This environmental program is in line with similar ones in other states.

Recreational fishing is alive and well in Western Australia as it is in other states. A valuable industry, the state's estimated 600 000 recreational fishers generate around $440 million annually.

For more details on any aspect of fishing in Western Australia, contact Fisheries Western Australia (Phone 08 9426 7333 or www.wa.gov.au/westfish).

INDEX

A Rostella Experience, Dilston Tas 265
A Taste of Byron, Byron Bay NSW 109
A Taste of Falls Creek, Falls Creek Vic 342
A Taste of Flinders Island, Whitemark, Flinders Island Tas 262
A Taste of South Australia Wine Tours, Adelaide SA 198
Aarli Bar, Broome WA 418
ABC Cheese Factory, Central Tilba NSW 115–16
Aboriginal Heritage Walk, South Yarra 300
AC Butchery, Leichhardt NSW 12–13
Acacia Ostrich Breeders, Albany WA 400
Ada at Strathvea, Healesville Vic 377
Adelaide SA 195–200
ADELAIDE AND SURROUNDS SA 193
Adelaide Central Market, Adelaide SA 196–7
Adelaide's Top Food and Wine Tours, Adelaide SA 198
Adventure Charters of Kangaroo Island, Kingscote SA 239
The Advocate Coffs Coast Food and Wine Festival, Coffs Harbour NSW 98
The Age Harvest Picnic at Hanging Rock, Woodend Vic 349
Agon Berry Farm, Pooraka SA 212–13
Aintree Almonds and Apiary, Bearii Vic 350
Airlie Beach Qld 188
Albany WA 400–1
Albany Farm Fresh Eggs, Redmond WA 414
Albany Farmers' Market, Albany WA 416
Albany Marron Farm and Nippers Cafe, Kalgan WA 407
The Albert Mill Restaurant and Pizzeria, Nairne SA 212
Albury NSW 110–12
Albury–Wodonga Wine and Food Festival, Albury–Wodonga Vic 342
Albury–Wodonga Wine and Food NSW 112
Aldinga Turkey Kitchen, McLaren Vale SA 209
Alexandra Vic 316
Alexandrina Cheese Co, Mount Jagged SA 211–12
Alice Springs NT 126–7
Allans Flat Vic 325
Allansford Vic 369
Allfresh Seafoods and Poultry, Warrnambool Vic 373
Allsun Farms, Goulburn NSW 38
The Almond and Olive Train, McLaren Flat SA 207–8
almonds
  South Australia
    Adelaide and Surrounds 207–9, 213–14
  Tasmania
    The North-West 288
Alpaca Country Shop and Cafe, Tichborne NSW 61
Alpaca Viande, Gloucester NSW 68–9
alpacas
  New South Wales
    Explorer Country and Riverina 61
    Hunter 68–9
Alpine Valleys Food and Wine Festival, Myrtleford Vic 342
Alstonville NSW 99
Amanda's on the Edge, Pokolbin NSW 74
Ameys Track Blueberries, Foster Vic 359
Amish Country Store, Nabiac NSW 93
Anakie Vic 301

Anchor Farm, Pyengana Tas 270
Andy's Bakery and Cafe, Westbury Tas 274
Ane's Cherrygrove, Young NSW 41
Angas Park Fruit Company Pty Ltd, Chiquita South Pacific Ltd, Angaston SA 217
Angas Park Fruit Company, Irymple Vic 352
Angaston Cottage Industries, Angaston SA 218
Angaston Gourmet Foods Cafe, Angaston SA 219
Angaston SA 217–20
Annapurna Wines and Cafe, Tawonga South Vic 338
Antiene NSW 66
antipasto
  New South Wales
    Blue Mountains and the Hawkesbury 23
    Explorer Country and Riverina 50, 58, 62
    North Coast 89
    South Coast and Illawarra 115
  South Australia
    Eyre Peninsula, Flinders Ranges and the Outback, 234
  Tasmania
    The North 265
  Western Australia
    The South-West 411
Apex Bakery, Tanunda SA 222
Apollo Bay Vic 369
Apple and Grape Harvest Festival, Stanthorpe Qld 187
Appledale Processors Cooperative, Orange NSW 59
apples
  New South Wales
    Blue Mountains and the Hawkesbury 20–1, 27, 28
    Capital Country and the Snowy Mountains 30–1, 33–4, 39
    Explorer Country and Riverina 59–60, 63
    New England and North-West 79, 83
    South Coast and Illawarra 119
  Queensland
    Toowoomba and the Southern Downs 181, 184–5
  South Australia
    Adelaide and Surrounds 199, 206, 215
    Barossa, 217
    Clare Valley and Yorke Peninsula, 226
    Limestone Coast 242, 245
  Tasmania 253
    Hobart and Surrounds 257
    The North 272
    The North-West 290, 292
    The South 275, 280–1, 284, 286
  Victoria
    Bays and Peninsulas 304
    Goldfields 312
    The Grampians 367
    Lakes and Wilderness 322
    Legends, Wine and High Country 329, 337
    Macedon Ranges and Spa Country 343, 346
  Western Australia
    Perth and Surrounds 398
    The South-West 414–15
Apricot Acres Farm, Bindoon WA 392
apricots
  Australian Capital Territory
    Canberra 30

INDEX

New South Wales
  Capital Country and the Snowy Mountains 34, 35, 37, 39
South Australia
  Barossa, 217
  Murray and Riverland 249
Tasmania 253
  Hobart and Surrounds 257
Western Australia
  Perth and Surrounds 392
  The South-West 402
Aqua Blue Marron Farm and Restaurant, Denmark WA 406
Araluen NSW 33
Ararat Vic 310
Arc-En-Ciel Rainbow Trout, Hanging Rock NSW 80
Archvale Trout Farm, Lithgow NSW 26
Arcoona Heritage Guesthouse, Deloraine Tas 264
Aridgold Farm, Alice Springs NT 126
Armidale NSW 79
Arriga Park Farmstay, Mareeba Qld 158
As Nature Intended, Belconnen ACT 31
Ascot Hotel, Rockhampton Qld 150
Ash Island NSW 66
Ashcrofts Restaurant, Blackheath NSW 22
Ashgrove Cheese, Elizabeth Town Tas 289
Asian Foods Australia, Cairns Qld 151
asparagus
  New South Wales
    Explorer Country and Riverina 49
    New England and North-West 83
  Western Australia
    Perth and Surrounds 392, 394
    The South-West 410, 417
Astrid Oysters, Ceduna SA 231
At The Star Restaurant, Narrandera NSW 59
Atherton Food Festival, Atherton Qld 166
Atherton Maize Festival, Atherton Qld 166
Atherton Market, Atherton Qld 166
Auburn SA 225
Auldstone Cellars, Taminick Vic 338
Aussie Fruit Boy, Mylor SA 212
AUSTRALIA'S CORAL COAST WA 426-7
AUSTRALIA'S NORTH-WEST WA 418-23
Australian Arid Lands, Port Augusta SA 229
Australian Country Life Festival, Great Western Vic 315
Australian Inland Wine Show, Swan Hill Vic 356
Australian Italian Festival, Ingham Qld 192
Australian Nougat Company, Eumundi Qld 173
Australian Oyster Festival, Ceduna SA 231
Australian Regional Food Store and Cafe, Pokolbin NSW 74-5
Avenel Vic 316
Avenel Produce Market, Avenel Vic 320
Avenue SA 241
Avenue Emus, Avenue SA 241
Avoca, Mildura Vic 353
Avoca Vic 310, 315
Avoca Beach NSW 44
The Avocado Farm Stall, Bobs Farm NSW 67
Avocado Festival, Duranbah NSW 109
avocados
  New South Wales
    Explorer Country and Riverina 49
    Hunter 67
    North Coast 85, 87-8
    Northern Rivers 106
    Outback and the Murray 111
    South Coast and Illawarra 122

Queensland
  Bundaberg, Fraser Coast and South Burnett 143
  Far North 160
  Sunshine Coast 173
  Townsville, Mackay and the Whitsundays 191
South Australia
  Adelaide and Surrounds 206, 209
Ayr Qld 188

Bacchus Marsh Vic 343
Back of Bourke Fruits, Bourke NSW 111
bacon *see* smoked smallgoods
Bad Habits Cafe, Daylesford Vic 344
Bairnsdale Farmers Market, Bairnsdale Vic 324
Baldivis WA 401-2
Balfours frog cakes 193
Balhannah SA 199-200
Ballandean Qld 181
bananas
  New South Wales 9
    North Coast 85, 87-8, 91, 93
    Northern Rivers 106
    South Coast and Illawarra 120, 122
  Northern Territory 125
  Queensland 186
    Far North 155, 159-60, 165
    Townsville, Mackay and the Whitsundays 191
  Western Australia
    Australia's Coral Coast 426
Bangalow NSW 99
The Bank Guest House and Tellers Restaurant, Wingham NSW 90, 97
The Bank Restaurant and Mews, Beechworth Vic 327
Bannisters Restaurant, Mollymook NSW 118
Banora Point Farmers' Market, Banora NSW 109
Barbushco Pty Ltd, Lorne NSW 91-2
Barilla Bay Tasmania, Cambridge Tas 276
Bark Hut Tourism Centre, Anaburroo NT 126-7
barley
  New South Wales
    Explorer Country and Riverina 54
  Victoria 295
  Western Australia 387
The Barn, McLaren Flat SA 208
Barnett's Rainbow Reach Oysters, South West Rocks NSW 95
BAROSSA SA 193
Barossa Daimler Tours, Rosedale SA 221
Barossa Farmers' Market, Angaston SA 217
Barossa Farmers' Market, Angaston SA 224
Barossa Olives, Truro SA 224
Barossa Valley Cheese Co, Angaston SA 218
Barossa Valley Produce, Tanunda SA 221-2
Barossa Vines Cellar Door, Tanunda SA 222-3
Barra Barra Pty Ltd, Kununurra WA 422
Barr-Vinum, Angaston SA 219
Bartholomew's Meadery, Denmark WA 405
Baskerville WA 391
Bass Vic 357
BASS STRAIT ISLANDS Tas 260-2
Batemans Bay NSW 113
Bathurst NSW 48
Bathurst Farmers' Markets, NSW 65
Bathurst Organic Markets, NSW 65
Batlow NSW 33-4
Batlow Fruit Cooperative Ltd, Batlow NSW 34
Battunga Country Growers' Market, Macclesfield SA 216
Bay of Fires Winery Cellar Door, Pipers River Tas 270

429

INDEX

Bayou Bill, Brooklyn NSW 44
BAYS AND PENINSULAS Vic 301-9
Bayside Meats, Sandy Bay Tas 256
Bazzani, Bendigo Vic 311
BD Farm Paris Creek Pty Ltd, Meadows SA 210
Beaconsfield Vic 357
Bearii Vic 350
Beaumont's Cafe, Rutherglen Vic 336
Beechworth Vic 325-7, 342
Beechworth Bakery, Beechworth Vic 325
Beechworth Bakery, Echuca Vic 351
Beechworth Country Craft Market, Beechworth Vic 342
Beechworth Harvest Festival, Beechworth Vic 342
Beechworth Provender, Beechworth Vic 325-6
beef
   Australian Capital Territory
      Canberra 32
   New South Wales 9
      Capital Country and the Snowy Mountains 36
      Explorer Country and Riverina 51-3, 55, 57, 63-4
      Hunter 66, 69, 70, 74, 76, 78
      New England and North-West 83
      North Coast 87-9, 92, 95, 97
      Northern Rivers 99-100, 102, 107
      Outback and the Murray 110
      South Coast and Illawarra 114, 121, 123
      Sydney and Southern Highlands 13, 15, 17
   Northern Territory
      Central Australia 128-9
      Top End 131
   Queensland 135
      Bundaberg, Fraser Coast and South Burnett 140-1, 143, 146
      Capricorn, Gladstone and the Outback 147-8, 150
      Far North 152, 154, 163
      Sunshine Coast 172, 173
      Toowoomba and the Southern Downs 181, 184, 186
      Townsville, Mackay and the Whitsundays 188
   South Australia 193
      Adelaide and Surrounds 201, 203, 210-11
      Clare Valley and Yorke Peninsula, 229
      Eyre Peninsula, Flinders Ranges and the Outback, 233
      Limestone Coast 242, 246
   Tasmania
      Bass Strait Islands 262
      Hobart and Surrounds 256
      The North 263, 268
      The North-West 288, 290, 293
      The South 275, 287
   Victoria
      Bays and Peninsulas 306
      Goldfields 310
      Goulburn Murray Waters 318
      The Grampians 365
      The Great Ocean Road 370
      Legends, Wine and High Country 328, 334
      Murray and the Outback 352-3
   Western Australia
      Australia's North-West 420
      The Outback 424
      Perth and Surrounds 389, 391, 393, 397
      The South-West 410, 412-13, 417
Beef on Barker, Casino NSW 109
Beenleigh Cane Festival, Beenleigh Qld 170
beer
   New South Wales
      New England and North-West 83

North Coast 96
South Coast and Illawarra 123
South Australia
   Adelaide and Surrounds 198, 203, 214
Tasmania
   Hobart and Surrounds 258-9
   The North 266, 268-9
   The South 277
Victoria
   Legends, Wine and High Country 327
   Phillip Island and Gippsland 363
Western Australia
   Australia's North-West 419-20, 423
Beerenberg Farm, Hahndorf SA 202
Bega NSW 113-14
Bega Cheese and Bega Cheese Heritage Centre, Bega NSW 113
Bega Dried Foods, Tathra NSW 120
Bega Valley Berry Wines, Cobargo NSW 116
Belford NSW 66
Belhus WA 392
Bella Restaurant, Chirnside Park Vic 375
Bellarine Estate, Bellarine Vic 301
Bellarine Vic 301
Bellbrae Harvest, Bellbrae Vic 302
Bellbrae Vic 302
Belli Park Qld 171
Bellingen NSW 85, 98
Bellingen Organic Markets, Bellingen NSW 98
Bellowing Bull Restaurant, Wollongbar NSW 107
Belltower Restaurant, Kingaroy Qld 143-4
Belstack Strawberry Farm, Kialla West Vic 317
Ben Furney Flour Mills, Dubbo NSW 50
Benalla Vic 310
Bendigo Vic 311, 315
Bendigo Easter Wine Festival, Castlemaine Vic 315
Bendigo Farmers' Market, Bendigo Vic 315
Benedictine Community of New Norcia, New Norcia WA 397
Bentley NSW 99-100
Berardo's Restaurant and Bar, Noosa Heads Qld 176
Bermagui NSW 114
Bermagui Seafood Supplies, Bermagui NSW 114
Berri SA 249-50
Berri Direct, Berri Ltd, Berri SA 249
Berridale NSW 34
Berriedale Tas 275
berries
   Australian Capital Territory
      Canberra 31
   New South Wales 9
      Blue Mountains and the Hawkesbury 20
      Capital Country and the Snowy Mountains 34, 36, 37, 41
      Explorer Country and Riverina 53, 61
      Hunter 75
      New England and North-West 79, 81-2
      North Coast 96
      South Coast and Illawarra 116, 121
      Sydney and Southern Highlands 15, 17, 18
   Queensland
      Toowoomba and the Southern Downs 181, 183-5
   South Australia 193
      Adelaide and Surrounds 210-12, 215
      Kangaroo Island, 240
   Tasmania 253
      Hobart and Surrounds 257-8
      The North 265

INDEX

The North-West 289
The South 276, 278, 282-5
Victoria
  Bays and Peninsulas 303, 307-8
  Goulburn Murray Waters 318-19
  The Grampians 367
  Lakes and Wilderness 321-2
  Legends, Wine and High Country 328, 330, 332, 335
  Macedon Ranges and Spa Country 347
  Phillip Island and Gippsland 358
  Western Australia 387
  The South-West 400, 409-10
The Berry Farm, Margaret River WA 409-10
Berry Good Cafe and Berry Farm, Drouin West Vic 359
Berry Sweet Australia and Tumbarumba Blueberries, Shepparton Vic 318-19
Berry Woodfired Sourdough Bakery, Berry NSW 114-15
Berry World, Timboon Vic 372
Berry NSW 114-15
Biboohra Qld 151
Big Apple Tourist Orchard, Bacchus Marsh Vic 343
The Big Apple-Spreyton Fruit and Meat Market, Spreyton Tas 292
The Big Banana, Coffs Harbour NSW 87
The Big Berry, Hoddles Creek Vic 378
Big Fish Restaurant, Shoal Bay NSW 95
The Big Orange, Berri SA 249
The Big Peanut, Tolga Qld 164
The Big Pineapple-Sunshine Plantation, Woombye Qld 178-9
Biji Bush Qubes, Hillston NSW 53
Billabong Boat Cruises, Longreach Qld 149
Billabong Produce, Jerilderie NSW 112
Billabong Restaurant, Taylors Arm NSW 95-6
Biloela Qld 147
Bilpin NSW 20-1, 27
The Bilpin Bite, Bilpin NSW 27
Bilpin Fruit Bowl, Bilpin NSW 21
Bilpin Springs Orchard, Bilpin NSW 20
Bimbadgen Blues, Pokolbin NSW 78
Binalong NSW 34
Bindoon WA 392
Birchs Bay Tas 275
Birdwood SA 200
Birdwood Wine and Cheese Centre, Birdwood SA 200
Birregurra Vic 370
biscuits *see* cakes, biscuits and pastries
Bistro Mont, Bowral NSW 15
Bit O'Heaven Orchard, Young NSW 41
Bittern Vic 302
Bittern Cottage, Bittern Vic 302
The Black Swan, Binalong NSW 34
blackberries
  Australian Capital Territory
    Canberra 31
  New South Wales
    Blue Mountains and the Hawkesbury 27
    Capital Country and the Snowy Mountains 34-5, 41
    Explorer Country and Riverina 53, 61
    New England and North-West 81
    South Coast and Illawarra 121
    Sydney and Southern Highlands 18
  Queensland
    Toowoomba and the Southern Downs 184
  South Australia
    Adelaide and Surrounds 212
  Victoria
    Goulburn Murray Waters 317

Legends, Wine and High Country 330
  Macedon Ranges and Spa Country 347
Blackbutt Country Markets, Blackbutt Qld 145
Blackheath NSW 21-2
Blackheath Continental Deli, Blackheath NSW, 21-2
Blackwood Cafe, Nannup WA 412
Blue Bell Inn, Sorell Tas 285
The Blue Fig, Sawtell NSW 94
BLUE MOUNTAINS 9, 20-9, 44
Blue Mountains Honey Company, Penrith NSW 12
Blue Mountains Paradise, South Bowenfels NSW 28
Blue Ox Berries, Oxley Vic 335
Blue Poles Cafe and Gallery, Byabarra NSW 87
Blue Pyrenees Estate, Avoca Vic 310
Blue Seafood and Char-grill, Mildura Vic 353-4
Blue View Olives, Landsborough Vic 367
Blue Wren Restaurant, Mudgee NSW 57
blueberries
  New South Wales
    Capital Country and the Snowy Mountains 41
    Explorer Country and Riverina 53
    New England and North-West 81
    Northern Rivers 99, 103
    South Coast and Illawarra 116, 121
    Sydney and Southern Highlands 18
  South Australia
    Adelaide and Surrounds 211-12, 215
  Tasmania
    The South 283
  Victoria
    Bays and Peninsulas 303
    Goulburn Murray Waters 318-19
    Legends, Wine and High Country 330, 335
    Macedon Ranges and Spa Country 347
    Phillip Island and Gippsland 359
  Western Australia
    The South-West 401
The Blueberry Patch, Mount Compass SA 211
Blyth SA 226
boab fruit
  Western Australia
    Australia's North-West 422-3
The Boathouse, Lakes Entrance Vic 322
Bobin NSW 85-6
Bobs Farm NSW 67
Bodalla NSW 115
Bondi Mall Organic Markets, Bondi Junction NSW 19
Bonville NSW 86
Boomerang by the Sea, King Island Tas 261
Boorhaman Vic 327
Boorowa NSW 35
Boort Vic 350
Borenore NSW 48
Boronia Farmers' Market, Boronia Vic 299
Boroondara Farmers' Market, East Hawthorn Vic 299
Borrodell on the Mount, Orange NSW 59-60
Bottle Tree Hill Organics, Murgon Qld 144-5
Bough House Restaurant, Uluru-Kata Tjuta National Park NSW 129
Boulevard Market, Homebush NSW 13
Boundary Park Olives, Tintinara SA 247
Bounty of the Sea Festival, Forster-Tuncurry NSW 98
Bourke NSW 111
Bowen Farmstay, Bowen Qld 188
Bowen Mountain NSW 22-3
The Bower Room, Byron Bay NSW 102
Bowra Hotel Bistro, Bowraville NSW 86-7

431

INDEX

Bowral Farmers' Market, Bowral NSW 19
Bowral NSW 15–16, 19
Bowraville NSW 86–7
Boyanup Farmers' Market, Boyanup WA 416
Boyanup WA 402
Boynton's Winery and Restaurant, Porepunkah Vic 336
Bradleys Head Bushfood Tours, The Rocks NSW 14
Braidwood NSW 35
The Bramble Patch Berry Gardens, Stanthorpe Qld 183–4
Bramblewood Fruit Wines, Belford NSW 66
Brambuk Living Aboriginal Cultural Centre, Halls Gap Vic 365–6
Branxton-Greta Farmers Market, Branxton NSW 78
Brass Monkey Season, Southern Downs Qld 187
bread
  Australian Capital Territory
    Canberra 31–3
  New South Wales
    Blue Mountains and the Hawkesbury 24, 26, 29
    Capital Country and the Snowy Mountains 34, 38, 42
    Central Coast 46
    Explorer Country and Riverina 50, 60–2
    Hunter 69, 74
    New England and North-West 81
    North Coast 94
    Northern Rivers 106
    South Coast and Illawarra 114–15, 117
  Northern Territory
    Central Australia 128
  South Australia
    Adelaide and Surrounds 196, 201, 203, 206, 211
    Barossa, 217, 219, 222–3
    Clare Valley and Yorke Peninsula, 228, 230
    Kangaroo Island, 238–9
  Tasmania
    The North 263, 272, 274
    The North-West 292
    The South 278, 287
  Victoria
    Bays and Peninsulas 302–3, 308
    Goldfields 310–11
    The Great Ocean Road 371
    Lakes and Wilderness 323
    Legends, Wine and High Country 325, 335–6
    Murray and the Outback 350–1, 353
    Phillip Island and Gippsland 361
    Yarra Valley and the Dandenong Ranges 375
  Western Australia
    Australia's North-West 422
    Perth and Surrounds 388, 397
The Breakout Brasserie, Cowra NSW 49
Briars Country Lodge and Inn, Bowral NSW 15
Brickworks Market, Torrensville SA 216
Bridgart's Orchard and Gourmet Preserves, Denmark WA 405
Bridge View Inn Bakery Cafe, Rylstone NSW 61
Bridgetown WA 402
Bridgewater SA 200
Bright Vic 327–8
Bright Berry Farm and Mt Buffalo Vineyard, Eurobin Vic 330
BRISBANE AND SURROUNDS Qld 135–8
Brisbane Food & Wine Month, Brisbane Qld 138
Brisbane Masterclass, Brisbane Qld 138
Bristowe Farm Hazelnuts, Mudgee NSW 55
Brittle Jacks, Orange NSW 60
Broke NSW 67

Brook Eden Vineyard, Lebrina Tas 269
Brooklyn NSW 44–5
Broome WA 418–21
Broome Courthouse Markets, Broome WA 418
Broome's Own, Broome WA 418
Brown Brothers Wine and Food Weekend, Milaw Vic 342
Brunswick WA 402
Bruny Island Tas 275–
Bryants Heritage Bakery Cafe, Goulburn NSW 37
Buckaneers for Seafood, Wynyard Tas 294
Buderim Ginger Ltd, Yandina Qld 179
buffalo
  New South Wales
    New England and the North-West 83, 90
    Outback and the Murray 110
  Northern Territory 125
    Central Australia 128
    Top End 131–2
  Queensland
    Far North 152–3
  South Australia
    Clare Valley and Yorke Peninsula, 227
    Eyre Peninsula, Flinders Ranges and the Outback, 234
  Victoria
    The Great Ocean Road 371, 374
    Yarra Valley and the Dandenong Ranges 382
  Western Australia
    Perth and Surrounds 389
Buffalo Brewery, Boorhaman Vic 327
Bulga NSW 67
Bull and Barrel Festival, Dardanup WA 416
Bullocks Wood Oven Eatery, Mount Barker SA 211
Bundaberg Qld 139–40
BUNDABERG, FRASER COAST AND SOUTH BURNETT Qld 139–46
Bundaberg Multicultural Food and Wine Festival, Bundaberg Qld 145
Bundaberg Rum Visitor Centre, Bundaberg Qld 139–40
Bundanoon NSW 16, 19
Bundanoon Village Nursery, Bundanoon NSW 16
Bundoora Park Farmers' Market, Bundoora Vic 299
Bunnyconnellen Olive Grove and Vineyard, Crows Nest Qld 182
Bunya Forest Gallery and Tearoom, Bunya Mountains Qld 181–2
Bunya Mountains Qld 181–2
Burnie Tas 288
Burnie Farmers' Market, Burnie Tas 294
Burning Beats Cafe, Kingaroy Qld 144
Buronga NSW 111
Burra SA 226
Burrawang NSW 16–17
bush foods
  Australian Capital Territory
    Canberra 32
  New South Wales
    Explorer Country and Riverina 53
    Hunter 66
    New England and North-West 80
    North Coast 85–6, 91–2
    Northern Rivers 104, 108
    Sydney and Southern Highlands 14, 16
  Northern Territory 125
    Central Australia 127–9
  Queensland
    Bundaberg, Fraser Coast and South Burnett 140
    Capricorn, Gladstone and the Outback 149

INDEX

Far North 152-4, 161
Toowoomba and the Southern Downs 181-2
South Australia
  Adelaide and Surrounds 197, 214
  Eyre Peninsula, Flinders Ranges and the Outback, 231, 233
Tasmania 253
Victoria
  Goldfields 312
  The Grampians 365-6
  Melbourne 300
Western Australia
  Australia's Coral Coast 426
  Australia's North-West 420-3
  The Outback 424
Bush Rock Cafe, Blackheath NSW 22
Bushranger Hotel, Collector NSW 35
butchers see meat
Butt's Gourmet Smokehouse, Albury NSW 110
Buxton Trout and Salmon Farm, Buxton Vic 375
Buxton Vic 375
Byabarra NSW 87
Byrne's Mill Restaurant, Queanbeyan NSW 40
Byron Bay Wholly Smoked Gourmet Foods, Byron Bay NSW 100
Byron Farmers' Market, Byron Bay NSW 109

Byron Bay NSW 100-2, 109
Cable Beach WA 421
Cabonne Country Honey, Mullion Creek NSW 58
Caboolture Qld 171
Cactus Cafe and Gallery, Wellington NSW 64
Cadbury Factory Tours, Claremont Tas 278
Cadi Jam Ora: First Encounters, Sydney NSW 14
Cafe Bon Ton, Leura NSW 25
Cafe Carlotta, Broome WA 418
Cafe Coco, Yamagen and Kingsfords Restaurants (Cairns International Hotel), Cairns NSW 152
Cafe de Railleur and Ye Olde Bicycle Shoppe, Bundanoon NSW 16
Cafe Sofala, Sofala NSW 62
Cafe Y, Lyndoch SA 220
Cairns Qld 151-4, 166
cakes, biscuits and pastries
  Australian Capital Territory
    Canberra 32
  New South Wales
    Blue Mountains and the Hawkesbury 23, 27
    Capital Country and the Snowy Mountains 35, 37
    Explorer Country and Riverina 49, 51, 54, 58
    Hunter 69, 77
    New England and North-West 79, 81, 82-3
    North Coast 89, 98
    Northern Rivers 100
    Outback and the Murray 111
    South Coast and Illawarra 114-15
  Northern Territory
    Central Australia 126
  Queensland
    Sunshine Coast 172
  South Australia
    Adelaide and Surrounds 206
    Barossa, 218-19, 222-3
    Clare Valley and Yorke Peninsula, 230
    Limestone Coast 241, 248
    Murray and Riverland 250
  Tasmania
    Bass Strait Islands 260, 262
    Hobart and Surrounds 255
    The North 272, 274
    The North-West 290, 292
  Victoria
    Bays and Peninsulas 308
    Goulburn Murray Waters 319
    The Grampians 367-8
    Legends, Wine and High Country 325-6, 337
    Melbourne 299
    Murray and the Outback 351
    Phillip Island and Gippsland 358, 361
    Yarra Valley and the Dandenong Ranges 384
  Western Australia
    Perth and Surrounds 388
    The South-West 406
Caldermeade Farm Cafe, Caldermeade Vic 357-8
Caldermeade Vic 357-8
Caloundra Qld 147, 171-2
Cambinata Extravaganza-Silver Service Dinner in the Shearing Shed, Kukerin WA 416
Cambridge Tas 276-8
Camden Produce Market, Camden NSW 27
camel
  Northern Territory 125
    Central Australia 127, 129
    Top End 131
Camelot Lavender Farm, Wallarobba NSW 77
Campaspe House, Woodend Vic 348
Campbell's Soups Australia, Lemnos Vic 317
CANBERRA ACT 9, 30-3, 35, 37
Canberra Fresh Fruit Market, Canberra City ACT 31
Canberra Growers' & Produce Market, Symonston ACT 43
Canberra Region Farmers' Market, Epic ACT 43
Candy Cow, Cowaramup WA 403
Canecutters Restaurant, Palm Cove Qld 161
Cape Lavender, Willyabrup WA 415
Cape Trib Exotic Fruit Farm and B&B, Cape Tribulation Qld 154
Cape Tribulation Qld 154
Capel's Bowral, NSW 15
Capparis, Gloucester NSW 69
CAPRICORN, GLADSTONE AND THE OUTBACK Qld 147-50
Caprino Farm, Herne Hill WA 395
Captain Starlight's, Paddy Dodger's Bistro, Flinders Range via Hawker SA 233
The Captains Catch, St Helens Tas 273
Cardinia Ranges Farmers' Market, Pakenham Vic 364
Carnarvon WA 426
Carnivale-Longest Lunch and Harvest of the Coral Sea, Port Douglas Qld 166
Carobana Confectionery, Korora NSW 91
Casalare Speciality Pasta Pty Ltd, Mirboo North Vic 362
Cascade Brewery Tours, South Hobart Tas 258-9
Cassilis Vic 321
Castlemaine Vic 312, 315
Castlereagh NSW 23
The Castle-Villa by the Sea, Phillip Island Vic 358
Catania Fruit Salad Farm Tours, Hanwood NSW 52
Catch a Crab Tour, Tweed Heads NSW 107
Cato's Restaurant and Bar, Noosa Heads Qld 177
Cavese Trattoria, Berry NSW 115
Caxton Street Seafood and Wine Festival, Brisbane Qld 138
The CB Alexander Agricultural College at Tocal, Paterson NSW 73-4
Ceduna SA 231-2
Ceduna Bakery Coffee Lounge, Ceduna SA 231-2

INDEX

Ceduna Clearwater Oysters, Ceduna SA 231
Ceduna Oyster Bar, Ceduna
The Cellar Restaurant, Pokolbin NSW 75-6
CENTRAL AUSTRALIA NT 126-30
CENTRAL COAST NSW 44-7
Central Queensland Multicultural Fair, Rockhampton Qld 150
Central Tilba NSW 115-16
Central Victoria Farmers' Market, Harcourt Vic 315
Central Victorian Yabby Farm, Heathcote Vic 346
Cha Cha Cha, Nundle NSW 82
Chapman Valley Wines, Nanson WA 427
Charella Farmstead Goat Dairy, Mudgeeraba Qld 168
Charlesworth Nuts, Marion SA 195
Charters Restaurant, Broome WA 418
The Chase Cafe, Flinders Chase SA 236-7
Chateau Barrosa, Lyndoch SA 220
Chatswood Organic Market, Chatswood NSW 19
Chatsworth Qld 172
cheese
  Australian Capital Territory
    Canberra 32
  New South Wales 9
    Blue Mountains and the Hawkesbury 22-4
    Capital Country and the Snowy Mountains 33, 35, 36, 38-40
    Explorer Country and Riverina 49-50, 54-5, 62
    Hunter 69, 72-4
    North Coast 85, 97
    Northern Rivers 105
    South Coast and Illawarra 113, 115-16
    Sydney and Southern Highlands 15, 17
  Queensland
    Bundaberg, Fraser Coast and South Burnett 142-4
    Far North 152, 156, 159
    Gold Coast 168
    Sunshine Coast 173-4, 178
  South Australia 193
    Adelaide and Surrounds 196, 201, 204, 206, 208, 210-15
    Barossa, 217-21, 223
    Kangaroo Island, 236-7, 239-40
    Limestone Coast 243-4
  Tasmania
    Bass Strait Islands 260-1
    Hobart and Surrounds 258
    The North 263-7, 269-70
    The North-West 288-9
    The South 275-8, 283-4, 287
  Victoria
    Melbourne 298-9
    Bays and Peninsulas 302, 306-7
    Goldfields 311
    The Great Ocean Road 369, 373-4
    Lakes and Wilderness 324
    Legends, Wine and High Country 326-8, 334-41
    Macedon Ranges and Spa Country 347-8
    Phillip Island and Gippsland 357, 360-4
    Yarra Valley and the Dandenong Ranges 375-8, 382-5
  Western Australia
    Perth and Surrounds 388-9, 393-9
    The South-West 403, 406-7, 409
The Cheese Factory Meningie's Museum Restaurant, Meningie SA 231
The Cheese Master Speciality Cheeses, Mount Gambier SA 244
The Cheese Store at Bowral 15

Cheese World, Allansford Vic 369
Cheesefreaks and Yarra Valley Fudge, Yarra Glen Vic 383-4
cherries
  New South Wales 9
    Blue Mountains and the Hawkesbury 20, 25
    Capital Country and the Snowy Mountains 33-4, 37, 41-2
    Explorer Country and Riverina 60, 63
    South Coast and Illawarra 116
  Queensland
    Toowoomba and the Southern Downs 182
  South Australia
    Adelaide and Surrounds 199, 205, 208, 212
  Tasmania
    The South 284
  Victoria
    Bays and Peninsulas 306
    Lakes and Wilderness 322
    Legends, Wine and High Country 328-30, 335
    Macedon Ranges and Spa Country 347
  Western Australia
    Perth and Surrounds 393, 398
The Cherry Patch, Glen Aplin Qld 182
Cherrybrook Cherry Farm, Mt Bruno Vic 335
Cherryhaven Orchards, Young NSW 41
Cheshunt Vic 329
chestnuts
  New South Wales
    Blue Mountains and the Hawkesbury 21
    Capital Country and the Snowy Mountains 33
    Explorer Country and Riverina 55, 60
  South Australia
    Adelaide and Surrounds 203-4
  Victoria
    Legends, Wine and High Country 326, 330
Chez Pok, Pokolbin NSW 75
chicken *see* poultry
Childers Qld 140
Childers Vic 358
Childers Multicultural Food, Wine and Arts Festival, Childers Qld 145
chillies
  Australian Capital Territory
    Canberra 30, 32
  New South Wales
    Blue Mountains and the Hawkesbury 21
    Explorer Country and Riverina 59
    Northern Rivers 99
    Outback and the Murray 112
  Queensland
    Sunshine Coast 180
  South Australia
    Adelaide and Surrounds 206
  Victoria
    Goldfields 310
  Western Australia
    The South-West 405
Chiltern Vic 329
Chiltern Honey Farm, Chiltern Vic 329
Chirnside Park Vic 375
Chittering WA 392
chocolate *see also* confectionary
  New South Wales
    Hunter 77
    North Coast 89, 93, 96
  Queensland
    Brisbane and Surrounds 136-7

INDEX

Far North 151–2
Gold Coast 169
South Australia
　Adelaide and Surrounds 197, 207, 216
　Clare Valley and Yorke Peninsula, 228
　Kangaroo Island, 237–8
　Limestone Coast 241, 248
　Murray and Riverland 250
Tasmania
　The North 267
　The North-West 290
　The South 278
Victoria
　Goldfields 314
Western Australia
　Perth and Surrounds 388, 393
Chocolate Sensations, Cairns Qld 151–2
Chris's Beacon Point Restaurant and Villas, Apollo Bay Vic 370
Christmas Hills Raspberry Farm, Elizabeth Town Tas 289
chutneys and pickles
　Australian Capital Territory
　　Canberra 32
　New South Wales
　　Blue Mountains and the Hawkesbury 24
　　Capital Country and the Snowy Mountains 35, 41–2
　　Explorer Country and Riverina 62
　　Hunter 67, 70
　　New England and North-West 83
　　North Coast 91, 94
　　Northern Rivers 100, 104
　Northern Territory
　　Central Australia 130
　Queensland
　　Far North 158
　　Toowoomba and the Southern Downs 182, 183, 185
　South Australia
　　Barossa, 218, 223
　　Clare Valley and Yorke Peninsula, 227–8
　　Eyre Peninsula, Flinders Ranges and the Outback, 234
　Tasmania
　　Bass Strait Islands 260–1
　　Hobart and Surrounds 256–7
　　The South 286–7
　Victoria
　　Melbourne 299
　　Bays and Peninsulas 302
　　Legends, Wine and High Country 326, 333–5
　　Macedon Ranges and Spa Country 344
　　Murray and the Outback 351–2
　　Phillip Island and Gippsland 364
　　Yarra Valley and the Dandenong Ranges 378
　Western Australia
　　Australia's North-West 422
　　The South-West 405, 415
cider
　Australian Capital Territory
　　Canberra 30, 32
　New South Wales
　　Blue Mountains and the Hawkesbury 21
　　Explorer Country and Riverina 60
　Queensland
　　Toowoomba and the Southern Downs 185
　South Australia
　　Clare Valley and Yorke Peninsula, 226
citrus *see also by type of fruit*
　New South Wales

Blue Mountains and the Hawkesbury 20, 23
Explorer Country and Riverina 48, 52
North Coast 88, 96
Outback and the Murray 111
Queensland
　Townsville, Mackay and the Whitsundays 191
South Australia
　Adelaide and Surrounds 201
　Murray and Riverland 249
Western Australia
　Australia's Coral Coast 426
　Perth and Surrounds 392
The Citrus Shop, Mildura Vic 352–3
Clare SA 227
CLARE VALLEY AND YORKE PENINSULA SA 193
Claremont Tas 278
Clearstream Olive Farm and Woolshed Cafe, Taggerty Vic 338
Clearview Farm Retreat, Warragul Vic 363
Cleaver's–The Organic Meat Company, Neutral Bay NSW 13
Clifford's Honey Farm, Kingscote SA 238
Cliffy's Emporium, Daylesford Vic 344
The Clipper Restaurant and Bar, Airlie Beach Qld 188
Clovely Estate, Moffatdale Qld 144
Cloverdene Dairy, Karridale WA 408
Clyde River Berry Farm, South Brooman NSW 121
Coal Valley Vineyard, Cambridge Tas 277
Coates Quality Smallgoods, Forth Tas 290
Cobargo NSW 116
Cobourg Peninsula NT 131
Cobram Vic 350–1
The Cockle Shuffle, SA 202
coffee
　New South Wales 9
　　Explorer Country and Riverina 51
　　North Coast 91
　　Northern Rivers 103, 106, 108
　Northern Territory
　　Top End 133
　Queensland 135
　　Brisbane and Surrounds 137
　　Far North 156–8, 164
　　Gold Coast 168
　South Australia
　　Limestone Coast 247
　Victoria
　　Yarra Valley and the Dandenong Ranges 381
The Coffee Works Cafe, Smithfield Qld 164
The Coffee Works Mareeba, Mareeba Qld 158
Coffin Bay
Coff's Coast Growers' Market, Coffs Harbour NSW 98
Coffs Harbour NSW 87–8, 98
Coffs Harbour Fishermen's Co-op, Coffs Harbour NSW 87
Coldstream Vic 376
Colemans at Kilikanoon, Penwortham SA 228
Coles Bay Tas 263
Colin James Fine Foods, Maleny Qld 174–5
Collector NSW 35–6
Colliers Chocolates, Sutton Grange Vic 314
Collingwood Childrens' Farm and Farmers' Market, Abbotsford Vic 299
Collits' Inn, Hartley Vale NSW 24
Comboyne NSW 88
Come 'N' Get It Restaurant, Merimbula NSW 118
Comedy and Food Festival, Biloela Qld 150
Commercial Fishermen's Cooperative Ltd, Tacoma NSW 46

435

INDEX

Commercial Fishermen's Cooperative Ltd, Gorokan NSW 45
Commercial Fishermen's Cooperative Ltd, Wickham NSW 97
confectionary see also chocolate; licorice
  New South Wales
    Explorer Country and Riverina 53-4
    Hunter 73, 77
    North Coast 91, 96
  Queensland
    Bundaberg, Fraser Coast and South Burnett 140
    Sunshine Coast 173, 179
  South Australia
    Adelaide and Surrounds 195-6
    Barossa, 217
  Tasmania 253
    Hobart and Surrounds 254-5, 257
    The North 266
  Victoria
    Murray and the Outback 350, 352
    Yarra Valley and the Dandenong Ranges 380, 383
  Western Australia
    Perth and Surrounds 395
    The South-West 403
Congewai NSW 67-8
The Coniston Dining Room, Romsey Vic 347
Conondale Qld 172
Conservation Hut Cafe, Wentworth Falls NSW 29
cooking classes and schools
  Australian Capital Territory
    Canberra 31-2
  New South Wales
    Blue Mountains and the Hawkesbury 28, 29
    North Coast 97
  Queensland
    Sunshine Coast 180
  South Australia
    Clare Valley and Yorke Peninsula, 228
    Limestone Coast 247
  Tasmania
    The North 268
  Victoria
    Phillip Island and Gippsland 361
  Western Australia
    Australia's North-West 419
Cooking Coordinates, Belconnen ACT 31-2
Cooktown Market, Cooktown Qld 166
Cooktown-Rossville Market, Cooktown Qld 166
Coolabine Farmstead Goats Cheeses, Kenilworth Qld 173-4
Coolac Festival of Fun Showcase, Coolac NSW 43
Cooma NSW 36
The Coonawarra Lavender Estate, Coonawarra SA 241
Coonawarra SA 241
Cooper Creek Homestay, Innamincka SA 233
Coopers Brewery Limited, Regency Park SA 198
The Coorong Fisherman, Meningie SA 243
Coowinga Qld 148
Copley SA 232
Coriole Vineyards, McLaren Vale SA 208
The Cork Street Cafe, Gundaroo NSW 37-8
Corowa Farmers' Market, Corowa Vic 342
Corralyn Emus, Wagin WA 424
Country Cuisine (Australia) Pty Ltd, Daylesford Vic 343-4
Country Fresh Eggs, Two Wells SA 229
Country Fresh Restaurant, Bangalow NSW 99
Country Style Meats, Garfield Vic 360
Courtyard Restaurant, Berry NSW 115

Cowaramup WA 403-5
Cowell Meat Service, Cowell SA 232
Cowell SA 232
Cowes Vic 358
Cowra NSW 9, 25, 49
Cowra Food and Wine Weekend, NSW 65
Cowra Region Farmers' Market, NSW 65
Cowra Wine Show, NSW 65
Crab'n'Oyster Cruise, Brooklyn NSW 45
crabs see seafood
Crackenback Cottage and Farm, Thredbo Valley NSW 40
Cradle Mountain Tas 288
Cradock SA 227-8
Cradock Hotel, Cradock SA 227-8
Crafers SA 200-1
Craigmoor Restaurant, Mudgee NSW 57
Crayhaven Aquaculture, North Arm Cove NSW 93
Crazy Chairs Restaurant and Gallery, Dungog NSW 68
Creative Chicken, Glenorchy Tas 254
crocodile
  Australian Capital Territory
    Canberra 32
  Northern Territory 125
    Central Australia 127, 129-30
    Top End 131-2
  Queensland
    Bundaberg, Fraser Coast and South Burnett 140
    Capricorn, Gladstone and the Outback 148, 150
    Far North 152, 156, 158, 162
    Townsville, Mackay and the Whitsundays 189
  Victoria
    The Grampians 366
    Legends, Wine and High Country 341
  Western Australia
    Australia's North-West 421
Crookwell NSW 37
Crows Nest Qld 182
Crustaceans on the Wharf, Palmerston NT 132
Crystal Waters Qld 172-3
Cullen Wines, Cowaramup WA 404
cumquats
  Australian Capital Territory
    Canberra 30
currants
  New South Wales
    Capital Country and the Snowy Mountains 34
Currency Creek Estate Wines, Currency Creek SA 201
Currency Creek SA 201
Currie, King Island Tas 260
Currumbin Qld 167
curry
  New South Wales
    Central Coast 44-5
    Explorer Country and Riverina 62
    Hunter 75
    South Coast and Illawarra 118
  Northern Territory
    Central Australia 127
  Victoria
    Lakes and Wilderness 323
  Western Australia
    Australia's North-West 420
    The South-West 414
custard apples
  New South Wales
    North Coast 96
    South Coast and Illawarra 122
  Queensland
    Far North 160

INDEX

Cygnet Tas 278-9
Cygnet River SA 236

Daintree Qld 154-5
Daintree Eco Lodge and Spa, Daintree Qld 154
Daintree Tea Company, Mossman Qld 160
Daintree Tea House Restaurant, Daintree Qld 154-5
dairy *see also* cheese
  Australian Capital Territory
    Canberra 31
  New South Wales 9
    Explorer Country and Riverina 54-5
    Hunter 72, 74
    North Coast 85, 87
    Northern Rivers 99-100
  Queensland 135
    Far North 152, 159
    Gold Coast 168
    Sunshine Coast 172-4
  South Australia
    Adelaide and Surrounds 201, 210, 213
    Kangaroo Island, 236-7
    Limestone Coast 244
  Tasmania
    The North 265
  Victoria 295
    Bays and Peninsulas 302
    Goulburn Murray Waters 319-20
    The Great Ocean Road 369-73
    Phillip Island and Gippsland 361-2
    Yarra Valley and the Dandenong Ranges 384-5
  Western Australia 387
    Australia's North-West 418
    The South-West 402-3, 408-9
Dairy Festival, Murgon Qld 145
Dan and Jude's Heritage Orchard, Flowerpot Tas 280
Dangerous Dan's Butchery, Macksville NSW 92
Daniel Alps at Strathlynn, Rosevears Tas 271
Daradgee Qld 155
Dare's Fresh Fruit and Vegetables, Wodonga Vic 340
Darley's, Katoomba NSW 24
Darrel's Gourmet Butchery, Gloucester NSW 69
Darwin NT 125, 132-3
Darwin Seafood Festival, Darwin NT 133
The Date Gardens, Alice Springs NT 126
dates
  New South Wales
    Explorer Country and Riverina 49
  Northern Territory
    Central Australia 126
  Queensland
    Capricorn, Gladstone and the Outback 148
David Medlow Chocolates, McLaren Flat SA 207
Daylesford Vic 343-5
Daylesford Macedon Market Day, Daylesford Vic 349
Days of Wine and Roses Festival, Canberra Region ACT 43
De Brueys Boutique Wines, Mareeba Qld 157-8
Deans Marsh Vic 370
Deeb's Kitchen at the Schoolmaster's House, Mudgee NSW 57
Deer-O-Dome (Aust), Albany WA 400-1
Delegate River Deer Farm, Berridale NSW 34, 36
Deloraine Showgrounds Market, Deloraine Tas 274
Deloraine Tas 264, 274
Denmark WA 405-6
Derby WA 421-2

desserts *see also* puddings
  New South Wales
    Blue Mountains and the Hawkesbury 22
    Explorer Country and Riverina 52, 55
    Hunter 74, 76
    New England and North-West 79
    North Coast 94
    South Coast and Illawarra 115
  Queensland
    Far North 151-2
  South Australia
    Adelaide and Surrounds 201, 203
  Tasmania
    Hobart and Surrounds 255
    The North 268, 273
    The South 286-7
  Victoria
    Goulburn Murray Waters 319
  Western Australia
    Perth and Surrounds 397
Deumer's Harcourt Valley Orchard, Harcourt Vic 312
Devonport Tas 289
Devonshire tea
  New South Wales
    Blue Mountains and the Hawkesbury 22
    Capital Country and the Snowy Mountains 41
    New England and North-West 81
    North Coast 96-7
    South Coast and Illawarra 113
  Queensland
    Bundaberg, Fraser Coast and South Burnett 142
    Far North 159
    Sunshine Coast 172
  South Australia
    Limestone Coast 247-8
  Tasmania
    The South 286-7
  Victoria
    Bays and Peninsulas 304
Dew's Meats Pty Ltd, Orroroo SA 233-4
di Lusso Estate, Mudgee NSW 56
Diamond Still Spring Water, Hobart Tas 254
Dilston Tas 265
Dingo Creek Rainforest Nursery, Bobin NSW 85-6
Discover Victoria, Tullamarine Vic 299
Dish Restaurant Raw Bar, Byron Bay NSW 101
diVine Cafe and Gourmet Deli, Penola SA 246
Dixons Creek Vic 376
DJ and CA Meek Butchers, Penola SA 246
DJ's Fish and Chips, Greenwell Point NSW 117
The Doncaster Small Luxury Hotel, Braidwood NSW 35
Donnybrook WA 406-7
Donnybrook Farmhouse Cheese, Donnybrook Vic 298
Doonkuna Orchard, Crookwell NSW 37
Doran's Fine Foods, Grove Tas 282
Dorrigo NSW 88-9
Dorrigo Woodfired Bakery, Dorrigo NSW 88-9
Dromana Vic 302
Dromana Bay Mussels, Safety Beach Vic 308
Drouin West Vic 359
Drouin Farmers' Market, Drouin Vic 364
Drysdale Vic 302-3
Drysdale Institute of TAFE, Launceston Tas 268
Dubbo NSW 50
Dubbo Farmers' Market, NSW 65
Ducat's Food Service, Shepparton Vic 319

437

INDEX

duck
  New South Wales
    Blue Mountains and the Hawkesbury 28
    Capital Country and the Snowy Mountains 34, 37, 40, 42
    Central Coast 45
    Explorer Country and Riverina 49, 52, 55, 60
    Hunter 68, 70, 72-5
    North Coast 94-5
    Northern Rivers 102, 104-5, 108
    South Coast and Illawarra 118, 121, 123
    Sydney and Southern Highlands 15-17
  Northern Territory
    Central Australia 127
  Queensland
    Bundaberg, Fraser Coast and South Burnett 143
    Toowoomba and the Southern Downs 186
    Townsville, Mackay and the Whitsundays 191
  South Australia
    Adelaide and Surrounds 201, 214
  Tasmania
    The South 284
  Victoria
    Goldfields 310, 312
    The Grampians 366
    The Great Ocean Road 370
    Legends, Wine and High Country 328, 336
    Macedon Ranges and Spa Country 345
    Murray and the Outback 351
  Western Australia
    Perth and Surrounds 392, 397
    The South-West 410
Dunalley Tas 279
Dunalley Fish Market, Dunalley Tas 279
Dunalley Waterfront Cafe, Dunalley Tas 279
Dungog NSW 68
Dunkeld Vic 365
Dunsborough WA 407
Duranbah NSW 102, 109
Dutton's Meadery, Manilla NSW 81
Dwellingup WA 393

East Maitland NSW 68
Ebenezer NSW 23
Ebor NSW 79
Echoes Boutique Hotel and Restaurant, Katoomba NSW 24-5
Echuca Vic 351
Eco Meats, Belconnen ACT 32
Eda-bull, Forbes NSW 51
Eden Gate Blueberry Farm, Albany WA 401
The Edge Restaurant, Coles Bay Tas 263
Edgecombe Brothers, Belhus WA 392
Edi Upper Vic 329-30
eels
  Tasmania
    The North 264
  Victoria
    Macedon Ranges and Spa Country 348
  Eels Australis, Deloraine Tas 264
eggplant
  New South Wales
    Explorer Country and Riverina 51
  Victoria
    Legends, Wine and High Country 339
eggs
  Australian Capital Territory
    Canberra 32

New South Wales
  Blue Mountains and the Hawkesbury 21-23, 25, 26, 29
  Central Coast 47
  Explorer Country and Riverina 48, 58, 62
  South Coast and Illawarra 120
  Sydney and Southern Highlands 16
Queensland
  Bundaberg, Fraser Coast and South Burnett 140
  Gold Coast 168
  Sunshine Coast 172
  Toowoomba and the Southern Downs 186
South Australia
  Barossa, 218
  Clare Valley and Yorke Peninsula, 226, 229-30
  Limestone Coast 242, 246
Victoria
  Bays and Peninsulas 305
  Goldfields 310-11
  Legends, Wine and High Country 332
  Macedon Ranges and Spa Country 344
  Phillip Island and Gippsland 358, 363
Western Australia
  Australia's North-West 420
  Perth and Surrounds 388, 392
  The South-West 414
Eleonore's at Chateau Yering, Yering Vic 384-5
Elevation Restaurant, Cooma NSW 36
Elizabeth Town Tas 289
Ellisfield Farm, Red Hill Vic 306
Elton's Brasserie, Mudgee NSW 58
Emerald Bank Heritage Farm, South Shepparton Vic 319
Emeu Inn Restaurant, B&B and Wine Centre, Heathcote Vic 312
Emma's Choice Jam Factory and Tearooms, Police Point Tas 283
emu
  Australian Capital Territory
    Canberra 32
    Capital Country and the Snowy Mountains 38
    South Coast and Illawarra 116
  Northern Territory
    Central Australia 127, 129-30
  Queensland
    Bundaberg, Fraser Coast and South Burnett 140
  South Australia
    Limestone Coast 241
    Murray and Riverland 249
  Tasmania 253
    Hobart and Surrounds 256
    The South 279
  Victoria
    The Grampians 366
    Legends, Wine and High Country 341
    Phillip Island and Gippsland 357
  Western Australia
    The Outback 424-5
    Perth and Surrounds 389, 399
Emu Park Lions Club Oktoberfest-Beer Festival, Emu Park Qld 150
Emu Ridge Eucalyptus, Kingscote SA 237-8
The Emu Shop, Falls Creek NSW 116
Enchanted Cottage Preserves, Mount Dandenong Vic 380
Enniskillen Orchard, Grose Vale NSW 23
The Epicurean Centre, Milawa Vic 334
Epicurean Tours, Padbury WA 389
Equitas Orchards, Young NSW 42
Erina Heights NSW 45

INDEX

Esca Bimbadgen, Pokolbin NSW 75
Escarpment Restaurant, Jabiru NT 127-8
Eschalot, Bowral NSW, 15-16
Esperance WA 424
Esse Restaurant, Kiama NSW 118
Essence Food and Wine, Devonport Tas 289
The Essential Ingredient, Sydney NSW 39
The Essential Ingredient, Tamworth NSW 83
Essential Ingredients on the Mall, Coffs Harbour NSW 88
Eulo Qld 148-9
Eulo General Store, Eulo Qld 148-9
Eumundi Qld 173
Eumundi Markets, Eumundi Qld 171, 173, 180
Eureka Farm, Scamander Tas 272
Eurobin Vic 330
EV Olives, Markwood Vic 332
Evelyn County Estate, Kangaroo Ground Vic 378
EXPLORER COUNTRY NSW 48-65
Explorers' Restaurant, Flinders Chase SA 237

Fairbrae Milk Co, Bentley NSW 99-100
Fairymead House Sugar Museum, Bundaberg Qld 140
Falcon WA 393
Falls Creek NSW 116
FAR NORTH Qld 151-66
Faraway Bay, The Bush Camp, Kununurra WA 423
Farm Follies and Hutton Vale Wines, Angaston SA 218
Farmers' Fresh and Seafood Markets, Brisbane Qld 138
Farmers' Market, Deloraine Tas 274
Farmgate at Statford Park, Wildes Meadow NSW 18
The Fat Cat Cafe, Yeoval NSW 64
Father Mac's Heavenly Puddings, Alstonville NSW 99
Fazzolari Olive Oils, Middle Swan WA 396
Feast of the Olive, lower Hunter NSW 73
Feast Restaurant, Avoca Beach NSW 44
Feast! Fine Foods, Mount Barker SA 210-11
Federal NSW 102-3
Fee and Me, Launceston Tas 267-8
Felons, Port Arthur Tas 283
Feral Brewing Company, Baskerville WA 391
Ferguson Australia Pty Ltd, Regency Park SA 195-6
Festivale, Launceston Tas 274
festivals
  New South Wales
    Blue Mountains and the Hawkesbury 27
    Capital Country and the Snowy Mountains 43
    Explorer Country and Riverina 65
    Hunter 78
    New England and North-West 84
    North Coast 98
    Northern Rivers 109
    Outback and the Murray 112
  Northern Territory
    Top End 133
  Queensland
    Brisbane and Surrounds 138
    Capricorn, Gladstone and the Outback 150
    Far North 166
    Gold Coast 170
    Toowoomba and the Southern Downs 187
  South Australia
    Adelaide and Surrounds 196, 216
    Barossa, 217
    Limestone Coast 247
  Tasmania
    The North 274
    The North-West 294
  Victoria
    Bays and Peninsulas 309
    Goldfields 315
    Goulburn Murray Waters 320
    The Grampians 368
    Lakes and Wilderness 324
    Legends, Wine and High Country 342
    Macedon Ranges and Spa Country 349
    Melbourne 299
    Phillip Island and Gippsland 364
    Yarra Valley and the Dandenong Ranges 385
  Western Australia
    Australia's North-West 423
    Perth and Surrounds 399
    The South-West 416
figs
  New South Wales
    Blue Mountains and the Hawkesbury 2
    Capital Country and the Snowy Mountains 35, 42, 42
    Central Coast 44
    Explorer Country and Riverina 49, 56
  Northern Territory
    Central Australia 126
  Queensland
    Capricorn, Gladstone and the Outback 148
    Toowoomba and the Southern Downs 181
  South Australia
    Adelaide and Surrounds 202, 204, 212
    Barossa, 217
  Victoria
    Goldfields 311
    Legends, Wine and High Country 326, 334
Figtree Retreat Olives and Farmstay, Mudgee NSW 56
Finley NSW 112
Fins Seafood Restaurant, Byron Bay NSW 101
Fiorelli's Restaurant and Bar, Port Douglas Qld 162
Fireside Inn Restaurant, Goulburn NSW 37
fish see also seafood
  New South Wales 9
    Blue Mountains and the Hawkesbury 23, 26, 27
    Capital Country and the Snowy Mountains 33, 35, 36
    Central Coast 44-7
    Explorer Country and Riverina 52, 59, 63
    Hunter 68-9
    New England and North-West 79-80, 83
    North Coast 85, 87, 93-7
    Northern Rivers 100-3, 107-8
    Outback and the Murray 110
    South Coast and Illawarra 113-14, 117, 121-2
    Sydney and Southern Highlands 12-14
  Northern Territory 125
    Central Australia 127-30
    Top End 131
  Queensland
    Brisbane and Surrounds 136
    Bundaberg, Fraser Coast and South Burnett 139
    Capricorn, Gladstone and the Outback 147-50
    Far North 151-4, 156-7, 161-4
    Sunshine Coast 175-7, 179-80
    Townsville, Mackay and the Whitsundays 188-92
  South Australia
    Adelaide and Surrounds 204, 210, 213
    Barossa, 219
    Clare Valley and Yorke Peninsula, 227-8
    Eyre Peninsula, Flinders Ranges and the Outback, 231-5

439

INDEX

fish *continued*
  South Australia *continued*
    Kangaroo Island, 236–7
    Limestone Coast 241–3
    Murray and Riverland 249–50
  Tasmania
    Bass Strait Islands 260–2
    Hobart and Surrounds 255–8
    The North 263–4, 267, 272–3
    The North-West 288–94
    The South 275, 279, 282–3
  Victoria
    Bays and Peninsulas 303, 305–9
    Goulburn Murray Waters 318
    The Grampians 365
    The Great Ocean Road 370–2
    Lakes and Wilderness 321–3
    Legends, Wine and High Country 341
    Macedon Ranges and Spa Country 344
    Phillip Island and Gippsland 360, 362
  Western Australia
    Australia's Coral Coast 426–7
    Australia's North-West 419–22
    Perth and Surrounds 393
    The South-West 406–8, 410, 412–15, 417
Fish Heads Cafe and Fine Foods, Umina Beach NSW 47
Fisheries on the Spit, Mooloolaba Qld 175
Fishermen's Pier Restaurant, Geelong Vic 303
Fishermen's Wharf Seafood, Ulladulla NSW 121–2
Fishermen's Wharf, Woy Woy NSW 47
Fishheads@Byron Restaurant, Byron Bay NSW 101
Fitzroy Inn, Mittagong NSW 17
The Five Islands Brewing Company, Wollongong NSW 123
Flaggy Rock Exotic Fruit Garden, Flaggy Rock Qld 189
Flaggy Rock Qld 189
Flair Restaurant, Erina Heights NSW 45
Fleurieu Peninsula SA 193, 199
Flinders Vic 303
Flinders Interstate Hotel, Whitemark, Flinders Island Tas 262
Flinders Island Tas 262
Flinders Island Bakery, Whitemark, Flinders Island Tas 262
Flinders Waterfront Restaurant, Gladstone Qld 149
Flinders Chase SA 236
Flooded Gums Restaurant, Bonville NSW 86
Florist Cafe, Cobram Vic 351
Flowerpot Tas 280
Flutes Restaurant, Willyabrup WA 415
Flying Fish, Port Elliot SA 213
The Flying Pig Deli at Redbank Winery, Redbank Vic 313
Food Barossa, Tanunda SA 222
Food Lovers Club, Moore River WA 416
Food Lovers, Belconnen ACT 32
Food Trail Tours—A Taste of the High Plains, Cairns Qld 153
Food Week, Orange NSW 65
Food Wine Friends, Bright Vic 336–7
Food, Wine and Jazz Festival, Smeaton Vic 349
Foodies' Dream Tour, Melbourne
Forbes NSW 51
Fordwich Grove, Broke NSW 67
Forest Fresh Marron, Pemberton WA 412–13
Forest Grove Olive Farm, Forest Grove WA 407
Forest Grove WA 407
Forgotten Fruits, Healesville Vic 377
Forth Tas 290
Foster Vic 359

Foster Seafood, Foster Vic 359
Fox Studios Farmers' Market, Moore Park NSW 19
Frangos and Frangos, Daylesford Vic 344
Frankland River Fine Wine, Food and Music Festival, Frankland WA 416
Franklin Manor, Strahan Tas 293
Fraser Island Qld 139, 140
Frederickton NSW 90
Fredo Famous Pies and Ice-creams, Frederickton NSW 90
Fremantle WA 390
French Island Vic 359–60
Fresh Provision Stores, Claremont WA 388
Freycinet Marine Farm, Coles Bay Tas 263
Frosty Mango, Mutarnee Qld 191
fruit *see also by type of fruit*
  Australian Capital Territory
    Canberra 30–3
  New South Wales
    Blue Mountains and the Hawkesbury 22–3
    Capital Country and the Snowy Mountains 39
    Explorer Country and Riverina 48–9, 52, 59–60, 63
    North Coast 96–7
    Northern Rivers 102
    Outback and the Murray 111–12
    South Coast and Illawarra 122
    Sydney and Southern Highlands 14, 16, 17, 19
  Northern Territory
    Top End 132
  Queensland
    Far North 152–6, 158–60, 162
    Sunshine Coast 174–5
    Townsville, Mackay and the Whitsundays 188–92
  South Australia
    Adelaide and Surrounds 196, 200, 202–3, 208–9, 214
    Barossa, 218–20
    Murray and Riverland 249
  Tasmania
    The North 270–1
    The North-West 290–1
    The South 284–5
  Victoria
    Bays and Peninsulas 304
    Goldfields 311
    Goulburn Murray Waters 316–18
    The Great Ocean Road 370–2, 374
    Lakes and Wilderness 322
    Legends, Wine and High Country 326, 328–9, 340–1
    Murray and the Outback 350–2
    Phillip Island and Gippsland 358, 363
    Yarra Valley and the Dandenong Ranges 375–8, 380–4
  Western Australia
    Perth and Surrounds 396
    The South-West 402
Fruit Ballad Country Wines, Quaama NSW 119
fruit wine
  New South Wales
    Hunter 66
    New England and North-West 82
    North Coast 89
    South Coast and Illawarra 119
  Queensland
    Bundaberg, Fraser Coast and South Burnett 139–40
    Far North 151, 153, 157–60, 164–5
    Toowoomba and the Southern Downs 185
  Victoria
    The Great Ocean Road 370
    Lakes and Wilderness 321

INDEX

Legends, Wine and High Country 325
Murray and the Outback 352
Yarra Valley and the Dandenong Ranges 377
Western Australia
The South-West 401, 402
The Fruitfarm Johnsonville, Johnsonville Vic 322
Fruitshack, Leeton NSW 54

Galloway Yabbies, Inman Valley SA 204–5
Gapsted Vic 330–1
Garden Restaurant, West Ballina NSW 107
Gardners Bay Tas 280
Garfield Vic 360
Garfield Berry Farm, Garfield Vic 360
Garfield Fish Farm, Garfield Vic 360
Gatton Qld 136
Gayndah Orange Festival, Gayndah Qld 145
Geelong Vic 303, 309
Geelong Farmers' Market, Geelong Vic 309
Gentle Annie Berry Gardens, Deans Marsh Vic 370
Georgies at the Gallery, Grafton NSW 103
Georgina's Restaurant, Benalla Vic 310
German Arms Hotel, Hahndorf SA 203
Gerringong NSW 116
Gerringong Gourmet Deli, Gerringong NSW 116
Gerroa NSW 116–17
Gidgegannup WA 393–4
Gigi's of Beechworth, Beechworth Vic 326
Gilberts Restaurant at Fruit Salad Farm, Marysville Vic 380
Gilgai Tavern, Kalbarri WA 426–7
Gillin's Butchery, Wellington NSW 64
ginger
  Australian Capital Territory
    Canberra 30
  Queensland 135
    Brisbane and Surrounds 137
    Sunshine Coast 171, 179, 180
  South Australia
    Adelaide and Surrounds 206
  Western Australia
    The South-West 411
Gippsland Food and Wine, Yarragon Vic 363–4
Gippsland Green Asparagus Tours, Koo-wee-rup Vic 361
Gip's Restaurant, Toowoomba Qld 186
Gipsy Point Vic 321
Gipsy Point Lodge and Restaurant, Gipsy Point Vic 321
Gisborne Vic 345–6
Gladstone Qld 149
The Gladstone Seafood Festival, Gladstone Qld 150
Gladysdale Apple and Wine Festival, Gladysdale Vic 385
Glaetzer's Blueberry Hill, Willunga SA 215
Glaziers Bay Tas 280
Glen Aplin Qld 182–3
Glen Huon Tas 281
Glen Innes NSW 79–80
Glengallan Seasonal Farmers' Markets, Allora Qld 187
Glenrowan Vic 331
Glenthompson Vic 365
Glo Glo's Restaurant, Latrobe Tas 291
The Globe Hotel, Albury NSW 110–11
The Globe Restaurant, Castlemaine Vic 312
Gloucester NSW 68–70
Gloucester Ridge Winery & Restaurant, Pemberton WA 413
Glover's of Auburn, Auburn SA 225
goat cheese see cheese
goats
  New South Wales
    Blue Mountains and the Hawkesbury 24, 25, 29

    Capital Country and the Snowy Mountains 38–40
    Explorer Country and Riverina 48, 54–5
    Hunter 69
    North Coast 95
  Queensland
    Bundaberg, Fraser Coast and South Burnett 142
    Capricorn, Gladstone and the Outback 150
    Gold Coast 168
    Sunshine Coast 173–4, 178
  South Australia
    Adelaide and Surrounds 202, 206, 215
  Tasmania
    Hobart and Surrounds 256
  Victoria
    Bays and Peninsulas 306
    Legends, Wine and High Country 328
  Western Australia
    Perth and Surrounds 394
    The South-West 403
Gold Coast Food and Wine Trails Qld 169
Gold Coast Signature Dish Competition, Main Beach Qld 170
GOLD COAST Qld 167–70
Golden Cow Dairy Education Centre, Tongala Vic 320
Golden Drop, Biboohra Qld 151
GOLDFIELDS Vic 309–15
Goldfields Greengrocers, Beechworth Vic 326
Good Living Growers' Markets, Pyrmont NSW 19
Goodwood SA 201
Goolwa SA 201
Goolwa Wharf Market, Goolwa SA 216
Goomeri Qld 141
Goomeri Pumpkin Festival, Goomeri Qld 145
Gooramadda, via Rutherglen Vic 331
Gooramadda Olives, Gooramadda Vic 331
Gordonvale Market, Gordonvale Qld 166
Gorman's Restaurant, Yamba NSW 108
Gorokan NSW 45
Gosford Fine Food Market, Mount Penang NSW 71
Goulburn NSW 37
GOULBURN MURRAY WATERS Vic 316–20
Gourlay's Sweet Shop, Launceston Tas 266
Gourmet Connection, Strathalbyn SA 214
Gourmet in Gundy, Goondiwindi Qld 187
Gourmet Safaris 14
Gourmet Taste Bud, Charnwood ACT 30
Gowrie Park Tas 290
Grafton NSW 103
Grafton Farmers' and Growers' Market, Grafton NSW 109
THE GRAMPIANS Vic 365–8
Grampians Grape Escape—The Food and Wine Festival, Halls Gap Vic 368
Grampians Lavender Patch, Stawell Vic 368
Grampians Pure Sheep Milk Products, Glenthompson Vic 365
Grandvewe Cheeses, Birchs Bay Tas 275
The Grange Cafe and Deli, Warragul Vic 363
Granite Belt Spring Wine Festival, Stanthorpe Qld 187
Grapefoodwine, Lake George NSW 39–40
grapefruit
  New South Wales
    Hunter 67
    Outback and the Murray 111
grapes see also wine
  New South Wales
    Outback and the Murray 111
  South Australia
    Murray and Riverland 249

441

INDEX

grapes *continued*
  Victoria
    Legends, Wine and High Country 329-30
  Western Australia
    Perth and Surrounds 394
Gray's Inn, Wollombi NSW 77
THE GREAT OCEAN ROAD Vic 369-74
Great Abalone Bake-off, Binalong Bay Tas 274
Great Bight Oysters, Ceduna SA 231
Great Lakes Great Produce Market, Forster NSW 98
Great Wine Tasting and Appreciation Expo, Bundanoon NSW 19
Green Farmhouse, Millicent SA 243
Green Grove Flour Mills, Junee NSW 46
Green Grove Organics—Junee Licorice and Chocolate Factory, Junee NSW 53-4
The Green Shed Bistro, Beechworth Vic 327
Greenmount Qld 183
Greenwell Point NSW 117
Gresford NSW 70
Greta NSW 70
Griffith NSW 51
Griffith Butchery and Bakery, Griffith ACT 32
Grong Grong NSW 52
Grose Vale NSW 23
Grove Tas 282
guavas
  New South Wales
    North Coast 96
    South Coast and Illawarra 122
Guesthouse Mulla Villa, Wollombi NSW 77-8
Gulf Buffalo, Warnertown SA 235
Gulf Seafoods Pty Ltd, Malvern SA 196
Gumeracha SA 202
Gumlu Capsicum Festival, Gumlu Qld 192
Gundaroo NSW 37-8
The Gunyah Restaurant at Paperbark Camp, Huskisson NSW 117
Guyra Lamb and Potato Festival, Guyra NSW 84
Guyra NSW 80
Gwydir Olives, Inverell NSW 80-1
Gympie Qld 173
Gympie Farm Cheese, Woolooga Qld 178

Hahndorf SA 202-3
Hahndorf Inn, Hahndorf SA 203
Hahndorf Venison, Hahndorf SA 202
Haigh's Chocolates Visitors Centre, Parkside SA 197
Hall NSW 38
Halls Gap Vic 365-6
ham *see* smoked smallgoods
Hamilton Vic 366
Hamilton Island Qld 189
Hamilton Strand Restaurant, Hamilton Vic 366
Hanging Rock NSW 80
Hanson's Swan Valley, Henley Brook WA 394
Hanuman Restaurant, Alice Springs NT 127
The Hanuman Restaurant, Darwin NT 132-3
Hanwood NSW 52
Harbourside Restaurant, Ulladulla NSW 122
Harcourt Vic 312, 315
Harcourt Applefest, Harcourt Vic 315
hare
  New South Wales
    New England and North-West 79
  South Australia
    Adelaide and Surrounds 202
  Tasmania
    Hobart and Surrounds 256

Harrietville Vic 331-2
Harrington NSW 90
Harrison and Sons Orchard, Araluen NSW 33
Harrison's Orchard, Pomonal Vic 367
Harry Black's Orchard, Wallaga Lake NSW 122-3
Hartley Vale NSW 24
Hartley Valley Teahouse, Little Hartley NSW 25
Hartzview Vineyard, Gardners Bay Tas 280
Harvest Corner Information and Craft, Minlaton SA 228
Harvest Home Country House Hotel, Avenel Vic 316
Harvey Cheese, Wokalup WA 398-9
Harveys Fish and Fun Park, Wodonga Vic 341
Hastings Farmers' Markets, Wauchope NSW 98
Hastings Harvest Picnic and Cultural Festival, Port Macquarie NSW 98
Hastings River Fishermens' Cooperative, Port Macquarie NSW 94
Hastings River Oyster Supplies, Port Macquarie NSW 94
Hastings Valley Olives, Wauchope NSW 97
Haven Glaze, Caboolture Qld 171
Havenhand Chocolates, Waikerie SA 250-1
Hawker SA 233
Hawker Bros Butchers, Stanthorpe Qld 184
HAWKESBURY 20-9
Hawkesbury Harvest Farmers' and Gourmet Food Market, Castle Hill NSW 19
Hay NSW 52-3
Hayman Island Qld 189
hazelnuts
  New South Wales
    Blue Mountains and the Hawkesbury 25
    Explorer Country and Riverina 55
  Victoria
    Phillip Island and Gippsland 359
Healesville Vic 377
Healesville Hotel, Healesville Vic 377
Heart of the Hills Market, Lobethal SA 216
Heathcote Vic 312, 346
Heathcote Wine and Food Festival, Heathcote Vic 315
Heathcote's Deep Winter Wine Dinner, Heathcote Vic 315
Heathfield SA 203-4
Henley Brook WA 394-5
Henry of Harcourt, Harcourt Vic 312
Henry's Winery Cafe, Wyanga Park Winery, Lakes Entrance Vic 323
Herb of Grace Organic Herbs, Cooma NSW 36
Herberton Market, Herberton Qld 166
Herbicious Delicious, Olinda Vic 380
herbs *see also* lavender
  Australian Capital Territory
    Canberra 31
  New South Wales
    Blue Mountains and the Hawkesbury 22-5, 28, 29
    Capital Country and the Snowy Mountains 36
    Explorer Country and Riverina 60
    Hunter 69-70, 73
    New England and North-West 83
    North Coast 86, 91
    Northern Rivers 102, 104-6
    South Coast and Illawarra 115, 123
    Sydney and Southern Highlands 12, 15, 16
  Northern Territory
    Central Australia 130
    Top End 132
  Queensland
    Brisbane and Surrounds 136
    Bundaberg, Fraser Coast and South Burnett 140, 144
    Far North 156

INDEX

Gold Coast 168
Sunshine Coast 171, 176, 180
Toowoomba and the Southern Downs 181
Townsville, Mackay and the Whitsundays 191
South Australia
   Adelaide and Surrounds 201-2
   Barossa, 218
   Clare Valley and Yorke Peninsula, 229
   Limestone Coast 241-4, 246
   Murray and Riverland 251
Tasmania
   The South 275, 286
Victoria
   Bays and Peninsulas 302
   Goldfields 311-12
   The Great Ocean Road 369
   Lakes and Wilderness 321
   Legends, Wine and High Country 326, 331, 334, 336
   Macedon Ranges and Spa Country 348
   Murray and the Outback 351, 355
   Yarra Valley and the Dandenong Ranges 378-9
Western Australia
   Australia's North-West 420
   The South-West 412
Herdsman Fresh Essentials, Churchlands WA 388-9
Herne Hill WA 395
Hervey Bay Qld 141
Hesket House, Romsey Vic 347
Hi Value Fruit and Berry Gardens, Thulimbah Qld 185
Hickinbotham of Dromana, Dromana Vic 302
Highbury Mountain—On Farm Cafe, Chatsworth Qld 172
Highland Heritage Estate Restaurant, Orange NSW 60-1
The Highland Restaurant, Cradle Mountain Tas 288
Hillfarm Spring Water, Borenore NSW 48
Hillsdale Orange Orchard, Bulga NSW 67
Hillston NSW 53
Hillview Herb Farm, Gloucester NSW 69
Hillwood Tas 265
Hillwood Strawberry Farm, Fruit Wine and Cheese Centre, Hillwood Tas 265
Hilton Masterclass, Brisbane Qld 138
Hindmarsh Island SA 204
Hinterland Country Markets, Tamborine Mountain Qld 170
HL and BJ Shapland, Albany WA 401
Hobart Tas 254-9
HOBART AND SURROUNDS Tas 254-9
Hobbitt Farm Goat Cheese, Jindabyne NSW 38-9
Hoddles Creek Vic 378
Hog Bay Apiary, Penneshaw SA 239
Home Hill Winery Restaurant, Ranelagh Tas 284
The Homestead Restaurant Plantation and Lodge, Tolga Qld 165
Hominy Bakery, Katoomba NSW 24
The Honey Place, Urunga NSW 96-7
honey
   New South Wales
      Blue Mountains and the Hawkesbury 23
      Capital Country and the Snowy Mountains 33
      Explorer Country and Riverina 48, 56, 58, 60, 62
      Hunter 69, 71
      New England and North-West 81
      North Coast 90, 96-7
      Northern Rivers 100, 104
      South Coast and Illawarra 116
      Sydney and Southern Highlands 12, 18
   Queensland
      Capricorn, Gladstone and the Outback 148
      Far North 158, 164

Gold Coast 167
Sunshine Coast 173, 178
Toowoomba and the Southern Downs 183
South Australia 193
   Barossa 218
   Clare Valley and Yorke Peninsula 225, 227
   Kangaroo Island 237-9
   Limestone Coast 241, 246, 248
   Murray and Riverland 250
Tasmania 253
   Bass Strait Islands 260-1
   Hobart and Surrounds 257-8
   The North 269
   The North-West 291
   The South 275, 287
Victoria
   Bays and Peninsulas 307
   Goldfields 310
   The Great Ocean Road 369
   Lakes and Wilderness 321
   Legends, Wine and High Country 326, 329, 333
   Murray and the Outback 350
Honeysuckle Cottage, Bowen Mountain NSW 22-3
Honeysuckle Markets, Newcastle NSW 71, 78
Hoochery Distillery, Kununurra WA 423
Hordern's Restaurant, Bowral NSW 16
horseradish
   South Australia
      Adelaide and Surrounds 205-6
The Hothouse Cafe, Bruny Island Tas 276
Hotson's Cherries, Chiltern Vic 329
Houghton SA 204
Houghton's Fine Foods, Mornington Vic 305
The House of Anvers, Latrobe Tas 290
Hudak's Bakery 15th Street Store, Mildura Vic 353
Hume Murray Food Bowl Farmers' Market, Milawa Vic 342
Hume Weir Trout Farm, Wodonga Vic 340
The Humpy, Tolga Qld 164-5
HUNTER 9, 66-78
Hunter Belle Cheese, Muswellbrook NSW 72
Hunter Grove, Hunter Olive Cooperative Ltd, Antiene NSW 66
Hunter Harvest Farmers' Market, Maitland NSW 78
Hunter Valley Cheese Company, Pokolbin NSW 66, 74, 76
Hunter Valley Day Tours, Paterson NSW 73
Hunter Valley Harvest Festival, Pokolbin NSW 78
Huntley NSW 53
Huntley Berry Farm, Huntley NSW 53
Huon Manor Bistro, Huonville Tas 282
Huon Valley Mushrooms, Glen Huon Tas 281
Huonville Tas 282
Huskisson NSW 117
hydroponic produce
   New South Wales
      Outback and the Murray, 112
      South Coast and Illawarra 119-20
   Queensland
      Far North 156
   Victoria
      Legends, Wine and High Country 326, 335

ice-cream
   New South Wales
      North Coast 89-90
      Northern Rivers 103, 105
      Sydney and Southern Highlands 15
   Northern Territory 125
      Central Australia 126

443

INDEX

ice-cream *continued*
  Queensland
    Bundaberg, Fraser Coast and South Burnett 140, 142
    Far North 155, 160
    Sunshine Coast 172, 175
    Toowoomba and the Southern Downs 183
    Townsville, Mackay and the Whitsundays 191
  South Australia
    Adelaide and Surrounds 211
    Barossa, 222
  Tasmania
    Hobart and Surrounds 254–5
    The North-West 290
    The South 275
  Victoria
    Goldfields 311
    The Grampians 367–8
    Lakes and Wilderness 321–2
    Legends, Wine and High Country 327
    Macedon Ranges and Spa Country 347–8
  Western Australia
    Australia's North-West 418
    The South-West 405
The Icecreamery, Chinatown, Broome WA 418
Indooroopilly Farmers' Market, Indooroopilly Qld 138
Inman Valley SA 204–5
Innamincka SA 233
Innes Boatshed, Batemans Bay NSW 113
Innisfail Qld 155
Inverell NSW 80–1
Irymple Vic 352
Island Olive Grove, Cambridge Tas 276–7
Island Produce Tasmania, South Hobart Tas 254–5
Island Pure, Cygnet River SA 236
Island Seafood, Kingscote SA 238
Italian Festa, Swan Hill Vic 356

J Boag & Son Brewing Ltd, Launceston Tas 266
Jabiru NT 127–8
The Jajjikari Cafe, Tennant Creek NT 128–9
jams and marmalades
  Australian Capital Territory
    Canberra 30–3
  New South Wales
    Blue Mountains and the Hawkesbury 22, 24
    Capital Country and the Snowy Mountains 35, 40–2
    Explorer Country and Riverina 49, 52, 61–2
    Hunter 67, 70–1, 77
    New England and North-West 81–2
    North Coast 86–7, 92–3, 96–7
    Northern Rivers 100, 104
    Sydney and Southern Highlands 12, 17
  Queensland
    Bundaberg, Fraser Coast and South Burnett 144
    Far North 158
    Sunshine Coast 172, 179
    Toowoomba and the Southern Downs 182, 183, 185
  South Australia
    Adelaide and Surrounds 202–5, 211–12, 215
    Barossa, 218–19, 223–4
    Clare Valley and Yorke Peninsula, 225, 227–8
    Kangaroo Island, 237–8
    Murray and Riverland 249
  Tasmania
    Hobart and Surrounds 255–7
    The North 271
    The North-West 289, 291
    The South 275, 282–7
  Victoria
    Bays and Peninsulas 302, 306–7
    Goldfields 310, 312
    Goulburn Murray Waters 317
    The Grampians 367–8
    The Great Ocean Road 369–70, 372
    Legends, Wine and High Country 326, 330, 341
    Macedon Ranges and Spa Country 343–4, 346–7
    Melbourne 299
    Murray and the Outback 352
    Phillip Island and Gippsland 360
    Yarra Valley and the Dandenong Ranges 375, 377, 378, 380, 382, 384
  Western Australia
    Australia's North-West 418, 422
    Perth and Surrounds 398
    The South-West 405, 411, 415
Jannei Goat Dairy, Lidsdale NSW 54–5
Jarrahdale WA 396
Jaspers Brush NSW 117
Jazz in the Vines, Pokolbin NSW 78
JD's Jam Factory, Young NSW 42
jellies *see* jams and marmalades
Jerilderie NSW 112
The Jetty Restaurant and Star Club Bar, Cowes Vic 358
Jills at Moorooduc Estate, Moorooduc Vic 305
Jindabyne NSW 38–9
Jindivik Smokehouse, Neerim South Vic 362
JJ's at the mouth of the Hawkesbury River, Brooklyn NSW 44
Joanna's Jams, Cascades Tas 255
Johnsonville Vic 322
Johnstone River Plantation Fruit Shop, Daradgee Qld 155
The Jolly Frog Restaurant, Wannunup WA 415
Jolly Jumbuck, Hay NSW 52–3
Jondaryan Qld 183
Jondaryan Woolshed, Jondaryan Qld 183
Joshua Creek Fruit Wines, Boyanup WA 402
Julie and Patrick's Restaurant and Hursey Seafoods, Stanley Tas 292–3
Junee NSW 53–4
Just Prunes, Kingsvale NSW 39

Kabuki by the Sea, Swansea Tas 273
Kaiwarra Food Barn and Cottages, Parndana SA 239
Kakadu Mango Winery, Palmerston NT 133
Kalangadoo Organic, Kalangadoo SA 242
Kalangadoo SA 242
Kalbarri WA 426–7
Kalgan WA 407
Kallangur Qld 136
Kalorama Vic 378
Kalorama Chestnut Festival, Kalorama Vic 385
Kangaroo Island Coast to Coast Tour, Kangaroo Island SA 240
kangaroo
  Australian Capital Territory
    Canberra 32, 34
  New South Wales
    Capital Country and the Snowy Mountains 36, 38
    Hunter 75
    Outback and the Murray 110
  Northern Territory
    Central Australia 127–30
  Queensland
    Bundaberg, Fraser Coast and South Burnett 140
    Far North 152–3, 156, 158, 162
  South Australia
    Adelaide and Surrounds 195, 208

INDEX

Eyre Peninsula, Flinders Ranges and the Outback, 232–4
Murray and Riverland 249–50
Tasmania
  The North 270
Victoria
  Goldfields 311
  The Grampians 366
  Legends, Wine and High Country 341
  Phillip Island and Gippsland 360
  Yarra Valley and the Dandenong Ranges 382
Western Australia
  Australia's North-West 421
  Perth and Surrounds 389
Kangaroo Ground Vic 378
Kangaroo Island SA 193
Kanmantoo SA 205
Kanmantoo Bacon & Quality Meats, Kanmantoo SA 205
Kapers, Gladstone Qld 149
Karambi Orchard, Bilpin NSW 20
Karingal Cafe, Toowoomba Qld 186
Karridale WA 408
Kate's Berry Farm, Swansea Tas 286
Katherine NT 128
Katialo Restaurant, Portarlington Vic 305–6
Katoomba NSW 24–5, 27
Kelly's Bakery, Korumburra Vic 361
Kelly's Humble Spud, Bright Vic 327
Kellybrook Cider Festival, Wonga Park Vic 385
Kendall NSW 90–1
Kenilworth Qld 173–4
Kenilworth Country Foods, Kenilworth Qld 174
Kenilworth Organic Olives, Belli Park Qld 171
Kennedy's Meats, Wodonga Vic 341
Kentish and Sons Pty Ltd, Mount Gambier SA 244
Kenton Valley SA 205
Kergunyah Vic 332
Kervella Cheese, Gidgegannup WA 393
Kettering Tas 282
Kialla West Vic 317
Kialla Pure Foods, Greenmount Qld 183
Kiama NSW 118
Kiewa Estate Olive Grove, Wodonga Vic 340
Kiewa Valley Meat Supply, Mt Beauty Vic 334
Kilcunda Lobster Festival, Kilcunda Vic 364
Kilkivan Qld 142
Kimberley Boab Kreations, Kununurra WA 422–3
King Island Tas 260–1
King Island Airport Kiosk, Currie, King Island Tas 260–1
King Island Bakery, Currie, King Island Tas 261
King Island Dairies, Loorana, King Island Tas 261
King Island Produce, Currie, King Island Tas 260
King River Cafe, Oxley Vic 336
King Trout Cafe, Pemberton WA 413
King Valley Olives, Cheshunt Vic 329
Kingaroy Cheese, Kingaroy Qld 142
Kingaroy Qld 139, 142–4
Kinglake Raspberries, Pheasant Creek Vic 380–1
Kinglake Raspberry Fair, Pheasant Creek Vic 385
Kings Choice, Hamilton Hill WA 389
Kingscliff Roadside Stalls, Kingscliff Qld 167
Kingscote SA 237–9
Kingscote IGA Friendly Grocer (Griffiths), Kingscote SA 238
Kingston SA 242–3
Kingsvale NSW 39
Kit and Kaboodle, Nabiac NSW 93
Kiwi Down Under Organic Farm and Teahouse, Bonville NSW 86

kiwifruit
  New South Wales
    Explorer Country and Riverina 53
  South Australia
    Adelaide and Surrounds 206
    Barossa, 217
  Western Australia
    The South-West 412
Knight's Meats Wholesale, Wagga Wagga NSW 62
Knockrow NSW 103
Knockrow Ridge Coffee, Knockrow NSW 103
Kojonup Country Kitchen, Kojonup WA 408
Kojonup WA 408
The Kookaburra Restaurant, Halls Gap Vic 366
Kookootonga, Mount Irvine NSW, 27
Kooljaman at Cape Leveque, Broome WA 420–1
Koolkuna Berries, Niangala NSW 81–2
Koonoomoo Vic 352
Koonwarra Vic 361
Koonwarra Fine Food and Wine Store, Koonwarra Vic 361
Kooragang City Farm, Ash Island NSW 66
Koorana Crocodile Farm, Coowonga Qld 148
Koorawatha Hydroponics, Sussex Inlet NSW 119–20
Koo-wee-rup Vic 361
Kore Farm Produce, Tea Gardens NSW 96
Koroit Vic 371
Koroit Tower Hill Gourmet Larder, Koroit Vic 371
Korora NSW 91
Korumburra Vic 361
Krondorf Road Cafe, Tanunda SA 223
Kroombit Park, Biloela Qld 147
Kuku Yalanji Dreamtime Walks, Mossman Gorge Qld 161
Kuniya Restaurant, Uluṟu-Kata Tjuṯa National Park NSW 129
Kununurra WA 422
Kuranda Qld 155–6, 166
Kuranda Home Made Tropical Fruit Ice Cream, Kuranda Qld 155
Kuranda Markets, Kuranda Qld 166
Kurrajong Hills NSW 25
Kydd's Butchery, Bega NSW 114
Kyneton Vic 346
Kyneton Country Fresh, Kyneton Vic 346
Kyneton Provender, Kyneton Vic 346
Kytren Fine Quality Goats Cheese, Gidgegannup WA 394

L'Ocean Fish and Chips, Lakes Entrance Vic 323
La Baracca Trattoria, Main Ridge Vic 304
La Buona Vita, Tanunda SA 223
La Fontaine, Hayman Island Qld 189
La Tartine, Somersby NSW 45–6
La Taverne du Naturaliste, French Island Vic 359–60
La Trattoria, Shepherds Flat Vic 348
Label and Table, Greenwood WA 391
Lacepede Seafood, Kingston SA 242–3
Lactos Cheese Tasting and Sales Centre, Burnie Tas 288
Laggan NSW 39
Laharum Vic 366
Lake George NSW 39–40
Lake House, Daylesford Vic 345
LAKES AND WILDERNESS Vic 321–4
Lakes Entrance Vic 322
Lakes Entrance Fish Tasting, Lakes Entrance Vic 324
Lakes Entrance Fishermens' Co-op, Lakes Entrance Vic 322
Lakeside Restaurant, Pemberton WA 413
lamb
  Australian Capital Territory
    Canberra 32

445

INDEX

lamb continued
  New South Wales 9
    Blue Mountains and the Hawkesbury 29
    Capital Country and the Snowy Mountains 36–8
    Explorer Country and Riverina 52–3, 55, 57, 59, 63
    Hunter 69, 74
    New England and North-West 79, 83
    North Coast 87, 96
    Northern Rivers 100
    Outback and the Murray 110
    South Coast and Illawarra 114
    Sydney and Southern Highlands 13
  Northern Territory
    Central Australia 128
  Queensland
    Bundaberg, Fraser Coast and South Burnett 140–1
    Toowoomba and the Southern Downs 181, 183, 186
  South Australia
    Adelaide and Surrounds 204, 210–11
    Barossa, 219
    Clare Valley and Yorke Peninsula, 225–9
    Kangaroo Island, 238–9
    Limestone Coast 242
  Tasmania
    Bass Strait Islands 262
    The North-West 290
    The South 285
  Victoria 295
    Goldfields 310
    The Grampians 366
    The Great Ocean Road 370
    Legends, Wine and High Country 326–7
    Macedon Ranges and Spa Country 344, 349
    Murray and the Outback 354
  Western Australia
    The South-West 415
Lambs Valley NSW 70–1
Lamont's Margaret River, Yallingup WA 417
Lamont's, Millendon WA 396–7
Lancefield & District Farmers' Market, Lancefield Vic 315
Landsborough Vic 367
Langhorne Creek SA 205–6
Lark Distillery Pty Ltd, Hobart Tas 255–6
Latrobe Tas 290–1
Launceston Tas 265–8, 274
Lauralla Guesthouse and Grapevine Restaurant, Mudgee NSW 58
Laurel Hill Berry Farm, Tumbarumba NSW 41
lavender
  New South Wales
    Explorer Country and Riverina 61
    Hunter 77
    Sydney and the Southern Highlands 18
  Queensland
    Bundaberg, Fraser Coast and South Burnett 142
  South Australia
    Adelaide and Surrounds 206–7
    Clare Valley and Yorke Peninsula, 225
    Limestone Coast 241, 247–8
  Victoria
    The Grampians 368
    Legends, Wine and High Country 331
    Macedon Ranges and Spa Country 346–8
    Melbourne 299
    Yarra Valley and the Dandenong Ranges 382
  Western Australia
    The South-West 415
Lavender Fest, Wandin Yallock Vic 385
Lavender Hue, Harrietville Vic 331–2

Lebrina Tas 268–9
Leederville WA 390
Leeton NSW 54
Leeuwin Estate Restaurant, Margaret River WA 410
The Left Bank Coffee and Foodbar, Swansea Tas 273
The Left Bank, Kilkivan Qld 142
Leichhardt Farmers' Markets, Leichhardt NSW 19
Lemnos Vic 317
lemon myrtle
  New South Wales
    North Coast 91–2
    Northern Rivers 100, 106
    Outback and the Murray 111
  Queensland
    Brisbane and Surrounds 137
    Bundaberg, Fraser Coast and South Burnett 140
    Toowoomba and the Southern Downs 182
Lenah Game Meats, Rocherlea Tas 271
Lennox Head NSW 103–4
Lenswood SA 206
Leongatha Vic 361–2
Lesley Black's, Hobart Tas 256–7
Letterbox, Terrigal NSW 46
Leura NSW 12, 25
Leura Falls NSW 26
Leura Butchery, Leura NSW 25
Lick the Spoon, Dorrigo NSW 89
licorice
  New South Wales
    Explorer Country and Riverina 53–4
    Sydney and Southern Highlands 12
  Queensland
    Sunshine Coast 176
Lidsdale NSW 54–5
Lifeboat Seafoods, Brooklyn NSW 44–5
Lighthouse Seafood Restaurant, Mackay Qld 190
The Lily Stirling Range Dutch Windmill, Stirling Range National Park WA 414–15
Lilydale Vic 379
Lime Grove, Narromine NSW 59
limes
  Australian Capital Territory
    Canberra 30
  New South Wales
    Explorer Country and Riverina 59
    North Coast 96
    Northern Rivers 100, 104, 105
  Northern Territory
    Central Australia 130
  Queensland
    Sunshine Coast 173, 180
  South Australia
    Adelaide and Surrounds 206
  Victoria
    Bays and Peninsulas 304
Limestone Coast Farmers Market SA 248
Limestone Coast Trout, Millicent SA 243
Limoncello Australia Pty Ltd, Goodwood SA 201
Limpinwood Teahouse, Limpinwood NSW 104
Limpinwood NSW 104
Lindegger Orchard, Crystal Waters Qld 172–3
Linden Tree Restaurant, Red Hill South Vic 307
Linga Longa Lunch Festival, Yarrawonga Vic 356
Linkes Central Meat Store, Nuriootpa SA 220–1
Lipscombe Larder, Sandy Bay Tas 257
liqueurs and spirits see also rum
  Queensland
    Far North 165
    Gold Coast 169
    Toowoomba and the Southern Downs 182–5

INDEX

South Australia
  Adelaide and Surrounds 201
Tasmania
  Hobart and Surrounds 255-5
  The South 280, 284
Western Australia
  Australia's North-West 423
  The South-West 405, 410
Lismore Rainbow Region Organic Market, Lismore NSW 109
Lismore NSW 104-5, 109
Lithgow NSW 26
Little Hartley NSW 26
L'Oasis Restaurant, Griffith NSW 51-2
Lobethal Bakery, Lobethal SA 206
Lobethal SA 206-7
lobster *see* seafood
The Local Harvest, Bilpin NSW 21
Local Producers' Market, Milawa Vic 342
Lockyer Discovery Tours, Gatton Qld 136
Lodge 241 Gallery Cafe, Bellingen NSW 85
Lolli Redini, Orange NSW 61
London Hill, Clare SA 227
The Lonely Palate Winery, Dorrigo NSW 89
Longreach Qld 149
Loorana, King Island Tas 261
The Loose Box Restaurant, Mundaring WA 397
Lorenzo's Diner, Wollongong NSW 123
Lorne NSW 91-2
Lorne Valley Macadamia Farm, Kendall NSW 90-1
Love at First Bite, Tilba Tilba NSW 120-1
Lovedale Long Lunch, Pokolbin NSW 78
Lower Deck, Hobart Tas 258
Loxton SA 250
LP Dutton Trout Hatchery, Ebor NSW 79
Luskintyre NSW 71
Lymington Tas 283
Lyndoch SA 220
Lyndoch Lavender Farm, Lyndoch SA 206-7
Lynton Fruit and Vegie Farm, Robigana Tas 271
Lynwood Cafe, Collector NSW 35-6

The Macadamia Castle, Knockrow NSW 103
macadamias
  New South Wales 9
    Capital Country and the Snowy Mountains 42
    Central Coast 47
    North Coast 87, 90-2
    Northern Rivers 99-100, 102-3, 106-7
  Queensland 135
    Brisbane and Surrounds 137
    Bundaberg, Fraser Coast and South Burnett 140
    Far North 153, 156
    Sunshine Coast 171, 173, 178-9
    Townsville, Mackay and the Whitsundays 191
  Western Australia
    The South-West 401-2
Macclesfield Vic 379
Macedon Grove Olives, Gisbourne Vic 345-6
MACEDON RANGES AND SPA COUNTRY Vic 343-8
Mackay Discovery Tours, Mackay Qld 190
Mackay Fish Market, Mackay Qld 189-90
Mackay Qld 189-90, 192
Macksville NSW 92
McLaren Flat SA 207-8
McLaren Vale SA 208-10
McLaren Vale Olive Groves, McLaren Flat SA 207
Maclean NSW 105, 109
Maclean Farmers' Market, Maclean NSW 109

MacNuts WA, Baldivis WA 401-2
McPhee's Meats, Burra SA 226
Macquarie Valley Wine & Food Festival, Dubbo NSW 65
Madge Malloys, Coles Bay Tas 264
Madura Tea Estates, Murwillumbah NSW 105-6
Maggie Beer's Farm Shop, Nuriootpa SA 221
Magill Estate Restaurant, Magill SA 199
Magnetic Island Qld 190-1
Magnetic Island Tropical Resort, Magnetic Island Qld 190-1
Magnetic Mango, Magnetic Island Qld 191
Magnolia Restaurant, Mount Victoria NSW 28
Magpie's Nest Restaurant, Wagga Wagga NSW 63
Mahalia Coffee, Robe SA 247
Main Beach Qld 167, 170, 170
Main Beach Farmers' Market, Main Beach Qld 170
Main Ridge Vic 303-4
Maitland NSW 71, 78
Malanda Dairy Centre, Malanda Qld 156
Malanda Qld 156
Maleeya's Thai, Porongurup WA 414
Maleny Qld 174-5
Mallee Fowl Restaurant, Berri SA 249
Mallyons on the Murray, Waikerie SA 251
Mammino Gourmet Ice-Cream, Childers Qld 140
Mandalong Grain-fed Lamb, Tamworth NSW 83
mandarins
  New South Wales
    Hunter 67
    Outback and the Murray 111
  Queensland
    Far North 160
Mandurah Quay Restaurant, Mandurah WA 396
Mandurah WA 396
mangos
  Australian Capital Territory
    Canberra 30
  New South Wales
    North Coast 96
    Northern Rivers 106
    South Coast and Illawarra 120
  Northern Territory 125
    Top End 133
  Queensland 135
    Bundaberg, Fraser Coast and South Burnett 139-40
    Far North 151-3, 158
    Toowoomba and the Southern Downs 186
    Townsville, Mackay and the Whitsundays 188, 191
  Western Australia
    Australia's Coral Coast 426
    Australia's North-West 418, 423
Manilla NSW 81
Manjimup WA 409
Manjimup Cherry Harmony Festival, Manjimup WA 416
Manjimup Farmers' Market, Manjimup WA 416
Manning Point NSW 92
Manning Valley River Cruises, Manning Point NSW 92
Mansfield Vic 332
Manyallaluk—The Dreaming Place, Katherine NT 128
Maple Street Co-op, Maleny Qld 174
Mareeba Qld 156-8, 166
Mareeba Coffee, Tichum Creek Coffee Farm, Mareeba Qld 156-7
Mareeba Markets, Mareeba Qld 166
Margaret River WA 409-11
The Margaret River Dairy Co, Metricup WA 411
Margaret River Dairy Company, Margaret River WA 409
Margaret River Farmers' Market, Margaret River WA 416
Margaret River Goat Cheese, Cowaramup WA 403

447

INDEX

Margaret River Hotel, Margaret River WA 410
Margaret River Venison, Margaret River WA 409
Margaret River Wine Region Festival, Margaret River WA 416
Margaret Riviera, Cowaramup WA 404
Marina Mirage Farmers' Markets, Marina Mirage Qld 170
marinades *see* sauces
Market at the Abbey, Warwick Qld 187
markets
  Australian Capital Territory
    Canberra 31, 43
  New South Wales
    Blue Mountains and the Hawkesbury 27
    Capital Country and the Snowy Mountains 33, 43
    Explorer Country and Riverina 65
    Hunter 78
    New England and North-West 84
    North Coast 92, 98
    Northern Rivers 109
    Sydney and Southern Highlands 13, 14, 19
  Northern Territory
    Top End 132–3
  Queensland
    Brisbane and Surrounds 138
    Bundaberg, Fraser Coast and South Burnett 145
    Far North 166
    Gold Coast 170
    Sunshine Coast 171, 173, 180
    Toowoomba and the Southern Downs 187
  South Australia
    Adelaide and Surrounds 196, 216
    Barossa, 217
    Limestone Coast 247
  Tasmania
    The North 274
    The North-West 294
  Victoria
    Bays and Peninsulas 309
    Goldfields 315
    Goulburn Murray Waters 320
    The Great Ocean Road 374
    Lakes and Wilderness 324
    Legends, Wine and High Country 342
    Macedon Ranges and Spa Country 349
    Melbourne 299
    Phillip Island and Gippsland 364
    Yarra Valley and the Dandenong Ranges 385
  Western Australia
    Australia's Coral Coast 426
    Australia's North-West 418, 423
    Perth and Surrounds 399
    The South-West 416
Markets at the Palms, Forster-Tuncurry NSW 98
Markwood Vic 332
marmalades *see* jams and marmalades
Maroondah Orchards, Coldstream Vic 376
Maroudas Olives, Thebarton SA 197
marron
  Western Australia
    Perth and Surrounds 395–7
    The South-West 406–8, 410, 412–14, 417
Martin's Orchard, Wandandian NSW 123
Mary Rose Restaurant, Denmark WA 406
Maryborough Heritage Markets, Maryborough Qld 145
Mary's Pasta Products, Moresby Qld 159
Marysville Vic 380
Matso's Broome Brewery, Broome WA 418–19
Mattisse, Hahndorf SA 203
Mauger's Meat, Burrawang NSW 16–17

Max's at Red Hill Estate, Red Hill South Vic 307
Mayfield Chocolates, Spring Hill Qld 136–7
mead
  New South Wales
    New England and North-West 81
  Tasmania
    Hobart and Surrounds 256
  Western Australia
    The South-West 405
Meadowbank Estate Vineyard and Restaurant, Cambridge Tas 277–8
Meadows SA 210
meat *see also* by type of meat
  Australian Capital Territory
    Canberra 32
  New South Wales
    Blue Mountains and the Hawkesbury 22, 24
    Capital Country and the Snowy Mountains 38, 42
    Explorer Country and Riverina 48, 51–3, 55, 57, 61–3
    Hunter 68–70, 76
    New England and North-West 79, 83
    North Coast 87–9, 92
    Outback and the Murray 110
    South Coast and Illawarra 114–15
    Sydney and Southern Highlands 12–13, 16–17
  Northern Territory
    Central Australia 127–30
  Queensland
    Bundaberg, Fraser Coast and South Burnett 140, 144
    Sunshine Coast 172
    Toowoomba and the Southern Downs 184
  South Australia 193
    Adelaide and Surrounds 196, 208–11, 215
    Barossa, 218–21
    Eyre Peninsula, Flinders Ranges and the Outback, 232–4
    Limestone Coast 246–7
  Tasmania
    Hobart and Surrounds 256
    The North 271
    The North-West 292
    The South 275, 285
  Victoria 295
    Bays and Peninsulas 302
    Legends, Wine and High Country 332, 341
    Murray and the Outback 354
    Phillip Island and Gippsland 360
  Western Australia
    Perth and Surrounds 389
Medicine Garden Australia, Lismore NSW 104
Medika Gallery, Blyth SA 226
Medowie NSW 92
Medowie Macadamia Growers, Medowie NSW 92
Megalong Valley NSW 26–7
Megalong Tearooms and Kiosk, Megalong Valley NSW 26–7
Melba's Chocolates, Woodside SA 216
MELBOURNE 9, 298–300
Melbourne Food and Wine Festival, Melbourne Vic 299
melons
  Northern Territory 131
  Queensland 186
  Western Australia 394
Melrose SA 233
Meningie SA 243
Merimbula NSW 118
Merlot Restaurant, Milawa Vic 333–4
Merrich Estate Olive Farm and Mediterranean Kitchen, Henley Brook WA 394–5

INDEX

Merricks North Vic 304
Merrijig Inn, Port Fairy Vic 372
Metricup WA 411
Michelin, Griffith NSW 52
Michel's Restaurant, South Townsville Qld 191
Mick's Bakehouse, Leeton NSW 54
Middle Swan WA 396
Midland Farmers Market, Midland WA 399
Midlands Aquaculture, Guyra NSW 80
Miellerie—The House of Honey, Woodbridge Tas 287
Mighty Murray Tart Trawl, Albury Vic 112
Mighty Murray Tart Trawl, Echuca Vic 356
Milawa Vic 333-4, 342
Milawa Cheese Factory Bakery and Restaurant, Milawa Vic 334
Milawa Cheese Shop, Carlton North Vic 299
Milawa Mustard Pty Limited, Milawa Vic 333
Mildura Vic 352-4
The Mill Cellar Door and Function Centre, Cowra NSW 49-50
The Mill Providore and Gallery, Launceston Tas 267
Millaa Millaa Qld 159
Millards Cottage, Ulladulla NSW 122
Millbrook Winery, Jarrahdale WA 396
Millendon WA 396-7
Millhouse Cafe and Chocolate Co, Dwellingup WA 393
Millicent SA 243
Millthorpe NSW 55
Mindil Beach Markets, Darwin NT 132
Mingling Waters, Nowa Nowa Vic 323
Minlaton SA 228
Minmore Farmstay, Kingaroy Qld 143
Minnucci's Green Gold, Huonville Tas 282
Mintaro SA 228
Miranda Wines, Rowland Flat SA 221
Mirboo North Vic 362
Mission Beach Qld 159
Mission Beach Banana Festival, Mission Beach Qld 166
Misty's Restaurant and Accommodation, Dorrigo NSW 89
Mitchelton Winery Restaurant, Nagambie Vic 318
Mittagong NSW 15, 17, 19
Mittagong Fruit Market, Mittagong NSW 19
MJ and VG Jennings, Mareeba Qld 157
Mocka's Pies, Port Douglas Qld 162
Moffatdale Qld 144
Moffat's Oyster Barn Restaurant, Swan Bay NSW 76
Mogo NSW 118
Mojo's on Wilderness, Rothbury NSW 76
Mole Creek Tas 291
Mollymook NSW 118
Mondo di Carne Pty Ltd, Inglewood WA 389
Mondo Organics, West End Qld 137
Montalto, Red Hill South Vic 307
Montrose Berry Farm, Sutton Forest NSW 18
Mooball NSW 105
Mooka Fishing Co, Falcon WA 393
Mooloolaba Qld 175-6
Mooral Creek NSW 92-3
Mooral Creek Farms, Mooral Creek NSW 92-3
Moorilla Estate Winery Restaurant, Berriedale Tas 275
Moorings, Port Lincoln SA 234
Moorooduc Vic 305
Mooroopna Friut Salad Day, Shepparton Vic 320
Mooroopna Vic 317
Moresby Qld 159
Morning Star Estate, Mount Eliza Vic 305
Mornington Vic 305
Mornington Wilderness Camp, Derby WA 421

Morpeth Honey Festival, Morpeth NSW 78
Morrall's Bakery and Cafe, Bourke NSW 111
Mossman Qld 160-1
Mossman Gorge Qld 161
Mount Barker SA 210-11
Mount Barker WA 411
Mount Beauty Vic 334-5
Mount Beauty Bakery-Pasticceria-Cafe, Mt Beauty Vic 335
Mount Bruno Vic 335
Mount Compass SA 211
Mount Cotton Qld 168, 170
Mount Dandenong Vic 380
Mount Egerton Vic 346-7
Mount Eliza Vic 305
Mount Gambier SA 244
Mount Hart Wilderness Lodge, Derby WA 422
Mount Horrocks, Auburn SA 225
Mount Hudson Strawberries, Lambs Valley NSW 70-1
Mount Irvine NSW 27
Mount Jagged SA 211-12
Mount Kembla NSW 119
Mount Lawley WA 390
Mount Lofty House, Crafers SA 200-1
Mount Magnet WA 424
Mount Markey Winery, Cassilis Vic 321
Mount Stirling Olives, Glen Aplin Qld 182-3
Mount Tamborine Qld 168, 170
Mount Tamborine Coffee Plantation and Fingerprint Gallery, Mount Tamborine Qld 168
Mount Tomah NSW 27-8
Mount Tomah Restaurant, Mount Tomah NSW 27-8
Mount Uncle Distillery, Walkamin Qld 165
Mount Victoria NSW 28
Mount Vincent NSW 71-2
Mount Zero Olives Cafe, Laharum Vic 366
Mountain Fine Fare, Leura NSW 12
Mountainview Hotel—King Valley, Whitfield Vic 340
The Mousetrap, Timboon Vic 373
Muddies Restaurant, Mackay Qld 190
Mudgee NSW 9, 14, 26, 55
Mudgee Gourmet @ Heart of Mudgee, Mudgee NSW 56
Mudgee Honey Company Pty Ltd, Mudgee NSW 56
Mudgee Regional Farmers' Markets, NSW 65
Mudgee Wine and Food Fair, Balmoral Beach NSW 19
Mudgee Wine Celebration, NSW 65
Mudgeeraba Qld 168, 170
Mudgeeraba Farmers' Market, Mudgeeraba Qld 170
Mugglestons General Store and Restaurant, Hahndorf SA 203
Mullion Creek NSW 58
Mundaring WA 397
Mungalli Creek Dairy and Out of the Whey Cafe, Millaa Millaa Qld 159
Munro's Prime Mallee Meats, Ouyen Vic 354
Murdering Point Winery, Silkwood East Qld 164
Mure's Fish Centre, Hobart Tas 257
Murgon Qld 144-5
Murgon Country Markets, Murgon Qld 145
Murgon Dairy Festival, Murgon Qld 145
Murgon Dairy Museum, Murgon Qld 144
Murrabit Country Market, Murrabit Vic 356
MURRAY AND THE OUTBACK NSW 350-6
Murray Goulburn Trading, Leongatha Vic 361-2
MURRAY REGION NSW 110-12
Murray Regional Food and Wine Diary NSW 111
Murray River Salt, Mildura Vic 352
Murwillumbah NSW 105-6, 109

449

INDEX

mushrooms
  Australian Capital Territory
    Canberra 31
  New South Wales
    Blue Mountains and the Hawkesbury 26, 28, 29
  Tasmania
    The North-West 291, 293
    The South 275, 281-2, 284
  Victoria
    Bays and Peninsulas 307
    Legends, Wine and High Country 334
Musk Vic 347
Musk Berry Farm, Musk Vic 347
The Mussel Boys Cafe Gallery, Taranna Tas 286
mustard
  Australian Capital Territory
    Canberra 32
  New South Wales
    Capital Country and the Snowy Mountains 40
    Explorer Country and Riverina 52, 63
    Sydney and Southern Highlands 18
  South Australia
    Adelaide and Surrounds 203, 206
    Clare Valley and Yorke Peninsula, 227
  Tasmania
    The North-West 291
    The South 284
  Victoria
    Legends, Wine and High Country 333-6
    Phillip Island and Gippsland 360
  Western Australia
    The South-West 411, 415, 415
Muswellbrook NSW 72
Mutarnee Qld 191
Mylor SA 212
Mytrleford Alpine Valleys Gourmet Weekend, Bright Vic 342
Mytrleford Produce Market, Myrtleford Vic 342

Nabiac NSW 93
Nagambie Vic 318
Nairne SA 212
Nambour Qld 176
Nannup WA 412
Nanson WA 427
Naracoorte SA 245
Nareeda Valley Olives, Vacy NSW 77
Narrandera NSW 59
Narromine NSW 59
Nashua NSW 106
National Cherry Festival, Young NSW 43
National Tomato Contest, Gunnedah NSW 84
National Wine Centre of Australia, Adelaide SA 197
Naturally Nichols, Sisters Creek Tas 292
Nautilus Floating Dockside Restaurant, Lakes Entrance Vic 323
Nautilus Restaurant, Port Douglas Qld 162-3
Neagles Rock Vineyard Restaurant, Clare SA 227
Nectarbrook Discovery Plantation, Port Augusta SA 234
nectarines
  New South Wales
    Blue Mountains and the Hawkesbury 20
    Capital Country and the Snowy Mountains 33-4, 39, 41-2
    Explorer Country and Riverina 60
    South Coast and Illawarra 121, 123
  Victoria
    Lakes and Wilderness 322

  Western Australia
    Perth and Surrounds 398
Neerim South Vic 362
Neila, Cowra NSW 49
Neil's Organics, Cairns Qld 152
Nelson Bay NSW 72
Nelson's Restaurant, Bridgetown WA 402
Netherhill Strawberry Farm, Kenton Valley SA 205
NEW ENGLAND AND NORTH-WEST NSW 79-84
New Norcia WA 397
New Norcia Woodfired Bakery, Subiaco WA 389
NSW Farmers' Association Farmers' Market, Liverpool NSW 19
Newcastle Fish Co-op, Newcastle NSW 44
Newmans Horseradish, Langhorne Creek SA 205-6
Newrybar NSW 106
Newslink Discover South Australia, Adelaide SA 197-8
Newton's Organic Prickleberry Farm, Whitfield Vic 339
Newtown Providores, Dubbo NSW 50
Ngoorabul Aboriginal Tourism, Glen Innes
Niangala NSW 81-2
Nimbin Organics, Nimbin NSW 106-7
Nimbin NSW 106-7
Nimmitabel Butchery, Cooom NSW 36
1918 Bistro and Grill, Tanunda SA 223
Nirvana Organic Produce, Heathfield SA 203-4
No 2 Oak St, Bellingen NSW 85
Nolan's Butchery, Mansfield Vic 332
Noosa Qld 176
Noosa Farmers' Markets, Noosaville Qld 180
Noosa Heads Qld 176-7
Noosa Sound Qld 177
Noosaville Qld 177
Norco Cooperative, South Lismore NSW 105
Norfolk Punch (Australia), Kendall NSW 91
Norland (Fig) Orchard, Borenore NSW 49
THE NORTH Tas 263-74
North Arm Cove NSW 93
North Queensland Gold Coffee Plantation, Mareeba Qld 157
North Rothbury NSW 72-3
North Star Inn Hotel, Melrose SA 233
North Tamborine Qld 169
North Torbay WA 412
North Walpole WA 412
THE NORTH-WEST Tas 288-94
Northbridge WA 390
NORTHERN RIVERS NSW 99-109
Northern Rivers Herb Festival, Lismore NSW 109
Northey Street Organic Market, Windsor Qld 138
Northside Produce Market, North Sydney NSW 19
Nowa Nowa Vic 323
Nucifora Tea, Innisfail Qld 155
Nundle Yabby Farm, Nundle NSW 82
Nundle NSW 82
Nuriootpa SA 220
The Nut Farm, Gympie Qld 173
nuts
  New South Wales
    Blue Mountains and the Hawkesbury 21, 25
    Capital Country and the Snowy Mountains 42
    Central Coast 47
    Explorer Country and Riverina 55
    New England and North-West 82-3
    North Coast 90-1
    Northern Rivers 99-100, 102-3, 106-7
  Queensland 135
    Brisbane and Surrounds 137
    Bundaberg, Fraser Coast and South Burnett 140, 143

INDEX

Far North 153, 156, 164
Sunshine Coast 171, 173, 178-9
Toowoomba and the Southern Downs 182, 186
Townsville, Mackay and the Whitsundays 191
South Australia
  Adelaide and Surrounds 195-6, 203-4, 207-8
  Barossa, 217
  Murray and Riverland 250
Victoria
  Legends, Wine and High Country 326, 328
  Murray and the Outback 350, 352-3
  Phillip Island and Gippsland 359
Western Australia
  The South-West 401-2
Nuts About Fruit, Renmark SA 250
Nutworks Macadamia Processors, Yandina Qld 179

The Oaks NSW 28-9
oats
  New South Wales
    Explorer Country and Riverina 54
  Victoria 295
  Western Australia 387
Oberon NSW 14, 25
The Ocean Front Brasserie, Coffs Harbour NSW 88
Ocean Grove Primary School Apple Fair, Ocean Grove Vic 309
Officer Organics, Beaconsfield Vic 357
oils
  Australian Capital Territory
    Canberra 32
  New South Wales
    Capital Country and the Snowy Mountains 41
  Queensland
    Bundaberg, Fraser Coast and South Burnett 143
  South Australia
    Adelaide and Surrounds 195-6
  Victoria
    Bays and Peninsulas 307
    Goldfields 311-12
    Yarra Valley and the Dandenong Ranges 376
  Western Australia
    The South-West 417
OiOi Oister Days, McLaren Vale SA 216
Old Bus Depot Markets, Kingston ACT 30, 33
The Old George and Dragon Restaurant, East Maitland NSW 68
Old Goldfields Orchard and Cider Factory, Donnybrook WA 406-7
The Old Mill Cafe and Restaurant, Millthorpe NSW 55
Old Zoo Cafe, Cable Beach WA 421
The Olde Apple Shed, Balhannah SA 199
Olinda Vic 380
Olio Bello Organic, Cowaramup WA 403-4
The Olive Nest, Mudgee NSW 55
The Olive Pit, Dunsborough WA 407
The Olive Shop, Milawa Vic 333
olives and olive oil
  Australian Capital Territory
    Canberra 31
  New South Wales 9
    Blue Mountains and the Hawkesbury 23, 28, 29
    Capital Country and the Snowy Mountains 42
    Explorer Country and Riverina 48-9, 51-2, 56-8, 60-3
    Hunter 66-7, 71, 74, 76, 77
    New England and North-West 79-81, 83-4
    North Coast 89, 97
    Northern Rivers 106

  Outback and the Murray 110-11
  South Coast and Illawarra 120
  Sydney and Southern Highlands 15, 18, 19
Queensland
  Bundaberg, Fraser Coast and South Burnett 139, 141-3
  Sunshine Coast 171
  Toowoomba and the Southern Downs 182-3, 186
South Australia
  Adelaide and Surrounds 197, 200, 202, 207-9, 211, 213-15
  Barossa, 219, 223-4
  Clare Valley and Yorke Peninsula, 225-7
  Kangaroo Island, 238
  Limestone Coast 247
  Murray and Riverland 249-50
Tasmania
  The South 276-7, 282, 284
Victoria
  Bays and Peninsulas 307
  Goldfields 311
  The Grampians 366-8
  The Great Ocean Road 371
  Legends, Wine and High Country 329, 331-4, 338-40
  Macedon Ranges and Spa Country 345-6
  Murray and the Outback 355
  Yarra Valley and the Dandenong Ranges 378
Western Australia
  Perth and Surrounds 394-7, 399
  The South-West 403-4, 407
Olivo, Byron Bay NSW 101
Olson Pheasant Farm, Swan Hill Vic 355
On The Inlet, Port Douglas Qld 163
On the Pier, Batemans Bay NSW 113
Orange NSW 9, 14, 59-61
Orange Farmers' Market, NSW 60, 65
Orange Festival, Bingara NSW 84
Orange Picking Day, Bingara NSW 84
oranges
  New South Wales
    Blue Mountains and the Hawkesbury 23, 26
    Explorer Country and Riverina 54
    Hunter 67
    New England and North-West 84
    Outback and the Murray 111
    Northern Territory 125
  South Australia
    Barossa, 217
    Murray and Riverland 249
  Victoria
    Goldfields 312
Orangeworld, Buronga NSW 111
O'Reilly's Orchard, Wirrabara Forest SA 230
Organic Energy, Griffith ACT 33
Organic Feast, Maitland NSW 71
Organic Food and Farmers' Market, Hornsby NSW 19
The Organic Market, Frenchs Forest NSW 19
The Organic Market, Stirling SA 213
Organic Pasta Shop, Balhannah SA 199-200
organic produce
  Australian Capital Territory
    Canberra 30-3
  New South Wales
    Blue Mountains and the Hawkesbury 21, 28
    Capital Country and the Snowy Mountains 35-6, 38
    Central Coast 46
    Explorer Country and Riverina 53-4, 60
    Hunter 66, 70-1, 76-8

451

INDEX

organic produce *continued*
  New South Wales *continued*
    New England and North-West 79
    North Coast 85-6, 89-90, 92-3
    Northern Rivers 100, 106
    South Coast and Illawarra 114-15, 118
    Sydney and Southern Highlands 12-14, 19
  Queensland
    Brisbane and Surrounds 136-7
    Bundaberg, Fraser Coast and South Burnett 144-5
    Far North 152, 157, 159
    Gold Coast 168
    Sunshine Coast 171-4, 176
    Toowoomba and the Southern Downs 183, 186
  South Australia
    Adelaide and Surrounds 197, 199, 203-4, 209-11, 213-15
    Clare Valley and Yorke Peninsula, 230
    Eyre Peninsula, Flinders Ranges and the Outback, 235
    Limestone Coast 242
    The North 268, 271, 274
    The North-West 290
    The South 278, 286-7
  Victoria
    Bays and Peninsulas 302-3, 305
    Goldfields 310
    Lakes and Wilderness 323
    Legends, Wine and High Country 331, 339
    Macedon Ranges and Spa Country 345
    Phillip Island and Gippsland 357, 360, 363
  Western Australia
    Perth and Surrounds 389
    The South-West 415
Organics on Eleventh, Mildura Vic 353
Orroroo SA 233-4
Oscar W's, Echuca Vic 351-2
Oss Eels, Skipton Vic 348
ostrich
  Australian Capital Territory
    Canberra 32
  Western Australia
    The South-West 400
Otherwood Farm Tours, Lenswood SA 206
Otway Harvest Festival, Geelong Vic 309
Otway Herbs, Apollo Bay Vic 369
Out and About Wine Tours, Bassendean WA 391
OUTBACK NSW 110-12
THE OUTBACK WA 424-5
Outback Pioneer BBQ and Bar, Uluṟu-Kata Tjuṯa National Park NSW 129
Outrigger Restaurant, Hamilton Island Qld 189
Ouyen Vic 354
Ovens Vic 335
Overlander's Steakhouse, Alice Springs NT 127
Oxley Vic 335
Oyster Cove Inn, Kettering Tas 283
The Oysterbeds, Coffin Bay SA 232
oysters
  New South Wales 9
    Central Coast 44-7
    Hunter 76
    North Coast 88, 92, 94-6
    Northern Rivers 101, 104, 107-8
    South Coast and Illawarra 113-14, 117-19, 121
    Sydney 12
  Northern Territory
    Central Australia 127
    Top End 132

Queensland
  Gold Coast 167
  Townsville, Mackay and the Whitsundays 189
South Australia
  Eyre Peninsula, Flinders Ranges and the Outback, 231-2, 235
Tasmania
  The North 263, 266, 272-3
  The South 276, 283-5
Ozone Seafront Hotel, Kingscote SA 238

Pacific Toyota Palm Cove Festival, Palm Cove Qld 166
Padthaway Homestead, Padthaway SA 245-6
Padthaway SA 245-6
Pallamallawa NSW 82-3
Palm Cove Qld 161-2, 166
Palm Grove Date Farm and Winery, Eulo Qld 148
Palmerston NT 133
Palmview Qld 177
Pambula NSW 119
pancakes
  New South Wales
    Blue Mountains and the Hawkesbury 29
    Capital Country and the Snowy Mountains 41
    South Coast and Illawarra 116
  South Australia
    Murray and Riverland 250
Parachilna SA 234
Paradise Estate Wines, Mission Beach Qld 159
Parap Fine Foods, Parap NT 133
Parap NT 133
Parker's Pies and Pastries, Rutherglen Vic 337
Parkes NSW 60, 61
Parklands Farmers' Markets, Parklands Showground Qld 170
Parndana SA 239
Passchendaele Verjuice, Edi Upper Vic 329-30
passionfruit
  Queensland
    Far North 163
  Western Australia
    Australia's Coral Coast 426
pasta
  Queensland
    Far North 159
  South Australia
    Adelaide and Surrounds 199
    Barossa, 221
  Victoria
    Goldfields 311
    Legends, Wine and High Country 328
    Phillip Island and Gippsland 362
    Yarra Valley and the Dandenong Ranges 377, 379, 381, 383
pastries *see* cakes, biscuits and pasties
Paterson NSW 73-4
Patrice's Table at Loxley on Bellbird Hill, Kurrajong NSW 25
Paupiettes, Lismore NSW 105
pawpaw
  New South Wales
    North Coast 87
    Northern Rivers 106
    South Coast and Illawarra 122
  Northern Territory
    Top End 132
  Queensland
    Far North 153, 160, 164
    Townsville, Mackay and the Whitsundays 192

452

INDEX

Western Australia
  Australia's Coral Coast 426
PCYC Markets, Griffith NSW 65
peaches
  New South Wales
    Blue Mountains and the Hawkesbury 20-1
    Capital Country and the Snowy Mountains 33-4, 39, 41-2
    Explorer Country and Riverina 49
    South Coast and Illawarra 119, 121, 123
  South Australia
    Adelaide and Surrounds 212
    Barossa, 217
    Murray and Riverland 249
  Victoria
    Lakes and Wilderness 322
  Western Australia
    Perth and Surrounds 398
Peanut Festival, Kingaroy Qld 145
The Peanut Van, Kingaroy Qld 143
peanuts
  Queensland
    Bundaberg, Fraser Coast and South Burnett 139, 143
    Far North 164
Pearl Luggers, Broome WA 421
pears
  New South Wales
    Blue Mountains and the Hawkesbury 20
    Capital Country and the Snowy Mountains 34, 37, 39
    Hunter 78
    South Coast and Illawarra 119-20
  Queensland
    Toowoomba and the Southern Downs 181
  South Australia
    Adelaide and Surrounds 199, 208, 212
    Barossa, 217
  Tasmania 253
    Hobart and Surrounds 257
  Victoria
    Goldfields 312
    The Grampians 367
    Legends, Wine and High Country 326, 328
    Macedon Ranges and Spa Country 347
  Western Australia
    Perth and Surrounds 398
Pecan Hill Tea Room—Museum, West Toodyay WA 398
pecans
  New South Wales 9
    New England and North-West 82-3
    North Coast 87
  Queensland
    Sunshine Coast 173
    Toowoomba and the Southern Downs 186
  Western Australia
    Perth and Surrounds 398
Pee Wee's at the Point, Darwin NT 132
Peel Estate Jazz, Baldivis WA 416
Pemberton WA 412-14
Pemberton Marron and Wine Festival, Pemberton WA 416
Pemberton Mill House Cafe, Pemberton WA 413
The Penguin Cafe on Bruny, Bruny Island Tas 275-6
Penguin Stop Cafe, Penneshaw SA 240
Peninsula Occasions, Rosebud Vic 308
Penna Lane Wines and Products, Penwortham SA 228
Penneshaw SA 239-40
Pennyroyal Raspberry Farm, Birregurra Vic 370
Penola SA 246-7

Penrith Valley Oranges, Castlereagh NSW 23
Penwortham SA 228
Peppermint Bay, Woodbridge Tas 287
Peppers on Queens, Ayr Qld 188
Perenti, Gloucester NSW 69-70
Perfect Poultry, Kojonup WA 408
persimmons
  New South Wales
    Blue Mountains and the Hawkesbury 21
  South Australia
    Adelaide and Surrounds 212
  Western Australia
    Perth and Surrounds 398
    The South-West 402
Perth Tas 269
Perth WA 388-91
PERTH AND SURROUNDS WA 388-99
Perth Royal Show, Claremont WA 399
pesto
  Australian Capital Territory
    Canberra 32
    Hunter 77
    North Coast 94
    South Coast and Illawarra 118
  South Australia
    Limestone Coast 246
Petaluma's Bridgewater Mill, Bridgewater SA 200
Peter Lehmann Wines, Tanunda SA 223
Peters and Sons Selected Meats and Delicatessen, North Albury NSW 110
Pettavel Winery and Restaurant, Waurn Ponds Vic 309
Petticoat Lane Herb Garden, Penola SA 246
Petuna Seafoods Pty Ltd, Launceston Tas 267
pheasant
  Victoria
    Murray and the Outback 355
Pheasant Creek Vic 380-1
Philip Rowe and Cathie Taylor, Childers Vic 358
PHILLIP ISLAND AND GIPPSLAND Vic 357-64
Phillip Island Farmers' Market, Churchill Island Vic 364
Pialligo Apples, Piallgo ACT 30-1
Pick-of-the-Crop Wine Expeditions, Mt Vincent NSW 71-2
Pick Your Own Fruit and Kimber Wines, McLaren Vale SA 208-9
The Pickled Sisters Cafe, Wahgunyah Vic 338-9
pickles *see* chutneys and pickles
Picnics at the Rocks, Yandina Qld 179-80
pie floaters 193
Pier Restaurant and Cocktail Bar, Hervey Bay Qld 141
Piers Restaurant, Portsea Vic 306
pies
  Australian Capital Territory
    Canberra 32
  New South Wales
    Blue Mountains and the Hawkesbury 26, 27, 28
    Capital Country and the Snowy Mountains 37, 41
    Explorer Country and Riverina 54
    North Coast 89-90, 96
    Outback and the Murray 111
    South Coast and Illawarra 115-16
    Sydney and Southern Highlands 17, 19
  Queensland
    Bundaberg, Fraser Coast and South Burnett 141
    Capricorn, Gladstone and the Outback 148
    Far North 162
  South Australia 193
    Adelaide and Surrounds 212
    Clare Valley and Yorke Peninsula, 230
    Eyre Peninsula, Flinders Ranges & the Outback, 231-2

453

INDEX

pies *continued*
Tasmania
Hobart and Surrounds 255
The South 276
Victoria
Goulburn Murray Waters 317
Legends, Wine and High Country 332, 337
Macedon Ranges and Spa Country 344
Phillip Island and Gippsland 361, 364
Western Australia
The South-West 406
pigeon
New South Wales
Explorer Country and Riverina 61
Hunter 73
Victoria
Goldfields 312
Legends, Wine and High Country 328
pineapples
New South Wales 9
Northern Rivers 106
Queensland
Brisbane and Surrounds 137
Bundaberg, Fraser Coast and South Burnett 139-40
Far North 160, 164
Sunshine Coast 171, 178-9
Townsville, Mackay and the Whitsundays 191
South Australia
Barossa, 217
Pipers of Penola, Penola SA 246-7
Pipers Brook Tas 269
Pipers River Tas 270
Pippies by the Bay, Warrnambool Vic 374
pistachios
New South Wales
Explorer Country and Riverina 63
pizza
Australian Capital Territory
Canberra 32
New South Wales
Capital Country and the Snowy Mountains 35, 38
Explorer Country and Riverina 58
New England and North-West 81
South Coast and Illawarra 115
Northern Territory
Central Australia 128
South Australia
Adelaide and Surrounds 214-15
Barossa, 220
Kangaroo Island, 239
Victoria
Bays and Peninsulas 304, 308
Western Australia
Australia's North-West 419
Plantation Lorna Macadamias, Newrybar NSW 106
Ploys at Tuross, Tuross Head NSW 121
plums
Australian Capital Territory
Canberra 30
New South Wales
Blue Mountains and the Hawkesbury 20-1
Capital Country and the Snowy Mountains 34, 41-2
Explorer Country and Riverina 52, 60
North Coast 91
Northern Rivers 100, 106
South Coast and Illawarra 119, 121
Sydney and Southern Highlands 16
Northern Territory
Central Australia 130

Queensland
Far North 164
South Australia
Adelaide and Surrounds 206
Tasmania 253
Hobart and Surrounds 257
Western Australia
Perth and Surrounds 398
The South-West 402
Plunkett Wines, Avenel Vic 317
Poacher's Pantry, Hall NSW 15, 38
Pogels Wood Cafe and Restaurant, Federal NSW 102-3
Pokolbin NSW 66, 74-6, 78
Pokolbin Creek Olives, Luskintyre NSW 71
Pokolbin Growers' Market, Pokolbin NSW 78
Police Point Tas 283
Pomona Eco Markets, Pomona Qld 180
Pomona Organic Markets, Pomona Qld 180
Pomonal Vic 367-8
Pomonal Berry Farm, Pomonal Vic 367
Pooraka SA 212-13
Porepunkah Vic 336
pork
Australian Capital Territory
Canberra 32
New South Wales
Blue Mountains and the Hawkesbury 25
Capital Country and the Snowy Mountains 36, 38
Explorer Country and Riverina 49, 51, 55, 64
Hunter 69
North Coast 96
Northern Rivers 102, 104-5, 107
Outback and the Murray 110
South Coast and Illawarra 114
Sydney and Southern Highlands 13, 17
Northern Territory
Central Australia 128
Queensland
Bundaberg, Fraser Coast and South Burnett 141, 143-4, 146
Far North 152
Sunshine Coast 172, 176
Toowoomba and the Southern Downs 181, 184
Tasmania
The North-West 290
Victoria 295
The Great Ocean Road 370
Legends, Wine and High Country 334, 336
Phillip Island and Gippsland 360
Yarra Valley and the Dandenong Ranges 375, 377, 379, 383
Western Australia
Perth and Surrounds 397
Porongurup WA 414
Port Arthur Tas 283
Port Augusta SA 229
Port Augusta SA 234
Port Campbell Vic 371
Port Douglas Qld 162-3, 166
Port Douglas Markets, Port Douglas Qld 166
Port Elliot SA 213
Port Elliot Market, Port Elliot SA 216
Port Fairy Vic 371-2
Port Lincoln SA 234-5
Port Macquarie NSW 94, 98
Port Willunga SA 213
Portabello's Cafe, Port Macquarie NSW 94
Portarlington Vic 305-6
Portsea Vic 306

454

INDEX

possum
　Australian Capital Territory
　　Canberra 32
　　Tasmania 253
　　The North 271
Possum Fruits Gourmet Foods, The Oaks NSW 28–9
Possum Shed Cafe & Craft Gallery, Westerway Tas 286–7
Possum's Cafe, Woolgoolga NSW 98
potatoes
　Australian Capital Territory
　　Canberra 33
　New South Wales
　　Capital Country and the Snowy Mountains 37, 39
　　Explorer Country and Riverina 52
　　New England and North-West 81, 84
　　Sydney and Southern Highlands 15, 17
　Queensland
　　Brisbane and Surrounds 136
　　Far North 156
　　Toowoomba and the Southern Downs 181
　South Australia
　　Limestone Coast 244
　Tasmania 253
　　Hobart and Surrounds 257
　　The North 270
　Victoria
　　Goldfields 313
　　Legends, Wine and High Country 327
　　Phillip Island and Gippsland 362–3
Potatoes, Paddock to Plate, Thorpdale Vic 362–3
Pottique Lavender Farm, Kingaroy Qld 142
poultry
　Australian Capital Territory
　　Canberra 32
　New South Wales
　　Blue Mountains and the Hawkesbury 26
　　Capital Country and the Snowy Mountains 35, 36, 38, 42
　　Explorer Country & Riverina 48–9, 52–3, 55, 61, 63
　　Hunter 69–71, 73
　　Northern Rivers 100, 102
　　Outback and the Murray 110
　　South Coast and Illawarra 118
　　Sydney and Southern Highlands 13
　Northern Territory
　　Central Australia 128, 130
　Queensland
　　Far North 152–3
　　Gold Coast 168
　　Sunshine Coast 172
　　Toowoomba and the Southern Downs 186
　South Australia
　　Adelaide and Surrounds 196, 210–11
　　Barossa, 217, 219, 221
　　Clare Valley and Yorke Peninsula, 227, 229
　　Kangaroo Island, 239
　Tasmania
　　Hobart and Surrounds 254
　　The North-West 290
　　The South 285
　Victoria 295
　　Goulburn Murray Waters 319
　　Legends, Wine and High Country 328, 334–6
　　Murray and the Outback 355
　　Phillip Island and Gippsland 360
　　Yarra Valley and the Dandenong Ranges 375–8, 383
　Western Australia
　　The South-West 408
Powerhouse Farmers' Markets, New Farm Qld 138

Prairie Hotel, Parachilna SA 234
prawns see seafood
preserves see jams and marmalades; sauces
Primex, Casino NSW 109
Primo Estate Wines, Virginia SA 230
Promised Land Tas 291
Proteco Pty Ltd, Kingaroy Qld 143
Proven-Artisan Breads and Pastries, Orange NSW 60
prunes
　New South Wales
　　Capital Country and the Snowy Mountains 39, 42
　　Explorer Country and Riverina 52
　　Hunter 73
Pub in the Paddock, Pyengana Tas 270–1
puddings see also desserts
　Australian Capital Territory
　　Canberra 30
　New South Wales
　　Capital Country and the Snowy Mountains 35
　　Northern Rivers 99, 104
　Tasmania
　　The North 272
　Victoria
　　Legends, Wine and High Country 337
The Pumpkin Pie Coffee Shop, Goomeri Qld 141
pumpkins
　New South Wales
　　New England and North-West 83
　Queensland
　　Brisbane and Surrounds 136
　　Bundaberg, Fraser Coast and South Burnett 141
　Western Australia
　　Australia's North-West 420
Pyengana Tas 270–1
Pyengana Dairy Company, Pyengana Tas 270
Pyrenees Pink Lamb and Purple Shiraz Country Race Meeting, Avoca Vic 315
Pyrenees Vignerons Gourmet Wine and Food Race Meeting, Avoca Vic 315

Quaama NSW 119
quail
　New South Wales
　　Explorer Country and Riverina 52
　　Hunter 72
　　Outback and the Murray 110
　　Sydney and Southern Highlands 17
　Tasmania
　　Hobart and Surrounds 256
　　The North 269
　　The North-West 291
　　The South 283
Quality Hotel Wangaratta Gateway, Wangaratta Vic 339
quandong
　South Australia
　　Clare Valley and Yorke Peninsula, 227–8
　　Eyre Peninsula, Flinders Ranges and the Outback, 232–4
Quandong Cafe and Bush Bakery, Copley SA 232
quarantine 149, 295, 387, 423
The Quarry Restaurant and Cellars, Cowra NSW 50
Queanbeyan NSW 40
Queen Victoria Markets, Melbourne Vic 299
Quills Restaurant, Bendigo Vic 311
quinces
　New South Wales
　　Capital Country and the Snowy Mountains 34, 35
　　Sydney and Southern Highlands 16

455

INDEX

quinces *continued*
South Australia
Barossa, 217
Victoria
The Grampians 367
Legends, Wine and High Country 326
Quinninup Farmers' Market, Quinninup WA 416

R Stephens Tasmanian Honey, Mole Creek Tas 291
rabbit
Australian Capital Territory
Canberra 32
New South Wales
Blue Mountains and the Hawkesbury 28
Explorer Country and Riverina 55, 57, 60
Hunter 69, 72
Outback and the Murray 110
Queensland
Gold Coast 168
South Australia
Barossa, 219
Tasmania
Hobart and Surrounds 256
Victoria
Legends, Wine and High Country 328, 332
Raeburn Orchards, Roleystone WA 398
Rae's on Wategos, Byron Bay NSW 101–2
Rainforest Foods, Byron Bay NSW 100
Rainforest Habitat, Port Douglas Qld 163
Rainforest Secrets, Mooball NSW 105
Rainforestation Nature Park, Kuranda Qld 156
Ranelagh Tas 284
Rankine's Landing Tavern, Hindmarsh Island SA 204
raspberries
Australian Capital Territory
Canberra 31
New South Wales
Capital Country and the Snowy Mountains 34–5
Explorer Country and Riverina 53, 61
New England and North-West 81
North Coast 96
Northern Rivers 103, 105
South Coast and Illawarra 116, 121
Sydney and Southern Highlands 18
Queensland
Toowoomba and the Southern Downs 184
South Australia
Adelaide and Surrounds 211, 215
Barossa, 224
Tasmania
The North 265, 272
The North-West 289
Victoria
Legends, Wine and High Country 330, 335
Macedon Ranges and Spa Country 347
Phillip Island and Gippsland 358
Western Australia
The South-West 412
Ravenshoe Market, Ravenshoe Qld 166
Red Bellies at the Pub, Nelson Bay NSW 72
Red Carp, Cowra NSW 49
The Red Crab Seafood Bar, Caloundra Qld 147
Red Hill Vic 306
Red Hill Cheese, Red Hill Vic 306
Red Hill Cool Stores, Red Hill South Vic 307
Red Hill South Vic 307–8
Red Ochre Grill, Alice Springs NT 127
Red Ochre Grill, Cairns Qld 152–3

Red Peppers Cafe, Launceston Tas 268
Red Rock Olive Oil, Pomonal Vic 367–8
Red Velvet Lounge and Firebird Bakehouse, Cygnet Tas 278–9
Redbank Vic 313
Redcliffe Farmers' Fresh and Seafood Market, Redcliffe Qld 138
Redfingers Cafe Bar and Grill, Coonawarra SA 242
Redlands Farmers' Market, Mount Cotton Qld 170
Redmond WA 414
The Reef House Restaurant, Palm Cove Qld 161–2
Rees Orchard, Mount Kembla NSW 119
Regional Fare, Yarra Glen Vic 383
Reilly's Winery and Restaurant, Mintaro SA 228
Relish Tasmania, Sydney NSW 13
relishes and processed vegetables
Australian Capital Territory
Canberra 32
New South Wales
Capital Country and the Snowy Mountains 35, 40
North Coast 94
Queensland
Brisbane and Surrounds 136
South Australia
Barossa, 218, 221, 223
Clare Valley and Yorke Peninsula, 227–8
Limestone Coast 246
Tasmania
The South 286–7
Victoria
Bays and Peninsulas 302
The Grampians 367
Legends, Wine and High Country 335
Macedon Ranges and Spa Country 343
Yarra Valley and the Dandenong Ranges 378, 380, 383–4
Western Australia
Perth and Surrounds 397
Renmark SA 250
Renner Springs NT 128
Renner Springs Desert Hotel Motel, Renner Springs NT 128
Restaurant at the Lilydale, Lilydale Vic 379
Restaurant Castalia, Yamba NSW 108
Restaurant Legall, Bathurst NSW 48
Restaurant Lurleens, Mt Cotton Qld 168
Restaurant Pizza Cafe Bella, Kingscote SA 239
Restaurant Q, Armidale NSW 79
rhubarb
New South Wales
Sydney and Southern Highlands 15
rice
New South Wales
Blue Mountains and the Hawkesbury 22
Explorer Country and Riverina 48, 54
Richmond Tas 284
Richmond Arms Hotel, Richmond Tas 284
Richmond Food and Wine Centre, Richmond Tas 285
Ricky Ricardo's, Noosa Sound Qld 177
The Right Food Group, Murwillumbah NSW 106
Ripe–Australian Produce, Sassafras Vic 381
Risby Cove Restaurant, Strahan Tas 294
Ristorante Leonardos, Albany WA 401
Riva Waterfront Restaurant, Noosa Qld 176
Riverboats, Jazz, Food and Wine Festival, Echuca and Moama Vic 356
Riverhouse Herb Farm, Wandandian NSW 123
RIVERINA NSW 9, 48–65

INDEX

Riverina Gourmet Gifts, Griffith NSW 51
Riverina Grove Pty Ltd, Griffith NSW 51
Riverina Olive Grove, Wagga Wagga NSW 62
Riverland Smallgoods, Waikerie SA 251
Rivers Cafe, Yarrambat Vic 384
Riverview Juices, Cobram Vic 350-1
Riviera Ice-Cream Parlour, Lakes Entrance Vic 322
RL Chapman and Sons, Silvan Vic 381
Robe SA 247, 248
Robe Village Fair, Robe SA 248
Robern Menz (Mfg) Pty Ltd, Glynde SA 196
Roberts at Peppertree, Pokolbin NSW 68, 75
Robertson NSW 17
The Robertson Pie Shop, Robertson NSW 17
Robertson Potatoes, Robertson NSW 17
Robigana Tas 271
Robinvale Vic 355
Robinvale Estate Olive Oil, Robinvale Vic 355
Robinvale Wines, Robinvale Vic 355
Rocherlea Tas 271
Rochford's Eyton, Coldstream Vic 376
Rockford-PS Marion, Tanunda SA 223
Rockhampton Qld 150
Rockpool Cafe, Stokes Bay SA 240
The Rocks Restaurant, South West Rocks NSW 95
Roleystone WA 398
Romsey Vic 347
Rosebud Vic 308
Rosedale SA 221
Rosevears Tas 271-2
Ross Tas 272
The Ross Village Bakery, Ross Tas 272
Rothbury NSW 76
Roundstone Winery and Cafe, Yarra Glen Vic 384
Rowland Flat SA 221
Royal Darwin Show, Darwin NT 133
Royal Mail Hotel, Dunkeld Vic 365
Ruffy Vic 318
Ruffy Produce Store, Ruffy Vic 318
rum
  Queensland 135
    Bundaberg, Fraser Coast and South Burnett 139, 144
Russell's Pizza, Willunga SA 215
Rustic Charm Restaurant, Wandin North Vic 382
Rusty's Markets, Cairns Qld 166
Rutherglen Vic 336-7
rye
  New South Wales
    Explorer Country and Riverina 54
Rylstone NSW 61-2
Rylstone Food Store, Rylstone NSW 62

Sadies, Pemberton WA 414
Safety Beach Vic 308
Safi Buluu, Flinders Vic 303
saffron
  Tasmania
    The South 280
Sage and Muntries Cafe, Mount Gambier SA 244
Sailors Falls Vic 313
St Helens Tas 272-3
St Ives Organic Market, St Ives NSW 19
St Leonards Vic 309
St Matthias Vineyard, Rosevears Tas 271-2
Salamanca Markets, Hobart Tas 257
Salami Festival, Yungaburra Qld 166
Salix Restaurant at Willow Creek Winery, Merricks North Vic 304

salmon *see also* fish
  Australian Capital Territory
    Canberra 32
  New South Wales
    Blue Mountains and the Hawkesbury 23
  Queensland
    Sunshine Coast 179
  South Australia
    Adelaide and Surrounds 210
    Barossa, 220
  Tasmania 253
    Hobart and Surrounds 257, 259
    The North 263, 273
    The North-West 291, 293
    The South 276, 282-3, 285, 287
  Victoria
    Goulburn Murray Waters 316
    Legends, Wine and High Country 341
    Yarra Valley and the Dandenong Ranges 375-7, 383-4
  Western Australia
    Australia's North-West 420
Salopian Inn, McLaren Vale SA 209-10
salt
  Victoria
    Murray and the Outback 352
Salt Water, Bermagui NSW 114
Salters at Saltram, Angaston SA 219
Salty Seas, St Helens Tas 272-3
San Remo Vic 362
San Remo Fishermen's Co-op, San Remo Vic 362
Sasha's of Bright and Sasha's Wine Bar, Bright Vic 328
Sassafras Vic 381
Sassi Cucina, Port Douglas Qld 163
sauces *see also* pickles, relishes and processed vegetables
  Australian Capital Territory
    Canberra 30, 32
  New South Wales
    Capital Country and the Snowy Mountains 35, 40-2
    Hunter 67, 77
    Northern Rivers 100
  Queensland
    Far North 151
    Toowoomba and the Southern Downs 182, 185
  South Australia
    Adelaide and Surrounds 212
    Barossa, 219-22
  Tasmania
    The South 286
  Victoria
    Goulburn Murray Waters 317
    Macedon Ranges and Spa Country 343, 346
    Melbourne 299
  Western Australia
    Perth and Surrounds 397
Sault Restaurant and Functions, Sailors Falls Vic 313
sausages
  New South Wales
    Capital Country and the Snowy Mountains 36
    Explorer Country and Riverina 64
    Hunter 69, 76
    New England and North-West 83
    North Coast 92
    Outback and the Murray 110
    South Coast and Illawarra 114, 115
    Sydney and Southern Highlands 13, 17
  Queensland
    Bundaberg, Fraser Coast and South Burnett 144
    Toowoomba and the Southern Downs 184

457

INDEX

sausages *continued*
  South Australia
    Adelaide and Surrounds 203
    Barossa, 217
  Victoria
    Legends, Wine and High Country 332
    Murray and the Outback 354
Sawtell NSW 94
Scamander Tas 272
Scarlett's Exotic Fruit and Veg, Burrawang NSW 17
Scenic Drive Strawberries, Koonoomoo Vic 352
Schmidt's Strawberry Winery, Allans Flat Vic 325
Schulz Butchers, Angaston SA 218-19
Scirocco Cafe Bar and Grill, South Townsville Qld 192
Seabelle Restaurant, Fraser Island Qld 140
seafood *see also* fish, shellfish
  New South Wales 9
    Capital Country and the Snowy Mountains 34-5
    Central Coast 44-7
    Explorer Country and Riverina 49, 51-2
    Hunter 68, 72, 76
    New England and North-West 80
    North Coast 87-8, 92, 94-7
    Northern Rivers 99-104, 107-8
    Outback and the Murray 110-11
    South Coast and Illawarra 113-14, 117-19, 121-2
    Sydney and Southern Highlands 12-15
  Northern Territory 125
    Central Australia 129-30
    Top End 131-3
  Queensland
    Brisbane and Surrounds 136
    Bundaberg, Fraser Coast and South Burnett 139-42
    Capricorn, Gladstone and the Outback 147-50
    Far North 151-4, 156-7, 161-3, 165
    Gold Coast 167
    Sunshine Coast 172, 175-7, 179-80
    Townsville, Mackay and the Whitsundays 188-92
  South Australia
    Adelaide and Surrounds 195-7, 204, 209, 213
    Barossa, 220
    Eyre Peninsula, Flinders Ranges and the Outback, 231-5
    Kangaroo Island, 236-7, 240
    Limestone Coast 241-6
  Tasmania
    Bass Strait Islands 260-2
    Hobart and Surrounds 255-8
    The North 263-4, 267-8, 272-3
    The North-West 288-94
    The South 275, 279, 282-3, 286
  Victoria
    Bays and Peninsulas 303, 305-9
    The Grampians 366
    The Great Ocean Road 373-4
    Lakes and Wilderness 322-3
    Legends, Wine and High Country 339, 341
    Murray and the Outback 353-4
    Phillip Island and Gippsland 358-9, 362-4
  Western Australia 387
    Australia's Coral Coast 426-7
    Australia's North-West 419-22
    The Outback 424
    Perth and Surrounds 388-9, 392-3, 396, 399
    The South-West 406-8, 410, 412-15
Seagrass Cafe, Huskisson NSW 117
Seagulls Restaurant, Townsville Qld 192
Seahaven Cafe and General Store, Gerroa NSW 116-17
Seasalt Restaurant, Terrigal NSW 46
Seashells Cafe, Harrington NSW 90
Selkirks Restaurant, Orange NSW 61
Sellicks Beach SA 213
Selzer Pastoral, Ovens Vic 335
Settlers House, York WA 399
Seven Mile Cafe Restaurant, Lennox Head NSW 103-4
Seven Spirit Bay Wilderness Lodge, Arnhem Land NT 131
Sevenhill SA 229
Shakey Tables, North Rothbury NSW 72-3
Shalom Sunday Markets, Bundaberg Qld 145
Shannonvale Tropical Fruit Wine Company, Mossman Qld 160
Shantell Restaurant, Dixons Creek Vic 376
shellfish *see* seafood
Shepherds Flat Vic 348
Shepparton Vic 318-19
Shields Orchard, Bilpin NSW 20-1
Shipard's Herb Farm, Nambour Qld 176
Shoal Bay NSW 95
Signal Point Cafe, Goolwa SA 201
Silks Brasserie, Leura NSW 25
Silkwood East
Silos Estate Winery and Restaurant, Jaspers Brush NSW 117
Silvan Vic 381
Silvan Estate Raspberries, Silvan Vic 382
Silver Bistro, Noosaville Qld 177
Simone's of Bright, Bright Vic 328
Simply Tomatoes, Boort Vic 350
Singleton NSW 76
Sirocco Restaurant, Cairns Qld 153
Sisters Creek Tas 292
Skillogalee Restaurant, Sevenhill SA 229
Skipton Vic 348
Slow Food Noosa, Pomona Qld 180
Smeaton Vic 313
Smithfield Qld 164
smoked smallgoods
  Australian Capital Territory
    Canberra 32
  New South Wales
    Capital Country and the Snowy Mountains 36, 38
    Explorer Country and Riverina 49-50, 62, 64
    Hunter 69-70
    Northern Rivers 100
    Outback and the Murray 110
    Sydney and Southern Highlands 13, 17
  Queensland
    Bundaberg, Fraser Coast and South Burnett 142-3, 145-6
  South Australia
    Adelaide and Surrounds 196, 202, 205, 210, 214
    Barossa, 217, 220-1, 223
    Limestone Coast 246-7
    Murray and Riverland 251
  Tasmania
    Hobart and Surrounds 256
    The North 264, 267, 271
    The North-West 290
  Victoria
    Goldfields 313
    Legends, Wine and High Country 326, 332, 334, 338
    Macedon Ranges and Spa Country 349
    Murray and the Outback 354
    Phillip Island and Gippsland 360, 362-3
Smokehouse Cafe, Hall NSW 38
Smokehouse Sorrento, Sorrento Vic 308

INDEX

Smoko's Big Shed Cafe and Fruit and Veg Market, Bright Vic 328
Smoky Bay SA 235
Smoky Bay Oyster Tours and Holmes Oysters, Smoky Bay SA 235
Snails Bon Appetite, Congewai NSW 67–8
snails
   New South Wales
      Hunter 68
Snowline Fruits, Stanley Vic 337
Snowy Mountains Regional Food Fair, Dalgety NSW 43
Snug Butchery, Snug Tas 285
Snug Tas 285
Sofala NSW 62
Solitary, Leura Falls NSW 25
Somersby NSW 45–6
Somerset Horticultural Farm, Mossman Qld 160
Sorell Tas 285
Sorell Fruit Farm, Sorell Tas 285
Sorrento Vic 308
Sounds of Silence, Uluru-Kata Tjuta National Park NSW 129–30
soups
   Victoria
      Goulburn Murray Waters 317–18
THE SOUTH Tas 275–87
South Australia Company Store and Kitchen, Angaston SA 219
The South Australian Olive Corporation Pty Ltd, Loxton SA 250
South Australian Tourism Commission, Adelaide SA 196
South Bowenfels NSW 28
SOUTH COAST AND ILLAWARRA NSW 113–23
South Coast Cheese, Bodalla NSW 115
South Gippsland Farmers' Market, Koonwarra Vic 364
South Shepparton Vic 319
South Townsville Qld 191
THE SOUTH-WEST WA 400–17
South West Rocks NSW 95
SOUTHERN HIGHLANDS 9, 12–19
Southern Highlands Olive Festival, Mittagong NSW 19
Southern Right Oysters, Ceduna SA 231
Southport Farmers' Market, Southport Qld 170
Spa Centre Meats, Daylesford Vic 344
SPC Ardmona Factory Sales, Mooroopna Vic 317–18
spices
   Australian Capital Territory
      Canberra 31
   New South Wales
      New England and North-West 83
      North Coast 86, 91
      Sydney and Southern Highlands 12, 16
   Queensland
      Far North 153
      Sunshine Coast 176
   Victoria
      Goldfields 313
      Murray and the Outback 352
The Spirit House-Restaurant and Cooking School, Yandina Qld 180
The Spirited Chef Foodstore and Comestibles, Beechworth Vic 326
Spoons Deli, Swan Hill Vic 356
Spreyton Tas 292
Spreyton Primary Apple Festival, Spreyton Tas 294
Spring Bay Seafoods Pty Ltd, Triabunna Tas 286
Spring Creek Cafe and Courtyard, Beechworth Vic 326
Spring Gully Olives, Goomeri Qld 141

Spring Hill Qld 136–7
Spring in the Valley, Swan Valley WA 399
Springs Smoked Seafoods, Mount Barker SA 210
Springvale Hills, Nashua NSW 106
The Spud Shed, Trafalgar Vic 363
SSS BBQ Barns (Stetson's Steakhouse and Saloon), Tamworth NSW 83
Stahmann Farms, Toowoomba Qld 186
Stahmann Farms Inc, Pallamallawa NSW 82–3
Stanley Tas 292–3
Stanley Vic 337
Stanley Bridge Tavern, Verdun SA 214
Stanley Seaquarium, Stanley Tas 292
Stanthorpe Qld 183–4, 187
The Stanthorpe Apple Shed, Thulimbah Qld 185
Stanthorpe in Season Farmers' Markets, Stanthorpe Qld 187
Stanthorpe Market in the Mountains, Stanthorpe Qld 187
Stanthorpe Showground Markets, Stanthorpe Qld 187
Stanthorpe Wine Centre, Thulimbah Qld 185
Star of Greece Cafe, Port Willunga SA 213
Staughton Vale Vineyard, Anakie Vic 301
Stawell Vic 368
Steenholdt's Organic Products, Cygnet Tas 278
Stefano's, Mildura Vic 354
Stellar Ridge Estate, Cowaramup WA 404
Stetson's Steakhouse and Saloon, Tamworth NSW 83
Stevens Quality Raspberries, Williamstown SA 224
Sticcádo Cafe, Yarragon Vic 364
Stillwater River Cafe, Restaurant and Wine Bar, Launceston Tas 268
Stingray Seafoods, North Hobart Tas 258
Stirling SA 213
Stirling Range WA 414–15
The Stock Exchange and Village Organic Farm, Conondale Qld 172
Stokes Bay SA 240
stone fruit
   Australian Capital Territory
      Canberra 31, 33
   New South Wales 9
      Blue Mountains and the Hawkesbury 20–1
      Capital Country and the Snowy Mountains 42
      New England and North-West 79
      Northern Rivers 106
      South Coast and Illawarra 116
   South Australia
      Clare Valley and Yorke Peninsula, 230
   Tasmania
      The North 272
   Victoria
      The Grampians 367
      Legends, Wine and High Country 328
   Western Australia
      Perth and Surrounds 392
Stonelea Country Estate, Alexandra Vic 316
Strahan Tas 293–4
Stratford Vic 324
Strathalbyn SA 214
strawberries
   New South Wales
      Capital Country and the Snowy Mountains 34–5, 41
      Explorer Country and Riverina 49, 53
      Hunter 70–1
      New England and North-West 79, 81
      North Coast 96
      South Coast and Illawarra 116, 120
      Sydney and Southern Highlands 16, 18

INDEX

strawberries *continued*
  Queensland
    Bundaberg, Fraser Coast and South Burnett 139–40
    Sunshine Coast 177
    Toowoomba and the Southern Downs 184
  South Australia
    Adelaide and Surrounds 202, 205, 212, 215
  Tasmania
    The North 265, 272
    The South 286
  Victoria
    Bays and Peninsulas 303–4, 306, 309
    Goulburn Murray Waters 317
    Legends, Wine and High Country 325
Strawberry Fields, Palmview Qld 177
Stringers Stores of Sorrento, Sorrento Vic 308
Stringybark Winery and Restaurant, Chittering WA 392
The Strudel Company, Huntingfield Tas 255
Subiaco WA 390
sugar cane
  New South Wales
    Northern Rivers 99
  Queensland 135, 186
    Bundaberg, Fraser Coast and South Burnett 139
    Far North 161–2, 164
    Townsville, Mackay and the Whitsundays 190
  Western Australia
    Australia's North-West 418
Summerland Olive Festival, Casino NSW 84
Sun Valley Australia Pty Ltd, Finley NSW 112
Suncream Homestyle Ice-Cream, Mooloolaba Qld 175
Sunny Creek Fruit and Berry Farm Vic 358
Sunny Ridge Strawberry Farm, Main Ridge Vic 303–4
Sunraysia Jazz, Food and Wine Festival, Mildura Vic 356
SunRice Country Visitors Centre, Leeton NSW 54
SunRice Festival, Leeton NSW 65
SUNSHINE COAST Qld 171–80
Suntralis Foods, Lonsdale SA 196
The Super Strawberry, Glen Innes NSW 79
Superbee Honey Factory, Tanawha Qld 178
Superbee Honeyworld, Currumbin Qld 167
Susie's Boutique Tours, Burnside SA 198
Sussex Barn, North Torbay (via Albany) WA 412
Sussex Inlet NSW 119–20
Sutton Forest NSW 18
Sutton Grange Vic 314
Swan Bay NSW 76
Swan Hill Vic 355–6
Swan Valley Central, Herne Hill WA 395
Swan Valley Eggs, West Swan WA 388
Swan Valley Winery Tours, Darlington WA 391
Swansea Tas 273
Swansea Tas 286
sweetcorn
  New South Wales
    Explorer Country and Riverina 52
Swickers Kingaroy Bacon Factory Pty Ltd, Kingaroy Qld 142–3
SYDNEY 9, 12–19, 44
Sydney Fish Market, Pyrmont NSW 13–14, 44, 47
Sydney Markets, Flemington NSW 14

The Table Guesthouse, Greta NSW 70
Tacoma NSW 46
Taggerty Vic 338
Talbot Yabbie Festival, Talbot Vic 315
Talinga Grove, Strathalbyn SA 214
Taltarni Avoca Cup, Avoca Vic 315
Tamarillo Fruit Farm, Main Ridge Vic 304

Tamborine Qld 169
Tamborine Mountain Country Markets, Mount Tamborine Qld 170
Tamborine Mountain Distillery, North Tamborine Qld 169
Tamborine Show, Mount Tamborine Qld 170
Taminick Vic 338
Tamworth NSW 83
Tanawha Qld 178
Tanja NSW 120
Tanja Olive Oil, Tanja NSW 120
Tanja's Beach Pavilion, Caloundra Qld 171–2
Tanunda SA 221–4
Tanunda's Nice Ice, Tanunda SA 222
Taranna Tas 286
Tarzali Qld 164
Tarzali Lakes Fishing Park, Tarzali Qld 164
Tasmania Distillery, Cambridge Tas 277
Tasmanian Honey Company, Perth Tas 269
Tasmazia and The Village of Lower Crackpot, Promised Land Tas 291
Tas-Saff, Glaziers Bay Tas 280
Tassie Blue, Lymington Tas 283
Taste Mount Barker Wine Cafe, Mt Barker WA 411
The Taste of Cowra, NSW 65
Taste of Tatura Wine and Food Festival, Tatura Vic 320
Taste of the Tropics Food Trails, Cairns Qld 154
Taste of the Valley, Swan Valley WA 399
Tastes of Prom Country, Foster Vic 364
Tastes of Rutherglen, Rutherglen Vic 342
Tasting Australia, Adelaide SA 216
The Tasting House, Richmond Tas 284
Tasting Mackay, Mackay Qld 192
Tastings at the Top, Cradle Mountain Vic 294
Tathra NSW 120
Tatong Farmers' Market, Tatong Vic 342
Tauondi Cultural Tours, Port Adelaide SA 198
Taverner's Products, Goodwood Tas 256
Tawonga South Vic 338
Taylors Arm NSW 95–6
tea
  New South Wales
    Northern Rivers 104–6
  Queensland 135
    Far North 155, 160, 164
Tea Gardens NSW 96
Teatrick Lavender Estate, Wolseley SA 247–8
Telegraph Farm, Esperance WA 424
Templestowe Farmers' Market, Templestowe Vic 385
Tennant Creek NT 128–9
Termite Fruit, Veg and Takeaway, Mareeba Qld 158
The Terrace Restaurant, Wahgunyah Vic 339
Terrace Seafood Restaurant, Maleny Qld 175
Terrigal NSW 46
Texas Qld 184–5
Thirlmere NSW 15
Thorn Park Country House, Sevenhill SA 229
Thorogood's Farmhouse Cider Cellar, Burra SA 226
Thorpdale Vic 362–3
Thredbo Valley NSW 40
3777 Restaurant, Healesville Vic 377
Three Snails Restaurant, Dubbo NSW 50
Thulimbah Qld 185
Thumm Estate Wines, Upper Coomera Qld 169
Thurlby Herb Farm, North Walpole WA 412
Tiaro Meats and Bacon, Tiaro Qld 145–6
Tiaro Qld 145–6
The Tide and Pilot Lower Deck and Upper Deck, Coffs Harbour NSW 88
Tilba Tilba NSW 120–1

INDEX

Timboon Vic 372
Timboon Fine Ice-cream, Port Campbell Vic 371
Tinglewood Wines and Puzzles, Denmark WA 406
Tintinara SA 247
Tizzana Winery, Ebenezer NSW 23
Toast to the Coast, Geelong Vic 309
Tokar Estate, Coldstream Vic 376
Tolga Qld 164-5
The Tomato Patch, Kallangur Qld 136
tomatoes
  Australian Capital Territory
    Canberra 31
  New South Wales
    Blue Mountains and the Hawkesbury 22
    Capital Country and the Snowy Mountains 38
    Explorer Country and Riverina 49, 51
    Hunter 67
    New England and North-West 84
    Outback and the Murray 112
    South Coast and Illawarra 120
  Northern Territory
    Central Australia 130
  Queensland
    Brisbane and Surrounds 136
    Toowoomba and the Southern Downs 186
  South Australia
    Clare Valley and Yorke Peninsula, 227
  Tasmania
    The South 286-7
  Victoria
    Goulburn Murray Waters 318
    Legends, Wine and High Country 326, 335-6
    Phillip Island and Gippsland 360, 364
Tongala Vic 320
Tonic, Millthorpe NSW 55
Tooperang Trout Farm, Mount Compass SA 211
Toot's Cafe, Manilla NSW 81
Toowoomba Qld 181, 186
TOOWOOMBA AND THE SOUTHERN DOWNS Qld 181-7
TOP END NT 131-3
Tory's Seafood Restaurant, Ulladulla NSW 122
Town Beach Cafe, Broome WA 420
Townsville Qld 192
TOWNSVILLE, MACKAY AND THE WHITSUNDAYS Qld 188-92
Trafalgar Vic 363
Treehouse Restaurant, Mossman Qld 161
Trentham Cliffs Vic 356
Trentham Estate, Trentham Cliffs Vic 356
Triabunna Tas 286
Tropical Fruit World, Duranbah NSW 102
Tropical Wines, Bundaberg Qld 139
trout
  New South Wales
    Capital Country and the Snowy Mountains 33, 35, 36, 40
    Explorer Country and Riverina 49, 62-3
    New England and North-West 79-80, 82-3
    Outback and the Murray 110
    South Coast and Illawarra 120-1
  Queensland
    Far North 163
    Toowoomba and the Southern Downs 181
  South Australia
    Adelaide and Surrounds 211
    Limestone Coast 243
  Tasmania
    The North 263, 269
    The North-West 290-1

  Victoria
    Goldfields 312-13
    Goulburn Murray Waters 316
    The Grampians 365
    The Great Ocean Road 370, 373-4
    Legends, Wine and High Country 326, 328, 334, 337, 340-1
    Macedon Ranges and Spa Country 344-5
    Phillip Island and Gippsland 357-8
    Yarra Valley and the Dandenong Ranges 375, 380, 382
  Western Australia
    The South-West 410, 412-13, 417
truffles
  Tasmania 253
  Western Australia 409
Trundle Bush Tucker Day, Trundle NSW 65
Trunk's Gourmet Meats, Singleton NSW 76
Truro SA 224
Tuckerberry Hill, Drysdale Vic 302-3
Tuileries, Rutherglen Vic 337
Tuki Trout Farm, Smeaton Vic 313
Tumbafest, Tumbarumba NSW 65
Tumbarumba NSW 41
Tumbeela Native Bushfoods, Verdun SA 214
Tumbling Waters Hydroponics, Malanda Qld 156
turkey
  Australian Capital Territory
    Canberra 32
  New South Wales
    Outback and the Murray 110
  South Australia
    Adelaide and Surrounds 208-9
  Victoria
    Goldfields 312
Tuross Head NSW 121
Tweed Heads NSW 107
Tweed Valley Banana Festival and Harvest Week, Murwillumbah NSW 109
27 Deakin, Mildura Vic 354
2 Fish Seafood Restaurant, Port Douglas Qld 163
2-Rice, Broome WA 420
Two Wells SA 229
Tyers (Traralgon) Farmers' Market, Tyers Vic 364

Uarah Fisheries, Grong Grong NSW 52
The Udder Cow Cafe, Comboyne NSW 88
Udder Delights, Lobethal SA 206
Ulladulla NSW 121-2
Ulladulla Oysters, Ulladulla NSW 121
Uluru-Kata Tjuta National Park NSW 129-30
Umina Beach NSW 47
UnWINEd in the Riverina, Griffith NSW 65
Upper Coomera Qld 169
Upper Deck, Hobart Tas 258
Upper Reach Winery and Cafe, Baskerville WA 392
Upstairs at Hollick, Coonawarra SA 242
Urunga NSW 96-7

Vacy NSW 77
Valley Nut Groves, Gapsted Vic 330-1
Valley of Armagh, Clare SA 227
Vasse Felix, Cowaramup WA 405
Vat 107, Margaret River WA 410
veal
  Australian Capital Territory
    Canberra 32

461

INDEX

veal *continued*
  New South Wales
    Explorer Country and Riverina 55
    Northern Rivers 107
    Outback and the Murray 110
    South Coast and Illawarra 123
    Sydney and Southern Highlands 13
  South Australia
    Adelaide and Surrounds 204, 208, 213
  Victoria
    Legends, Wine and High Country 328
  Western Australia
    Perth and Surrounds 389
Veg Out Farmers' Market, St Kilda Vic 299
The Veg Shed-Kester's Apples, Naracoorte SA 245
vegetables *see also by type of vegetables*
  Australian Capital Territory
    Canberra 31–3
  New South Wales
    Blue Mountains and the Hawkesbury 22, 23
    Capital Country and the Snowy Mountains 37, 38
    Explorer Country and Riverina 49, 52, 63
    Hunter 68, 70
    Outback and the Murray 112
    Sydney and Southern Highlands 14–17, 19
  Northern Territory 125
    Top End 132
  Queensland
    Brisbane and Surrounds 136
    Far North 152–5, 158
    Sunshine Coast 174–5
    Toowoomba and the Southern Downs 181
  South Australia
    Adelaide and Surrounds 196, 200, 202–3, 208–9, 212, 214
    Barossa, 218–20
  Tasmania
    The North 270–1
    The North-West 290–1
  Victoria
    Goulburn Murray Waters 316
    The Great Ocean Road 369, 374
    Legends, Wine and High Country 326, 328, 340–1
    Murray and the Outback 351
    Phillip Island and Gippsland 358–60, 363
    Yarra Valley and the Dandenong Ranges 375, 381, 384
  Western Australia
    Perth and Surrounds 396
    The South-West 401–2, 412
venison
  Australian Capital Territory
    Canberra 32
  New South Wales 9
    Blue Mountains and the Hawkesbury 26
    Capital Country and the Snowy Mountains 34, 36
    Explorer Country and Riverina 49, 55, 57, 60–1
    Hunter 73
    South Coast and Illawarra 123
  Northern Territory 125
    Top End 131
  Queensland
    Toowoomba and the Southern Downs 184–5
    Townsville, Mackay and the Whitsundays 189
  South Australia
    Adelaide and Surrounds 201–2, 208, 213–14
    Barossa, 223
    Clare Valley and Yorke Peninsula, 225

  Tasmania
    Hobart and Surrounds 256
    The North 271
    The North-West 289
    The South 279, 282–5
  Victoria
    Legends, Wine and High Country 332, 339, 341
    Phillip Island and Gippsland 357, 360
    Yarra Valley and the Dandenong Ranges 376, 382, 384
  Western Australia
    The Outback 424
    Perth and Surrounds 389
    The South-West 400–1, 404, 409–11, 415
Verandah Restaurant, Tamborine Qld 169
Verdun SA 214
Verity Prunes, Young NSW 42
Victor Harbor SA 215
Victor Harbor Country Market, Victor Harbor SA 216
Victor Harbor Winery, Victor Harbor SA 215
Victorian Olive Groves, Bendigo Vic 311
Victorian Strawberry Fields, Silvan Vic 381
Victory Hotel, Sellicks Beach SA 213
Vie Bar, Main Beach Qld 167
Villa Gusto-La Dolce Vita, Bright Vic 328
Vincenzo's at the Big Apple, Stanthorpe Qld 184
vinegars
  Australian Capital Territory
    Canberra 30, 32
  New South Wales
    Blue Mountains and the Hawkesbury 21
    Hunter 67, 74, 77
    New England and North-West 81–2
  Queensland
    Toowoomba and the Southern Downs 183–4
  South Australia
    Adelaide and Surrounds 208
  Victoria
    Lakes and Wilderness 321
    Legends, Wine and High Country 331
    Macedon Ranges and Spa Country 344
    Yarra Valley and the Dandenong Ranges 380
Vines Cafe and Bar, Ararat Vic 310
Vines Restaurant of the Yarra Valley, Warranwood Vic 383
Vineyard Balcony, Gumeracha SA 202
Vineyard Cafe and Cottages, Ballandean Qld 181
Vintners Bar and Grill, Angaston SA 220
Virginia SA 230
Voyager Estate Winery, Margaret River WA 410

Waddingtons at Kergunyah, Kergunyah Vic 332
Wa-De-Lock Cellar Door, Stratford Vic 324
Wagga Wagga NSW 54, 62–3
Wagga Wagga Farmers' Markets, NSW 65
Wagga Wagga Winery, Wagga Wagga NSW 63
Wagin WA 424
Wahgunya Vic 338–9
Wahroonga Dairy Goat Farm, Victor Harbor SA 215
Waikerie SA 250–1
Walkabout Apiaries, Milawa Vic 333
Walkamin Qld 165
wallaby
  Australian Capital Territory
    Canberra 32
  Tasmania 253
    Bass Strait Islands 262
    Hobart and Surrounds 256
    The North 271

INDEX

Wallaga Lake NSW 122-3
Wallarobba NSW 77
Wallendbeen NSW 63
Wallington Vic 309
Wallington Berries, Wallington Vic 309
Wallington Strawberry Fair, Wallington Vic 309
walnuts
  New South Wales
    Blue Mountains and the Hawkesbury 21
    Explorer Country and Riverina 55
  Victoria
    Legends, Wine and High Country 330
  Western Australia
    The South-West 411
Walsh's Country Kitchen, Boorowa NSW 35
Wandandian NSW 123
Wandin North Vic 382
Wandin Yallock Vic 382
Wangaratta Vic 339, 342
Wangaratta Jazz Festival, Wangaratta Vic 342
Wanneroo Local HarvestFarmers' Market, Wanneroo WA 416
Wannunup WA 415
Warnertown SA 235
Warragul Vic 363
Warranwood Vic 383
Warratina Lavender Farm, Wandin Yallock Vic 382
Warrawee Orchard, Chesnut Vic 329
Warrnambool Vic 373
Warrnambool Trout Farm, Warrnambool Vic 373
Warwick Band Centre Markets, Warwick Qld 187
Warwick Town Hall Carpark Markets, Warwick Qld 187
wasabi
  Tasmania 253
    The North-West 289
water (mineral)
  Victoria
    Macedon Ranges and Spa Country 349
water (spring)
  Tasmania
    Hobart and Surrounds 254
water chestnuts
  New South Wales
    Sydney and Southern Highlands 16
Waterfront Restaurant, Tea Gardens NSW 96
Wauchope Ice-cream Van, Wauchope NT 125
Wauchope NSW 97, 98
Waurn Ponds Vic 309
Waves Port Campbell, Port Campbell Vic 371
Weindorfers, Gowrie Park Tas 290
Wellington NSW 64
Wellington Farmers' Market, Sale Vic 364
Wentworth Falls NSW 27, 29
West Ballina NSW 107
West End Qld 137
West Eyre Shellfish, Ceduna SA 231
West Toodyay WA 398
Westbury Tas 274
Western Australian Wine and Food Exhibition, Perth WA 399
Westerway Tas 286-7
Westhaven Dairy, Launceston Tas 265
Westhaven Yacht Charters, Strahan Tas 294
Westoby Banana Plantation, Carnarvon WA 426
Weston's Walnuts, Eurobin Vic 330
Wharf Restaurant, Broome WA 420
Wharf Restaurant, Derby WA 422

wheat
  New South Wales 9
    Explorer Country and Riverina 50, 53-4
    New England and North-West 79
  Queensland
    Toowoomba and the Southern Downs 183
  Victoria 295
    Murray and the Outback 355
    Phillip Island and Gippsland 362
  Western Australia 387
    The Outback 424
    The South-West 400
Wheeler's Oysters, Pambula NSW 119
Whirrakee Restaurant and Wine Bar, Bendigo Vic 311
Whispers from Provence, Kalorama Vic 378
White Cottage Herbs, Glenrowan Vic 331
White Fisheries Pty Ltd, St Leonards Vic 309
White Gums Restaurant and Arnguli Flame Grill, Uluṟu-Kata Tjuṯa National Park NT 130
White Rocks Farm, Brunswick WA 402-3
Whitemark, Flinders Island Tas 262
Whitfield Vic 339
The Wholefood Garden, Gresford NSW 70
Wholefoods Cooperative Ltd, Geelong Vic 303
Wiberforce NSW 29
Wickham NSW 97
Wildbite, Byron Bay NSW 100
Wildes Meadow NSW 18
Wildlife Wonderland, Bass Vic 357
Wilga Vale Venison, Texas Qld 184-5
Wilgro Orchard Roadside Stall, Batlow NSW 33-4
Willabrand, Houghton SA 204
Williamstown SA 224
Willmullie Farm, Echuca Vic 351-2
Willow Cafe, Mogo NSW 118
Willowbrae Chevre Cheese, Wiberforce NSW 29
Willowvale Mill, Laggan NSW 39
Willunga Farmers' Market, Willunga SA 216
Willunga SA 215
Willyabrup WA 415
Wily Trout Vineyard, Hall NSW 38
wine see also fruit wine
  Australian Capital Territory
    Canberra 30
  New South Wales 9
    Blue Mountains and the Hawkesbury 23, 25, 26
    Capital Country and the Snowy Mountains 35, 36, 38, 39
    Explorer Country and Riverina 51-2, 58-9, 62-3
    Hunter 66, 71-3, 77
    New England and North-West 81
    North Coast 89
    South Coast and Illawarra 117, 119
    Sydney and Southern Highlands 19
  Queensland
    Bundaberg, Fraser Coast and South Burnett 142-4
    Far North 165
    Toowoomba and the Southern Downs 181-2, 185
  South Australia 193
    Adelaide and Surrounds 197-201, 204, 213, 215
    Barossa 217-18, 220,-4
    Clare Valley and Yorke Peninsula, 225, 229-30
    Kangaroo Island, 239-40
  Tasmania 253
    Bass Strait Islands 260
    Hobart and Surrounds 257
    The North 268-72
    The North-West 288-9
    The South 275-8, 280, 284

INDEX

wine continued
  Victoria
    Bays and Peninsulas 301-2, 305-7, 309
    Goldfields 313
    Goulburn Murray Waters 317
    The Great Ocean Road 369, 372
    Lakes and Wilderness 321, 323, 324
    Legends, Wine and High Country 325-42
    Macedon Ranges and Spa Country 344
    Murray and the Outback 351, 355
    Phillip Island and Gippsland 357, 363
    Yarra Valley and the Dandenong Ranges 376, 378, 382, 384
  Western Australia
    Australia's Coral Coast 427
    Australia's North-West 423
    Perth and Surrounds 389, 391-2, 394
    The South-West 400, 402, 404-7, 409-11, 413
Wine and Food Affair, Stanthorpe Qld 187
WINE AND HIGH COUNTRY Vic 325-42
The Wine and Truffle Company, Manjimup WA 409
Wine Coast Avocado Farm, McLaren Vale SA 209
Wineglass Bar and Grill, Mudgee NSW 58
The Winery Cafe, Pipers Brook Tas 269
Wineworks Downunder, Yungaburra Qld 165
Wingham NSW 97
Winkiku, Uluṟu-Kata Tjuṯa National Park NT 130
Wirrabara Old Bakery, Wirrabara SA 230
Wirrabara SA 230
Wirrabara Forest SA 230
Wirraninna Ridge Apple and Cider Vinegar, Bilpin NSW 21
Wise Vineyard Restaurant, Dunsborough WA 407
Wisharts at the Wharf, Port Fairy Vic 371-2
Wodonga Vic 340-1
Wogarno Station, Mount Magnet WA 424
Wokalup WA 398-9
Wollombi NSW 77, 78
Wollombi Markets, Wollombi NSW 78
Wollongbar NSW 107
Wollongong NSW 123
Wolseley SA 247-8
Wombah Coffee Plantation, Woombah NSW 108
Woodbridge Tas 287
Woodend Vic 348
Woodside SA 216
Woodstock Coterie, McLaren Flat SA 208
Woolgoolga NSW 98
Wooli NSW 107-8
Wooli Oyster Supply, Wooli NSW 107-8
Woolooga Qld 178
Woombah NSW 108
Woombye Qld 178-9
World's Longest Lunch, Albury-Wodonga 112
Woy Woy NSW 47
Wursthaus at Olivers, Launceston Tas 267
Wycliffe Well NT 130
Wycliffe Well Holiday Park, Wycliffe Well NT 130
Wynyard Tas 294
Wynyard Farmers' Market, Wynyard Tas 294
Wynyard Wharf Seafoods, Wynyard Tas 294

Xanadu Margaret River Restaurant and Cellar Door, Margaret River WA 411

yabbies
  New South Wales
    Hunter 73
    New England and North-West 82
    North Coast 93
    Northern Rivers 107
  South Australia 193
    Adelaide and Surrounds 204
    Clare Valley and Yorke Peninsula, 227
    Limestone Coast 245
  Victoria
    Goldfields 312-14
    The Great Ocean Road 373
    Legends, Wine and High Country 339
    Macedon Ranges and Spa Country 345-6
    Murray and the Outback 352
    Yarra Valley and the Dandenong Ranges 382
  Western Australia
    Australia's Coral Coast 427
Yabbies Restaurant, Paterson NSW 73
Yachtaway Cruising Holidays, Port Lincoln SA 235
Yackandandah Vic 341
Yackandandah General Store, Yackandandah Vic 341
Yallingup WA 417
Yamba NSW 108
Yandilla Mustard Seed Oil, Wallendbeen NSW 63
Yandina Qld 179-80
Yarr Valley Farmers' Market, Yering Vic 385
Yarra Glen Vic 383-4
YARRA VALLEY AND THE DANDENONG RANGES Vic 375-85
Yarra Valley Dairy, Yering Vic 385
Yarra Valley Free Range Pork, Macclesfield Vic 379
Yarragon Vic 363-4
Yarramalong NSW 47
Yarramalong Macadamia Nut Farm, Yarramalong NSW 47
Yarraman Bacon Factory, Yarraman Qld 146
Yarraman Qld 146
Yarrambat Vic 384
Yelland Yabbies, Naracoorte SA 245
Yeoval NSW 64
Yering Vic 384-5
Yoey's Traditional Fine Foods, Mount Gambier SA 244
yoghurt
  New South Wales
    Explorer Country and Riverina 54
  Queensland
    Far North 152, 153, 159
    Gold Coast 168
    Sunshine Coast 174
  South Australia
    Adelaide and Surrounds 210
    Kangaroo Island, 236-7
  Victoria
    The Grampians 365
York WA 399
Yorktown Organics, Yorktown Tas 274
Yorktown Tas 274
Young NSW 9, 13, 35, 41-3
Yulefest, Katoomba NSW 27
Yum Yums, Mossman Qld 160
Yungaburra Market, Yungaburra Qld 166
Yungaburra Qld 165, 166
Yuulong Lavender Estate, Mt Egerton Vic 346-7

Zest Restaurant, Nelson Bay NSW 72
Zilch—Food store and cafe, Wodonga Vic 341
Zouch Cafe-Restaurant, Young NSW 42-3